Dynamics of Magnetic Fluctuations in High-Temperature Superconductors

NATO ASI Series

Advanced Science Institutes Series

A series presenting the results of activities sponsored by the NATO Science Committee, which aims at the dissemination of advanced scientific and technological knowledge, with a view to strengthening links between scientific communities.

The series is published by an international board of publishers in conjunction with the NATO Scientific Affairs Division

A	**Life Sciences**	Plenum Publishing Corporation
B	**Physics**	New York and London
C	**Mathematical and Physical Sciences**	Kluwer Academic Publishers Dordrecht, Boston, and London
D	**Behavioral and Social Sciences**	
E	**Applied Sciences**	
F	**Computer and Systems Sciences**	Springer-Verlag
G	**Ecological Sciences**	Berlin, Heidelberg, New York, London,
H	**Cell Biology**	Paris, and Tokyo

Recent Volumes in this Series

Volume 240—Global Climate and Ecosystem Change
edited by Gordon J. MacDonald and Luigi Sertorio

Volume 241—Applied Laser Spectroscopy
edited by Wolfgang Demtröder and Massimo Inguscio

Volume 242—Light, Lasers, and Synchrotron Radiation: A Health Risk Assessment
edited by M. Grandolfo, A. Rindi, and D. H. Sliney

Volume 243—Davydov's Soliton Revisited: Self-Trapping of Vibrational Energy in Protein
edited by Peter L. Christiansen and Alwyn C. Scott

Volume 244—Nonlinear Wave Processes in Excitable Media
edited by Arun V. Holden, Mario Markus, and Hans G. Othmer

Volume 245—Differential Geometric Methods in Theoretical Physics: Physics and Geometry
edited by Ling-Lie Chau and Werner Nahm

Volume 246—Dynamics of Magnetic Fluctuations in High-Temperature Superconductors
edited by George Reiter, Peter Horsch, and Gregory C. Psaltakis

Volume 247—Nonlinear Waves in Solid State Physics
edited by A. D. Boardman, M. Bertolotti, and T. Twardowski

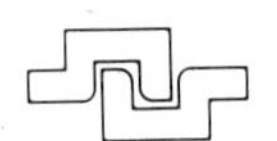

Series B: Physics

Dynamics of Magnetic Fluctuations in High-Temperature Superconductors

Edited by

George Reiter

University of Houston
Houston, Texas

Peter Horsch

Max-Planck-Institut für Festkörperforschung
Stuttgart, Germany

and

Gregory C. Psaltakis

University of Crete
and Research Center of Crete
Heraklion, Greece

Plenum Press
New York and London
Published in cooperation with NATO Scientific Affairs Division

Proceedings of a NATO Advanced Research Workshop on
Dynamics of Magnetic Fluctuations in High-Temperature Superconductors,
held October 9-14, 1989,
in Aghia Pelaghia, Crete, Greece

Library of Congress Cataloging-in-Publication Data

NATO Advanced Research Workshop on Dynamics of Magnetic Fluctuations in High-Temperature Superconductors (1989 : Hagia Pelagia, Greece)
Dynamics of magnetic fluctuations in high-temperature superconductors / edited by George Reiter, Peter Horsch, and Gregory C. Psaltakis.
p. cm. -- (NATO ASI series. Series B, Physics ; v. 246)
"Proceedings of a NATO Advanced Research Workshop on Dynamics of Magnetic Fluctuations in High-Temperature Superconductors, held October 9-14, 1989, in Aghia Pelaghia, Crete"--T.p. verso.
"Held within the program of activities of the NATO Special Program on Condensed Systems of Low Dimensionality, running from 1983 to 1988 as part of the activities of the NATO Science Committee"--
"Published in cooperation with NATO Scientific Affairs Division."
Includes bibliographical references and index.
ISBN 0-306-43810-0
1. High temperature superconductors--Congresses. 2. Valence fluctuations--Congresses. I. Reiter, George. II. Horsch, Peter. III. Psaltakis, Gregory C. IV. North Atlantic Treaty Organization. Scientific Affairs Division. V. Special Program on Condensed Systems of Low Dimensionality (NATO) VI. Title. VII. Title: Magnetic Fluctuations in High-Temperature Superconductors.
VIII. Series.
QC611.98.H54N36 1989
537.6'23--dc20 91-10370
CIP

A Division of Plenum Publishing Corporation
233 Spring Street, New York, N.Y. 10013

Printed in the United States of America

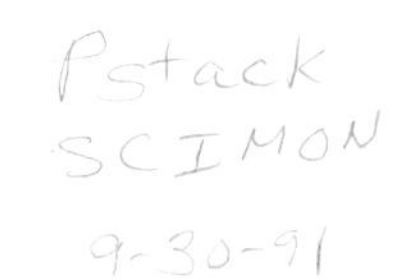

SPECIAL PROGRAM ON CONDENSED SYSTEMS OF LOW DIMENSIONALITY

This book contains the proceedings of a NATO Advanced Research Workshop held within the program of activities of the NATO Special Program on Condensed Systems of Low Dimensionality, running from 1985 to 1990 as part of the activities of the NATO Science Committee.

Other books previously published as a result of the activities of the Special Program are:

Volume 148 INTERCALATION IN LAYERED MATERIALS
edited by M. S. Dresselhaus

Volume 152 OPTICAL PROPERTIES OF NARROW-GAP LOW-DIMENSIONAL STRUCTURES
edited by C. M. Sotomayor Torres, J. C. Portal, J. C. Maan, and R. A. Stradling

Volume 163 THIN FILM GROWTH TECHNIQUES FOR LOW-DIMENSIONAL STRUCTURES
edited by R. F. C. Farrow, S. S. P. Parkin, P. J. Dobson, J. H. Neave, and A. S. Arrott

Volume 168 ORGANIC AND INORGANIC LOW-DIMENSIONAL CRYSTALLINE MATERIALS
edited by Pierre Delhaes and Marc Drillon

Volume 172 CHEMICAL PHYSICS OF INTERCALATION
edited by A. P. Legrand and S. Flandrois

Volume 182 PHYSICS, FABRICATION, AND APPLICATIONS OF MULTILAYERED STRUCTURES
edited by P. Dhez and C. Weisbuch

Volume 183 PROPERTIES OF IMPURITY STATES IN SUPERLATTICE SEMICONDUCTORS
edited by C. Y. Fong, Inder P. Batra, and S. Ciraci

Volume 188 REFLECTION HIGH-ENERGY ELECTRON DIFFRACTION AND REFLECTION ELECTRON IMAGING OF SURFACES
edited by P. K. Larsen and P. J. Dobson

Volume 189 BAND STRUCTURE ENGINEERING IN SEMICONDUCTOR MICROSTRUCTURES
edited by R. A. Abram and M. Jaros

Volume 194 OPTICAL SWITCHING IN LOW-DIMENSIONAL SYSTEMS
edited by H. Haug and L. Banyai

Volume 195 METALLIZATION AND METAL-SEMICONDUCTOR INTERFACES
edited by Inder P. Batra

Volume 198 MECHANISMS OF REACTIONS OF ORGANOMETALLIC COMPOUNDS WITH SURFACES
edited by D. J. Cole-Hamilton and J. O. Williams

SPECIAL PROGRAM ON CONDENSED SYSTEMS OF LOW DIMENSIONALITY

Volume 199 SCIENCE AND TECHNOLOGY OF FAST ION CONDUCTORS
edited by Harry L. Tuller and Minko Balkanski

Volume 200 GROWTH AND OPTICAL PROPERTIES OF WIDE-GAP II–VI LOW-DIMENSIONAL SEMICONDUCTORS
edited by T. C. McGill, C. M. Sotomayor Torres, and W. Gebhardt

Volume 202 POINT AND EXTENDED DEFECTS IN SEMICONDUCTORS
edited by G. Benedev, A. Cavallini, and W. Schröter

Volume 203 EVALUATION OF ADVANCED SEMICONDUCTOR MATERIALS BY ELECTRON MICROSCOPY
edited by David Cherns

Volume 206 SPECTROSCOPY OF SEMICONDUCTOR MICROSTRUCTURES
edited by Gerhard Fasol, Annalisa Fasolino, and Paolo Lugli

Volume 213 INTERACTING ELECTRONS IN REDUCED DIMENSIONS
edited by Dionys Baeriswyl and David K. Campbell

Volume 214 SCIENCE AND ENGINEERING OF ONE- AND ZERO-DIMENSIONAL SEMICONDUCTORS
edited by Steven P. Beaumont and Clivia M. Sotomayor Torres

Volume 217 SOLID STATE MICROBATTERIES
edited by James R. Akridge and Minko Balkanski

Volume 221 GUIDELINES FOR MASTERING THE PROPERTIES OF MOLECULAR SIEVES: Relationship between the Physicochemical Properties of Zeolitic Systems and Their Low Dimensionality
edited by Denise Barthomeuf, Eric G. Derouane, and Wolfgang Hölderich

PREFACE

This NATO Advanced Research Workshop was held at a time when there was little consensus as to the mechanism for high temperature superconductivity, in the context of a world undergoing major changes in its political alignments and sense of the possibility for the future. It was characterized by generosity in the sharing of our uncertainties and speculations, as was appropriate for both the subject matter and the context. The workshop was organized, of necessity around the experimental work, as is this volume. Where the theoretical work is directly relevant to particular experiments, it is included in the appropriate sections with them.

Most of the participants felt strongly that magnetic fluctuations played an important role in the mechanism for high T_c, although with the exception of the μSR work reported by Luke showing results inconsistent with the anyon picture, and the work on flux phases by Lederer, the mechanism remained an issue in the background.

A major focus was the phenomenological interpretation of the NMR data. Takigawa interprets his data on ^{17}O and ^{63}Cu in terms of a single spin fluid model, supported by Rice, Bulut and Monien. Berthier argues that one needs separate degrees of freedom on the oxygen and copper to describe all the data including that from ^{89}Y. Mehring presents data on ^{205}Tl that could naturally be interpreted in terms of two fluids, points out that the two spin fluids ought to interact, and that the data could as well be described by a single spin fluid with the right fluctuation spectrum. Emery prefers to describe the situation in terms of coupled spin fluids, and Mezei cautions that even a low density of impurities may dominate the NMR relaxation in real systems.

The dramatic reduction in the NMR widths below T_c appears to require a gap opening in the fluctuation spectrum, but the neutron scattering data of Tranquada and

Shirane, and of Rossat-Mignod, clearly show spectral intensity much below any calculation of the gap energy and apparently extending to zero frequency.

There were some areas of consensus. The systematic variation of T_c with density of doped holes is demonstrated by Uemura et al. and Ansaldo, using μSR. The picture of the ground state of the Heisenberg model being Neel ordered, with well defined spin wave excitations, as in the calculations of Becher and Reiter and of Grempel, is confirmed by the neutron scattering measurements reported by Mook. The existence of quasiparticles in the t-J model is a theme in the works of Prelovsek, of Stephan and Horsch, and of Gunn, although Emery and Long point out that there are important differences with two band models.

Taken as a whole, the reader will find here an overview of the subject of the dynamics of magnetic fluctuations that we expect will be useful for anyone seeking an understanding of the physics of the high T_c materials from the perspective that these fluctuations matter for the mechanism, or that they are interesting in and of themselves.

We would like to thank the Texas Center for Superconductivity at the University of Houston, and the Mitos Corporation at the University of Crete. Their clerical and logistical support made a major contribution to the smooth working of the conference.

G. Reiter

P. Horsch

G. Psaltakis

CONTENTS

A. Neutron Scattering

B. NMR

C. μSR

D. Raman Scattering

E. Photoemission

F. Macroscopic Fluctuations

G. Transport Properties

H. General Theory

I. One Band Hubbard Models

J. Heisenberg Model Dynamics

K. Quasiparticles and Magnetic Fluctuations

L. Magnetic phases

ANTIFERROMAGNETIC SPIN FLUCTUATIONS IN CUPRATE SUPERCONDUCTORS

J. M. Tranquada and G. Shirane

Physics Department
Brookhaven National Laboratory
Upton, NY 11973

Introduction

In this paper we will discuss the results of some neutron scattering studies of antiferromagnetic spin fluctuations in the superconductors $YBa_2Cu_3O_{6+x}$ and $La_{2-x}Sr_xCuO_4$. We begin by discussing spin wave measurements in the antiferromagnetic phases of these compounds and comparisons between experimental and theoretical results for the spin-$\frac{1}{2}$ Heisenberg model in two dimensions. Next, recent studies of antiferromagnetic excitations in metallic and superconducting phases are described. In particular, the topics of incommensurate scattering, the temperature dependence of the spin susceptibility, and the existence of spin fluctuations at temperatures below the superconducting transition temperature will be covered. Finally, we consider the connection between neutron scattering measurements and studies of nuclear spin relaxation rates obtained by nuclear magnetic resonance. The paper concludes with a short summary.

Insulating Phases

Magnetic Order in $YBa_2Cu_3O_{6+x}$

Before discussing spin waves in antiferromagnetic $YBa_2Cu_3O_{6+x}$, it may be helpful to review the phase diagram shown in Fig. 1. (The discussion here will be kept short; further details are available elsewhere.[1-3]) At $x = 0$ the crystal exhibits a simple Néel structure involving the $Cu(2)^{2+}$ ions in the CuO_2 layers [see Fig. 2(a)]. The two-fold coordinated $Cu(1)^{1+}$ ions are nonmagnetic, and so do not participate in the magnetic structure. When oxygen is added, it goes into the Cu(1) layer converting some Cu^{1+} to Cu^{2+}(Ref. 4). Beyond $x = 0.2$, the added oxygens begin to form O–Cu(1)–O chain segments,[5] which results in a low density of O $2p$ holes. The Néel temperature stays constant below $x = 0.2$,[2] but begins to decrease beyond that point as a very small density of holes enters the CuO_2 planes and causes some disorder. The increased disorder is also reflected in the decrease in the average ordered moment observed at low temperature with increasing x.[2,6]

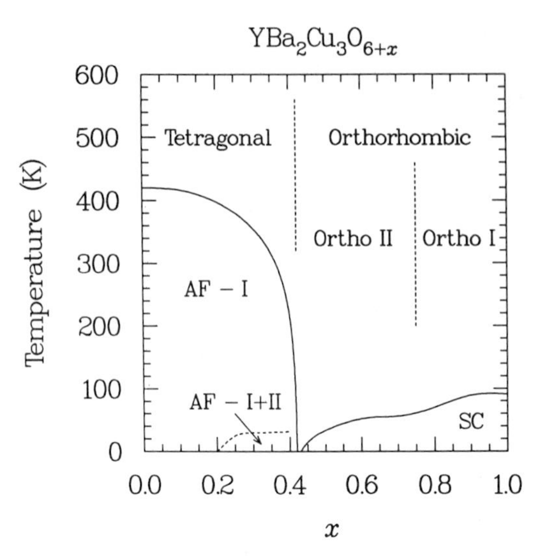

Fig. 1. Phase diagram of $YBa_2Cu_3O_{6+x}$. The positions of the phase boundaries with respect to x depend on sample preparation and treatment.

The $Cu(1)^{2+}$ ions which are formed by adding oxygen would like to couple ferromagnetically to the CuO_2 layers,[7] but they are frustrated by the antiferromagnetic coupling between the planes. Ferromagnetic coupling across the Cu(1) layer would lead to a doubling of the unit cell along the c axis, and would result in a significant hyperfine field at the Cu(1) sites. Such a magnetic structure is induced at low temperature by replacing as little as 1% of the Cu(1) sites with Fe.[8] Without doping, such long-range order does not occur.[8] A study[9] of magnetic susceptibility revealed a Curie-like contribution with an effective moment that increases with x up to $x \sim 0.3$, but which decreases sharply for temperatures below $\sim$ 30 K. Monte Carlo simulations of the magnetic lattice including some magnetic $Cu(1)^{2+}$ sites give a reasonable description of this behavior.[9] The average ordered moment measured by neutron diffraction is also observed to decrease below $\sim$ 30 K for $x \gtrsim 0.2$.[2] In a crystal with $x \approx 0.3$ we have observed excess inelastic magnetic scattering which peaks near 30 K; below this temperature diffuse elastic 2D scattering appears.[10] Rossat-Mignod *et al.*[11] have argued that the low-temperature transition is due entirely to localization of holes in the CuO_2 layers. Alternatively, we have suggested[10] that the transition involves local ferromagnetic coupling of $Cu(1)^{2+}$ ions to the planes; such defects in the magnetic order of the planes should provide good sites for holes to localize.

Spin Waves in Antiferromagnetic $YBa_2Cu_3O_{6+x}$

We have made extensive measurements[10] of spin waves in a large crystal ($\sim$ 0.5 cm^3) of $YBa_2Cu_3O_{6+x}$ with $x \approx 0.3$ and $T_N = 260$ K. Most of the measurements were made at 200 K, where little scattering is expected from $Cu(1)^{2+}$ ions. For an antiferromagnet, the long-wavelength spin-wave dispersion is given by $\omega = cq$, where c is the spin-wave velocity and q is the wave vector measured relative to a magnetic Bragg peak. The dispersion is dominated by the in-plane superexchange $J_{\|}$ between nearest-neigbor Cu atoms. As a result, if we hold the energy transfer ΔE $(= \hbar\omega)$ constant and scan the wave vector across the 2D rod [scan A in Fig. 2(b)], then we expect to see spin-wave peaks at $q_{\|} = \pm\omega/c$, where $q_{\|}$ is the component of the AF wave vector perpendicular to the 2D rod, and $c = \sqrt{2}J_{\|}a$. Examples of such scans, measured at three different excitation energies, are shown in Fig. 3. Because of the coarse spec-

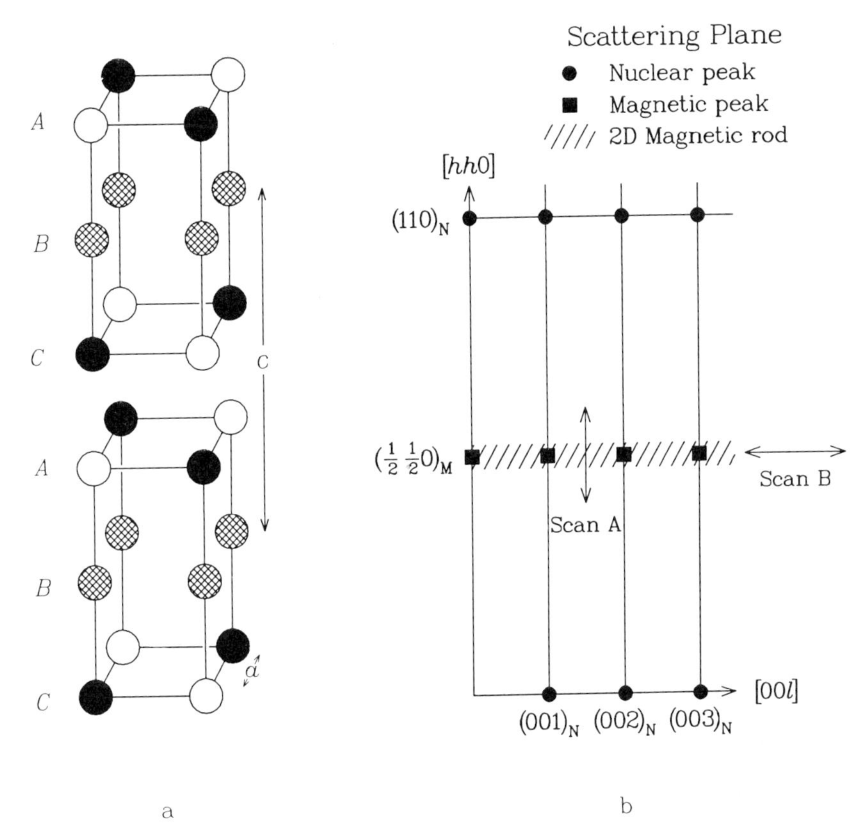

Fig. 2. (a) Magnetic spin arrangement in $YBa_2Cu_3O_{6+x}$ with x near zero. Cross-hatched circles represent nonmagnetic Cu^{1+} ions, while solid and open circles indicate antiparallel spins at Cu^{2+} sites. (b) Reciprocal space (hhl) zone. The hatched line along $(\frac{1}{2}, \frac{1}{2}, l)$ is the magnetic rod for two-dimensional scattering, and A and B indicate scans across and along the rod, respectively.

trometer resolution relative to the extremely steep dispersion, the spin-wave peaks are not resolved. The solid lines are fits made using the standard Heisenberg-model spin-wave cross section (with no damping) and taking into account the spectrometer resolution function. The amplitude was adjusted for each data set, but the overall variation was less than 10%. Equally good agreement was obtained for fits to a range of measurements, indicating that the spin waves are well described by the Heisenberg model. The value of $J_{\parallel}$ obtained from the fits in Fig. 3, 80^{+60}_{-30} meV, is rather imprecise, but it is consistent with the somewhat larger value extracted from Raman scattering studies[12] and from higher-resolution neutron scattering measurements.[13]

For a perfectly 2D system, the spin-wave intensity should not vary significantly as the wave vector is scanned perpendicular to the planes [scan B in Fig. 2(b)]. However, as shown in Fig. 4, the spin-wave intensity measured in such a scan in $YBa_2Cu_3O_{6.3}$ is strongly modulated. There are two reasons for this modulation. The first is that because there are two CuO_2 layers per unit cell separated along the c axis by an arbitrary distance zc, the Cu atoms do not form a Bravais lattice. As a result, the spin-wave modes are split into acoustic and optical branches.[14] The second reason

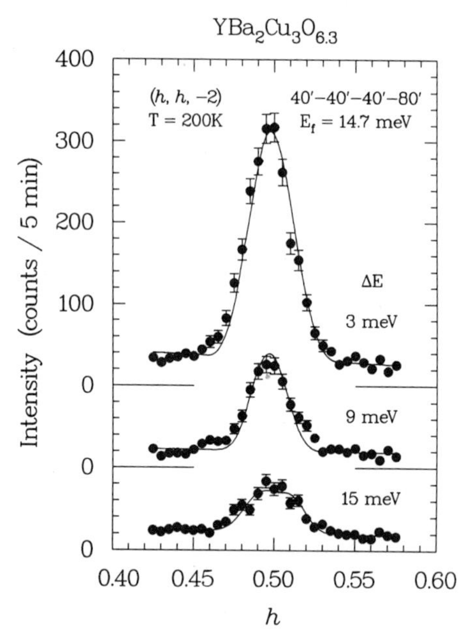

Fig. 3. Several constant ΔE scans of type A [see Fig. 2(b)] across the 2D magnetic rod in a $YBa_2Cu_3O_{6+x}$ single crystal at 200 K. The solid lines are fits to the data as discussed in the text. From Ref. 10.

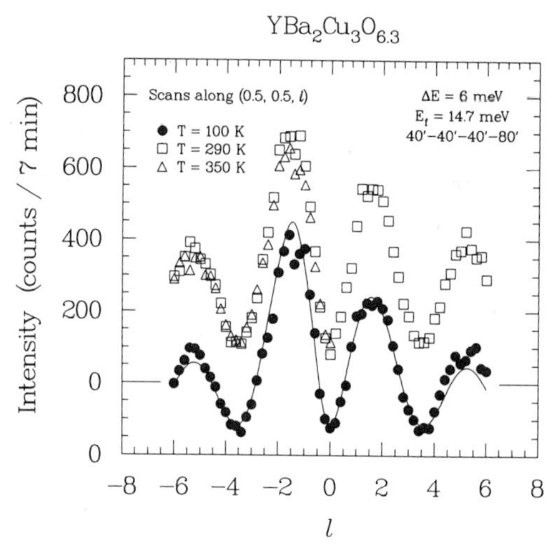

Fig. 4. Constant ΔE scan of type B [see Fig. 2(b)] along the 2D rod with $\Delta E = 6$ meV for $YBa_2Cu_3O_{6.3}$ at several temperatures. The modulation is due to the inelastic structure factor. The solid line is a fit. From Ref. 10.

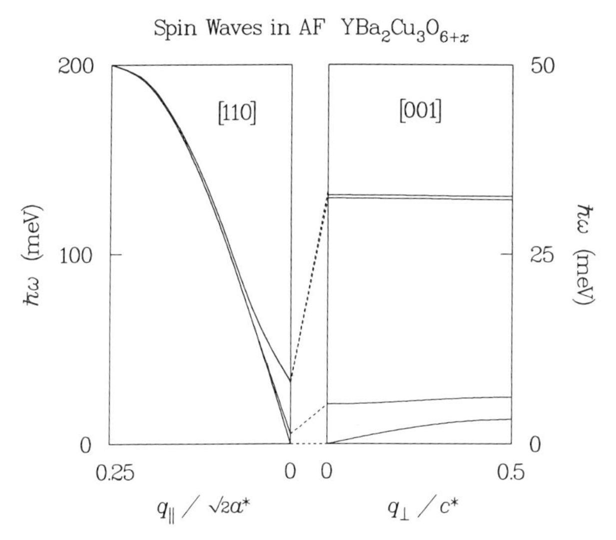

Fig. 5. Schematic diagram of spin-wave dispersion in $YBa_2Cu_3O_{6+x}$. Note that the energy scales for the two panels differ by a factor of four. The 2D nature of the magnetic interactions makes the dispersion very large along $q_{\parallel}$ but extremely weak along $q_{\perp}$. From Ref. 10.

is that because of a reasonably strong coupling $J_{\perp 1}$ between nearest-neighbor layers, presumably due to direct exchange along the *c* axis between neighboring Cu atoms, the optical modes are at energies greater than 30 meV. Assuming $J_{\parallel} = 80$ meV, we obtain the limit $J_{\perp 1} \gtrsim 2$ meV. Thus, at 6 meV one observes only acoustic modes modulated by the appropriate inelastic structure factor. The modulation provides a useful signature for correlations within the bilayers. In particular, it is found that the modulation survives above T_N, indicating that the bilayers remain strongly correlated in the absence of long-range order.

To complete the picture of spin-wave dispersion it is necessary to take into account the weak coupling $J_{\perp 2}$ between next-nearest-neighbor CuO_2 planes, separated by a CuO_x layer. This interaction causes a weak dispersion as a function of the wave vector component $q_{\perp}$ perpendicular to the planes. There is also a weak XY-like anisotropy of the in-plane exchange which causes a splitting of the acoustic and optical modes (each of which would otherwise be doubly degenerate). A schematic diagram of the spin-wave dispersion is shown in Fig. 5. The values of $J_{\perp 2}$ and the exchange anisotropy are on the order of $10^{-4} \times J_{\parallel}$. The overall picture we have obtained for spin-wave dispersion in $YBa_2Cu_3O_{6+x}$ is quite similar to that determined by Rossat-Mignod and coworkers.[11,13]

Spin Fluctuations in La_2CuO_4

Studies of the magnetic structure and spin waves in doped and undoped La_2CuO_4 have been reviewed by Birgeneau and Shirane.[15] More recently, Yamada *et al.*[16] have performed an extensive inelastic neutron scattering study on a La_2CuO_4 crystal with $T_N = 245$ K at temperatures up to 520 K. They compared their measurements with the theoretical formula for the dynamical structure factor $S(\mathbf{q}, \omega)$ determined by Chakravarty and coworkers[17,18] in their analysis of the 2D, spin-$\frac{1}{2}$ Heisenberg model. Inelastic scattering measurements above T_N were well described

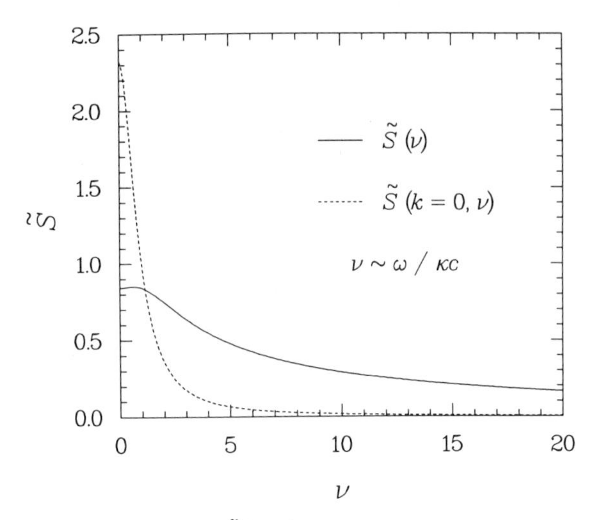

Fig. 6. The scaling function $\tilde{S}(k, \nu)$ for $k = 0$, and also integrated over k, plotted vs. ν. The region of critical scattering is $\nu \lesssim 1$.

by the theoretical formula when the experimentally determined correlation length was used. The only discrepancy was an excess elastic component observed experimentally whose origin was attributed to magnetic defects. The temperature dependence of the correlation length was also in good agreement with the theoretical result.

As the correlation length ξ grows and the Néel temperature is approached from above, one expects to observe a critical slowing down of the spins. This effect should manifest itself as a sharp peak in $S(\mathbf{q}, \omega)$ at $\mathbf{q} = 0$ having a width $\Gamma/\hbar \approx \kappa c$, where $\kappa = 1/\xi$. For a 2D Heisenberg system, this quasielastic peak evolves into the antiferromagnetic Bragg rod at $T = 0$. Because of the very strong superexchange within an insulating CuO_2 layer, ξ is several hundred anstroms even at 300 K, and correspondingly $\Gamma \sim 1$ meV.[19] Capellmann and coworkers[20] have argued that since the expected quasielastic scattering has not been directly observed in neutron scattering measurements, the picture of fixed, localized moments on the Cu atoms implicit in applications of the Heisenberg model does not properly characterize the CuO_2 layers above T_N. However, the theoretical form of $S(\mathbf{q}, \omega)$ given by Tyč *et al.*[18] and tested by Yamada *et al.*[16] does contain the expected quasielastic component. Why is this component not obvious in the neutron scattering measurements?

The dynamical structure factor of Tyč *et al.*,[18] appropriate at low temperatures and frequencies, can be written as

$$S(q, \omega) = \omega_0^{-1} S_0 \tilde{S}(k, \nu), \tag{1}$$

where $\nu \equiv \omega/\omega_0$, $k \equiv q/\kappa$, and

$$\omega_0 \equiv \kappa c \sqrt{k_B T / A J}, \tag{2}$$

with $A = 0.944$. To analyze the quasielastic component it is sufficient to consider the scaling function $\tilde{S}$. The quasielastic regime in which spin fluctuations are overdamped corresponds to $\omega < \omega_0$ (i.e. $\nu < 1$). Figure 6 shows the function $\tilde{S}(k = 0, \nu)$, which clearly has most of its weight below $\nu = 1$. If the neutron scattering measurements were performed with infinite resolution, then one should indeed observe

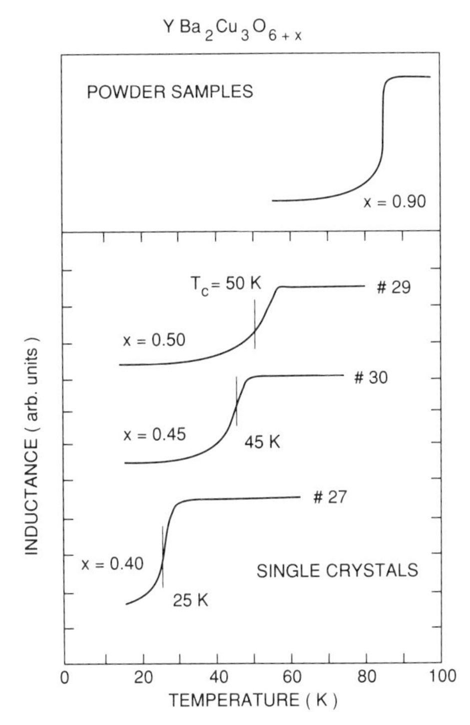

Fig. 7. T_c determination for $YBa_2Cu_3O_{6+x}$ crystals by ac induction method. Top panel shows typical data for powder samples obtained by the identical setup. From Ref. 22.

a strong quasielastic peak. Instead, however, the experiments have been performed with a coarse resolution, so that $S(q, \omega)$ is averaged over a large region of q. A more appropriate quantity to compare with experiment is

$$\tilde{S}(\nu) = \int dk^2 \, \tilde{S}(k, \nu). \tag{3}$$

As shown in Fig. 6, this function has its weight spread over a much wider range of ν, consistent with experiment. While the quasielastic contribution is sharply peaked in ν, it occupies only a small area in phase space, and hence can only be resolved by high resolution measurements. Indeed, Mook reported at this meeting that Aeppli[21] has observed the quasielastic peak in a La_2CuO_4 crystal using cold neutrons.

Metallic Phases

Superconducting $YBa_2Cu_3O_{6+x}$

Our initial attempts to observe inelastic magnetic scattering in superconducting crystals of $YBa_2Cu_3O_{6+x}$ were unsuccessful.[10] Fortunately, Sato and Shamoto at the Institute for Molecular Science in Japan have been able to grow several very large crystals (~ 1 cm^3) from which it has been possible to observe spin fluctuation scattering. Crystals with $x = 0.4$, 0.45, and 0.5 were initially studied at Riso National Laboratory in Denmark.[22] The superconducting transition temperatures were determined by ac susceptibility measurements (see Fig. 7); the T_c values are 25 K,

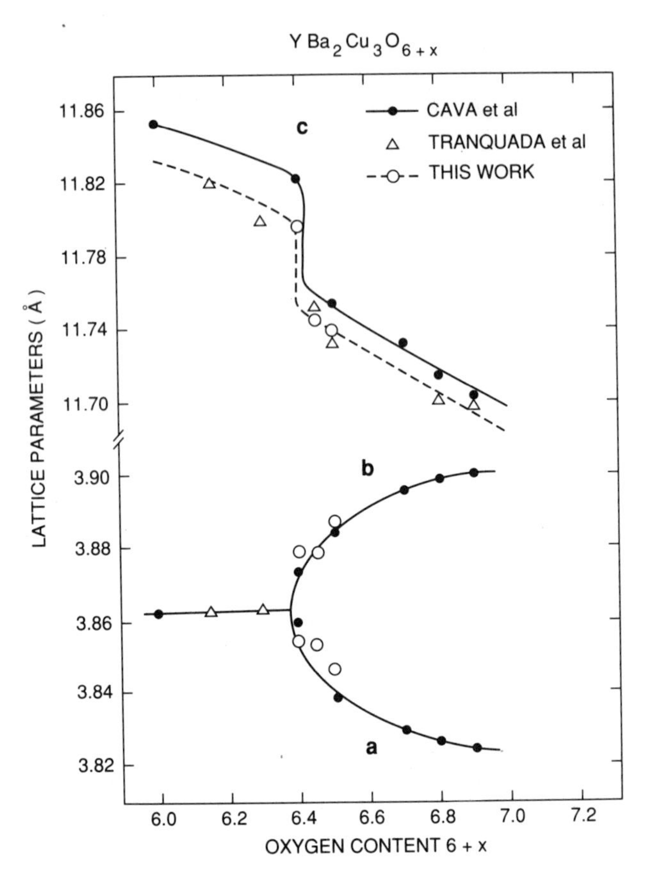

Fig. 8. Lattice constants at room temperature for $YBa_2Cu_3O_{6+x}$ crystals. Solid circles are for powder samples studied by Cava *et al.* (Ref. 23), triangles are for crystals in an earlier study by Tranquada *et al.* (Ref. 10), and the open circles are for the same crystals characterized in Fig. 7. From Ref. 22.

45 K, and 50 K, respectively. The oxygen contents of the crystals were inferred from the room temperature lattice parameters (see Fig. 8) by comparing with results for powder samples by Cava *et al.*[23] No magnetic Bragg peaks were observed in any of these crystals, suggesting that the density of mobile holes in the CuO_2 planes is sufficient in each to prevent the development of long-range order. However, the $x = 0.4$ and 0.45 crystals are quite near to the orthorhombic-tetragonal boundary, where in powder samples T_c and the Meissner fraction are found to drop off sharply. A study of muon spin rotation at $T < 0.1$ K in powder samples by Kiefl *et al.*[24] has found evidence for static magnetic order for $x \lesssim 0.5$ even in the orthorhombic phase. No static order was observed for samples with $x \gtrsim 0.5$ and $T_c \gtrsim 50$ K. Those results suggest that the magnetic correlation length in our $x = 0.5$ crystal should be much shorter than that in either the $x = 0.4$ or 0.45 crystal. As we will show, this is indeed the case.

Inelastic scans at an excitation energy of 3 meV measured on all three crystals at both 11 K and 300 K are shown in Fig. 9. At 11 K one observes strong magnetic scattering peaked on the antiferromagnetic rod in the $x = 0.4$ and 0.45 samples; however, there is a drastic reduction in cross section for the $x = 0.5$ crystal. The

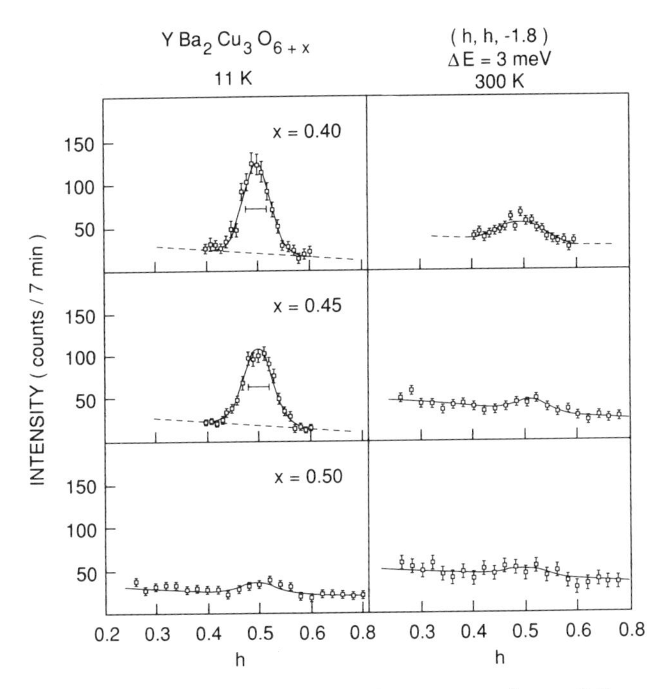

Fig. 9. Comparison of inelastic magnetic cross sections of three superconducting crystals of $YBa_2Cu_3O_{6+x}$ [scans of type A in Fig. 2(b)]. All three crystals have similar volumes. Notice the sudden decrease of intensities between $x = 0.45$ and $x = 0.50$. From Ref. 22.

intensity is considerably reduced at 300 K for the crystals with lower oxygen content, while for the $x = 0.5$ sample one can only say that the cross section did not increase. Further measurements on the $x = 0.45$ sample indicated that the cross section at 3 meV decreases monotonically with increasing temperature.

What do these limited observations tell us? The differential scattering cross section can be written in general as[25]

$$\frac{d^2\sigma}{d\Omega dE} \sim S(\mathbf{q},\omega) \sim [n(\omega)+1]\chi''(\mathbf{q},\omega), \tag{4}$$

where $\chi''(\mathbf{q},\omega)$ is the imaginary part of the generalized susceptibility, and $n(\omega) = [\exp(\hbar\omega/k_BT) - 1]^{-1}$ is the Bose factor. If χ'' is independent of temperature, as in an antiferromagnet below T_N, then the temperature dependence of the cross section comes just from the Bose factor, which increases monotonically with temperature. If the cross section is observed to decrease with increasing temperature, as in an antiferromagnet above T_N,[16] then the susceptibility must be changing with temperature. If we assume that the magnetic moments and the interactions between them are constant, then the most likely source of the temperature dependence is the spin-spin correlation length. The increasing intensity and relatively large cross section at low temperature observed in the $x = 0.4$ and 0.45 crystals suggest that the correlation length is growing significantly with decreasing temperature. Such a behavior is consistent with the μSR observations of static order at very low temperature in samples with similar oxygen concentrations. For the $x = 0.5$ crystal it is necessary to study the energy dependence of the cross section before drawing any conclusions.

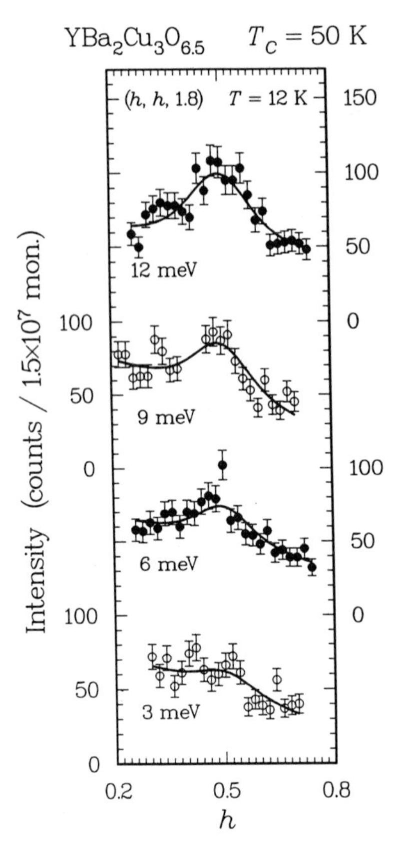

Fig. 10. Constant-E scans across the 2D antiferromagnetic rod at $\Delta E = 3$, 6, 9, and 12 meV and $T = 12$ K measured on the $YBa_2Cu_3O_{6.5}$ crystal. Solid lines are fits using Eq. (5) as discussed in the text. From Ref. 26.

To study the ω dependence of the magnetic cross section, the experiments were continued at Chalk River Nuclear Laboratories.[26] Scans measured on the $x = 0.5$ crystal at $T = 12$ K for several different excitation energies are shown in Fig. 10. Although the peak is fairly broad in q, the cross section grows with increasing energy to a level comparable with the $x = 0.3$ antiferromagnetic sample. The ω dependence is more clearly illustrated in Fig. 11, where it is compared with similar measurements on the $x = 0.45$ crystal. To understand the difference in behavior between the two samples, the measurements were fit using the dynamical structure factor appropriate for paramagnetic scattering[25]:

$$S(\mathbf{q},\omega) = \frac{\hbar\omega}{1 - e^{-\hbar\omega/k_BT}} \frac{A}{\kappa^2 + q^2} \left[\frac{\Gamma}{(\hbar\omega - \hbar\omega_{\mathbf{q}})^2 + \Gamma^2} + \frac{\Gamma}{(\hbar\omega + \hbar\omega_{\mathbf{q}})^2 + \Gamma^2} \right] . \quad (5)$$

This formula is quite similar to that used by Tyč *et al.*[18] if one sets $\Gamma/\hbar \equiv \kappa c$. Adjusting the parameters c, κ, and A by trial and error, it was found that a reasonable fit could be obtained to the data with $\hbar c = 100$ mev-Å. The values of the inverse correlation length for the $x = 0.45$ and 0.5 crystals were 0.025 Å^{-1} and 0.3 Å^{-1}, respectively. This order of magnitude difference in κ for the two samples appears to be consistent with μSR results concerning the disappearance of static order.[24] It is clear that the magnetic scattering in the $T_c = 50$ K sample is coming from a highly

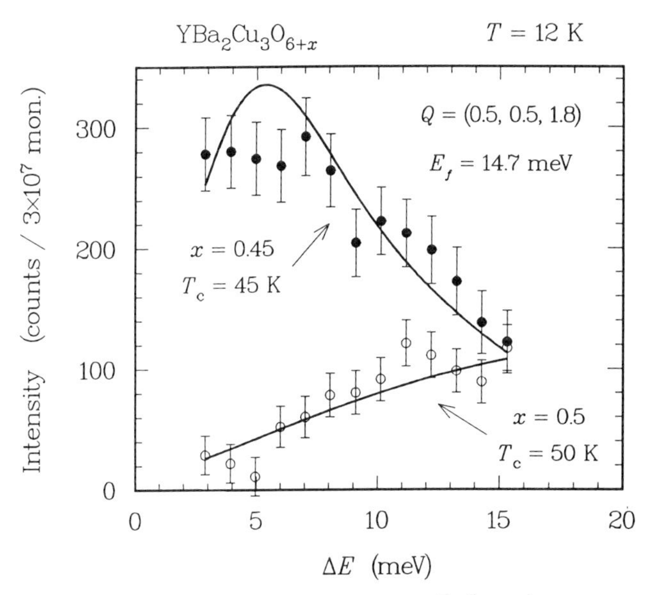

Fig. 11. Constant-**Q** scans measured at $\mathbf{Q} = (\frac{1}{2}, \frac{1}{2}, 1.8)$ and $T = 12$ K on the $x = 0.45$ (solid circles) and $x = 0.5$ (open circles) crystals. A constant background value has been subtracted from each set of data. The solid lines are fits using Eq. (5) as discussed in the text. From Ref. 26.

doped region, since the energy dependence of the cross section is inconsistent with a long correlation length.

In the analysis described above, the possibility of an incommensurate scattering component was ignored. As we will discuss in the next section, it appears that in well-doped crystals of $La_{2-x}Sr_xCuO_4$ the inelastic magnetic scattering peaks at points off of the 2D rod.[27,28] Some form of incommensurate scattering is expected on the basis of theoretical models for dilute holes in a 2D Heisenberg antiferromagnet, in which the spins are found to be canted in a spiral around a hole.[29,30] Explicit predictions for incommensurate inelastic scattering have been obtained by Bulut *et al.*[31] from an RPA analysis of the single-band Hubbard model with weak correlations. In the latter model the incommensurability takes the form of stripe domains. One might expect that the positions of incommensurate peaks are determined by the hole density, while the width of the peaks, and their intensity, depends on the inverse correlation length. Such a picture has been used to analyze energy-integrated measurements of 2D magnetic scattering in a crystal of $La_{1.89}Sr_{0.11}CuO_4$.[32] Although no splitting of the inelastic features has been observed in $YBa_2Cu_3O_{6+x}$, the fairly constant q width as a function of ω and T suggests that Eq. (5) is not really adequate for fitting all of the measurements on the $x = 0.4$, 0.45, and 0.5 samples. Hence, the parameter values obtained in fitting the data may be in error in an absolute sense. Nevertheless, we believe that the conclusion concerning an order of magnitude difference in the correlation lengths is correct.

The results shown in Fig. 11 are important for several reasons. First of all, they are direct evidence that antiferromagnetic correlations survive in metallic $YBa_2Cu_3O_{6+x}$. Secondly, the measurements were performed well below T_c, and hence they demonstrate that the antiferromagnetic spin fluctuations are not frozen

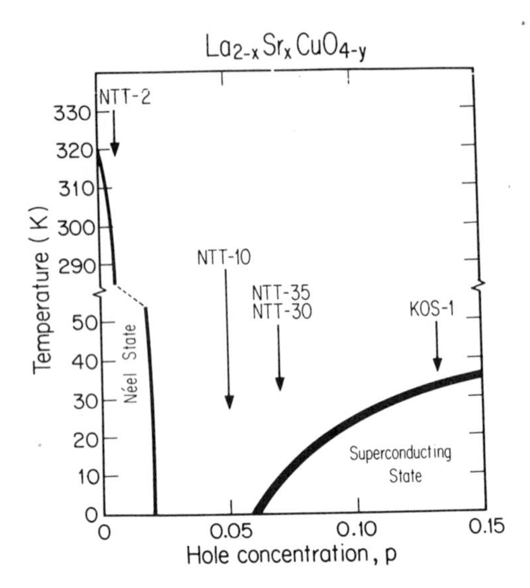

Fig. 12. Phase diagram of $La_{2-x}Sr_xCuO_{4-y}$ after Thurston *et al.* (Ref. 27). Hole concentration p is approximately equal to Sr content x.

out in the superconducting state. We note that Rossat-Mignod *et al.*[11] have obtained similar and complementary results on a crystal with $x = 0.45$ and $T_c = 35$ K. From an unsuccessful search for a magnetic cross section in a powder sample of $YBa_2Cu_3O_7$, Brückel *et al.*[33] concluded that no magnetic moments were present in the material. The present analysis suggests, alternatively, that the negative results at low energy may have been due to a very short correlation length. Measurements on large crystals with high oxygen concentrations are necessary to test this idea.

Spin Fluctuations in $La_{2-x}Sr_xCuO_4$

Figure 12 shows a phase diagram of $La_{2-x}Sr_xCuO_4$, in which the characteristics of several crystals recently studied by neutron scattering are indicated. For some of these samples, the initial experiments[32,34] involved measurement of the energy-integrated magnetic scattering using the two-axis method. The measurements were originally analyzed assuming the q dependence of the cross section to be a single Lorentzian with a half-width of κ, the inverse of the 2D correlation length. Because of the appearance of two peaks in the two-axis scans of crystal NTT-30,[32] the data were reanalyzed assuming a cross section consisting of two Lorentzians with peaks on either side of the 2D rod and each having a half-width of κ. From the latter analysis it was determined that the correlation length (measured at low temperature) in crystals NTT-10, 30, and 35 is approximately 18 ± 6 Å. However, because of problems with contamination of the elastic channel in the two-axis measurements, further work has focussed on the inelastic cross section.

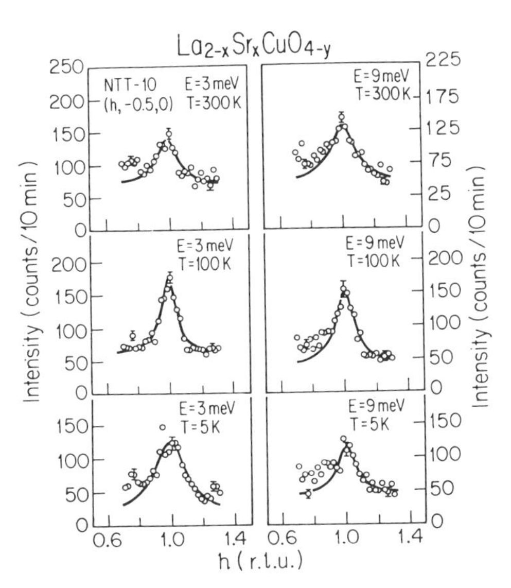

Fig. 13. Temperature dependence of the 3 and 9 meV scattering in crystal NTT-10. The spectrometer was set in the E_f-fixed mode at 14.7 meV and the collimation was 40′-40′-40′-80′. From Ref. 27.

Thurston *et al.*[27] recently studied the temperature dependence of the inelastic cross section in the crystals NTT-10 ($x = 0.06$) and NTT-35 ($x = 0.11$). The latter crystal, as grown, exhibited a broad superconducting transition centered around 7 K. Magnetization measurements of a piece of the crystal indicated a large Meissner fraction; however, a muon spin rotation study of NTT-10 and a sample prepared under identical conditions to NTT-35 found evidence for static magnetic order below 5 K.[35] Figure 13 shows examples of scans across the 2D rod measured on NTT-10 at two different excitation energies (3 and 9 meV) for temperatures from 5 to 300 K. A single peak is observed, with an amplitude and width which change relatively little with temperature and energy transfer. The peak intensity integrated over Q is shown as a function of temperature in Fig. 14 for NTT-10 and NTT-35. For both crystals the integrated intensity shows a relatively weak temperature dependence. As discussed in the previous section, the fact that the temperature dependence does not follow the Bose factor $n(\omega) + 1$ indicates that the susceptibility is temperature dependent. The static magnetic order observed by μSR also suggests that the correlation length must be increasing with decreasing temperature.

An exceptionally good single crystal of $La_{1.85}Sr_{0.15}CuO_4$, denoted KOS-1, was grown recently by Tanaka and Kojima.[36] As indicated by the resistivity measurements shown in Fig. 15, the crystal has a T_c of 33 K. Magnetization measurements on the entire sample (6 mm diameter, 17 mm long) imply a flux exclusion of $80 \pm 20\%$. The effects of trapped flux in the superconducting state can be seen directly in neutron depolarization measurements, as shown in the insert of Fig. 15. The magnetic inelastic cross section measured on this crystal is compared with that of an undoped sample in Fig. 16. The scattering from KOS-1 is quite broad compared to the undoped

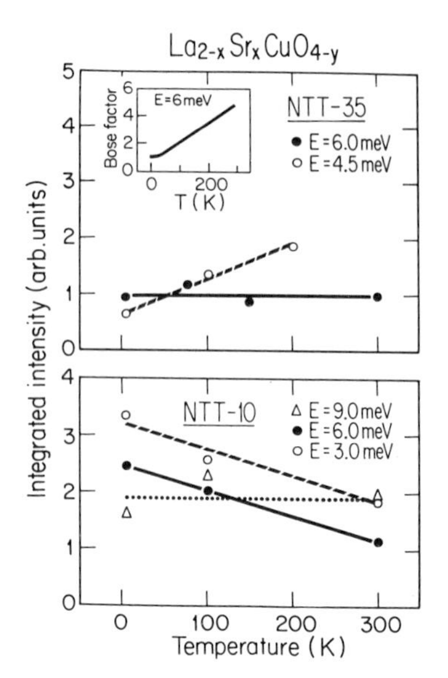

Fig. 14. Integrated ($\int dQ$) intensity vs. temperature for NTT-10 and NTT-35. The data in this figure were taken at a variety of different experimental conditions; consequently the magnitude of the intensities cannot be compared at different energy transfers and between the two samples. From Ref. 27.

sample, and it exhibits a double-peaked structure, as also observed for NTT-30 and NTT-35.[27,32] Nevertheless, taking into account differences in crystal volumes, there is little change in the integrated intensity with doping.

Of particular interest are the data on temperature dependence of the integrated intensities at energies of 6 and 12 meV shown in Fig. 17. (Note that the intensity data for each energy are normalized to one at an arbitrary temperature.) The 6 meV signal increases with temperature to $\sim$ 120 K, where it seems to saturate. In contrast, the 12 meV data appear to be temperature independent. This behavior was initially interpreted[28] as evidence for a gap in the spin fluctuation spectrum. One difficulty with such an interpretation is that the low-energy inelastic cross section never goes to zero. Our results[26] for $YBa_2Cu_3O_{6.5}$ can be explained without invoking a gap, and this suggests that the interpretation of the data in Fig. 17 be reconsidered. It appears that the 6 meV data follow the Bose factor at low temperature; the Bose factor at 12 meV does not deviate significantly from 1 below 120 K. From such a behavior one can infer that the susceptibility is roughly constant at low temperature, presumably because the correlation length has saturated. The temperature independent behavior at higher temperatures suggests that there the correlation length is decreasing. Rather than indicating a gap in the spin fluctuations associated with superconductivity, the latter interpretation leads to the conclusion that antiferromagnetic excitations remain unaffected by Bose condensation. Similar interpretations in terms of the temperature

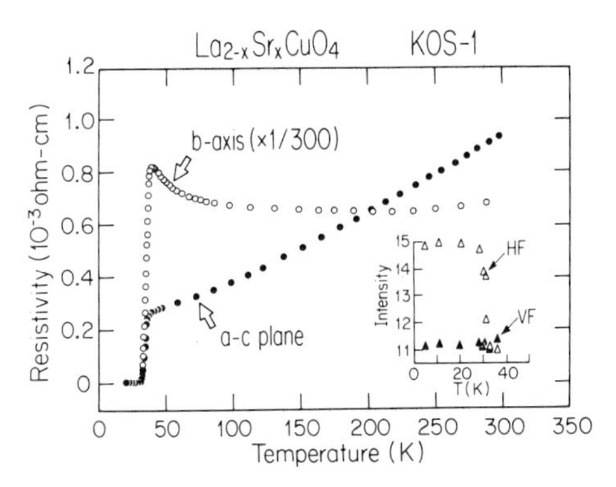

Fig. 15. Resistivity, parallel and perpendicular to the CuO_2 planes, of KOS-1 crystal (Ref. 36). Insert shows spin flip neutron intensity (Ref. 28) with the applied field vertical (VF) and horizontal (HF).

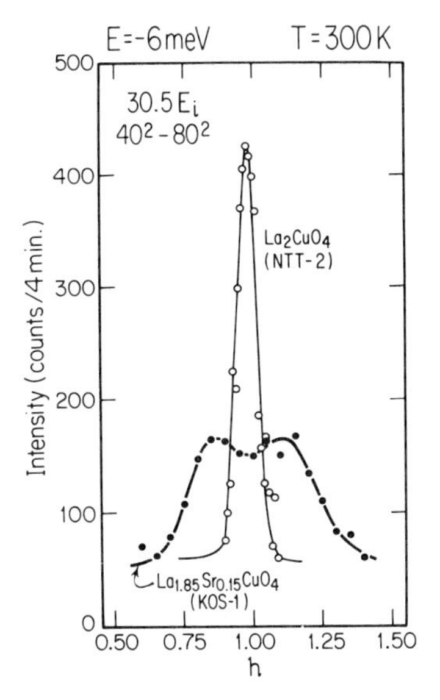

Fig. 16. Direct comparison of magnetic excitations of La_2CuO_4 and $La_{1.85}Sr_{0.15}CuO_4$. From Ref. 28.

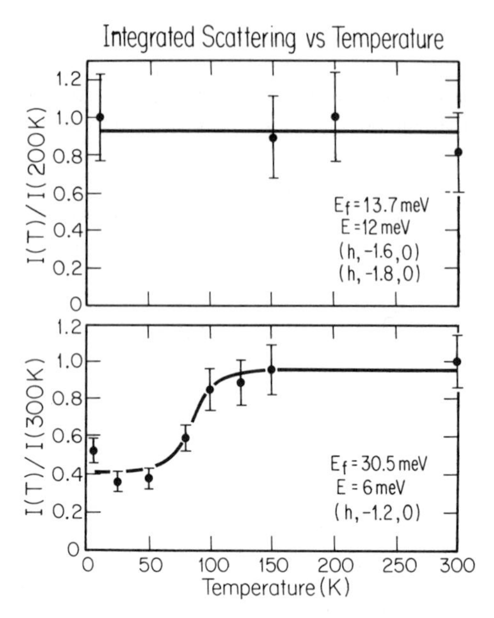

Fig. 17. Temperature dependence of normalized integrated intensities at energies of 6 meV and 12 meV for $La_{1.85}Sr_{0.15}CuO_4$ (Ref. 28). Solid lines are guides to the eye.

dependence of the spin susceptibility have been given by Monien and coworkers[37] and by Bulut *et al.*[31]

Neutron Scattering vs. Nuclear Magnetic Resonance

As indicated by Eq. (4), the differential scattering cross section for neutrons is proportional to the imaginary part of the dynamical susceptibility. In principle one can measure the susceptibility at any point in $(\mathbf{Q}, \omega)$-space by neutron scattering. In practice, however, the measurements require very large single crystals and are limited by resolution and background scattering. Experimentally, it is difficult to identify intrinsic scattering which does not show structure in $\mathbf{Q}$, ω, or T. As a result, neutron scattering studies of the copper oxide superconductors have focussed on the region of reciprocal space around the 2D antiferromagnetic rod where spin waves are observed in the insulating phases.

As emphasized by a number of speakers at this workshop, the spin-lattice relaxation rate measured by nuclear magnetic resonance spectroscopy is also sensitive to the dynamical susceptibility. The relaxation rate is proportional to an average of the susceptibility over the Brillouin zone, weighted by a form factor and the inverse frequency, all in the limit $\omega \to 0$. The form factor depends on the atomic site probed, so one obtains different measures of the susceptibility by looking at Cu and O, for example. While relaxation rate measurements contain less information than the neutron scattering cross section, they probe with high precision the low frequency regime which is quite sensitive to the opening of gaps in the spin fluctuation spectrum. Thus, NMR and neutron scattering are complementary techniques which can be combined to put strong constraints on theoretical models.

NMR measurements of the relaxation rate at the Cu(2) site in metallic $YBa_2Cu_3O_{6+x}$ provide evidence for antiferromagnetic spin fluctuations,[38,39] consistent with the neutron scattering results discussed here. Yasuoka and coworkers[40] have observed a systematic variation of the relaxation rate as a function of doping in superconducting $La_{2-x}Sr_xCuO_4$ and $YBa_2Cu_3O_{6+x}$, with a break in the temperature dependence at a characteristic temperature between 100 and 200 K, well above T_c. They interpreted the decrease of the relaxation rate below the characteristic temperature as evidence for suppression of antiferromagnetic fluctuations at low temperature. Alternatively, the temperature dependence of the relaxation rate[39,40] has been explained in terms of a temperature-dependent correlation length[37,41] or more generally in terms of the temperature dependence of the susceptibility.[31] Below T_c the relaxation rates at both the oxygen and copper sites drop off sharply.[42] Rice[43] has argued that this behavior indicates that all of the electronic spins are pairing up below T_c, so that few free spins remain at low temperature. Our neutron scattering results contradict this conclusion—it appears that antiferromagnetic spin fluctuations are relatively unaffected by the superconducting transition. It follows that the abrupt change in the susceptibility at T_c must be occurring in a different region of reciprocal space.[42] While it may be difficult to find the right signal, it would be of interest to attempt a more complete survey of the 2D Brillouin zone using neutron scattering.

Summary

We have discussed a number of neutron scattering studies of antiferromagnetic spin fluctuations in insulating and metallic $YBa_2Cu_3O_{6+x}$ and $La_{2-x}Sr_xCuO_4$. For the insulating phases, the Heisenberg Hamiltonian appears to be appropriate for describing the spin dynamics both below and above T_N. The magnetic correlations are highly two-dimensional, although bilayer correlations are important in $YBa_2Cu_3O_{6+x}$. Recent experiments have shown that dynamical antiferromagnetic spin correlations survive in the metallic phases and even in the superconducting state. Much work remains to more completely characterize these compounds, and we look forward with extreme anticipation to the return of neutrons to Brookhaven.

Acknowledgments

The work discussed in this paper is the result of collaborations and discussions with many scientists. We would especially like to acknowledge J. Als-Nielsen, W. J. L. Buyers, R. J. Birgeneau, H. Chou, V. J. Emery, P. Gehring, and G. Reiter. A major part of the work was supported by the U.S.-Japan Cooperative Neutron Scattering Program. Research at Brookhaven National Laboratory is supported by the Division of Materials Sciences, U.S. Department of Energy, under Contract No. DE-AC02-76CH00016.

References

1. J. M. Tranquada, in *Earlier and Recent Aspects of Superconductivity*, edited by J. G. Bednorz and K. A. Müller (Springer-Verlag, Berlin, to be published).
2. J. Rossat-Mignod, P. Burlet, M. J. Jurgens, C. Vettier, L. P. Regnault, J. Y. Henry, C. Ayache, L. Forro, H. Noel, M. Potel, P. Gougeon, and J. C. Levet, J. Phys. (Paris) **49**, C8-2119 (1988).
3. S. K. Sinha, in *Studies of High Temperature Superconductors, Vol. 4*, edited by A. V. Narlikar (Nova Science, New York, to be published).
4. J. M. Tranquada, S. M. Heald, A. R. Moodenbaugh, and Y. Xu, Phys. Rev. B **38**, 8893 (1988).
5. J. L. Hodeau, C. Chaillout, J. J. Capponi, and M. Marezio, Solid State Commun. **64**, 1349 (1987); Y. Nakazawa and M. Ishikawa, Physica C **158**, 381 (1989).
6. J. M. Tranquada, A. H. Moudden, A. I. Goldman, P. Zolliker, D. E. Cox, G. Shirane, S. K. Sinha, D. Vaknin, D. C. Johnston, M. S. Alvarez, A. J. Jacobson, J. T. Lewandowski, and J. M. Newsam, Phys. Rev. B. **38**, 2477 (1988).
7. Y. Guo, J.-M. Langlois, and W. A. Goddard III, Science **239** (1988); A. H. Moudden, G. Shirane, J. M. Tranquada, R. J. Birgeneau, Y. Endoh, K. Yamada, Y. Hidaka, and T. Murakami, Phys. Rev. B **38**, 8720 (1988).
8. H. Lütgemeier, R. A. Brand, Ch. Sauer, B. Rupp, P. M. Meuffels, and W. Zinn, in *Proc. of the International M^2S-HTSC Conference*, edited by N. E. Phillips, R. N. Shelton, and W. A. Harrison, to be published in Physica C.
9. W. E. Farneth, R. S. McLean, E. M. McCarron, III, F. Zuo, Y. Lu, B. R. Patton, and A. J. Epstein, Phys. Rev. B **39**, 6594 (1989).
10. J. M. Tranquada, G. Shirane, B. Keimer, S. Shamoto, and M. Sato, Phys. Rev. B **40**, 4503 (1989).
11. J. Rossat-Mignod, L. P. Regnault, M. J. Jurgens, C. Vettier, P. Burlet, J. Y. Henry, and G. Lapertot, in *Proc. of the Santa Fe Conf. on the Physics of Highly Correlated Electrons Systems,* to be published in Physica B.
12. K. B. Lyons, P. A. Fleury, L. F. Schneemeyer, and J. V. Waszczak, Phys. Rev. Lett. **60**, 732 (1988).
13. C. Vettier, P. Burlet, J. Y. Henry, M. J. Jurgens, G. Lapertot, L. P. Regnault, and J. Rossat-Mignod, Physica Scripta **T29**, 110 (1989).
14. M. Sato, S. Shamoto, J. M. Tranquada, G. Shirane, and B. Keimer, Phys. Rev. Lett. **61**, 1317 (1988).
15. R. J. Birgeneau and G. Shirane, in *Physical Properties of High Temperature Superconductor*, edited by D. M. Ginsberg (World-Scientific, Singapore, 1989).
16. K. Yamada, K. Kakurai, Y. Endoh, T. R. Thurston, M. A. Kastner, R. J. Birgeneau, G. Shirane, Y. Hidaka, and T. Murakami, Phys. Rev. B **40**, 4557 (1989).
17. S. Chakravarty, B. I. Halperin, and D. R. Nelson, Phys. Rev. B **39**, 2344 (1989).
18. S. Tyč, B. I. Halperin, and S. Chakravarty, Phys. Rev. Lett. **62**, 835 (1989).
19. D. R. Grempel, Phys. Rev. Lett. **61**, 1041 (1988).
20. H. Capellmann, lecture at this meeting; O. Schärpf, H. Capellmann, T. Brückel, A. Comberg, and H. Passing, to be published.
21. G. Aeppli *et al.*, unpublished.
22. G. Shirane, J. Als-Nielsen, M. Nielsen, J. M. Tranquada, H. Chou, S. Shamoto, and M. Sato, Phys. Rev. B (to be published).
23. R. J. Cava, B. Batlogg, K. M. Rabe, E. A. Rietman, P. K. Gallagher, and L. W. Rupp, Jr., Physica C **156**, 523 (1988).

24. R. F. Kiefl *et al.*, Phys. Rev. Lett. **63**, 2136 (1989).
25. W. Marshall and R. D. Lowde, Rep. Prog. Phys. **31**, 705 (1968).
26. J. M. Tranquada, W. J. L. Buyers, H. Chou, T. E. Mason, M. Sato, S. Shamoto, and G. Shirane, submitted to Phys. Rev. Lett..
27. T. R. Thurston, R. J. Birgeneau, M. A. Kastner, N. W. Preyer, G. Shirane, Y. Fujii, K. Yamada, Y. Endoh, K. Kakurai, M. Matsuda, Y. Hidaka, and T. Murakami, Phys. Rev. B **40**, 4585 (1989).
28. G. Shirane, R. J. Birgeneau, Y. Endoh, P. Gehring, M. A. Kastner, K. Kitazawa, H. Kojima, I. Tanaka, T. R. Thurston, and K. Yamada, Phys. Rev. Lett. **63**, 330 (1989).
29. A. Aharony, R. J. Birgeneau, A. Coniglio, M. A. Kastner, and H. E. Stanley, Phys. Rev. Lett. **60**, 1330 (1988).
30. B. I. Shraiman and E. D. Siggia, Phys. Rev. Lett. **62**, 1564 (1989); A. Mauger and D. L. Mills, Phys. Rev. B **40**, 4913 (1989).
31. N. Bulut, D. Hone, D. J. Scalapino, and N. E. Bickers, preprint UCSBTH-89-35.
32. R. J. Birgeneau, Y. Endoh, Y. Hidaka, K. Kakurai, M. A. Kastner, T. Murakami, G. Shirane, T. R. Thurston, and K. Yamada, Phys. Rev. B **39**, 2868 (1989).
33. T. Brückel, H. Capellmann, W. Just, O. Schärpf, S. Kemmler-Sack, R. Kiemel, and W. Schaefer, Europhys. Lett. **4**, 1189 (1987).
34. R. J. Birgeneau, D. R. Gabbe, H. P. Jenssen, M. A. Kastner, P. J. Picone, T. R. Thurston, G. Shirane, Y. Endoh, M. Sato, K. Yamada, Y. Hidaka, M. Oda, Y. Enomoto, M. Suzuki, and T. Murakami, Phys. Rev. B **38**, 6614 (1988).
35. Y. J. Uemura and B. Sternlieb, unpublished.
36. I. Tanaka and H. Kojima, Nature **337**, 21 (1989).
37. H. Monien, lecture at this meeting; D. Pines, in *Proc. of the Santa Fe Conf. on the Physics of Highly Correlated Electrons Systems,* to be published in Physica B.
38. M. Horvatić, P. Ségransan, C. Berthier, Y. Berthier, P. Butaud, J. Y. Henry, M. Couach, and J. P. Chaminade, Phys. Rev. B **39**, 7332 (1989); C. H. Pennington, D. J. Durand, C. P. Slichter, J. P. Rice, E. D. Bukowski, and D. M. Ginsberg, Phys. Rev. B **39** (1989).
39. W. W. Warren, Jr., R. E. Waldstedt, G. F. Brennert, R. J. Cava, R. Tycko, R. F. Bell, and G. Dabbagh, Phys. Rev. Lett. **62**, 1193 (1989).
40. H. Yasuoka, T. Imai, and T. Shimizu, in *Strong Correlations and Superconductivity,* edited by H. Fukuyama, S. Maekawa, and A. P. Malozemoff (Springer-Verlag, Berlin, 1989); T. Imai, T. Shimizu, H. Yasuoka, Y. Ueda, K. Yoshimura, and K. Kosuge, submitted to Phys. Rev. B.
41. B. S. Shastry, Phys. Rev. Lett. **63**, 1288 (1989).
42. P. C. Hammel, M. Takigawa, R. H. Heffner, Z. Fisk, and K. C. Ott, Phys. Rev. Lett. **63**, 1992 (1989).
43. T. M. Rice, Physica Scripta **T29**, 72 (1989).

NEUTRON SCATTERING MEASUREMENTS OF THE MAGNETIC EXCITATIONS OF HIGH-TEMPERATURE SUPERCONDUCTING MATERIALS

H. A. Mook,[1] G. Aeppli,[2,6] S. M. Hayden,[3,6] Z. Fisk,[4] and D. Rytz[5]

1. Oak Ridge National Laboratory, Oak Ridge, Tennessee 37831
2. AT&T Bell Laboratories, Murray Hill, New Jersey 07974
3. Institut Laue-Langevin, 38042 Grenoble, France
4. Los Alamos National Laboratory, Los Alamos, New Mexico 87545
5. Hughes Research Laboratory, Malibu, California 90265
6. Risø National Laboratory, Roskilde, Denmark

INTRODUCTION

Since the discovery of superconductivity in $(La_{1-x}Ba_x)_2CuO_4$ by Bednorz and Müller[1] in 1986, there has been an enormous interest in all aspects of this family of materials. Our measurements have used neutron scattering techniques to study the magnetic properties of these materials. The measurements were made at the Institut Laue-Langevin in Grenoble, France, and at the Risø National Laboratory at Roskilde, Denmark. A number of studies of these types of materials have previously been made at other laboratories, particularly Brookhaven National Laboratory.[2] Many of our measurements have been made on a different energy scale than the earlier measurements, and different aspects of the problem have been addressed. Where the experiments have overlapped, good agreement has generally been found.

We have chosen to address the problem by first using high-energy neutrons so that the fundamental properties of the excitations such as the spin wave velocity and lifetimes can be isolated. We then used lower energy neutrons to cover a wide energy scale. Indeed, particularly for La_2CuO_4, we have used incident neutrons varying in energy from 0.3 eV to 4 meV. This required experiments on a number of different neutron spectrometers. We find very different phenomena for pure La_2CuO_4 than when the material is doped with Ba. Indeed, very different behavior is found for only slightly different concentrations of Ba and O in the material. We will thus limit our discussion to two materials, La_2CuO_4 and $(La_{0.975}Ba_{0.025})_2CuO_4$. The first of these is an antiferromagnetic insulator, while the second shows metallic

Dynamics of Magnetic Fluctuations in High-Temperature Superconductors
Edited by G. Reiter *et al.*, Plenum Press, New York, 1991

behavior for T > 30 K and is a spin glass. We have chosen not to discuss the superconductor directly at this time, but rather to give as much information as possible about the insulator and the metal which show extraordinarily interesting phenomena in their own right.

EXPERIMENTAL CONSIDERATIONS

The $(LaBa)_2CuO_4$ structure consists of Cu-O planes separated from each other by a sufficient distance that the dominant magnetic interactions are those within the planes. At temperatures above about 500 K, La_2CuO_4 is tetragonal, but at room temperature, the material is orthorhombic. Our Ba doped sample is also orthorhombic, but the transition from tetragonal to orthorhombic takes place slightly above room temperature. The lattice constants of our La_2CuO_4 sample are a = 5.375 Å, b = 13.156 Å, and c = 5.409 Å, while the Ba doped sample has lattice constants of a = 5.357 Å, b = 13.165 Å, and c = 5.399 Å. Our sample of La_2CuO_4 undergoes an antiferromagnetic transition to a three-dimensional ordered state at about 295 K so that the exchange interaction between planes is sizable, even though it is still small compared to the exchange coupling in the planes. Figure 1 shows a phase diagram of the $(La_{1-x}Ba_x)_2CuO_4$ system as determined by Fujita et al.[3] We see that the composition of our Ba doped sample is such that it is as far as possible from the Néel state without reaching the composition that shows superconductivity.

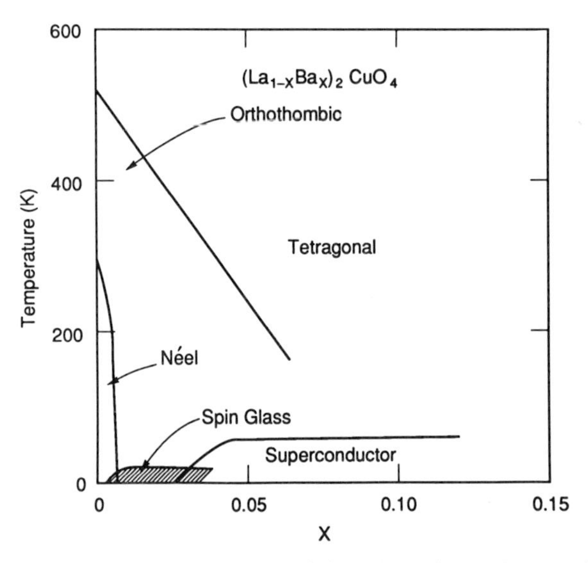

FIG. 1. Phase diagram of $(La_{1-x}Ba_x)_2CuO_4$ as determined by X ray, magnetic susceptibility, and resistance measurements.

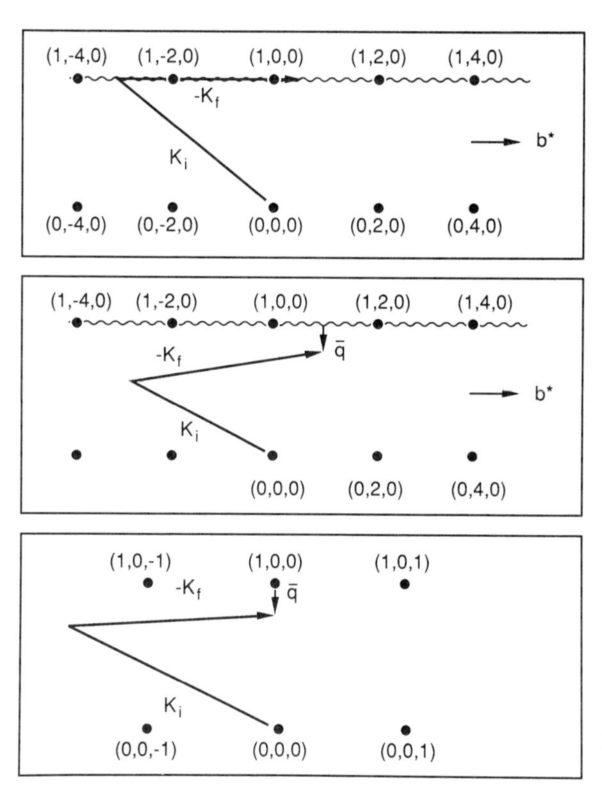

FIG. 2. Reciprocal lattice diagram showing different scattering geometries for La_2CuO_4. The wavy line shows the two-dimensional fluctuations.

Because the dominant magnetic interactions are in the planes, the scattering of interest is independent of the direction between planes and thus occurs in rods along b*. Figure 2 shows the reciprocal lattice where the wavy line represents the rod of scattering that passes through the magnetic Bragg point (1,0,0) and extends along b*. Since the material is orthorhombic and the crystal consists of random domains, there are really two rods of scattering slightly displaced from each other. This doesn't alter the experimental situation greatly, but since our experiments are done with high resolution, it must be taken into account carefully in the data analysis. The experiment is done by bringing neutrons of wave vector $\bar{K}_I$ onto the sample and scattering them along direction $\bar{K}_F$. The length of $\bar{K}_I$ is determined by the setting of the monochromator crystal of the triple-axis spectrometer, while the length of $\bar{K}_F$ is determined by the analyzer angle. If the analyzer is removed, then all lengths of $\bar{K}_F$ are possible. The integrated scattering along the rod can thus be measured by setting $-\bar{K}_F$ along the rod, as shown in the top diagram of Fig. 2, since the distance to the rod is then the momentum transfer $\bar{K}$ which is given by $\bar{K}_I$-$\bar{K}_F$. The rod of scattering in La_2CuO_4 was first discovered by Shirane et al.[4] using this technique. In our measurements we want to determine the size of the energy transfer so that the analyzer crystal is used. In this case, the measurement can take place as shown in the second diagram in Fig. 2. The momentum of the excitation is determined by the distance from the rod given by $\bar{q}$ and the excitation energy by the different lengths of $\bar{K}_I$ and $\bar{K}_F$. Another possibility is to mount the sample crystal so that the b* axis is up and the rods of scattering are perpendicular to the scattering plane as shown in the third diagram of Fig. 2. The momentum transfer of the excitation is then still given by the distance to the rod $\bar{q}$ and the energy transfer by $\bar{K}_I$ and $\bar{K}_F$. However, this scattering arrangement has the great advantage that the length of $\bar{q}$ is very insensitive to the out-of-plane components of $\bar{K}_I$ and $\bar{K}_F$. This means that neutron beams with a large vertical divergence can be used while still keeping good resolution in the momentum and energy transfer. Large focusing monochromator and analyzer crystals can thus be used, greatly increasing the neutron count rate.

In order to obtain sufficient signal to noise, large high-quality crystals are needed. The La_2CuO_4 sample was constructed from crystals obtained from melts containing excess CuO. The sample volume used for the highest energy measurements was about 8 $cm.^3$ The Ba doped crystals were grown by top-seeded solution growth from melts containing excess CuO and varied in size from about 0.3–1.2$cm.^3$ Scanning microprobe analysis showed that the Ba is uniformly distributed throughout the crystal.

EXPERIMENTAL RESULTS FOR La_2CuO_4

The excitations at high energy for La_2CuO_4 were measured at the hot source triple-axis spectrometer IN1 at the Institut Laue-Langevin. The neutron moderator for this spectrometer is a graphite block kept at a temperature of 2000 K and thus provides a high neutron flux for energies up to about 1 eV. Figure 3 shows measurements made with energy transfers varying from 140–30 meV. The solid lines are least-squares fits in which a model dispersion surface is convoluted with the spectrometer resolution. The parameters obtained from the best fit with the model dispersion surface give a spin wave velocity of $\hbar c = 0.78 \pm 0.04$ eV with zero spin wave damping and a linear spin wave dispersion. Measurements made at 5 K give the slightly larger value of $\hbar c = 0.85 \pm 0.03$ eV. For a near-neighbor Heisenberg model, this would result in an antiferromagnetic exchange coupling constant of 0.16 eV. The dashed line shows the calculated result of a measurement if the spin wave velocity is infinite and gives a measure of the spectrometer resolution.

At small q the neutron scattering cross section takes on the rather simple form given by

$$\frac{\partial^2 \sigma}{\partial \Omega \partial \omega} = \frac{k_i}{k_f} A_q[(n(\omega)+1)\delta(\omega - cq)+n(\omega)\delta(\omega+cq)] \ , \tag{1}$$

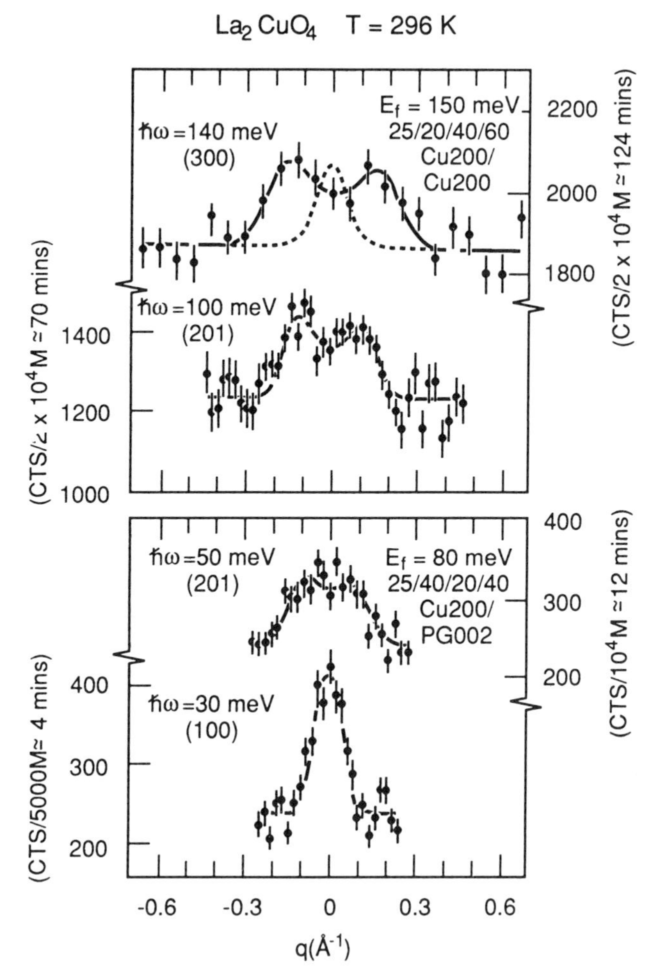

FIG. 3. High-resolution measurements of the magnetic excitations of La_2CuO_4 for energy transfers up to 140 meV.

where

$$n(\omega) = (\exp(\hbar\omega/kT)-1)^{-1} ,$$

and A_q is proportional to $1/q$.

Thus, for classical spin wave theory, the overall amplitude varies as $1/\omega\ (n(\omega)+1)$.

By using other triple-axis spectrometers with lower energy incident neutrons, the cross section has been determined over the energy range from 140–5 meV. These measurements will be dealt with in detail in Ref. 5; therefore, we will not discuss them at this time. The results of the measurements are shown in Fig. 4. The solid line is the function $1/\omega(n(\omega)+1)$, and we see that it gives an excellent representation of the data over two orders of magnitude of intensity. Since the spin-spin correlation length ξ is rather long at 290 K, these measurements are all in the regime where $q\xi$ is large compared to 1.

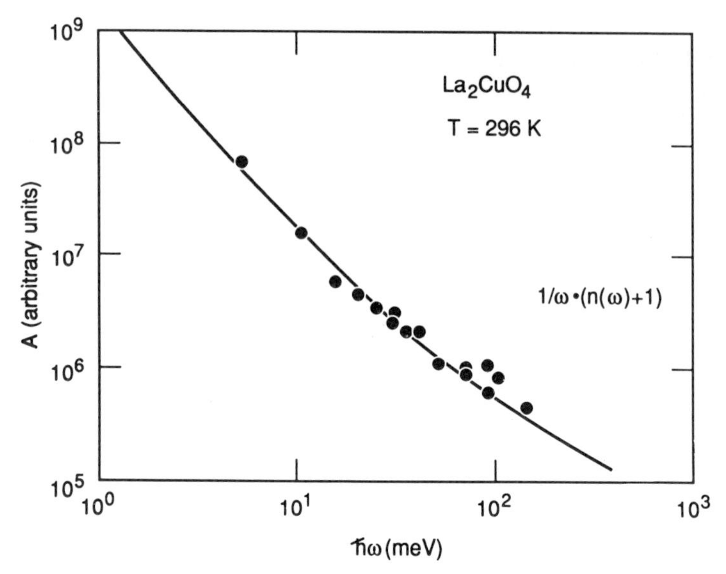

FIG. 4. Measurements of the spin wave intensity for La_2CuO_4 for a wide range of energy transfers. The solid line is the calculated intensity derived from classical spin wave theory.

At low energies where q gets smaller, the finite size of the correlation length will come into play, and the spin wave cross section will have to be modified to take this into account. In order to examine this regime, we need to employ a different energy scale, and the experiment was moved to the cold source triple-axis spectrometer at Risø in Denmark. We have observed no effects that cannot be explained by classical antiferromagnetism (no quantum effects), so we can model our cross section in a way similar to that used for the classical three-dimensional antiferromagnet $RbMnF_3$.[6] Using a function $C(q,\omega)$, as in Ref. 6, the cross section then becomes proportional to

$$C(q,\omega) = \frac{1}{q^2 + \xi^{-2}} \left[\frac{\Gamma}{\Gamma^2 + (\omega - cq)^2} + \frac{\Gamma}{\Gamma^2 + (\omega + cq^2)} \right] , \tag{2}$$

where Γ is the spin wave damping.

In this formula we have measured the spin wave velocity, so we only need to determine ξ and Γ. Earlier measurements[1] have shown that ξ/a is about 200 Å for La_2CuO_4 at 290 K. Our measurements suggest that ξ/a may be somewhat longer than this for our sample; however, since the exact number is not critical for our purposes, we will utilize the value of 200 Å.

We have never observed any finite Γ for the temperatures used in our measurements; however, we will include the possibility of spin wave damping by using Γ as a free parameter.

We will only consider the point q = 0 here, since obviously qξ is smallest at this point. We can then calculate the response of our neutron spectrometer using equation (2) and the known resolution of our spectrometer. It then turns out that if high resolution is used, the spin wave cross section is rather flat as a function of energy in the region of interest between -5 and 10 meV. Figure 5 shows a calculation of the spin wave cross section for different values of Γ when the spectrometer resolution is good (0.5 meV). If anything, the cross section has a minimum at energy E equal to zero, which may be surprising since we know the spin wave energy is zero at q = 0. It is an unusual feature of finite correlation range systems that the density of scattering states is rather featureless on a fine energy scale in the region near q = 0. Knowing this, we can exploit this fact to examine the cross section in the region near q = 0.

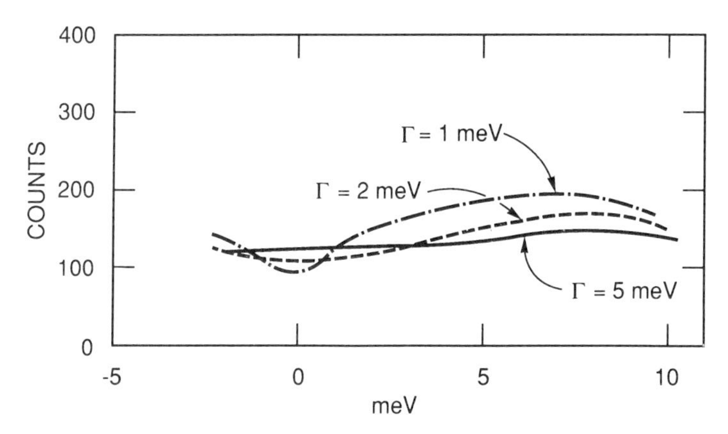

FIG. 5. Calculations of the spin wave intensity for a constant q scan for q = 0. For high resolution, the scattering is rather featureless, even when spin wave damping is considered.

Figure 6 shows a constant q scan at q = 0 for La_2CuO_4 taken at a temperature of 320 K. As can be seen from the background scan taken off the magnetic rod, there is some extra scattering at very low energies from the cryostat and other sources that falls within the resolution width of the spectrometer. In addition, there is scattering that is centered at zero energy from the sample that falls well outside the spectrometer resolution width. This results from quasielastic scattering or fluctuations in the order parameter of the magnetic system. Under all of this scattering there is the rather flat scattering that stems from the spin wave excitations. The size of this can be established from the difference in the scattering off and on the rod. The data can be fitted to a quasielastic form

$$C(0,\omega) = \text{cost.} \frac{1}{q^2 + \xi^{-2}} \frac{\Gamma}{\omega^2 + \Gamma^2} \quad . \tag{3}$$

For T = 320 K and ξ/a =200 Å, we obtain $\hbar\,\Gamma = 1.8 \pm 0.5$ meV. We then have a situation not unlike the classical antiferromagnet $RbMnF_3$, spin waves at the larger q values and spin relaxation at low q. La_2CuO_4, of course, has two-dimensional interactions, so we expect differences from the three-dimensional system.

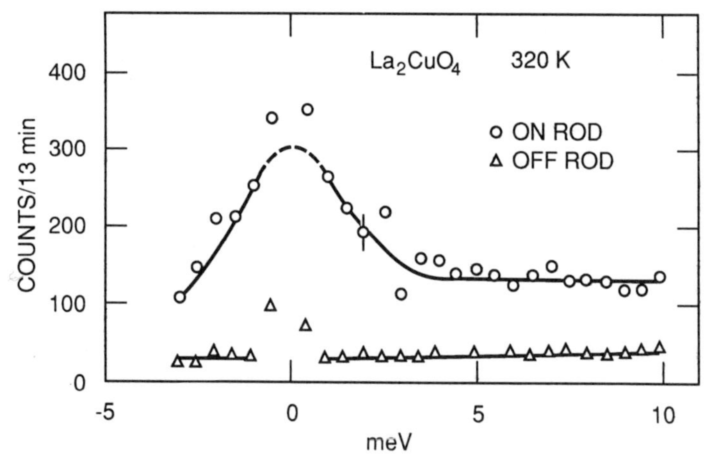

FIG. 6. Constant q scan at q = 0 for La_2CuO_4 using high resolution. Quasielastic scattering is clearly observed around zero energy transfer. The data taken off the scattering rod show the background intensity.

A recent and very important contribution to the field of magnetism in materials like La_2CuO_4 is the theoretical solution of the two-dimensional Heisenberg antiferromagnet.[7–10] Indeed, La_2CuO_4 appears to be a prototypical two-dimensional Heisenberg antiferromagnet. The calculations predict that the quasielastic peak width should be given by

$$\hbar\,\Gamma \simeq \hbar\,\xi^{-1} \left(\frac{T}{2\pi\,\rho_S} \right)^{1/2} \quad . \tag{4}$$

We know all the elements of this equation since ρ_s can be related to the exchange energy, c is the spin wave velocity, and ξ the correlation length. Inserting these values gives 1.7 meV in excellent agreement with our results.

Between the measurements and the calculations, we are acquiring a good understanding of the magnetic excitations in La_2CuO_4. In fact, the dynamics seem to be rather straightforward in the sense that we have not found it necessary to include quantum effects. The dynamics at high q are governed by the spin waves and at low q by the correlation length. We will see that doping with Ba dramatically alters this situation.

EXPERIMENTAL RESULTS FOR $(La_{0.975}Ba_{0.025})_2CuO_4$

We will not discuss the high-energy transfer measurements for $(La_{0.975}Ba_{0.025})_2CuO_4$ at this time. High-energy results for the Ba doped system are covered in Refs. 11 and 12. Instead, we will continue where we left off with La_2CuO_4 with low energies and high resolution to study the spin freezing phenomena. The first consideration for the doped material is the loss of three-dimensional long-range order. To prove that three-dimensional order is no longer present in the Ba doped sample, we must examine the (1,0,0) antiferromagnetic peak. The top part of Fig. 7 shows a measurement of this using the cold neutron source at Risø. At 183 K, a scan through the antiferromagnetic Bragg peak shows some very weak scattering that comes from second order processes and multiple scattering. At 9.25 K, the scattering is stronger, but the extra intensity comes from the wings on the peak, rather than the center. In fact, the extra scattering is no stronger than elsewhere on the two-dimensional rod shown by the result at (1,0.3,0) given in the bottom part of Fig. 7. Three-dimensional order thus no longer is present as all the scattering can be accounted for by two-dimensional fluctuations.

The second important consideration is the temperature dependence of the correlation length. We find that the correlation length does change with temperature for temperatures above 120 K, but that it remains constant for temperatures below this value. This is a very different situation than for La_2CuO_4. The temperature dependence of the magnetic phenomena in La_2CuO_4 is a result of the changing correlation length. Since the correlation length is fixed in the doped material, other considerations determine the behavior of the magnetic response.

Before discussing the dynamics, we will first consider the spin freezing in $(La_{0.975}Ba_{0.025})_2CuO_4$. The quantity of interest in this regard is the Edwards-Anderson order parameter which is given by

$$\lim_{t\to\infty}\left[< S_i(t)\, S_i(0) >\right]_{av} \quad .$$

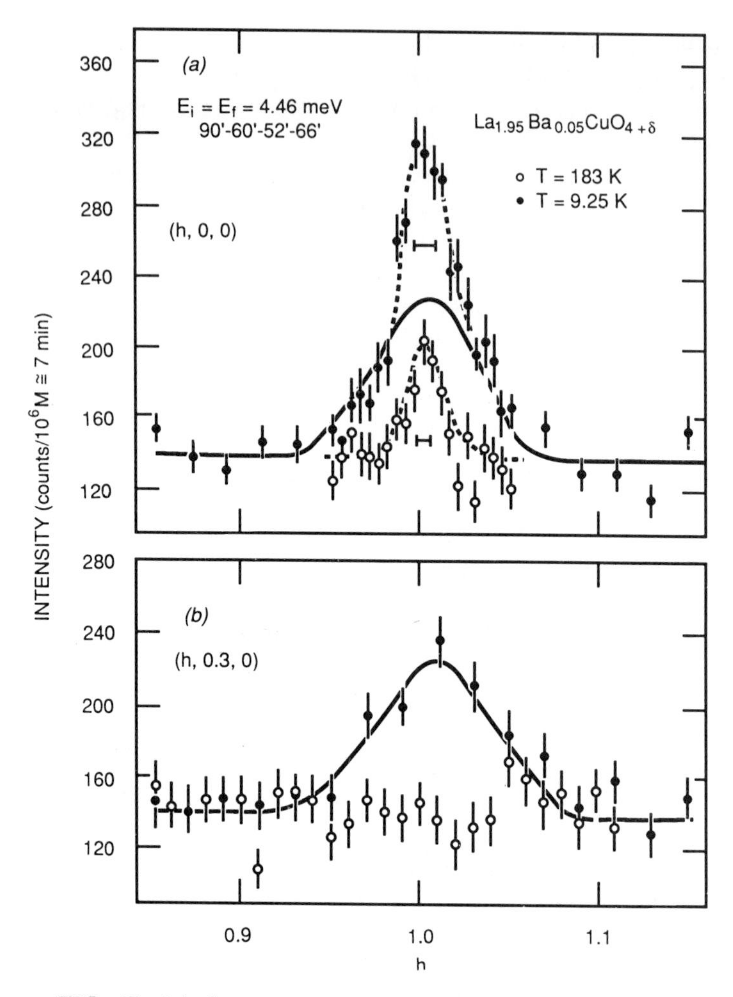

FIG. 7. (a) shows a scan through the antiferromagnetic Bragg reflection position at 183 and 9.25 K. At the lower temperature, a small increase in intensity appears but no more than is found elsewhere on the magnetic rod of scattering, as is shown in (b).

For most systems, this is a hard quantity to measure. Because it is a single-site correlation function, one has to sum over all of reciprocal space to determine it. Also, because it must be determined for long times, very high-energy resolution is generally needed. However, for two-dimensional freezing, all scattering is limited to the rod of scattering, and thus the volume of reciprocal space which must be investigated is severely restricted. In fact, all of the scattering can be collected by setting the scattering geometry as shown in the second part of Fig. 2 and scanning through the rod. In this case, of the three integrals to cover all reciprocal space, only one has to be done by hand. The intensity in the scan over the rod has to be summed, but the vertical resolution of the spectrometer takes care of the out-of-plane integral. The third integral is effectively done by the system itself as the sum along the rod is determined by the two-dimensional properties of the system.

The problem of having sufficient energy resolution to sample only elastic scattering on a very small scale is also simplified for our case. The spectrometer resolution of about 0.1 meV, given by the scattering configuration used with the cold neutron source, samples freezing on a temperature scale of about a degree. Because of the large energy scale of the excitations for the magnetic system, this resolution is sufficient. Figure 8 shows the result of the measurement demonstrating spin freezing beginning at about 30 K. The dashed line shows the background level obtained at 183 K.

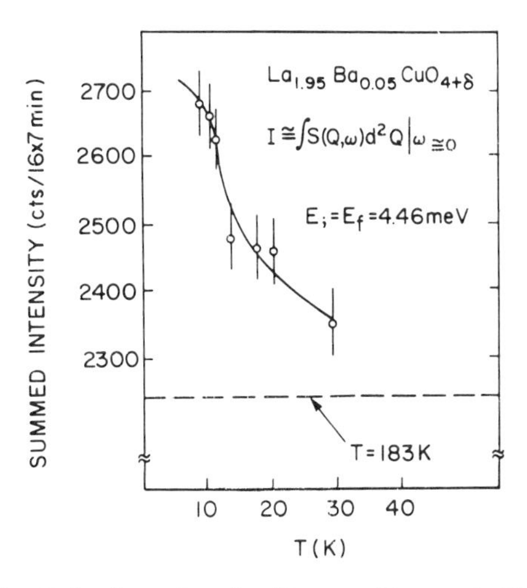

FIG. 8. Spin freezing in $(La_{0.975}Ba_{0.025})_2CuO_4$. The measurement gives a good approximation to the Edwards-Anderson order parameter for the material.

Now that we have established the spin freezing temperature and the temperature dependence of the magnetic correlation length, we can examine the behavior of the magnetic excitations. High-resolution measurements of the temperature dependence of the magnetic scattering are shown in Fig. 9. Rather, energy independent scattering is found at T = 300 K. There are a couple of high points near zero energy transfer that stem from the elastic scattering from the sample and cryostat, as shown from the background scan off the rod. The energy independent scattering results from the spin wave excitations as shown earlier for La_2CuO_4. The situation is basically unchanged for a temperature of 120 K. However, at 36 K, a noticeable quasielastic peak is observed that results from the dynamics of spin freezing.

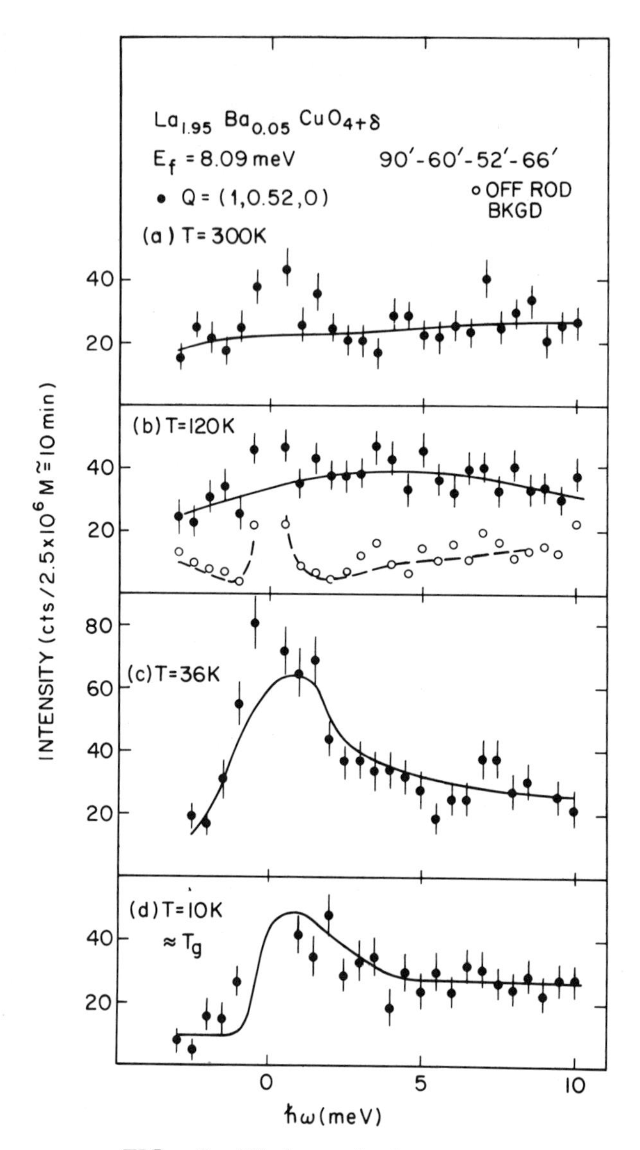

FIG. 9. High-resolution measurements of the spin dynamics of $(La_{0.975}Ba_{0.025})_2CuO_4$ at four temperatures. For temperatures above and below the spin freezing temperature, the spectra reflect the spin wave scattering which is rather flat on this scale. A quasielastic peak is observed for 36 K, which is near the spin freezing temperature.

This quasielastic scattering largely disappears at 10 K as the spin system is essentially frozen at this temperature. We remember that the magnetic correlation length is constant for temperatures lower than 120 K, so that the change in spin dynamics is governed by other considerations. Indeed the change in the dynamics results from the change in the susceptibility of the electronic system as the temperature is changed.

We have, therefore, a situation very different than for La_2CuO_4. The intensities of the magnetic excitations are no longer given by a simple formula like (1). However, the cross section always contains the term $n(\omega)$ which stems from very general principles. Thus, in determining the temperature dependence of the cross section, the influence of the term $n(\omega)$ must be taken into account. The fact that the cross section for the doped material does not scale in temperature like $n(\omega)$ means that the electronic character of the system changes with temperature.

CONCLUSIONS

Neutron scattering measurements with large, high quality single crystals are now yielding detailed information about the magnetism in La_2CuO_4 and $(LaBa)_2CuO_4$. In order to get a clear picture for the magnetic response of these materials, high-resolution measurements have been made on different energy scales. High-energy measurements using the hot source moderator were used to establish the spin wave velocity and lifetimes. Low-energy neutrons from the cold source were used to study the quasielastic behavior and spin freezing.

Our measurements for La_2CuO_4 are consistent with a classical antiferromagnetic response independent of quantum corrections. However, the neutron scattering from La_2CuO_4 is different in many respects from that of other materials. These differences are the manifestation of a two-dimensional magnetic system with a finite correlation length and large exchange interactions. There is still more to be done on La_2CuO_4, especially temperature dependent effects; however, between the measurements and the theory, we are beginning to get a good understanding of this material.

The situation is quite different for the Ba doped material. The measurements we have made on $(La_{0.975}Ba_{0.025})_2CuO_4$ have elucidated the spin freezing process for this material. However, in this case the temperature dependence of the spin dynamics is not determined by a changing correlation length or the standard type of critical slowing down. The magnetic response is directly affected by the underlying change in the electronic susceptibility. It would be helpful to have more theoretical guidance in this regime. Obviously the electronic response of such systems is very important, because with a very slight increase in doping, high-temperature superconductivity is achieved.

ACKNOWLEDGEMENTS

This research was supported in part by the Division of Materials Sciences, U.S. Department of Energy under contract DE-AC05-84OR21400 with Martin Marietta Energy Systems, Inc.

REFERENCES

1. J. G. Bednorz and K. A. Müller, Z. Phys. B 64:189 (1986).
2. R. J. Birgeneau and G. Shirane, p. 151 in: "Physical Properties of High-Temperature Superconductors 1," D. M. Ginsberg, ed., World Scientific, Singapore (1989).
3. F. Fujita, Y. Aoki, Y. Maeno, J. Sakurai, H. Fukuba, and H. Fujii, Jpn. J. Appl. Phys. 26:L368 (1987).
4. G. Shirane, Y. Endoh, R. J. Birgeneau, M. A. Kaster, Y. Hidaka, M. Oda, M. Suzuki, and T. Murakami, Phys. Rev. Lett. 59:1613 (1987).
5. S. M. Hayden, G. Aeppli, H. A. Mook, Z. Fisk (to be published).
6. H. Y. Lau, L. M. Corliss, A. Delapalme, J. M. Hastings, R. Nathens, and A. Tucciarone, Phys. Rev. Lett. 23:1225 (1969).
7. S. Chakravarty, B. Halperin, and D. Nelson, Phys. Rev. Lett. 60:1057 (1988).
8. D. R. Grempel, Phys. Rev. Lett. 61:1041 (1988).
9. S. Chakravarty, B. Halperin, D. Nelson, Phys. Rev. B 39:2344 (1989).
10. S. Tyč, B. Halperin, and S. Chakravarty, Phys. Rev. Lett. 62, 835,(1989)
11. G. Aeppli, S. M. Hayden, H. A. Mook, Z. Fisk, S.-W. Cheong, D. Rytz, J. P. Remeika, G. P. Espinosa, and A. S. Cooper, Phys. Rev. Lett. 62:2052 (1989).
12. G. Aeppli, S. M. Hayden, H. A. Mook, Z. Fisk, and D. Rytz, (to be published).

NEUTRON SCATTERING STUDY OF THE SPIN DYNAMICS IN $YBa_2Cu_3O_{6+x}$

J. Rossat-Mignod, L.P. Regnault, J.M. Jurguens* , P. Burlet ,
J.Y. Henry, G. Lapertot

Centre d'Etudes Nucléaires, DRF/SPh-MDN, 85 X
38041 Grenoble Cedex, France

C. Vettier

Institut Laue Langevin, 156 X, 38042 Grenoble Cedex, France

INTRODUCTION

Since the discovery of superconductivity in lamellar copper oxide materials a huge amount of experimental and theoretical works has been performed[1] but there is not yet a consensus on the physical mechanism of superconductivity. In order to clarify the physics involved accurate single crystal experiments must be performed. In this context, the neutron scattering technique plays an important role because both spatial and temporal spin fluctuations can be probed. A detailed investigation of the (La-Sr)$_2$CuO$_4$ system has been carried out at Brookhaven[2] and large antiferromagnetic (AF) spin correlations have been found above $T_N \sim 240$ K and in Sr-doped samples. The AF ordering was found to be very sensitive to both the Sr content[3] and the oxygen stoechiometry : T_N can be as high as 300 K[4] when oxygen p-holes are no longer present in CuO_2 planes.

At the Centre d'Etudes Nucléaires de Grenoble we have focused our effort on the $YBa_2Cu_3O_{6+x}$ system. This system has the advantage that, by changing only the oxygen content, we can investigate successively different interesting regimes. An AF ordering within CuO_2 planes was discovered in the insulating compound $YBa_2Cu_3O_6$[5,6] which persists up to $T_N = 415$ K[5]. We have undertaken systematic magnetic scattering studies of $YBa_2Cu_3O_{6+x}$ compounds as a function of the oxygen content using single crystal samples in order to undertake a detailed investigation of the magnetic phase diagram. Preliminary single crystal results where reported at the Interlaken Conference[7] and more detailed results at ICM conference in Paris[8]. These results have shown, as powder experiments[9], that the long range 3D-AF ordering disappears suddenly for $x \sim 0.40$.

These experiments allow us to define five different states : the AF-state ($x < 0.2$), the AF insulating states with a small amount of holes in CuO_2 planes ($0.25 < x < 0.40$), the metallic state ($0.4 < x < 0.5$) and the two superconducting states with $T_c = 60$ K ($x \sim 0.66$) and 92 K ($x \sim 0.95$). Successfully, at the CEN-Grenoble we initiated a programme to grow large single crystals of $YBa_2Cu_3O_{6+x}$. Single crystals of about 0.2 cm^3 were obtained, which have allowed us to perform inelastic neutron scattering experiments. Up to now we have investigated only the first three regimes and in this paper we will give a brief account of neutron scattering studies of the spin dynamics in samples with x = 0.15, 0.37, 0.45. Preliminary results on x = 0.15, 0.37 samples were presented at the EPS conference at Nice[10] and at the Stanford conference[11]. Results on the superconducting sample with x = 0.45 were first presented at the

---

* On leave from Kamerlingh Onnes Laboratory, Leiden, The Netherlands.

Santa Fe conference[12]. Moreover a general overview of our results was given at the International Seminar on High Temperature Superconductors in Dubna[13].

The paper is arranged as follows. The experimental conditions and the sample preparation are described first, then the investigation of the magnetic phase diagram will be summarized. The spin dynamics in the purely AF-state (x = 0.15), in the AF-state with a small amount of holes (x = 0.37) and in the metallic superconducting state (x = 0.45) will be presented successively. The obtained results will be discussed and some conclusions will be established.

EXPERIMENTAL AND SAMPLES

Neutron scattering experiments were performed on single crystals using three-axis spectrometers, mainly IN8, at the Institut Laue Langevin.

Ge(111) or graphite (002) crystals were used as monochromators and graphite (002) as analyzer. One or two pyrolytic graphite filters were employed to suppress higher-order contamination. High neutron energy transfer experiments were performed with Cu(111) or Cu(220) as monochromator and graphite (004) or Cu(111) as analyzer. Collimations of the neutron beams were chosen such as to optimize intensity and resolution.

The single crystal sample (~ 0.40 cm^3) was mounted in a standard ILL cryostat with the (110) and (001) axes in the scattering plane.

Large single crystals have been successfully grown at the Centre d'Etudes Nucléaires de Grenoble. These single crystals have a good mosaic spread (smaller than one degree). Moreover, they have the great advantage that the oxygen content can easily be changed from x = 0 to x = 0.99. Actually, this makes the $YBa_2Cu_3O_{6+x}$ system easier to investigate because the complete range can be covered, from the AF to the superconducting state by using the same crystal. A well defined procedure has been used for the intercalation of oxygen in Cu(1) planes. The oxygen stoechiometry was better than x = 0.01 and the accuracy in the determination of x was less than x = 0.02.

MAGNETIC PHASE DIAGRAM[8,14]

$YBa_2Cu_3O_6$[5,6] orders with an antiferromagnetic structure described by a wave vector $\vec{k}$ = (1/2, 1/2, 0), i.e. corresponding to the magnetic unit cell ($a\sqrt{2}$, $a\sqrt{2}$, c) shown in Fig. 1. The absence of intensity at (1/2, 1/2, 0) substantiates the AF coupling between Cu^{+2} moments in the Cu(2) Bravais sublattices at (00z) and (00$\bar{z}$) and confirms the non-magnetic Cu^{+}state of Cu(1) site at (0,0,0). The AF direction lies

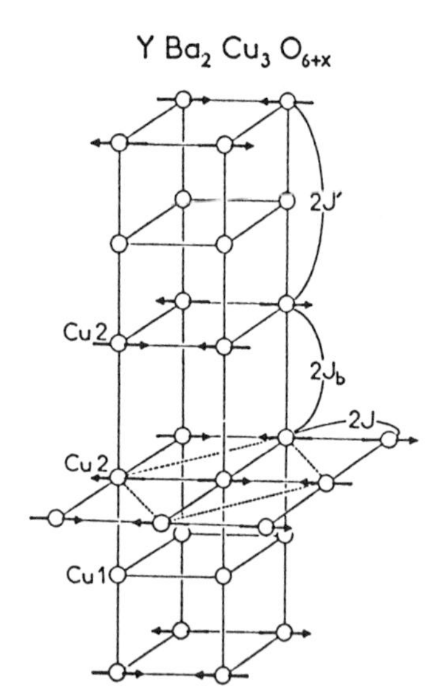

Fig. 1 . Antiferromagnetic structure of $YBa_2Cu_3O_{6-x}$

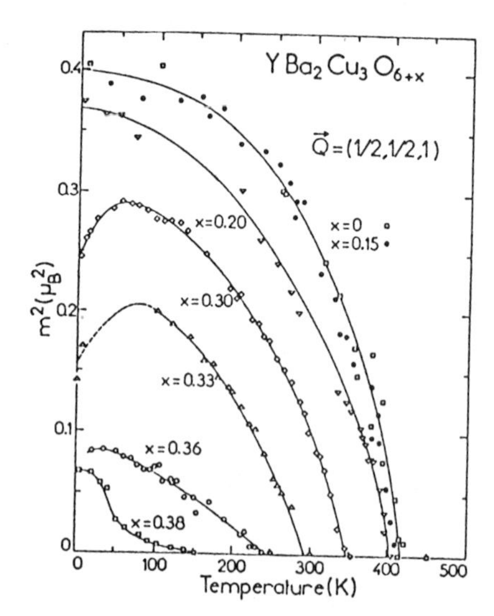

Fig. 2 . Intensity of the magnetic Bragg peak (1/2, 1/2, 1) as a function of temperature of $YBa_2Cu_3O_{6+x}$ for various oxygen contents. Intensities have been normalized in order to reflect the square of the ordered moment.

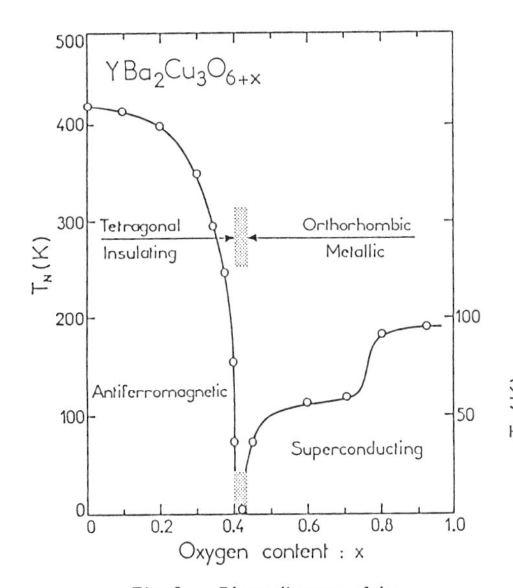

Fig. 3 . Phase diagram of the $YBa_2Cu_3O_{6+x}$ system

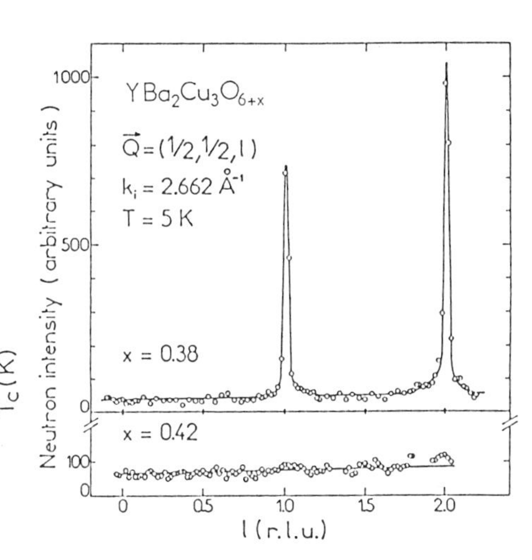

Fig. 4 . Elastic scans along (1/2, 1/2, l) for $YBa_2Cu_3O_{6+x}$ with x = 0.38 and 0.42.

within the basal plane, as established unambiguously by magnetic intensity measurements on single crystals[7]. Single crystal magnetic intensity measurements[8] yield a low temperature moment value $m_0 = 0.64 \pm 0.03$ μ_B. An accurate study as a function of temperature indicates a second order phase transition with an ordering temperature $T_N = 415$ $\pm$ 5 K and a critical exponent of the staggered magnetization $\beta = 0.25 \pm 0.03$ typical of a quasi-2D XY AF system[15] (see Fig. 15).

Crystals with various oxygen contents where investigated in detail as a function of temperature. The temperature dependence of magnetic Bragg peak intensities is reported on a normalized scale in Fig. 2.

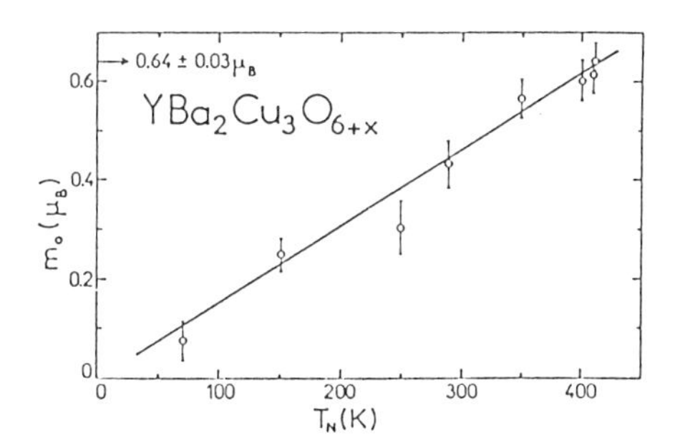

Fig. 5 . Value of the low temperature ordered moment as a function of the ordering temperature for $YBa_2Cu_3O_{6+x}$.

From these data, the magnetic phase diagram given in Fig. 3 has been deduced. For increasing oxygen contents up to x = 0.20 the AF structure remains unchanged, m_0 and T_N staying constant. For x > 0.20 both T_N and m_0 decrease, rather smoothly up to x = 0.35 and more abruptly above resulting in a sharp disappearance of the long range AF ordering at $x_c = 0.41 \pm 0.02$. The steepness of the magnetic - nonmagnetic transition is evidenced in Fig. 4 which indicates a 3D-AF ordering for x = 0.38 (T_N = 150 K) and the absence of ordering for x = 0.42. This result is a direct proof of the high homogeneity of the oxygen content of our samples.

A striking result is the linear relationship between the low temperature 3D-ordered moment m_0 and the 3D-ordering temperature T_N for any oxygen content smaller than x_c (see Fig. 5). As we shall see later the 3D-AF order is destroyed very quickly by a small amount of p-holes ($n_h \leq 2\%$) which create a static disorder around them within the Cu(2) planes. Two typical behaviours emerge: at low temperatures a reentrant-type behaviour is observed and close to the critical oxygen concentration the 3D order parameter exhibits a very unusual temperature dependence. This will be discussed in more details in section 5.

Moreover, for all samples studied up to x = 0.38, only the reflections (1/2, 1/2, *l*) with *l* integer were observed. No other magnetic peaks could be found down to the lowest temperatures. In particular, no magnetic intensity appears at (1/2, 1/2, 0) and (1/2, 1/2, 1/2) which indicates that the Cu(1) sites carry no long range ordered moment and that the bi-layers remain always antiferromagnetically coupled.

High magnetic field measurements on a sample with x = 0.35 have established that the AF-direction within the (a, b) plane lies along (100) or equivalently along (010).

SPIN DYNAMICS IN THE PURE AF-STATE[10,11,12,13]

As it is rather difficult to reach the fully deoxygenated limit we have prepared a sample with the composition $YBa_2Cu_3O_{6.15}$. With such an oxygen content no hole is created within CuO_2 planes, this is supported by the large value of T_N = 410 K. Below this temperature only copper spins in Cu(2) sites of the CuO_2 layers develop a 3D-long range AF ordering. Down to the lowest temperature, no order is found for Cu(1) spins. Therefore when investigating the spin wave spectrum, the most important exchange interactions are the strong in-plane superexchange interaction J, the direct exchange coupling between the two CuO_2 sheets, J_b, and the coupling between bi-layers, J', corresponding to the indirect exchange via Cu ions in the Cu(1) site. Cu^{2+} being in a S = 1/2 ground state the planar anisotropy results from the anisotropy of the interaction $\Delta J = J^{\perp} - J^{//}$, between in-plane and out-of-plane spin components, according to the exchange Hamiltonien.

$$H = -\sum_{ij} J_{ij}^{//} S_i^z S_j^z + J_{ij}^{\perp}\left(S_i^x S_j^x + S_i^y S_j^y\right)$$

We must notice that with this definition the copper pair interaction is actually 2J (see Fig. 1).

Taking into account the AF-coupling between the two Bravais sub-lattices $\vec{m}_1$, in (00z) and $\vec{m}_2 = -\vec{m}_1$ in (001-z) and the value of the AF-ordering wave vector $\vec{k}$ = (1/2, 1/2, 0), the spin wave spectrum has been calculated from an usual spin wave theory. As shown in Fig. 6, we expect along the (q, q, 0) direction a single excitation energy, $\hbar\omega(q) = 4J \sin 2\pi q$, over almost the whole Brillouin zone which extends up to 4J (0.35 - 0.4 eV). However near the zone center and the zone boundary the acoustic mode is well separated from the optical one.

The energy, $\hbar\omega(\vec{q}) = c_o|\vec{q}|$, of the acoustic mode near the zone center defined the spin wave velocity $c_o = 4J\, a/\sqrt{2}$. Using the value a = 3.85 Å we have c_o (meV.Å) ~ J(K). At the zone boundary, the planar anisotropy gives rise to an energy gap $4J\sqrt{2\Delta J/J}$ where $\Delta J = J^{\perp} - J^{//}$. The optical mode is expected at the energy $4J\sqrt{J_b/J}$ both at the zone boundary and zone center.

Along the (0, 0, q) direction a very weak energy dispersion is expected because only the weak coupling J' is involved ; at the zone boundary (0, 0, 1/2) the energy gap is $4J\sqrt{J'/J}$.

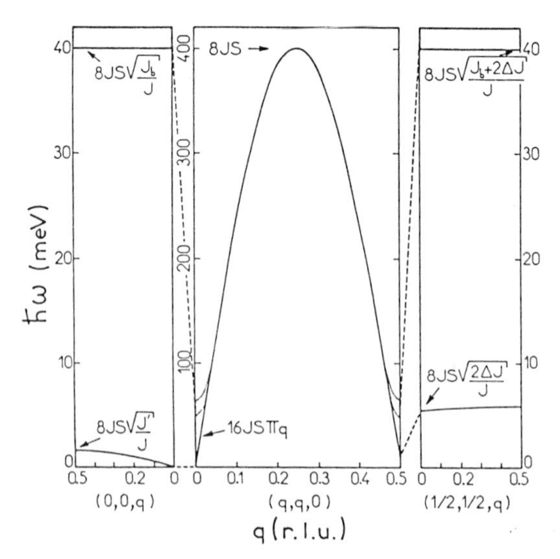

Fig. 6 . Theoretical spin wave spectrum of $YBa_2Cu_3O_{6+x}$

Energy scans for scattering vectors $\vec{Q}$ = (1/2, 1/2, *l*) clearly show a double peak structure (see Fig. 7). The absence of the high energy peak for large *l* value establishes that the low energy part of the spectrum can be assigned to excitations of in-plane spin components and the high energy part to excitations of the out-of plane spin component.

The strong intensity modulation observed in q - scans along (0,0,*l*) (Fig. 8) with the absence of scattering at *l* = 0 and 3.5 is a signature of the acoustic nature of the mode.

Therefore the optical mode must have a maximum intensity at *l* = 0 and 3.5, but, as show in Fig. 9, there is no detectable inelastic signal up to 40 meV indicating that the optical mode has an energy larger than 50 meV.

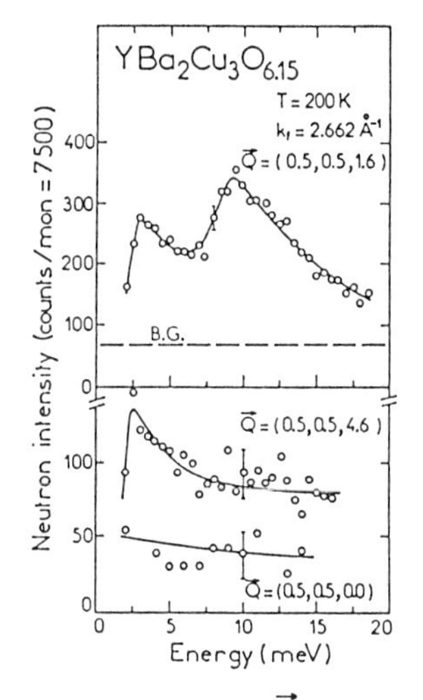

Fig. 7 . Energy scans for $\vec{Q} = (1/2, 1/2, \ell)$ with $l = 1.6$ and 4.6 at $T = 200$ K for $YBa_2Cu_3O_{6.15}$

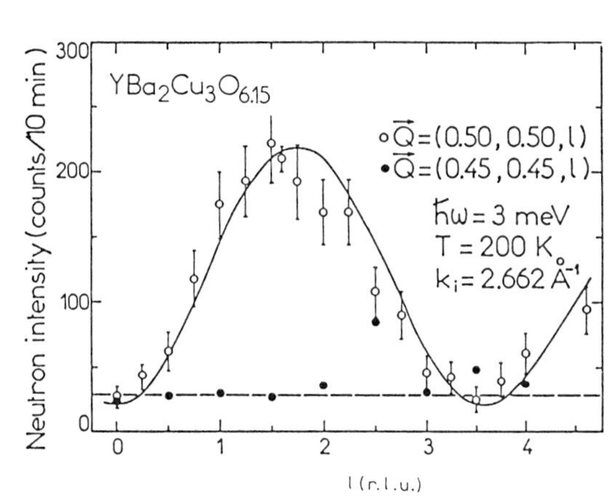

Fig. 8 . Q-scans along the magnetic rod at $\hbar\omega = 3$ meV and $T = 200$ K for $YBa_2Cu_3O_{6.15}$ showing the intensity modulation typical for an acoustic mode.

Q-scans performed at constant energy transfers up to $\hbar\omega = 50$ meV give a single peak ; the q-width up to $\hbar\omega = 15$ meV is close to the experimental resolution, only around 30 meV a q-width value twice the experimental resolution ($\Delta q_{res.} = 0.030$ r.l.u) has been achieved (see Fig. 10). The deconvolution of these data yields the dispersion curve, near the zone boundary, reported in Fig. 11. An extremely large value for the spin wave velocity $c_o = 1.0 \pm 0.05$ eVÅ is deduced, yielding an in-plane Cu pair interaction 2J = 2000 K.

The energy gap of out-of-plane excitation is $\Delta^z \sim 5$ meV resulting from an XY anisotropy exchange interaction $\Delta J = J^{\perp} - J^{//} = 10^{-4}$ J.

The low energy part of the spectrum, typical of excitations of in-plane spin components, exhibits a dispersion along the (0,0,q) direction with a maximum value $\hbar\omega$ (0, 0, 0.5) = 1.6 meV at the zone boundary. Therefore the coupling J' between the bi-layers is extremely small $J' = 10^{-5}$ J, establishing that YBa_2CuO_6 is a very good quasi-2D antiferromagnet. High resolution energy scans near the zone center cannot give evidence for a gap. If it exists, the zone center energy gap must be smaller than 0.2 meV yielding a negligible Ising-type anisotropy within the basal plane.

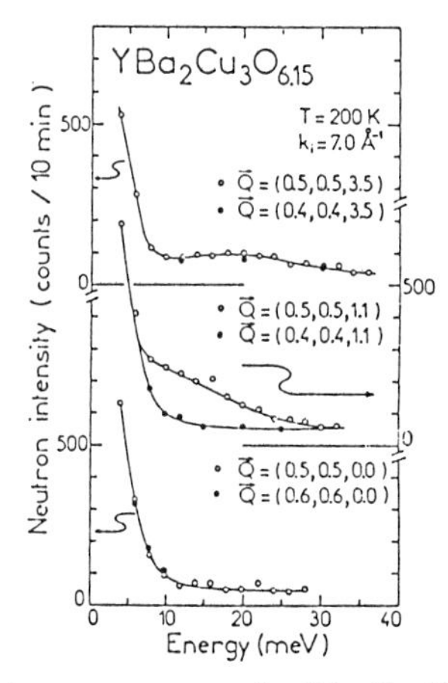

Fig. 9 . Energy scans for $YBa_2Cu_3O_{6.15}$ at $T = 200$ K.

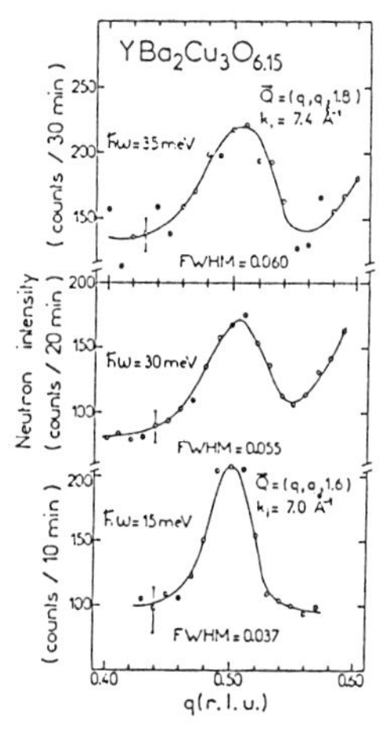

Fig. 10 . Q-scans across the magnetic rod for energy transfers $\hbar\omega = 15, 30$ and 35 meV for $YBa_2Cu_3O_{6.15}$

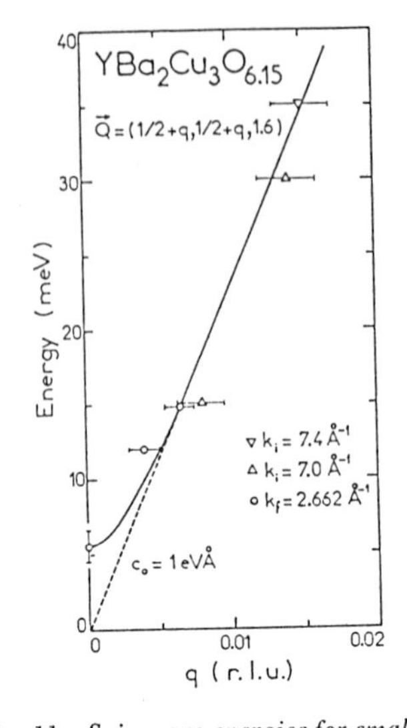

Fig. 11 . Spin wave energies for small wave vectors around the AF Bragg peak of the purely AF sample $YBa_2Cu_3O_{6.15}$. The dotted line corresponds to a spin wave velocity c_o = 1 eVÅ.

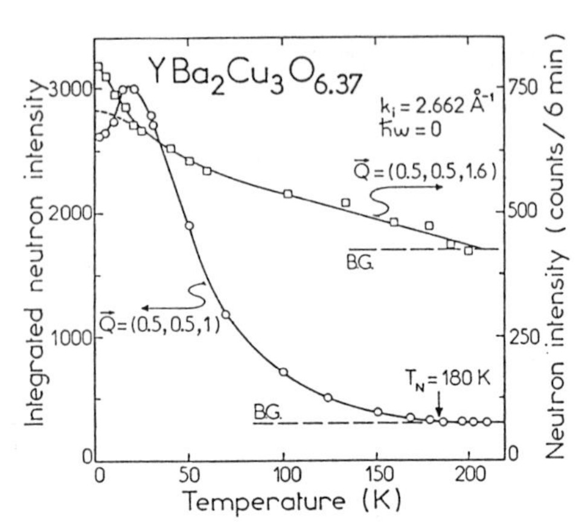

Fig. 12 . Intensity of the magnetic Bragg peak (1/2, 1/2, 1) and of the magnetic rod at $\vec{Q}$ = (1/2, 1/2, 1.6) as a function of temperature for $YBa_2Cu_3O_{6.37}$

Moreover, as no optical mode has been detected up to 50 meV, the coupling between the two Cu(2) layers is rather large : $J_b > 10^{-2}$ J, but must be smaller than 10^{-1}J.

Clearly these results establish that $YBa_2Cu_3O_{6+x}$ is a S = 1/2 bi-layer Heisenberg antiferromagnet with a very weak XY anisotropy (10^{-4}) and inter-bi-layer coupling (10^{-5}). While the XY anisotropy is weak, it clearly affects the pseudo critical exponent (ß = 0.25).

INFLUENCE OF P-HOLES ON THE AF-ORDER AND THE SPIN DYNAMICS[11, 12,13]

A detailed study has been carried out on a sample close to the critical oxygen concentration with x = 0.37. This sample develops an AF-order below T_N = 180 K (see Fig. 12) and no sign of superconductivity was detected[16]. The low temperature value of the 3D-ordered moment, $m_o = 0.3\ \mu_B$, is reduced because p-holes induce some static disorder in the moment direction within the Cu(2)-planes, as shown by the additional elastic magnetic scattering existing along the (1/2, 1/2, l) rod (see Fig. 12-13). The q-width of the (1/2, 1/2, l) rod yields a value Γ_q = 0.015 r.l.u., i.e. an in-plane correlation length ξ = 7.5 a.

The q-scan along the magnetic rod, reported in Fig. 14, clearly indicates that coexist both well-defined magnetic Bragg peaks and magnetic Bragg rods (within the energy resolution $\Delta E \sim 0.5$ meV) with an intensity modulated along l. The comparison of the intensity at low-T shows that there is as much intensity in the Bragg peak than in the Bragg rod indicating that the value of the local moments is not reduced but that p-holes induce, within Cu(2) planes, a disorder in moment directions. However the modulation of the rod intensity implies that the AF-coupling between the two Cu(2) layers is not affected by this in-plane disorder which is a result of the large value of the coupling between the layers, J_b. Only the stacking of the bi-layers is perturbed.

Moreover, as shown in Fig. 12, the magnetic Bragg peak intensity exhibits a quite unusual temperature dependence indicating that the 3D-ordering cannot built up in an usual manner. We suspect than in addition to thermal fluctuations there exist also quantum fluctuations induced by the hoping of p-hole magnetic polarons.
The influence of hole motion on the AF-phase transition can also be observed on the critical behaviour. While the experiments are not very accurate we can see in Fig. 15 that the value of the critical exponent ß, which describes the T-behaviour of the 3D-order parameter, increases from 0.25 up to 0.5

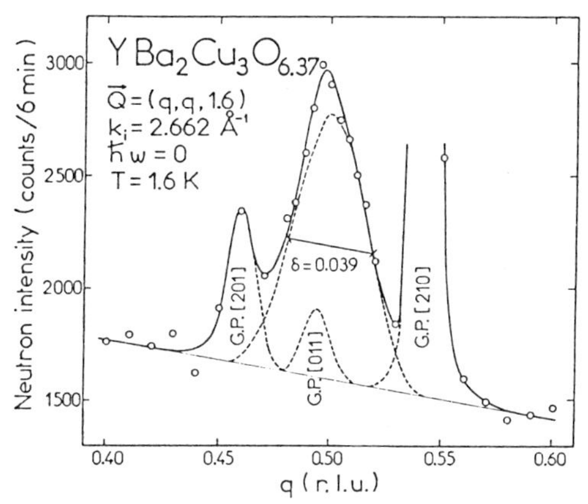

Fig. 13 . *Q-scan across the magnetic rod for* $\vec{Q} = (q, q, 1.6)$ *at* $T = 1.6$ K *and* $\hbar\omega = 0$ *for* $YBa_2Cu_3O_{6.37}$. *Additional peaks are due to the green phase.*

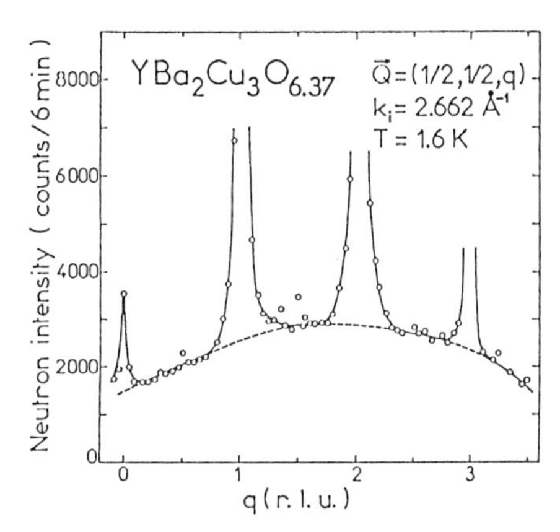

Fig. 14 . *Q-scan along the magnetic rod for* $\vec{Q} = (1/2, 1/2, q)$ *at* $T = 1.6$ K *showing the intensity modulation due to the AF-coupling between the two Cu(2) layers.*

when the amount of holes increases yielding even to the unusual behaviour found close to the critical hole concentration.

The hole concentration can be estimated from the in-plane correlation length because for small hole concentrations (n_h) we can expect ξ to be the distance between holes because the size of magnetic defects or polarons is smaller than the distance between polarons, then $n_h = 1/\xi^2$. So in the x = 0.37 sample a hole concentration $n_h = 1.8\%$ can be deduced indicating that the critical hole concentration is $n_h^c \approx 2\ \%$ in good agreement with results found in the $(LaSr)_2\ CuO_4$ system[2,3]. So a quite general result is that about 2% of holes in CuO_2 planes prevent any 3D-AF ordering to develop down to the lowest temperature. Within a Cu(2)-plane magnetic polarons are randomly distributed but, as the AF-coupling between the two Cu(2)-planes is not affected, they are antiferromagnetically coupled along the c-axes and, holes pile up to form a kind of bipolarons.

Another important result is the observation of a reentrant behaviour at low temperature (see Fig. 12) that we ascribe to hole localization. The Bragg peak intensities are depressed and transferred into the magnetic

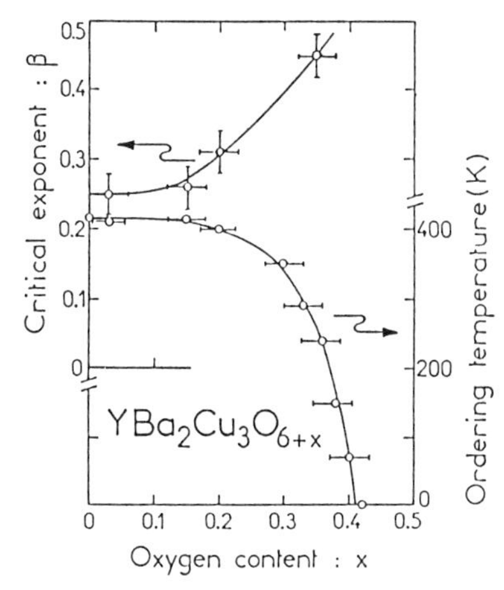

Fig. 15 . *Critical exponent* β *of the staggered magnetization as a function of the oxygen content x for the system* $YBa_2Cu_3O_{6+x}$.

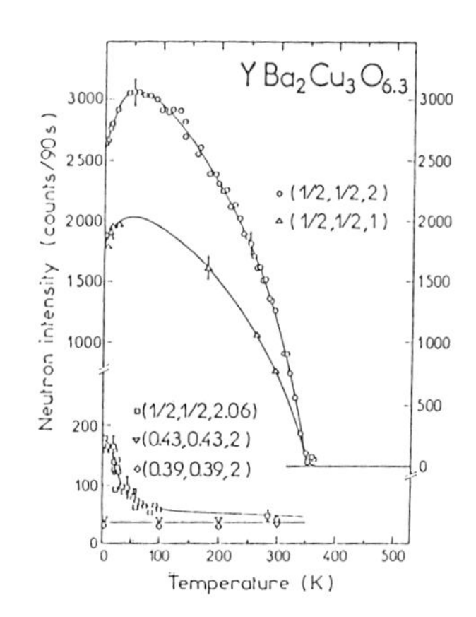

Fig. 16 . *Intensity of magnetic Bragg peaks (1/2, 1/2, 1) and (1/2, 1/2, 2) and of the magnetic rod at* $\vec{Q} = (1/2, 1/2, 2.06)$ *as a fucntion of temperature for* $YBa_2Cu_3O_{6.30}$.

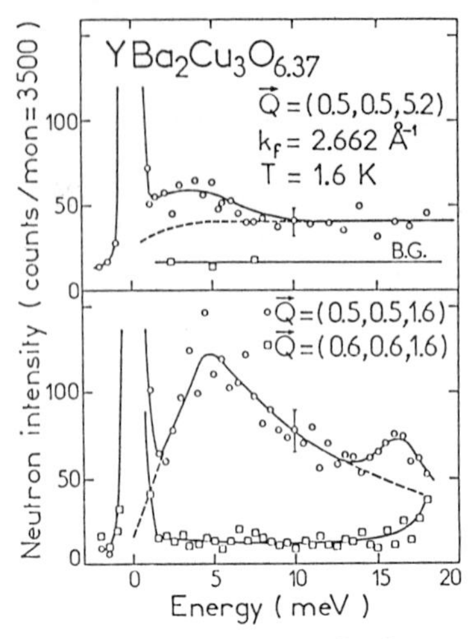

Fig. 17 . Energy scans in $Yba_2Cu_3O_{6.37}$ at $\vec{Q}$ = (1/2, 1/2, 1.6) showing the existence of the out-of plane fluctuations. At $\vec{Q}$ = (1/2, 1/,2 5.2), the overdamped in-plane fluctuations (XY) dominate the scattering.

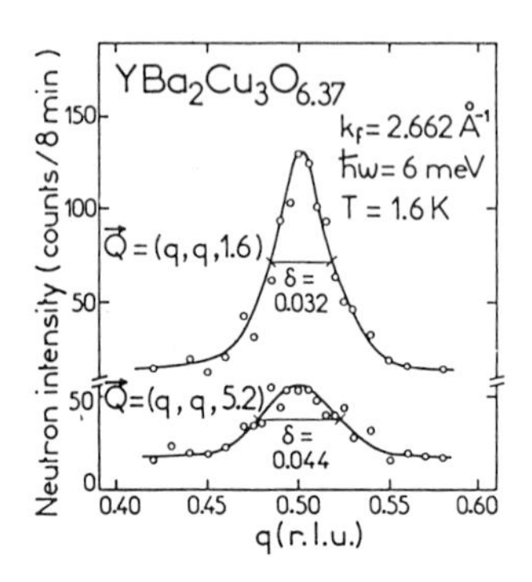

Fig. 18 . Q-scans in $YBa_2Cu_3O_{6.37}$ at ħω = 6 meV showing the importance of XY and Z spin fluctuations on the q-width.

rods indicating an additional lost of the AF-coupling in the stacking of the bi-layers. It must be noticed that this reentrant temperature T_r is depending on the amount of holes. Larger is the amount of holes, smaller is T_r:T_r = 15 K for x = 0.37 and T_r = 50 K for x = 0.30 (see Fig. 16). So T_r appears to vanish at the critical hole concentration which seems to rule out a spin glass behaviour. It must be underligned that a quite similar behaviour was found in an off-stoechiometric La_2CuO_4 crystal[3].

A detailed inelastic neutron scattering study has been performed on the sample $YBa_2Cu_3O_{6.37}$. Typical energy scans are reported in Fig. 17. They clearly show that at T = 1.6 K a small amount of holes (~ 1.8%) strongly modifies the spin dynamics. The analysis of the scattering intensity as a function of $\vec{q}$ along the (0,0,*l*) direction allows us to separate the contributions of in-plane and out-of-plane spin components. Excitations associated with the out-of-plane spin component have a propagative character and the anisotropy gap is reduced by a factor two (Δ^Z = 2.5 meV). However, excitations associated with in-plane spin components are overdamped (diffusive behaviour) with a characteristic energy Γ_ω ~ 17 meV. Such a difference in behaviour between out-of-plane and in-plane components is another signature of the static disorder of the direction of the copper moments within the basal plane induced by p-holes. In this energy range (ħω < 20 meV) the wave length of the excitation ($q < \Gamma_q$) is larger than the correlation length and spin waves cannot propagate. Only excitations of in-plane components are overdamped we conclude that the static disorder concerns only the in-plane components indicating that local static moments remain in the basal plane.

Q-scans performed at energy transfers of 6 and 12 meV show single broadened peaks, as show in Fig. 18, (Γ_q = 0.032 r.l.u. at ħω = 6 meV) indicating a softening of the spin wave velocity. However q-scans at higher energies, around 30 meV, do not show any indication for a double peak arising from the spin waves with wave vectors $+\vec{q}$ and $-\vec{q}$. The main reason of this result is that in-plane and out-of-plane excitations behave differently. In order to separate these two contributions the same q-scans have been performed around both $\overline{Q}$ = (1/2, 1/2, 1.6) and (1/2, 12, 5.2) for energy transfers of 6 and 12 meV, at higher energies the contribution of the in-plane spin component is dominant (see Fig. 17). The obtained widths are larger for the in-plane component, at ħω ~ 6 meV Δq^Z = 0.030 r.l.u. and Δq^{xy} = 0.052 r.l.u. ($\Delta q_{res.}$ = 0.017 r.l.u.).

The deconvolution of the data is reported in Fig. 19 ; for the out-of-plane spin component no important damping was found, whereas a very large damping (Γ_ω = 17 meV) was determined for the in-plane contribution. The obtained results clearly indicate that the renormalization of spin wave energies is q-dependent and reaches a factor two at small wave vectors ($q < \Gamma_q$). A spin wave velocity c = 0.45 ± 0.05 eVÅ can be deduced indicating a large reduction by a small amount of p-holes. While it is not possible to get experimental data we can anticipate that the renormalization is negligible for $q >> \Gamma_q$ which means

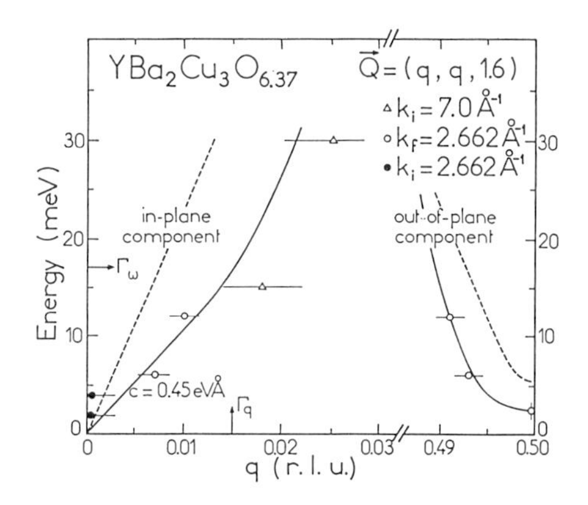

Fig. 19. Excitation energies of in-plane and out-of-plane spin components for the sample $YBa_2Cu_3O_{6.37}$. The renormalization of excitation energies is large at small q as shown by the comparison with the undoped AF sample (dotted line).

that the Cu-Cu superexchange coupling J is not affected. It is worth noting that the damping $\Gamma_\omega = 17$ meV is related to $\Gamma_q = 0.015$ r.l.u. by the simple relation $\Gamma_\omega = c.\ 2.3\ \Gamma_q$ (the factor 2.3 transforms Γ_q in r.l.u. to $Å^{-1}$) as can be seen in Fig. 19.

Therefore the main effect of oxygen p-holes, at low temperatures, is to produce some local static disorder, i.e. some kind of magnetic polarons. As the disorder consists mainly in a spin rotation within the basal plane, these polarons strongly disturb the propagation of excitations of in-plane spin components and reduce the magnetic stiffness. Actually the magnetic stiffness is likely to vanish when the hole concentration reaches the critical value $n_h^c \approx 2\ \%$ at $x_c \sim 0.41$.

THE METALLIC SUPERCONDUCTING REGIME[12]

In order to understand how the superconductivity develops we decided first to investigate samples with oxygen contents in the superconducting region but close to the border line (see Fig. 3). First a sample with an oxygen content $YBa_2Cu_3O_{6.45}$ has been investigated. For such a composition no trace of 3D-AF ordering has been found down to T = 1.6 K, however a.c. susceptibility measurements[16] have given evidence for a very sharp ($\Delta T_c \sim 2$ K) superconducting transition at $T_c = 35$ K. Therefore, as shown in Fig. 3, superconductivity appears just above the critical oxygen concentration $x_c = 0.41$ and T_c increase sharply with x up to a plateau of about 60 K[17].

Typical energy and q-scans, measured at low temperatures (T = 5 K) in the superconducting state, are reported in Fig. 20 and 21, respectively. A clear magnetic scattering is observed around the AF scattering vector. However, the comparison with the sample x = 0.37, shows that an important decrease of the correlation length occurs when the hole concentration increases. The q-width, $\Delta q \sim 0.105$ r.l.u., practically independent of the energy transfer, yields a value $\Gamma_q = 0.050$ r.l.u (0.11 $Å^{-1}$), i.e. a correlation length $\xi = 2.2$ a. Energy scans clearly indicate that propagative spin excitations do not exist any more. High energy scans show a broad excitation spectrum with a maximum around 8-10 meV and which extends, with appreciable intensity, up to about 30-40 meV. So in the superconducting state there exist only short range dynamical antiferromagnetic correlations. In particular there is no sign of any incommensurate magnetic scattering both elastic and inelastic.

Moreover, a careful analysis shows that the low energy part of the spectrum is depressed for $T<T_c$ and no more magnetic scattering remains below an energy transfer of about 2 meV (see Fig. 20). A detailed investigation has been performed as a function of temperature. A typical example is reported in Fig. 22 which shows the scattering intensity measured at $\vec{Q} = (1/2, 1/2, 1.6)$ for energy transfers of 2 and 6 meV as a function of temperature. For $\hbar\omega > 5$ meV the intensity is almost temperature independent whereas for low energy transfers the magnetic scattering gradually decreases when cooling from 60 K down to 10 K. We can conclude that superconductivity suppresses low energy magnetic excitations of copper spins.

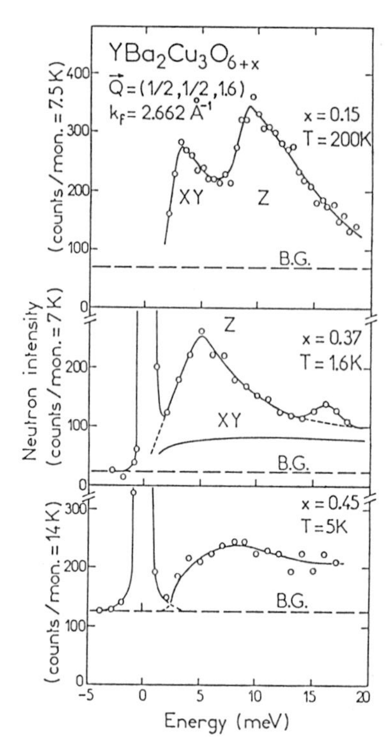

Fig. 20. *Energy scans performed at $\vec{Q} = (1/2, 1/2, 1.6)$ for $YBa_2Cu_3O_{6+x}$ with $x = 0.15$, 0.37 and 0.45. The contribution of in-plane (XY) and out-of-plane (Z) excitations is shown.*

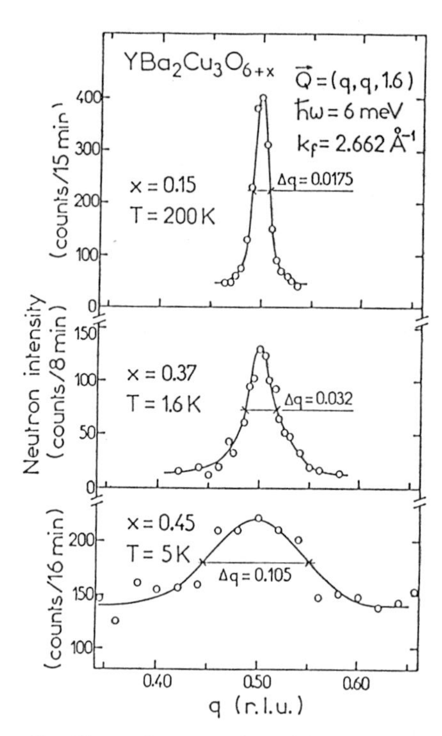

Fig. 21. *Q-scans performed for an energy* transfer $\hbar\omega = 6$ meV for $YBa_2Cu_3O_{6+x}$ *with $x = 0.15$, 0.37 and 0.45. The q-width due to the resolution is $\Delta q_{res} = 0.017$ r.l.u.*

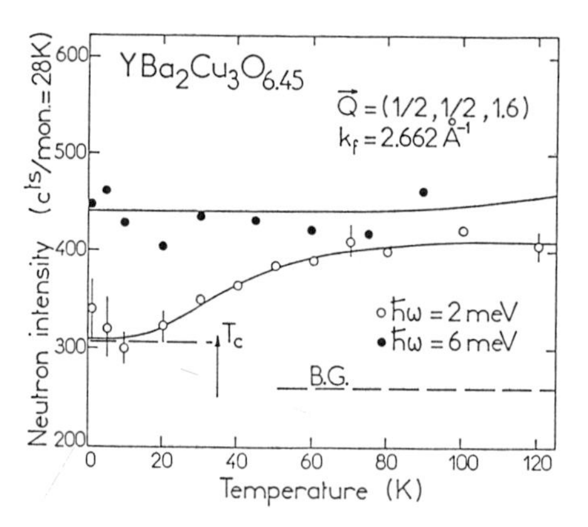

Fig. 22. *Magnetic intensity as a function of temperature measured at $\vec{Q} = (1/2, 1/2, 1.6)$ for energy transfers $\hbar\omega = 2$ and 6 meV for $YBa_2Cu_3O_{6.45}$.*

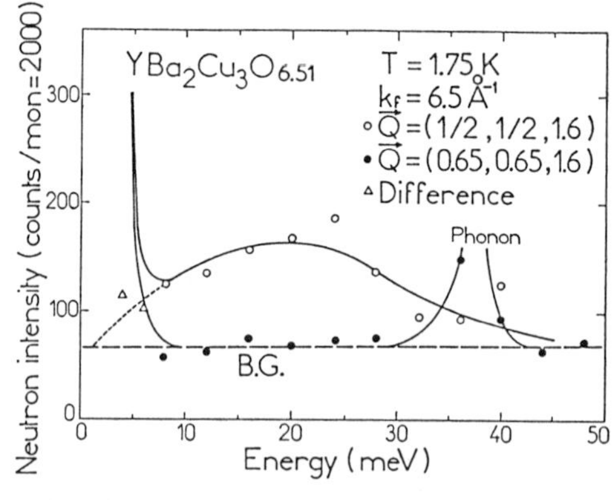

Fig. 23. *Energy scans for $YBa_2Cu_3O_{6.51}$ at $T = 1.75$ K.*

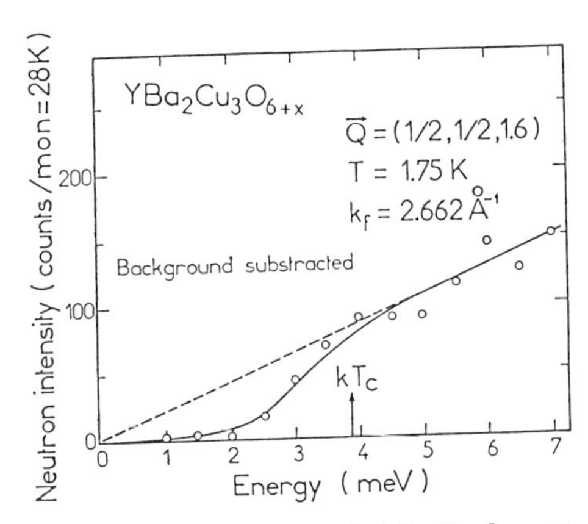

Fig. 24 . Low energy scan for $YBa_2Cu_3O_{6.51}$ at $T = 1.75$ K, the background has been substracted.

Very recently a new composition, with $x = 0.51$, was investigated and all these unusual results were confirmed with a better accuracy. The sample is superconducting with $T_c = 45$ K[16]. Q-scans give the same correlation length ($\Delta q = 0.10$ r.l.u. ~ 2.2 a) as for $x = 0.45$. Energy scans show also a broad spectrum extending up to about 40 meV but with a maximum at higher energy around 20 meV (see Fig. 23). The low energy part of the magnetic scattering, reported in Fig. 24, clearly shows the strong depression ; no intensity exists below about 2 meV. It must be noted that it is not really a gap because the intensity is depressed progressively from about 5 to 2 meV which is an energy range much larger than the energy resolution of about 1 meV. The temperature dependence of the scattering at low energy transfers shows that above T_c (T = 50 K), the intensity depression do not exist any more. The magnetic scattering (Im χ (q,ω)) recovers a linear behaviour as a function of the energy transfer.

Therefore in both samples with $x = 0.45$ and 0.51, in the superconducting state, the excitation spectrum of copper spins is strongly affected for energy transfers of the order and smaller than T_c and even completely washed out for $\hbar\omega < 2$ meV. This result is in agreement with NMR experiments (see as example[18]) indicating that $1/T_1T$ drops down below T_c. $1/T_1T$ being proportional to Im $\chi(q,\omega)/\omega$ is nothing than the slope of the magnetic scattering as a function of energy. So the anomalous shape found for the magnetic scattering at low energies instead of a well defined magnetic gap can explain the power law behaviour of $1/T_1T$ as a function of temperature instead of an exponential behaviour for a BCS-type gap.

DISCUSSION AND CONCLUDING REMARKS

The neutron scattering results presented above for the $YBa_2Cu_3O_{6+x}$ system are not yet complete enough, especially in the superconducting state, to get definitive information for the understanding of the mechanism of superconductivity. However a few conclusions can already be drawn and a comparison will be done with the $(LaSr)_2CuO_4$ system.

1. **The AF-state with no hole in CuO_2 planes is well described by a quasi 2D-Heisenberg model with a weak XY-anisotropy :**

- Experiments on La_2CuO_4[3] have established that above T_N^{3D} the correlation length varies exponentially with temperature. This result was well explained theoretically by Chakravarty et al.[19] who showed that

$$\xi/a = C_\xi e^{2\pi\rho_s/kT}$$

with $2\pi\rho_s = AJ$ (A ~ 1.9) and $C_\xi \sim 0.5$

- The 3D-AF ordering temperature is slightly larger in $YBa_2Cu_3O_6$ (415 K) than in La_2CuO_4 (~ 300 K) and the same conclusion for the low-T ordered moment $m_o = 0.64 \pm 0.3$ μ_B and 0.55 ± 0.05 μ_B[3,4], respectively. This difference may be accounted for by the fact that $YBa_2Cu_3O_6$ is a bi-layer system whereas La_2CuO_4 is a single layer system. In particular spin wave theory yields quantum spin reduction factors $\alpha = 0.27$ and 0.40 for double and single layer system, respectively[20]. Therefore we can expect a low-T ordered moment $m_o = g\mu_B S(1-\alpha)$. With $g = 2.12$ the obtained values 0.77 μ_B and 0.64 μ_B, respectively, are larger than the experimental ones which is an indication of an additionnal moment reduction by covalency. We can estimate this covalency reduction factor of about 15 - 20%. Therefore we are expecting a spin density on oxygen p-orbitals not larger than 10% of the copper $3d_{x^2-y^2}$ orbitals (in a planar square coordination Cu has two neighbouring oxygen atoms). This conclusion is in agreement with spin density measurements by polarized neutron diffraction[21] on the sample $YBa_2Cu_3O_{6.15}$ which gives at T = 30 K and H = 4.6 T an induced moment for Cu(2) site $m_{Cu2} = (0.65 \pm 0.05).10^{-3}$ μ_B and for $O_{2,3}$ site $m_{O2,3} = (0.05 \pm 0.05).10^{-3}$ μ_B.

- Spin excitations are well described by the spin wave theory. The in-plane spin wave velocity in $YBa_2Cu_3O_6$ is very large, $c_o = 1.0 \pm 0.05$ eVÅ ; a value 15% smaller was found in La_2CuO_4 (0.85 ± 0.05 eVÅ)[22]. The classical spin wave theory yields an effective Cu-pair superexchange integral $J^{eff}_{Cu\text{-}Cu} = 2\,J = 2000$ K. However for a S = 1/2 spin system quantum corrections have to be performed to get the true value $J_{Cu\text{-}Cu} = J^{eff}_{Cu\text{-}Cu}/Z_c$ (for S = 1/2, Z_c = 1.17[19]). So the value of the in-plane exchange integral, which enters in the (t, J) model as example, is $J_{Cu\text{-}Cu}$ = 1 700 K = 0.15 eV. It must be noted that this experimental value is quite well explained by ab initio cluster calculations by Sawatzky who estimated a value of about 0.12 - 0.17 eV[23].

- $YBa_2Cu_3O_{6+x}$ is a bi-layer system ($10^{-2} < J_b/J < 10^{-1}$) with a very good quasi-2D character ($J'/J = 10^{-5}$) and a very weak XY-anisotropy ($\Delta J/J = 10^{-4}$).

- The 3D-ordering temperature can be estimated from the following relation:

$$kT_N^{3D} \approx 2\,J'\,S^2\,(\xi/a)^2$$

using the above expression for ξ/a we get :

$$T_N^{3D} = T_N^{MF}\,\frac{A}{\mathrm{Ln}\,(J/J')} \quad \text{with} \quad T_N^{3D} = 2\,J = J_{Cu\text{-}Cu}$$

With the experimental values $J_{Cu\text{-}Cu}$ = 1700 K, A ~ 1.9 and J/J' = 10^5 the estimated value $T_N^{3D} \approx 300$ K is smaller than the observed one, this discrepancy could be accounted for by the bi-layer character.

- The large value of the in-plane exchange interaction accounts quite well for the low value of the spin susceptibility. From single crystal measurements on a x = 0.15 sample we estimate the spin susceptibility of Cu(2) of about 6-7 10^{-5} emu/mol Cu[24]. As the low-T susceptibility scales with J ($\chi_o \propto 1/J$) we can get a theoretical estimate of about $5.2.10^{-5}$ emu/mol. Cu by taking the value found in $Cu(C_5H_5NO)_6\,(BF_4)_2$[25] (χ_o = 0.04 emu/mol) which has a T_N^{MF} = 2.2 K (T_N^{3D} = 0.64 K).

2. The AF ordering in CuO_2 planes is strongly affected by a small amount of p-holes :

- Holes in CuO_2 planes induce a static disorder in the moment direction by producing a magnetic polaron. It must be emphasized that magnetic Bragg peaks coexist together with rods of magnetic scattering. Therefore the order is characterized by a mean AF-component which has a 3D-ordering and a rather disordered component, with an in-plane correlation length which decreases when the hole concentration increases. The disorder consists mainly in a rotation of the local moment direction within the basal plane, the value of copper moments remaining unchanged. This behaviour could be similar to that predicted by Shraiman and Siggia[26] which have found that the quantum motion of the hole results in a long-range dipolar twist of the AF-direction. However, we want to emphasize that we have not found any trace of the predicted spiral phase. The magnetic correlations remain always maximum at the AF-scattering vector (1/2, 1/2, 0). So magnetic polarons remain disordered within the basal plane down to the lowest temperatures. However, we have shown that the AF-coupling between the two Cu(2)-layers leads to a piling up of two polarons, i.e. of two holes.

- A critical hole concentration $n_h^c \approx 2\%$ is large enough to prevent any 3D-AF ordering to build up. This critical value must be related to the coupling J' between bi-layers and the hole mobility.

- The low-T reentrant behaviour is associated with the localization of holes. The reentrant temperature T_r decreases when the hole concentration increases and appears to vanish at the critical hole concentration. For $T>T_r$ the hole motion induces quantum spin fluctuations which reduce T_N^{3D} and affect the critical behaviour.

- The linear relationship between the 3D-ordered moment m_o and the ordering temperature T_N can be understood within the assumption that the local moment value remains constant. Then only the molecular field is reduced by the moment orientation disorder and it scales with m_o like T_N^{MF}, i.e. T_N^{3D}.

- A small amount of p-holes strongly modifies the spin dynamics : excitations of the out-of-plane component remain propagative in character whereas those associated with in-plane components are strongly overdamped. This result is a consequence of the disordered orientation of the local moments within the basal plane.

- Holes give rise to a strong renormalization of the low energy spectrum. For $n_h \sim 1.8\%$ the spin wave velocity and the anisotropy gap is renormalized by a factor two. The renormalization is q-dependent and decreases as q becomes larger than Γ_q $(1/\xi)$. This result is quite similar to the renormalization of the energy spectrum calculated by Grempel[27] for the disordered phase of the 2D-Heisenberg antiferromagnet.

- At the critical hole concentration $n_h^c \approx 2\%$ the spin wave velocity vanishes together with the spin-stiffness constant $\rho_s = c^2\chi_\perp(0)$ because the susceptibility remains almost unchanged. Moreover the activation energy of the polaron motion is also expected to vanish at n_h^c.

3. Dynamical antiferromagnetic correlations persist in the superconducting state and superconductivity depresses the low energy part of the excitation spectrum :

- In both samples $YBa_2Cu_3O_{6+x}$ with x = 0.45 (T_c = 35 K) and x = 0.51 (T_c = 45 K) there is no trace of 3D-AF ordering or even for a static component. There exist only dynamical magnetic correlations centered around the AF-scattering vector (1/2, 1/2, 0). No evidence for incommensurate spin fluctuations has been found, in contrast with results on the KOS-1 sample $La_{1.85}Sr_{0.15}CuO_4$ (T_c = 33 K)[28].

- The q-width of the magnetic scattering is energy independent. So the imaginary part of the dynamical susceptibility $\chi(q,\omega)$ can be written as :

$$\mathrm{Im}\,\chi\,(q,\omega) = \chi(q)F(\omega)\quad \frac{\Gamma_q}{\Gamma_q^2 + q^2}$$

- For $T<T_c$ the low energy part of the spectrum is strongly reduced and is no more detectable for energy transfers smaller than 2 meV. The unusual shape of the energy spectrum may explain the power law T-behaviour of $1/T_1T$.

- Up to 200 K the correlation length does not exhibit any temperature dependence indicating that at this temperature we are still in the quantum fluctuation regime. A good estimate of the characteristic energy is given by the maximum of the energy spectrum (~ 20 meV for x = 0.51). So the fundamental assumptions in the paper of Millis et al.[29] to explane the nuclear relaxation in high T_c superconductors are questionable.

- Even for $T>T_c$ the energy spectrum has not a Lorentzian shape. However, more quantitative experiments are required to confirm this conclusion.

4. The analysis of the magnetic phase diagram and neutron scattering data gives valuable information on the charge transfer mechanisms when oxygen is introduced in Cu(1) plane:

As a function of the oxygen content x five different regimes can be distinguished.

i) for 0 < x < 0.20, the filling of oxygen in Cu(1) planes induces electron transfers only within the Cu(1) plane, no holes are transferred to Cu(2) planes. The CuO_2 planes contain no 2p-oxygen holes whereas Cu(1) planes contain a mixture of Cu^+, Cu^{2+} and oxygen holes.
ii) for 0.20 < x < 0.40 only a small amount of 2p-holes are created in Cu(2) planes, only 2% for x = 0.40.
iii) for 0.40 < x < 0.50 suddenly a large amount of p-holes (10-15%) are transferred into the Cu(2) planes yielding a metallic state and superconductivity is appearing.
iv) for x > 0.50 two superconducting states build up with $T_c \sim 60$ K (0.50 < x < 0.70) and $T_c \sim 90$ K (x > 0.80). An amount of p-holes of about 25% is expected in Cu(2) planes.

REFERENCES

1. Proceedings of the International Conference on High Temperature Superconductors, Interlaken, ed. J. Müller and J.L. Olsen, Physica C135-155, 1988.

2. R.J. Birgeneau and G. Shirane in "Physical properties of high temperature superconductors, ed. D.M. Ginsberg (World Scientific Publishing).

3. Y. Endoh, K. Yamada, R.J. Birgeneau, D.R. Gabbe, H.P. Jensen, M.A. Kastner, C.J. Peters, P.J. Picone, T.R. Thurston, J.M. Tranquada, G. Shirane, Y. Hidaka, M. Oda, Y. Enomoto, M. Suzuki, T. Murakami, Phys. Rev. B37, (1988) 7443.
   R.J. Birgeneau, D.R. Gabbe, H.P. Jensen, M.A. Kastner, P.J. Picone, T.R. Thurston, G. Shirane, Y. Endoh, M. Sato, K. Yamada, Y. Hidaka, M. Oda, Y. Enomoto, M. Suzuki, T. Murakami. Phys. Rev. B 38 (1988) 6614.

4. K. Yamada, E. Kudo, Y. Endoh, Y. Hidaka, M. Oda, M. Suzuki, T. Murakami, Solid State Commun 64, 753 (1987).

5. J. Rossat-Mignod, P. Burlet, M.J. Jurgens, J.Y. Henry, C. Vettier, Physica C152, (1988) 19.

6. J.M. Tranquada, D.E. Cox, W. Kannmann, A.H. Moudden, G. Shirane, M. Suenaga, P. Zolliker, D. Vaknin, S.K. Sinha, Phys. Rev. Lett. 60, (1988) 156.

7. P. Burlet, C. Vettier, M.J. Jurgens, J.Y. Henry, J. Rossat-Mignod, H. Noel, M. Potel, P. Gougeon, J.C. Levet, Physica C153-155, (1988) 1115.

8. J. Rossat-Mignod, P. Burlet, M.J. Jurgens, L.P. Regnault, J.Y. Henry, C. Ayache, L. Forro, C. Vettier, H. Noel, M. Potel, M. Gougeon and J.C. Levet, J. Physique 49 (1988) 2119.

9. J.M. Tranquada, A.H. Moudden, A.I. Goldman, P. Zolliker, D.E. Cox, G. Shirane, S.K. Sinha, D. Vaknin, D.C. Johnston, M.S. Alvarez, J. Jacobson, Phys. Rev. B38, (1988) 247.

10. C. Vettier, P. Burlet, J.Y. Henry, M.J. Jurgens, G. Lapertot, L.P. Regnault and J. Rossat-Mignod, Physica Scripta, T29 (1989) 110.

11. J. Rossat-Mignod, L.P. Regnault, M.J. Jurgens, C. Vettier, P. Burlet, J.Y. Henry, G. Lapertot, Physica C., 162-164 (1989) 1269.

12. J. Rossat-Mignod, L.P. Regnault, M.J. Jurgens, C. Vettier, P. Burlet, J.Y. Henry, G. Lapertot, Proceeding of the International Conference on the Physics of Highly Correlated Electron Systems, Santa Fe, Physica B., 1990.

    P. Burlet, L.P. Regnault, M.J. Jurgens, C. Vettier, J. Rossat-Mignod, J.Y. Henry, G. Lapertot, Proceeding of the European Conference on Low Dimensional Conductors and Superconductors, Dubrovnic, 1989.

13. J. Rossat-Mignod, J.X. Boucherle, P. Burlet, J.Y. Henry, M.J. Jurgens, G. Lapertot, L.P. Regnault, J.Schweizer,
    Proceeding of the International Seminar on High Temperature Superconductivity, Dubna 1989, to be published by World Scientific.

14. M.J. Jurgens, P. Burlet, C. Vettier, L.P. Regnault, J.Y. Henry, J. Rossat-Mignod, H. Noel, M. Potel, P. Gougeon, J.C. Levet. Physica B, 156-157 (1989) 846.

15. L.P. Regnault and J. Rossat-Mignod, Phase transitions in quasi 2D-planar magnets, in Magnetic properties of layered transition metal compound, L.J. de Jongh and R.D. Willet, under press (1990).

16. M. Couach, private communication.

17. R.J. Cava et al, Physica C, 153-155 (1988) 560.

18. C. Berthier, Y. Berthier, P. Butaud, M. Horvatic, Y. Kitaoka, P. Segransan.
    To be published in this proceeding.

19. S. Chakravarty, B.I. Halperin, D.R. Nelson
    Phys. Rev. B 39 (1989) 2344.

20. R. Navarro, L.J. de Jongh
    Physica B 98 (1979) 1

21. J.X. Boucherle, J.Y. Henry, M.J. Jurgens, J. Rossat-Mignod, J. Schweizer, F. Tasset.
Physica C, 162-164(1989) 2052.

22. G. Aeppli, S.M. Hayden, H. Mook, Z. Fisk, S.W. Cheong, D. Rytz, J.P. Remeika, G.P. Espinosa, A.S. Cooper.
Phys. Rev. Letters, 62 (1989) 2052.

23. G.A. Sawatzky
To be published in this proceeding.

24. M.J. Jurgens
Ph. D. thesis, University of Leiden, to be published.

25. L.J. de Jongh
Solid state commun. 65 (1988) 963.

26. B.I. Shraiman, E.D. Siggia.
Phys. Rev. Letters, 61 (1988) 467.

27. D. Grempel.
To be published in this proceeding.

28. G. Shirane, R.J. Birgeneau, Y. Endoh, P. Gehring, M.A. Kastner, K. Kitazawa, H. Kogima, I. Tanaka, T.R. Thurston and K. Yamada.
Physical Review Letters 63 (1989) 330.

39. A.J. Millis, H. Monien, D. Pines.
Preprint.

DISORDERED, LOW ENERGY COMPONENT OF THE MAGNETIC RESPONSE IN BOTH ANTI-FERROMAGNETIC AND SUPERCONDUCTING Y-Ba-Cu-O SAMPLES

F. Mezei

Hahn-Meitner-Institut
Pf. 390128
D-1000 Berlin 39, Germany

INTRODUCTION

Neutron scattering lends itself as a prominently powerful and direct tool to the investigation of magnetic phenomena, under the assumption that the magnetic scattering effects can be unambiguously identified and separated from an often stronger background of non-magnetic signal. This is the case in particular with relatively small magnetic effects, such as in high T_c superconductors. Magnetic Bragg peaks and excitations around them can be well identified by their localized character in the reciprocal space. On the other hand, the hardly q dependent diffuse scattering from magnetic disorder can only be identified by the use of polarization analysis, which implies a dramatic loss of neutron intensity, i.e. sensitivity. This latter kind, much more limited studies performed by now are complementary to the single crystal work described in other contributions in this volume and they provide evidence for the existence of a disorder type, relatively low frequency range contribution to the total magnetic response in Y-Ba-Cu-O compounds. This "impurity" kind magnetism is expected to also manifest itself in μSR and NMR experiments.

EXPERIMENTAL

As of today the best neutron flux conditions in neutron spin polarization analysis can be achieved at relatively long neutron wavelengths (4-6 Å, i.e. low neutron energies of 5-3 meV) by the application of supermirror optical devices[1] as opposed to the conventional magnetic crystals. The main feature of these devices is that, in contrast to the crsytals, they are not energy selective. This allows us, on the one hand, to use a relatively broad wavelength band to increase the flux, but, on the other hand, it makes us lose most information on the inelasticity of the scattering, i.e. on the frequency of the observed magnetic fluctuations. Thus, similarly to neutron diffraction work without energy analysis, we have a "constant scattering angle" situation (cf. Fig. 1) as opposed to the more familiar and favourable "constant momentum transfer **q**" situation of triple-axis spectroscopy. In the detector (which also includes the polarization analyser device) we record simultaneously various outgoing neutron momenta **k'**, consequently various **q** vectors and neutron energy changes. The uncertainty of the direction of **q** essentially influences the interpretation of the polarization analysis data, since in a "paramagnetic", i.e. macroscopically magnetically isotropic sample (e.g. a polycrystalline antiferromagnet) the polarization **P'** of the magnetically

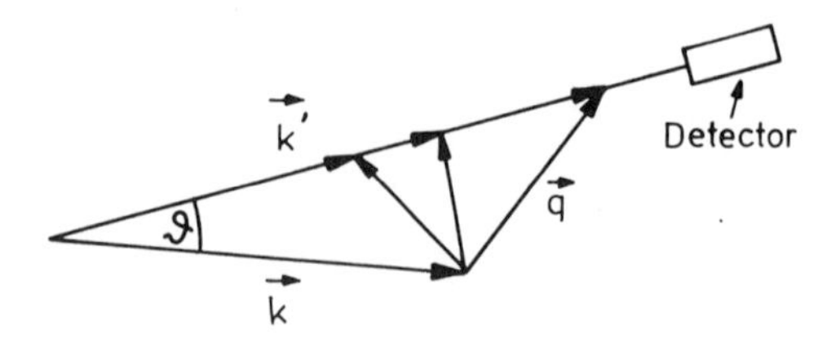

Fig. 1. "Constant scattering angle" experimental configuration showing a few possible scattering triangles with different **q** vectors.

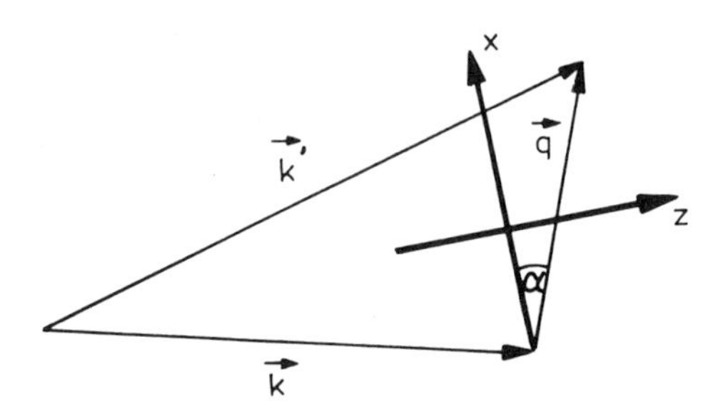

Fig. 2. Choice of reference directions in "three directional polarization analysis" in order to check the inelasticity of the scattering.

scattered beam is parallel to **q**, as given by the well known Halpern-Johnson relation:

$$\mathbf{P'} = -\mathbf{q}(\mathbf{Pq})/|\mathbf{q}|^2 = \mathbf{P'}(\mathbf{P}) \tag{1}$$

where **P** is the incoming beam polarization. The polarization of the scattered beam is independent of the direction of **q** for nuclear or nuclear spin scattering. The characteristic **q** dependence (1) is used to identify magnetic scattering effects. However, since in our case the incoming beam direction and the position of the detector only determines the scattering plane (say horizontal) but not the direction of **q** within this plane, we have to apply a special trick, the so called "three directional polarization analysis" method, which has been first introduced by the Leningrad group[2] and somewhat later, independently, at the Institut Laue-Langevin (ILL) in Grenoble[3]. This method consists of determining the spin flip (sf) and non-spin-flip scattering intensities for 3 mutually perpendicular incoming beam polarization **P** directions **x**, **y** and **z**. It is preferential to make one of these, say **y**, vertical, i.e. in any case perpendicular to **q**. Thus in view of (1) $\mathbf{P'}(\mathbf{y})=0$. Now, if the angle between **q** and **x** is named α, we have $|\mathbf{P'}(\mathbf{x})| = -\cos^2\alpha$ and $|\mathbf{P'}(\mathbf{z})| = -\sin^2\alpha$. Thus the sum of the scattered beam polarizations measured in the 3 directions is -1, independently of the direction of **q**. If, in addition, we choose e.g. **z** to be the bisector of the scattering angle ϑ, (cf. Figs. 1 and 2) we also can obtain some information on the inelasticity of the scattering purely from the polarization analysis measurement, as first pointed out by Maleev[4]. Namely, for purely elastic scattering now $\mathbf{q} \parallel \mathbf{x}$, and hence $\mathbf{P'}(\mathbf{z})=0$, while inelasic scattering contributes to both $\mathbf{P'}(\mathbf{x})$ and $\mathbf{P'}(\mathbf{z})$, cf. Fig. 2, and the ratio $|\mathbf{P'}(\mathbf{x})|/|\mathbf{P'}(\mathbf{z})|$ is related to the effective inelastic linewidth. (More details can be found in Refs. 3 and 5.)

Another experimental aspect is particularly relevant for the present case. The two strongest contributions to the elastic non-magnetic

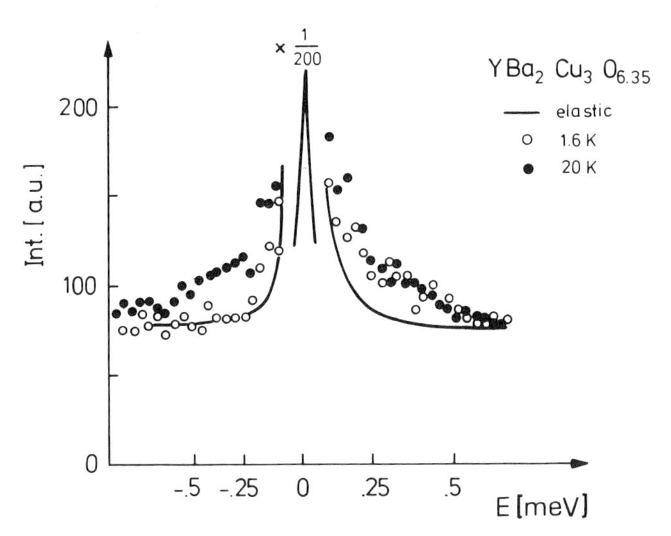

Fig. 3. Time-of-flight spectra measured in an Y-Ba-Cu-O powder sample at low temperatures. The line indicates the instrumental resolution function and the background measured with a V sample. The data were obtained by adding the spectra in all detectors between 0.4 and 2 $Å^{-1}$, with the exception of those detectors at or in the immediate vicinity of a Bragg peak.

background we have established, viz. both multiple Bragg scattering and nuclear spin incoherent scattering on H atoms in the sample, are much stronger than the magnetic signal. The first one contributes to the non-spin-flip channel, the second one primarily to the spin-flip channel. Since in our case this background was 2x smaller in the spin-flip channel, this channel only was used for data collection. With the **y** being vertical, spin-flip counts taken with this incoming polarization, $I_{sf}(y)$ give the polarization of the non-magnetic background, and the magnetic signal M is thus given as:

$$\frac{1}{2} P_o M = (I_{sf}(\mathbf{x})-I_{sf}(\mathbf{y})) + (I_{sf}(\mathbf{z})-I_{sf}(\mathbf{y})), \qquad (2)$$

where P_o is the overall polarization efficiency of the instrument. For sufficiently elastic scattering the first term only is non-zero in the sum on the right hand side, as it was found in basically all data sets taken. This sum gives 1/2 of the total magnetic scattering intensity, the other half contributes in a similar fashion to the non-spin-flip channel, which was only monitored to check proper experimental conditions. The neutron counts were converted into absolute cross sections by using the integrated intensity of the (001) powder diffraction line determined in the same polarization analysis configuration. The strong intensity of this Bragg peak was also used to verify that the polarization efficiency of the spectrometer was identical for all three incoming polarization directions. A uniformity of better than 10^{-3} was established in several control runs, which implies that the systematic error of our data is negligible compared to the statistical one. Typical counting rates for about 40 g Y-Ba-Cu-O powder samples on the IN11 Neutron Spin Echo spectrometer at the ILL are: 300 c/min spin-flip, 600 c/min non-spin-flip counts, compared to the magnetic intensity of 20 c/min.

The samples were prepared by the superconductivity group at the Central Research Institute for Physics (KFKI) in Budapest using the ususal techniques[6]. The polycristalline pellets were crashed into powder, in order to avoid orientational anisotropy. The samples were kept in He atmosphere in sealed Al containers in order to avoid the absorbtion of water from the air, and keep the background H scattering as low as possible. This precaution is vital, but some amount of H was found to be always present in the samples, even after high temperature heat treatment in controlled atmosphere. Our samples contained on the whole the equivalent of about 12 % per formula (or 200 ppm in weight) H, probably mostly on the surface of the grains. A sample left in the open air would absorb within a few days several times this amount. It is likely, that most samples actually used in various experiments contain substantial unvoluntary amount of H. The O composition of our samples was determined by Rietveldt analysis of neutron diffraction data and we estimate the error to be 0.02. This is consistent with the results of weight analysis during fabrication. The x=6.35 sample was found to be antiferromagnetic with an ordered moment of 0.4 ± 0.1 μ_B at low temperatures and the x=6.6 sample superconducting with T_c=47 K. This latter is somewhat smaller than the expected 55 K plateau, which might be due to concentration fluctuations. In any case, by neutron diffraction the sample was found to be in at least 90 % of the volume ortorombic.

RESULTS

In the first stage of this study[7] the existence of low energy diffuse magnetic scattering has been established in both antiferromagnetic and superconducting $YBa_2Cu_3O_x$ powder samples with x=6.15, 6.35 and 6.6. The present results concern the wavenumber q dependence of this quasielastic magnetic scattering cross section in the two latter samples at selected temperatures. The scattering was found little inelastic in its polarization analysis behaviour (less than some 0.5 meV). The inelasticity was also checked (without polarization analysis) on the IN5 time-of-flight spectrometer at the ILL, as shown in Fig. 3. The observed inelastic tail can tentatively be identified with the magnetic scattering, (normal phonon scattering does not peak at zero energy) and the variation between 1.6 and 20 K is compatible with the change of the Bose factor. Comparing the integrated inelastic intensity with the magnetic one as determined by polarization analysis we find that the major part of the magnetic scattering is contained in the resolution broadened elastic line, thus the effective inelastic linewidth is about 0.2 meV. In addition, in a complementary Neutron Spin Echo scan at T=2 K about 60 % of the magnetic signal was found to be frozen on the energy scale of 0.5 μeV, which suggests that the disordered magnetism is of spin glass character at this temperature. To complete the picture, the inelastic linewidth was also checked on the triple-axis spectrometer IN20 using both polarization and energy analysis, the latter with a resolution of 1 meV. Within the limit of this resolution no broadening of the qausielastic magnetic scattering was found up to room temperature. Note, that the detector counting rate on this spectrometer was found to be two times less than that on the NSE spectrometer IN11. Therefore the IN20 data could sensibly only be used to check the inelasticity taking advantage of the fact that the backgroud becomes some two orders of magnitude smaller on leaving the elastic channel (cf. Fig. 3). All the inelastic measurements have been performed on the x=6.35 antiferromagnetic sample. The IN5 time-of-flight data also showed that there is no substantial q dependence in the inelasticity between 0.2 and 2 $Å^{-1}$.

Having established that the magnetic diffuse scattering is little inelastic (either by inelastic scans, or by the polarization analysis check described above), we can drop the second term on the right hand side of eq. (2). Thus determining the magnetic signal from $I_{sf}(\mathbf{x})$ and $I_{sf}(\mathbf{y})$ only we achieve the same statistical accuracy in one quarter of data collection time, and the result is an underestimate of the magnetic intensity if the quasielastic assumption were unjustified. For independent paramagnetic spins the quasielastic magnetic scattering cross section per magnetic atom can be given as

$$\frac{d\sigma}{d\Omega} = \frac{2}{3} g^2 f^2 \mu_{eff}^2 \tag{3}$$

where $g=0.269\times10^{-12}$ cm/μ_B, f is the magnetic structure factor (for Cu^{++} f^2 is expected[8] to drop smoothly from 1 to .75 between q=0 and 1.5 Å^{-1}) and μ_{eff}^2 is the effective squared magnetic moment well known from the interpretation of susceptibility data. For ideal, spin only magnetism with exact half-integer spin values this latter quantity is 4S(S+1), i.e. 3 for spin 1/2.

Our main results, the quasielastic diffuse magnetic scattering cross section at various wavenumbers and temperatures are shown in Figs. 4 and 5. Technically it is rather delicate and time consuming to perform this kind of polarization analysis experiments on a superconducting sample, essentially because of the influence of eventual trapped magnetic fields (even as small as 1 Øe) on the polarization of the neutron beam. Therefore most of the low temperature data on the x=6.6 sample were collected above T_c. The main features of these results are (a) a roughly q independent cross section between 0.4 and 1.35 Å^{-1}, characteristic of isolated spins, (b) the enhanced cross section towards q=0 indicates the existence of a ferromagnetic short range order compatible with an Ornstein-Zernicke correlation length of about 5 Å, and (c) little temperature dependence.

DISCUSSION

In contrast to the bulk susceptibility, finite q neutron scattering data are not essentially influenced by the presence of very small amounts of magnetic impurity phases. The q dependences shown in Figs. 4 and 5 imply that we have to do with isolated spins with slight ferromagnetic type correlations with a correlation length comparable to the size of the unit cell. Thus, this magnetic scattering suggests the presence of some kind of atomic impurity type magnetism inside the bulk of the sample matrix. Comparing our results to those of the only available similar studies[9-11] performed on the D7 polarization analysis instrument at the ILL on samples from two very different origins, we can observe the systematic presence of diffuse magnetic scattering, although its strength fluctuates from one study to another, as summarized in Fig. 6. The experiments in Refs. 9-11 had no access to the small q range, therefore, the ferromagnetic correlations could not possibly be observed. In the figure the average cross sections between q=0.5 and 1.35 Å^{-1} are compared, and on the right hand scale also expressed in terms of the equivalent concentration of full spin 1/2 Cu spins required to produce the same amount of magnetic scattering. The data in Ref. 10 were taken at higher temperatures, above the Neel temperature, but since the antiferromagnetic fluctuations of the Cu(2) (in plane) spins above the Neel temperature are known to be rather localized in the q space and very inelastic, the diffuse scattering observed in that work has to be of the same nature as the one we have put to evidence.

The most plausible source of the present quasielastic magnetic response is the magnetic nature of some of the Cu(1) (in chain) atoms

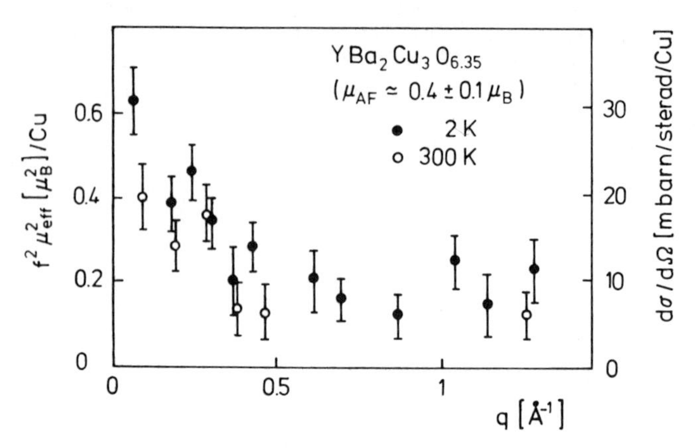

Fig. 4. Quasielastic diffuse magnetic neutron scattering cross section in an Y-Ba-Cu-O powder sample as determined by neutron spin polarization analysis on the IN11 Neutron Spin Echo Spectrometer at ILL. The spectrometer was used in the "three directional polarization analysis" mode without (spin echo) energy analysis. The scale on the left hand side was calculated by using eq. (3). The Neel temperature of this sample is expected to be just above room temperature.

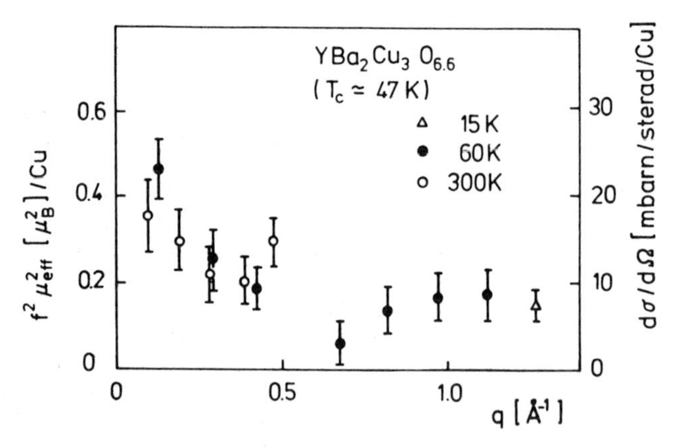

Fig. 5. Quasielastic diffuse magnetic neutron scattering cross section in a superconducting Y-Ba-Cu-O powder sample, determined as in Fig.4. Most low temperature data were taken just above T_c in order to avoid experimental difficulties with trapped fields. The one data point below T_c confirms the coexistence of magnetic disorder scattering and superconductivity, as also reported in Ref. 7.

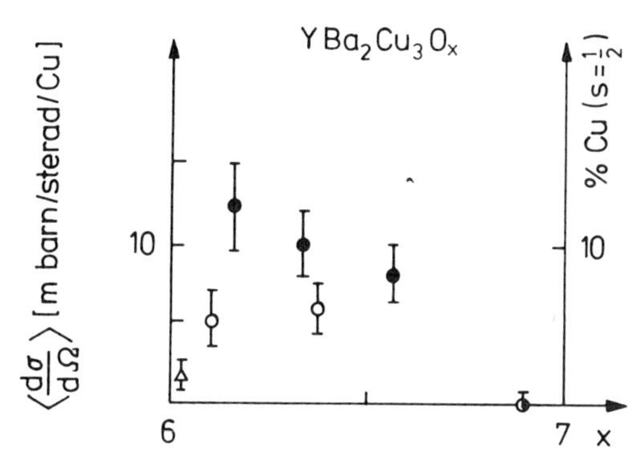

Fig. 6. Comparison of the average low temperature diffuse quasielastic magnetic cross sections for various $YBa_2Cu_3O_x$ samples. The average refers to the q range of 0.5 to 1.35 Å^{-1}. On the right hand side the equivalent fraction of free 1/2 spin Cu atoms is also shown, which would give the same cross section, cf. eq. (3). Filled circles: present work, half-filled circle: Ref. 9, open circles: Ref. 10, triangle: Ref. 11.

In the intermediate oxygen concentration range in question there has to be an amount of O disorder in the chain positions, i.e. different Cu(1) atoms experience different chemical environments. Since the O disorder can be very different from one sample to the other (e.g. shorter or longer filled and empty chain sections), this can offer an explanation for the variation of the magnetism from one sample to another. The existence of paramagnetic Cu(1) spins is also supported by neutron diffraction studies of single crystal samples in high magnetic fields, showing a strong magnetic polarizability on these sites[12]. Of course, one cannot fully exclude the existence of additional foreign impurity atoms in the lattice, e.g. quasielastic scattering similar to the one in Fig. 3. has been reported upon introduction of Co impurities[13]. It is worth mentioning that the x=6.05 sample (Ref. 11) contains an order of magnitude less H than the others.

The small but marked ferromagnetic short range order observed below 0.4 Å^{-1} can either be explained by some ferromagnetic coupling between the paramagnetic spins we are concerned with or by some kind of covalency effect, i.e. magnetic polarization induced by the localized spin in its neighbourhood. Since we cannot be sure that all of the diffuse scattering comes from the Cu(1) sites, it is premature to argue that this ferromagnetic short range order is a characteristic, intrinsic feature of the Y-Ba-Cu-O matrix, but this certainly is a possibility, too.

In sum, there is broad, concurring evidence that in real Y-Ba-Cu-O samples the magnetic response contains, beside the antiferromagnetism and high energy antiferromagnetic fluctuations on the Cu(2) plane sites[14], a low energy quasielastic disordered paramagnetic type component too, which is at least partially related to the Cu(1) sites. The fluctuation rate of the diffuse component is in the range of 100 GHz, therefore, its weight in the local (i.e. integrated over q) spectral density function becomes at frequencies smaller than this rate (thus including the NMR frequencies) some 2 orders of magnitude higher than that of the contribution of the high energy Cu(2) antiferromagnetic fluctuations, even if the paramagnetic response is equivalent to a few percent of the full spin 1/2 magnetism only. On the one hand, this would imply that the NMR relaxation of the magnetic Cu(1) sites is so rapid, that no NMR signal can be observed at all on these sites. On the other hand, the very high spectral density of the magnetic fluctuations on these sites might contribute to an appreciable degree to the NMR relaxation of the O and Cu(2) sites, as well as of the non-magnetic Cu(1) sites. The similarity of the NMR relaxation rates observed[15] for the Cu(1) and Cu(2) sites at both x=6.6 and 6.9 might be just related to a common contribution of this type. In this respect it is rather strange, that no Cu(1) magnetism has been observed in the neutron scattering work at x=6.9 (cf. Fig 6 and Ref. 9), since the high Cu(1) NMR relaxation rate cannot be explained without substantial magnetic fluctuations in the chains. In μSR the most apparent signature of the present findings could be the appearence of a spin glass phase at low temperatures, coexisting with the AF order or with superconductivity, as suggested by the Neutron Spin Echo observation of freezing in the spin dynamics.

One of the simple logistics problems of the study of the magnetism of high T_c superconductor systems is the fact that virtually every experiment has been performed on a different sample. Since the sample quality remains a factor of uncertainty in any case, at least because of the varying degree of O disorder even at perfect purity and uniformity, it is rather hard to reliably compare the various bits of information. In particular, the two neutron scattering techniques applied by now, triple-axis spectroscopy and low energy diffuse scattering with polarization analysis are complementary in the type of information they were able to provide (high energy ordered response vs. low energy disordered response). Furthermore, while the first method requires single crystals (by now rather limited in size), the second one needs big sample volumes only, and therefore no sample could be studied by both methods by now. The eccessive speed which characterizes the research in this field is not favourable for performing various type of experiments by different groups on the same, well characterized samples. Nevertheless an important next step should be the combined NMR, μSR and both kind of (i.e. both low and high energy) neutron scattering study of the magnetic response in a few high quality samples of various O concentrations.

ACKNOWLEDGEMENT

The author is indebted to G. Hutiray and L. Mihaly of KFKI, Budapest for providing the samples, to B. Farago, P. Frings and D. Kearley of ILL and C. Lartigue of Hahn-Meitner-Institut for their help in the experiments on IN11, IN20 and IN5 spectrometers at ILL, to J. Pannetier of ILL and I. Abraham of Technische Universität Berlin for their help in the neutron diffraction Rietveldt analysis determination of the sample compositions.

REFERENCES

1. F. Mezei, in: "Use and development of Low and Medium Flux Research Reactors", O.K Harling, L. Clark, P. von der Hartd, eds. Supplement to Atomenergie-Kerntechnik Vol. 44, Karl Thiemig, München (1984) p. 735.
2. B.P Toperverg, V.V. Runov, A.G. Gukasov, Phys. Lett. 71A:289 (1979)
3. F. Mezei and A.P. Murani, J. Mag. Mag. Mat. 14:211 (1979)
4. S.V. Maleev, Soviet Phys. JETP Lett. 2:338 (1966).
5. F. Mezei, in: "Neutron Spin Echo", F. Mezei, ed., Springer Verlag, Heidelberg (1980) p. 21.
6. H. Kuzmany, M. Matus, E. Faulgues, S. Pekker, Gy. Hutiray, E. Zsoldos and L. Mihàly, Solid State Comm. 65:1343 (1988).

7. F. Mezei, B. Farago, C. Pappas, Gy. Hutiray, L. Rosta and L. Mihàly, Physica C 153-155:1669 (1988).
8. J. Akimitsu and Y. Ito, J. Phys. Soc. Japan 40:1621 (1976).
9. T. Brückel, H. Capellman, W. Just, O. Schärpf, S. Kemmler-Sack, R. Kiemel and W. Schäfer, Europhys. Lett. 4:1189 (1987).
10. T. Brückel, K.U. Neumann, H. Capellmann, O. Schärpf, S. Kemmler-Sack, R. Kiemel and W. Schäfer, J. de Physique, suppl. 49:C8-2155 (1988).
11. H. Capellmann and O. Schärpf, Z. Phys. B, in press; H. Capellmann, in this volume.
12. B. Gillon, D. Petitgrand, A. Delapalme, G. Collin and P. Schweiss, Physica C, in press, and J. Rossat-Mignod, in this volume.
13. A.J. Dianoux et al. in: "Proc. of Workshop on High Temp. Supercond.", Dubna, July 1989, World Scientific (Singapore) in press.
14. See contributions by H. Mook, J. Rossat-Mignod, G. Shirane and J.M. Tranquada in this volume.
15. See contributions by C. Berthier and M. Takigawa in this volume.

COPPER AND OXYGEN NMR STUDIES ON THE MAGNETIC PROPERTIES OF $YBa_2Cu_3O_{7-y}$

Masashi Takigawa

Los Alamos National Laboratory

Los Alamos, NM 87545

ABSTRACT

Microscopic magnetic properties of the CuO_2 layers in $YBa_2Cu_3O_{7-y}$ have been investigated from Cu and O NMR experiments on the $y \simeq 0$ (T_c=92K) and y=0.37 (T_c=62K) materials. The Knight shift at the planar Cu and the planar oxygen sites are found to be proportional to a common spin susceptibility χ_s which depends on temperature and oxygen-content, strongly supporting a single component spin model for the CuO_2 planes. In the y=0.37 material, χ_s shows a significant reduction with decreasing temperature in the normal state. The nuclear relaxation rate ($1/T_1$) in the $y \simeq 0$ material, particularly different behaviors at Cu and O sites, can be accounted for by the hyperfine coupling of Cu and O nuclei to an antiferromagnetically correlated single spin system. Quite different behaviors of $1/T_1$ were observed in the y=0.37 material, which might be due to combined effects of antiferromagnetic correlations and a temperature-dependent spin susceptibility.

1. INTRODUCTION

In the high-T_c Cu oxides, the magnetism and the nature of spin fluctuations in the two dimensional CuO_2 layers are of central importance in understanding the normal and superconducting properties of these materials. In this paper, we discuss the microscopic magnetism of the CuO_2 layers in $YBa_2Cu_3O_{7-y}$ system from both the static and the dynamical point of view based on the ^{63}Cu and ^{17}O NMR results. The parent compound $YBa_2Cu_3O_6$ is an antiferromagnetic insulator where each planar Cu has 2+ valence ($3d^9$ configurations). When more oxygen is added, filling the chain O(1) sites, electrons are removed from both the chain Cu(1) sites and the Cu(2)O_2 planes. There are strong spectroscopic evidences that planar holes are of primarily O-2p character.[1-3] The long range antiferromagnetic order in the CuO_2 planes is easily destroyed by a small number of these plane holes. At higher doping, these doped holes are responsible for the metallic conduction and the superconductivity. Therefore, it is important to know the details of the doped hole state and the magnetic interaction between doped holes and Cu d-spins.

There are a wide variety of models to describe the doped hole states. If the doped holes go into the O-2pσ orbitals, a strong antiferromagnetic exchange is expected between Cu d and oxygen hole spins. Zhang and Rice argued that a doped hole spin whose orbital state is spread over square coordinated oxygen sites will form a local singlet with the the central Cu d-spin.[4] This local singlet then moves through the lattice in a similar way as a hole (an empty site) in a single band Hubbard model. Thus the transport carriers do not have spin degrees of freedom in this single band picture. Magnetism is associated only with Cu d-spins except for covalency effects which produce finite spin density at the oxygen sites. Emery and Reiter, on the other hand, argued that a mobile hole rather has spin 1/2 resulting from the superexchange between an oxygen and its two Cu neighbors.[5] Therefore, a mobile hole carries both charge and spin, a

Dynamics of Magnetic Fluctuations in High-Temperature Superconductors
Edited by G. Reiter *et al.*, Plenum Press, New York, 1991

significant fraction of the latter resides in the oxygen sites. This picture leads to the possibility that the oxygen holes have distinct spin degrees of freedom (different spin dynamics) from the Cu d-spins. If oxygen $2p\pi$ orbitals are occupied by the doped holes, these holes will have distinct spin degrees of freedom from Cu d-spins because these two states do not hybridize. NMR is a particularly useful probe to this question since we can obtain selective information about the magnetic properties at a specific Cu or O site.

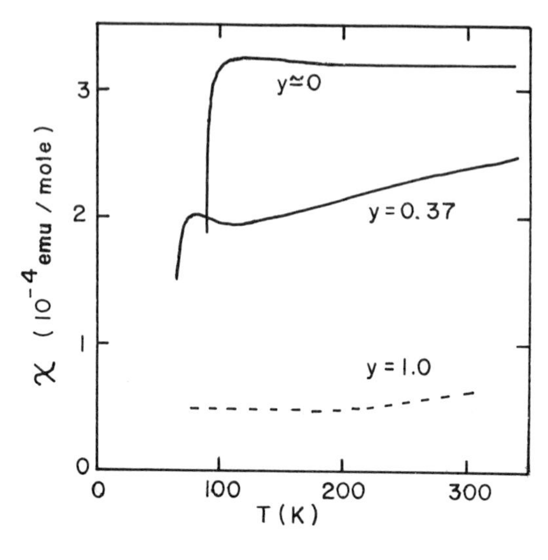

Fig. 1. Magnetic susceptibility of $YBa_2Cu_3O_{7-y}$. The y=0.37 data was taken on the same sample as used in the NMR experiment, which was obtained by reducing the y~0 sample. y=1.0 data is taken from ref. 6.

The magnetic susceptibility (χ) of $YBa_2Cu_3O_{7-y}$ is quite unusual[6] as shown in Fig. 1. In the antiferromagnetic phase, χ is nearly t-dependent. χ increases with increasing oxygen content. For y=0.37 (T_c=62K), χ shows a reduction with decreasing temperature. At higher doping (y~0, T_c=92K), χ becomes T-dependent again. Johnston has analyzed similar data on $(La_{1-x}Sr_x)CuO_4$ systems based on two band picture, where χ is the sum of the contribution from localized 2D Cu moments and Pauli like term due to mobile holes.[7] The validity of such analysis should be examined by the Knight shift measurements which detect the spin susceptibility at each site separately.

Finally, we are interested in the spin dynamics. In the undoped insulator, the ground state and the elementary excitations as well as the spin dynamics at finite temperatures are well described by 2-dimensional Heisenberg model for s=1/2.[8,9] It is not clear yet how the dynamical behavior changes as we go into the metallic and superconducting phase. Measurements of the nuclear relaxation rate will provide useful information about the nature of low frequency spin fluctuations. This paper is organized as follows. The sample preparation and NMR experiments are described in section 2. Section 3 is devoted to the discussion of the static magnetic properties based on the Cu and O Knight shift results in the y~0 and the y=0.37 samples. In section 4, we discuss the spin dynamics in both samples based on the data of Cu and O nuclear spin relaxation rate.

2. EXPERIMENTAL

The powder samples of $YBa_2Cu_3O_{7-y}$ used in this experiment were prepared as follows. First, powder sample of fully oxygenated $YBa_2Cu_3O_{7-y}$ was made by standard ceramic method of solid state reaction. ^{17}O isotope was then introduced by annealing the powder at 670 C in 45% ^{17}O atmosphere. Zr getting technique was employed as described by Cava et al.[10] to make the y=0.37 sample. A proper amount of fully oxygenated powder and Zr foil were sealed in a quartz tube and annealed at 490 C for 50 hours followed by slow cooling to room temperature in 100 hours. Such a low temperature annealing and slow cooling seems to be important to obtain good homogeneity and oxygen ordering. The value of y was determined from iodometric titration and weight gain of the Zr foil. We use a label y~0 for the fully oxygenated sample, although the actual value of y is around 0.05. The y~0 (y=0.37) sample shows almost 100% (about 70%) shielding magnetization at 10 G. NMR measurements were made on the oriented powder sample as described in ref. 11. The methods of measurements of Cu and O Knight shift and nuclear spin relaxation rate ($1/T_1$) were already described elsewhere.[11-15]

3. Cu AND O KNIGHT SHIFT

In this section we discuss the static magnetic properties of the CuO_2 layers revealed by Knight shift measurements on the planar Cu and O sites. the results in the y~0 material were already discussed in earlier publications[11-14] and we review a few important conclusions.

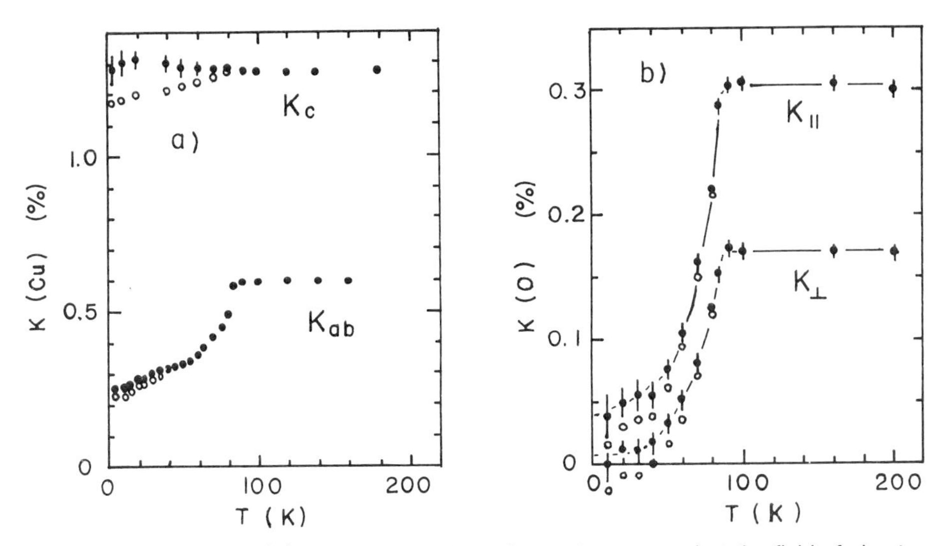

Fig. 2. Cu (a) and O (b) Knight shift in the y~0 sample measured at the field of about 7tesla. Solid (open) circles are the data with (without) correction for the field due to diamagnetic supercurrent. $K_{||}$ ($K_{\perp}$) at the O sites indicates the Knight shift with the magnetic field along (perpendicular to) the Cu-O bond axis, i.e., $K_{||}$ for the O(2) sites is equal to K_a and $K_{||}$ for the O(3) sites is equal to K_b.

Cu Knight shift in the y~0 (T_c=92K) material Figure 2a) shows the T-dependence of the Knight shift at the planar Cu(2) sites in the y~0 sample measured at 85MHz (about 7 tesla). The Knight shift is T-independent in the normal state as in the magnetic susceptibility. In the superconducting state, K_{ab} decreases rapidly below T_c, whereas K_c hardly changes. In the superconducting (vortex) sate, the screening diamagnetic current produces additional local field at the nuclear sites. The solid (open) symbols show the Knight shift with (without) the correction for this field. This correction is based on the comparison of the shift of the ^{17}O resonance field with the bulk magnetization. The correction for K_c was found to be much smaller than the original estimation in ref. 12, in agreement with the measurements of Barrett et al.,[16] who made this correction using Y resonance. Generally, the observed Knight shift consists of the spin and orbital (Van Vleck) parts, $K = K_{spin} + K_{orb}$. We assume that only K_{spin} depends on temperature below T_c and K_{spin}=0 at T=0. The validity of this assumption is discussed in ref. 12. K_{orb} is then determined from the value of Knight shift at T=0 in each direction. Virtually unchanged K_c below T_c indicates very small $K_{spin,c}$ in the normal state. The magnitude and the large anisotropy of K_{orb} thus determined are well accounted for by a commonly accepted crystal field level scheme of Cu 3d-states.[11,17]

We are mainly interested in the magnitude and anisotropy of K_{spin} in the normal state. Since K_{spin} is axially symmetric around the c-axis, K_{spin} follows the orientational dependence,

$$K_{spin}(\theta) = K_{iso} + K_{ax}(3\cos^2\theta - 1) \qquad (1)$$

where θ is the angle between the external field and the c-axis. From the data in Fig. 2a), we obtain K_{iso} = 0.23 ± 0.03 %, K_{ax} = -0.12 ± 0.02 %. A similar result has been obtained by Barrett et al.[16] (K_{iso} = 0.20 ± 0.02 %, K_{ax} = -0.10 ± 0.01 %). The negative sign of K_{ax} is consistent with the hyperfine field from spin density in the Cu-3d(x^2-y^2) state. The anisotropy of this field is due to the dipolar field which is strongly modified by spin-orbit coupling. One can easily see that the spin density distribution on the $d(x^2-y^2)$ state produces negative dipolar field (i.e. in opposite direction to the external field) when spin is polarized along the c-axis and positive field when spin polarization is in the ab-plane.

The positive sign of K_{iso}, however, is quite unusual. In most of the 3d transition metal compounds including CuO[18], the isotropic hyperfine field originates from core polarization effect which gives a negative field of the order of -100 KOe/μ_B.[19] The positive K_{iso} could be due to the transferred hyperfine coupling to oxygen hole spins,[12,20]

$$K_{spin,i} = A_d^i \chi_d/\mu_B + B_h \chi_h/\mu_B \qquad (i = c \text{ or } ab), \qquad (2)$$

where χ_d and χ_h is the susceptibility of the Cu d- and O hole spins and the second term gives rise to the positive isotropic shift. Mila and Rice, however, have shown that the mixing between Cu 4s states and nearest neighbor Cu $d(x^2-y^2)$ state via the intermediate O 2p states leads to a significant isotropic positive transferred hyperfine field and explicit coupling to the oxygen holes (the B_h term in eq. 2) is not required.[17]

$$K_{spin,i} = (A_d^i + 4B)\, \chi_d/\mu_B \quad . \qquad (3)$$

The factor 4 is the number of nearest neighbor Cu sites. The positive K_{iso} results from the 4B term dominating over the negative core polarization field included in A_d^i.

Since B_h and B terms in eq. (2) and (3) are isotropic, K_{ax} is coupled only to the planar Cu d-spins via on site hyperfine parameters, $K_{ax} = (A_d^c - A_d^{ab})\chi_d/3$. $A_d^c - A_d^{ab}$ can be estimated with reasonable accuracy. Using $A_d^c - A_d^{ab} = -220 \pm 20$ kOe/μ_B[17,20] and from the value of K_{ax}, we obtain

$$\chi_d = (9.2 \pm 1.5) \times 10^{-5} \text{ emu/mole Cu(2)} \qquad (4)$$

<u>O Knight shift in the y~0 (T_c=92 K) material</u> A similar analysis was made on the oxygen Knight shift.[13,14] We discuss only the result son the planar O(2,3) sites. The T-dependence of the Knight shift in the ab plane is shown in Fig. 2b). Assuming again that $K_{spin}=0$ at T=0, the values of K_{spin} at the O(2) sites in the normal state are obtained, $K_{spin,a} = 0.26 \pm 0.01$ %, $K_{spin,b} = 0.16 \pm 0.01$ %, $K_{spin,c} = 0.16 \pm 0.03$ %. For the O(3) sites which have the nearest Cu neighbor along the b-axis, the values of $K_{spin,a}$ and $K_{spin,b}$ are interchanged. K_{spin} at O(2,3) sites is uniaxial around the Cu-O bond axis and expressed by eq. (1) with $K_{iso} = 0.19 \pm 0.02$ % and $K_{ax} = 0.033 \pm 0.003$ %, θ being the angle between the external field and the Cu-O bond axis. Similar results have been obtained by Horvatic et al.[21] K_{ax} at O(2,3) sites is due to the dipolar field from the spin density on O-2p orbitals. As shown in Fig. 3, the dipolar field is positive (negative) when spin polarization is along (perpendicular to) the lobe of the 2p orbital. Therefore, the positive sign of K_{ax} (i.e. positive dipolar field when spin polarization is along the Cu-O bond axis) indicates that the spin density resides on the 2pσ orbital.

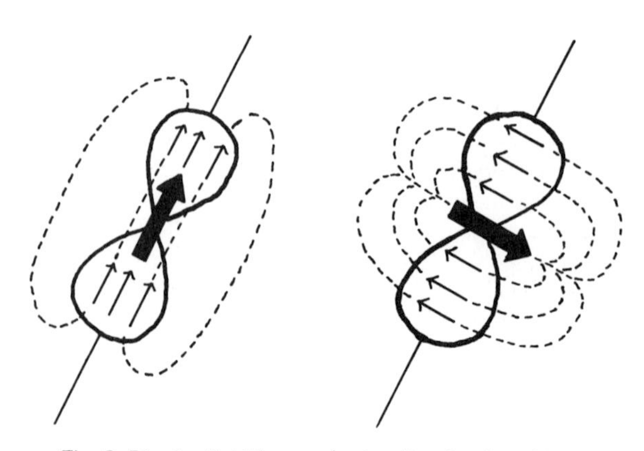

Fig. 3 Dipolar field from spin density of a 2p orbital.

K_{ax} is directly related to the spin susceptibility of O-2p states. $K_{ax} = A_p\chi_p$. Using $A_p = 90$ kOe/μ_b,[13] we obtain $\chi_p = (2.1 \pm 0.2) \times 10^{-5}$ emu/mole O. By comparing this value with eq.(4) and considering that there are two oxygen sites in a unit square, we conclude that in the $Cu(2)O_2$ plane, about 70% of the spin density resides on Cu $d(x^2-y^2)$ states and 30% on the O $2p\sigma$ states. (The spin density on the O 2s state which is responsible for K_{iso} is estimated to be about an order of magnitude smaller than that on the 2p states.[21] Because of the high hyperfine field from 2s state (6000 kOe), such a small spin density gives rise to large K_{iso}.) It is important to know how much of the observed χ_p is actually associated with doped holes. A finite spin density on the oxygen sites is expected even for the undoped insulator (y=1) due to covalency effect, as has been observed in many magnetic insulators. Such spin density, however, does not behave independently from Cu d-spins. Only the spin density associated with the doped holes could have distinct degrees of freedom. An important clue to this problem has been provided by the Knight shift results in the y=0.37 sample.

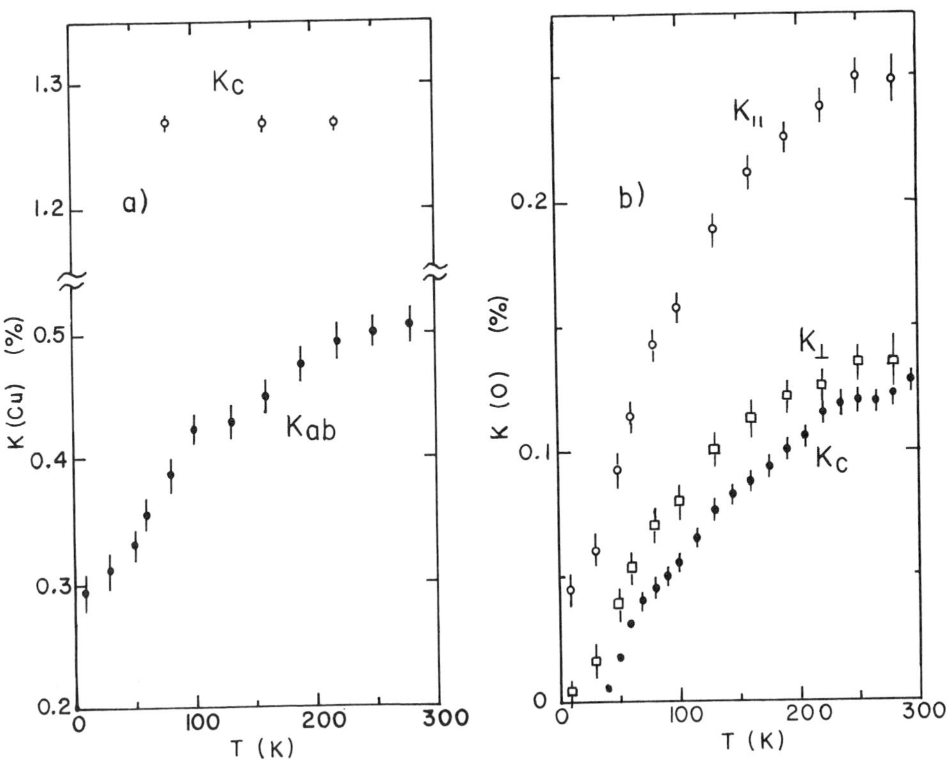

Fig.4. Cu (a) and O (b) Knight shift in the y=0.37 sample measured at the field of 7 tesla. No correction is made for the field produced by diamagnetic supercurrent below T_c.

Cu and O Knight shift in the y=0.37 (T_c=62 K) material The T-dependences of the Knight shift at the planar Cu and O sites in the y=0.37 (T_c=62 K) sample are shown in Fig. 4 a) and b). In contrast to the y~0 material, the various principal components of Knight shift except K_c(Cu) show significant T-dependence in the normal state as does the bulk susceptibility. K_c(Cu) remains unchanged from y~0 compound, further supporting that K_c(Cu) is entirely of orbital origin. Since K_c(Cu) is T-independent, K_{ab}(Cu) has the same T-dependence as K_{ax}(Cu). Thus the spin part of K_{ab}(Cu) is proportional to the d-spin susceptibility χ_d (eq. (3)). Similar results of Cu Knight shift have been reported by Shimizu et al.[18] and Walstedt et al.[22] Now we look at the oxygen data. It is noticed that the difference between $K_\perp$ and K_c is roughly T-independent, indicating that K_{spin} at the O sites is uniaxial as in the y~0 material. We define as $K_{ax} = (K_{||} - K_\perp)/3$, $K_{iso} = (K_{||} + 2K_\perp)/3$. It should be emphasized again that the spin part of K_{ax} results from the spin density on the O-2p states and that K_{iso} is due to the spin density on the O-2s states.

Our major finding is that these various components of Cu and O Knight shift follow the same T-dependence. More precisely, all these components (K_{ax}, K_{iso}, K_c for oxygen and K_{ab} for Cu) are coupled to a common t-dependent spin susceptibility χ_s as

$$K_i(T) = A_i\chi_s(t) + K_i(O) \tag{5}$$

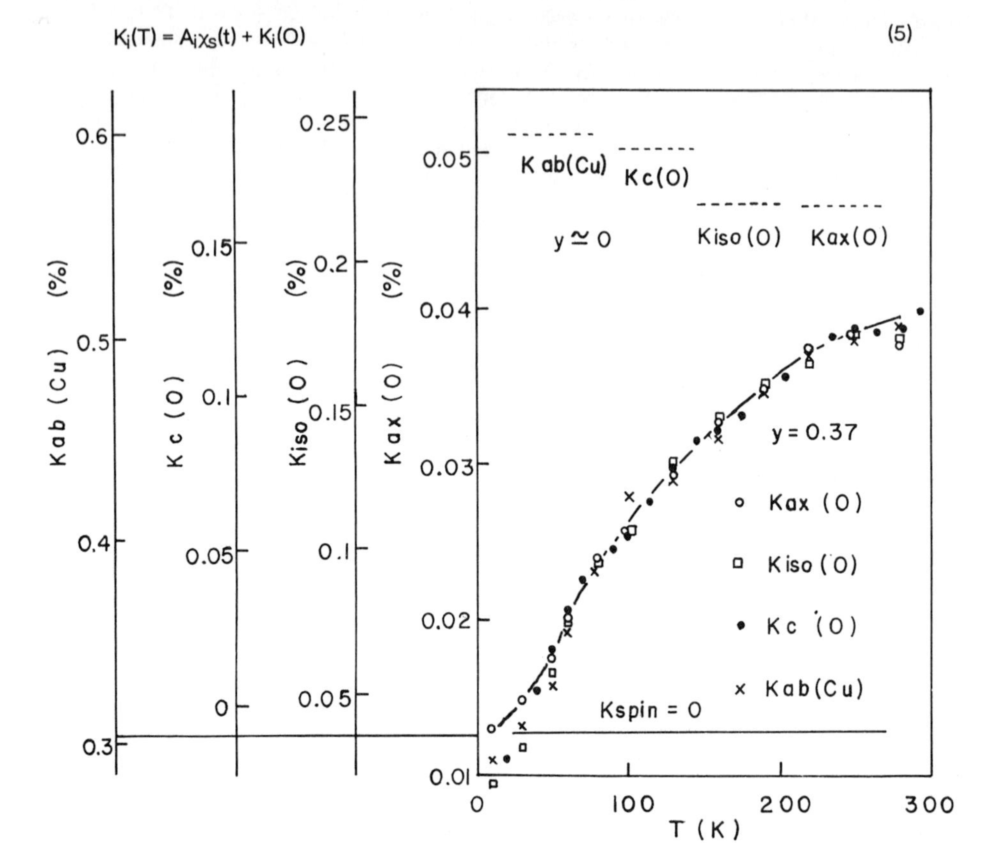

Fig. 5. K_{ax}, K_{iso} and K_c at the O sites and K_{ab} at the Cu sites in the y=0.37 sample are plotted against temperature with different vertical scales as shown on the left. All the data above T_c are on a single curve (solid line). The orbital parts of these components are shown by the line K_{spin}=0. The values of these Knight shift components in the y$\simeq$0 sample (T-independent) are also shown by dashed lines.

This is demonstrated in Fig. 5 where K_i's are plotted with different scales corresponding to different choice of A_i and $K_i(O)$. The values of $K_i(O)$ (K_i at T=0 where χ_s=0) are determined as follows. The effect of diamagnetic current below t_c should be cancelled for K_{ax} at the O sites since this is the difference between two Knight shift components with the same field direction (H$\perp$c). Therefore, $K_{ax}(O)$ is determined simply by extrapolating the data to T=0. A_iA_{ax} and $K_i(O)$ for other components are determined by fitting the data of $K_{ax}(T)$ and $K_i(T)$ above T_c to the form, $K_i(T) = (K_{ax}(T) - K_{ax}(O))A_i/A_{ax} + K_i(O)$. The obtained values of $K_i(O)$ ($K_{ax}(O)$ = 0.013, $K_{iso}(O)$ = 0.036, $K_c(O)$ = -0.006 % for oxygen and $K_{ab}(O)$ = 0.30 % for Cu) will represent the orbital parts and are indeed largely equal to the orbital Knight shift estimated in the y$\simeq$0 material.[12,13,16] The values of $K_i(O)$ are indicated by the line K_{spin}=0 in Fig. 5. We can see in the plot of Fig. 5 that all the data points in the normal state are essentially on a single curve. Systematic deviation seen below T_c is probably due to local field by diamagnetic current.

The relations in eq. (5), particularly those for K_{ax} at O and K_{ab} at Cu, indicate that both χ_d and χ_p have the same T-dependence, strongly suggesting that the spin density on the Cu-3d and the O-2p states behave as parts of a single spin system. Moreover, the same relation can be extended to the y$\simeq$0 material, as shown by the plot of K_i in the y$\simeq$0 sample in the same scale in Fig.5. This indicates that the ratio χ_d/χ_p does not change from y=0.37 to y$\simeq$0. These results are entirely consistent with the observation by Alloul et al.[23] that the Knight shift at the yttrium (Y) sites has the same T- and y (oxygen content)-dependence as the bulk susceptibility, if we notice that Y nuclei are dominantly coupled to the spin on the O-2p states[24] and the bulk spin susceptibility is dominated by Cu d-spin. It is quite impressive that this behavior is observed in a wide range of y ($0 \leq y \leq 0.6$)[23] covering the insulating phase were the number of doped holes in the CuO_2 plane must be very small (a few percent). This indicates that χ_d/χ_p is constant over such a wide rnagle of hole (carrier) concentration. If the doped holes contribute to the spin susceptibility, they will contribute more to χ_p than χ_d. Therefore, y-independence of χ_d/χ_p suggests that the doped holes or charge carriers do not have spin degrees of freedom and observed ξ_p is simply due to the covalency effect which does not much depend on y. Thus all the data of Cu, O and Y Knight shift fit quite naturally to the picture of single spin component which mainly resides on the Cu sites.

This situation allows us to extract the unique T-dependence of the spin susceptibility χ_s of the CuO_2 planes as shown by the line in Fig. 5. χ_s shows strong reduction with decreasing temperature in a wide t-range above T_c. This apparent suppression of spin excitations in the reduced T_c material is in clear contrast to the Pauli-like T-independent susceptibility in the fully oxygenated material. It should be noted that a similar suppression of susceptibility has been observed in $(La_{1-x}Sr_x)_2CuO_4$ systems with the highest T_c (~35K).[7,25] The peculiar T-dependence of χ_s seems to be a common feature of the "intermediate-T_c" materials. We do not see such a strong reduction in the bulk susceptibility (Fig. 1) even after correcting for the diamagnetic and orbital susceptibility. The data shows a slight Curie-Weiss contribution probably from an impurity phase. Also there must be contribution from the CuO chains whose T-dependence is not known. Therefore, we have not tried to make quantitative comparison between Knight shift and bulk susceptibility data.

4. Cu and O NUCLEAR SPIN-LATTICE RELAXATION RATE

Measurement of nuclear spin-lattice relaxation rate ($1/T_1$) provides useful information about the nature of low frequency spin fluctuations. $1/T_1$ is the transition probability between nuclear spin Zeeman levels caused by fluctuations of the hyperfine field and quite generally be expressed in terms of the dynamical spin susceptibility,[26]

$$\frac{1}{T_1} = \frac{\gamma_n^2 k_B T}{2\mu_B^2} \sum_q |A_q|^2 \frac{\mathrm{Im}\ \chi(q,\omega_o)}{\omega_o}$$

$$A_q = \sum_q A_i \exp(i\vec{q}\vec{r}_i) \qquad , \qquad (6)$$

where γ_n is the nuclear gyromagnetic ratio and A_i is the hyperfine coupling between a nuclear spin and electron spin at a site $\vec{r}_i$. ω_o is the nuclear Larmor frequency which is much smaller than the characteristic energy of electronic spin system. If we assume a Lorentzian frequency spectrum having a width Γ_q for $\mathrm{Im}\chi(q,\omega)/\omega$,

$$\mathrm{Im}\chi(q,\omega_o)/\omega_o = \pi\chi(q)/\Gamma_q \qquad , \qquad (7)$$

where $\chi(q)$ is the static q-dependent susceptibility. For a Fermi liquid without strong magnetic correlation, both $\chi(q)$ and $1/\Gamma_q$ is roughly q-independent and is given by the density of state ρ_o. Therefore, $\Sigma\ \mathrm{Im}\chi(q,\omega_o) \sim \pi h \mu_B^2 \rho_o^2 \sim \pi h \chi_s^2/\mu_B^2$, where χ_s is the uniform spin susceptibility. For non-interacting electrons, we have an exact relation, $\Sigma\ \mathrm{Im}\chi(q,\omega_o) \sim \pi h \chi_s^2/(2\mu_B^2)$. Furthermore, if the hyperfine coupling is local (i.e. and A_q is q-independent) and isotropic, we have a universal relation between $1/(T_1T)$ and the spin Knight shift (Korringa relation),

$$\frac{1}{T_1 T K^2} = \frac{\pi h \gamma_n^2 k_B}{\mu_B^2} \equiv S \qquad . \qquad (8)$$

The way this relation is modified by the electron-electron interaction has been studied by the random phase approximation (RPA).[26] The enhancement of the Korringa product $(1/)T_1T\,K^2)$ relative to the non-interacting value S is given by[26]

$$\frac{1}{T_1 T K^2 S} = \frac{\langle (\chi(q)/\chi_o(q))^2 \rangle}{(\chi_s/\chi_o(0))^2} \qquad (9)$$

where $\chi_o(q)$ is the static susceptibility for the non-interacting system and < > means a certain average over the Fermi surface. From eq. (9), we expect $1/(T_1T\ K^2\ S)$ to be less than 1 when $\chi(q)$ is enhanced mostly around q~0 (nearly ferromagnetic) and greater than 1 when the enhancement of $\chi(q)$ occurs around a finite value of q far from the zone center. Although eq. (9) is specific to RPA, this qualitative conclusion would be valid for a general itinerant electron system.

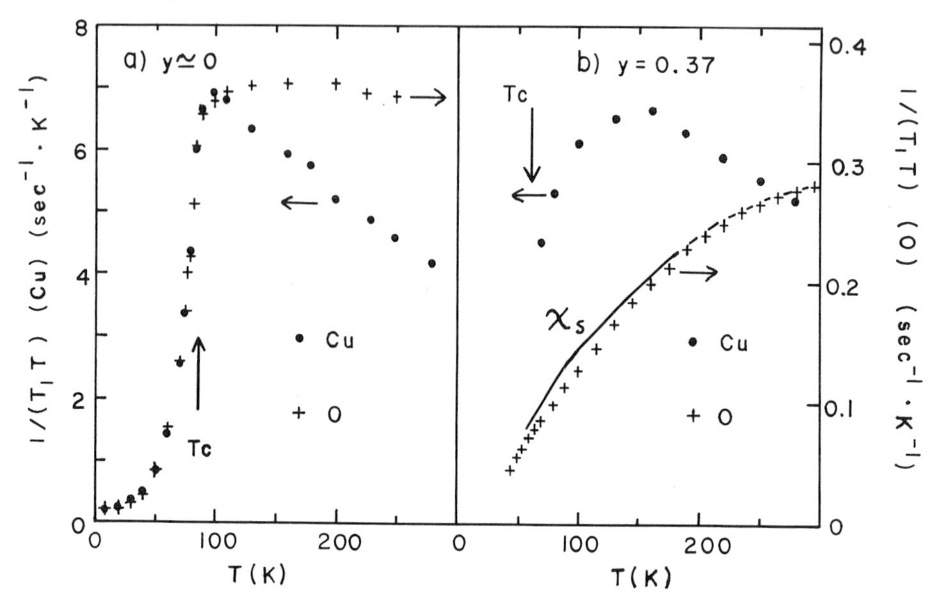

Fig. 6. Temperature dependences of $1/(T_1T)$ at the Cu (left scale) and O (right scale) sites in the y~0 sample (from ref. 15) (a) and the y=0.37 sample (b). $1/T_1$ was measured at the field of about 7 tesla along the c-axis. T-dependence of the spin susceptibility in the y=0.37 sample deduced from the plot in Fig. 5 is also shown in an arbitrary unit in b).

<u>y~0 (T_c=92K) material</u> The temperature dependences of $1/(T_1T)$ at the Cu and O sites in the y~0 sample (data from re. 15) are shown in Fig. 6 a). The cur results are in agreement with the data by other groups.[27-29] One can see that $1/(T_1T)$ at the Cu and the O sites show quite different T-dependence above about 110 K. $1/(T_1T)$ (O) is T-independent as should be expected from T-independent χ_s. $1(T_1T)$ (Cu) is decreasing with increasing temperature and its magnitude is about 20 times larger than $1(T_1T)$ (O) at 110 K. Below about 110 K, $1(T_1T)$ at both sites show the same t-dependence. $1(T_1T)$ decreases very rapidly below T_c and we do not see the "Hebel-Slichter" peak,[30] which has been usually observed just below T_c in conventional superconductors. It has been sometimes argued that the absence of this peak suggests anisotropic d-wave pairing which removes the singularity in the density of quasiparticle states.[28] However, this is in contradiction to other results such as the T-dependence of the penetration depth[31] which supports s-wave pairing. Alternative explanation in s-wave model has been proposed by Kuroda and Varma.[32] This involves a T-dependent pair breaking which is effective only in the vicinity of T_c.

Now we will focus on the normal state results. It is not obvious how to estimate the enhancement of the Korringa constant $1/(T_1TK_2S)$ since the hyperfine coupling is non-local for both Cu and O and highly anisotropic for Cu. A reasonable way to estimate this enhancement is described in ref. 15. We find that[15] $1/(T_1T\ K^2)$ at Cu at 110 K is enhance by a factor of 11 compared with what one would expect for a non-interacting electrons. This enhancement factor at the O sites is much smaller, about 1.4. A somewhat more realistic estimation was made by Millis et al.,[33] which gives the enhancement factor 15 for Cu and 2.5 for O.

According to the previous discussion, the large enhancement of $1(T_1T)$ at Cu implies significant enhancement of $\chi(q)$ a some finite $q=q_o$. The T-dependence of $1/(T_1T)$ at Cu can be explained by Curie-Weiss like increase of $\chi(q_o)$ with decreasing temperature in the normal state. What is then responsible for the Korringa-like behavior at the O site? A possible explanation may be based on a two band model, where O holes have distinct spin dynamics from Cu d-spin (Cox and Tree[34]). The Knight shift results, however, supports a model of single spin component which mainly resides on the Cu sites.

In a single spin component model, the hybridization between O-2s and Cu-3d states is the dominant source for the transferred hyperfine field at the O sites, which is isotropic and written as

$$H_{hf} = A(\vec{S_i} + \vec{S}_{i+1}) = \sum_q A_q \vec{S_q}$$

$$A_q = A(1 + e^{iq_x a}) \qquad , \qquad (10)$$

where $\vec{S_i}$ and $\vec{S}_{i+1}$ are the two nearest neighbor Cu spins. This hyperfine interaction now contains q-dependent coupling constant (form factor) A_q. (See. eq. 6.) This coupling constant becomes O at the antiferromagnetic (AF) wave vector $Q=(\pi,\pi)$. Therefore, if $\chi(q)$ has a peak at (π,π) $(q_o=Q)$, or equivalently, of there is a strong AF correlation among Cu spins, this will enhance only the Cu relaxation rate. Since O relaxation is dominated by spin fluctuations in a broad q-space around the zone center (q=0) where $\chi(q)$ is T-independent and not much enhanced, $1(T_1T)$ (O) will show Korringa-like behavior, in agreement with experiment.

This qualitative explanation has been put into more quantitative scheme in several theoretical papers. Shastry[35] has discussed the relation between $1/(T_1T)$ (Cu) and AF correlation length ξ based on the dynamical scaling hypothesis. Bulut et al.[36] have used RPA to calculate Cu and O relaxation rate in 2D tight binding Hubbard model and reproduced the characteristic feature observed in the experiment. Millis, Monien and Pines[33] have developed a phenomenology to give a consistent account of the whole body of NMR data on the $y \simeq 0$ material. They proposed a model dynamical susceptibility which consists of a sharp peak around $Q=(\pi,\pi)$, whose width is related to the AF correlation length ξ, and a broad constant background over entire q-space. They obtained good fit to the data of the Cu and O relaxation rate by assuming T-dependence of ξ as $\xi(T)^2 = \xi(O)^2\, T_x/(T + T_x)$ with $T_x \simeq 120$K and $\xi(O)$ being 3~4 times the lattice constant ($\simeq$15Å).

<u>y=0.37 (T_c=62K) material</u> The nuclear relaxation behavior in the y-0.37 sample is quite different from that n the $y \simeq 0$ sample, as shown in Fig. 6 b). The data of $1/(T_1T)$ at Cu is in agreement with the data by Yasuoka et al.[37] A striking feature is a broad peak of $1/(T_1T)$ (Cu) around 160K. The behavior above this temperature is similar to that in the $y \simeq 0$ sample and may be understood as a result of the growth of AF correlation with decreasing temperature. Reduction of $1/(T_1T)$ (Cu) below 160K, a temperature well above T_c, is a novel feature not seen in the $y \simeq 0$ material. Warren et al. discussed this as a possible evidence of the superconducting precursor effect.[38] Yasuoka et al. suggested a certain kind of gap opening in the spin excitation spectrum at these temperature.[37]

$1/(T_1T)$ at the O sites, on the other hand, shows a monotonic decrease with decreasing temperature in a wide t-range in the normal state. Moreover, the T-dependence of $1/(T_1T)$ at the O sites is found to be almost the same as that of the spin susceptibility extracted from the Knight shift measurements (shown by the solid line in Fig. 6 b)).

$$\frac{1}{(T_1T)(O)} \propto \chi_s(T) \qquad (11)$$

This is in contradiction to the argument by Alloul et al.[23] that $1/(T_1T)$ at the Y sites is proportional to the square of K_{spin} at Y. We found, however that this discrepancy is resolved and proportionality between $1/(T_1T)$ and K_{spin} at the Y sites is obtained simply by taking the chemical shift at the Y sites to be 170ppm instead of 300ppm given by Alloul et al.

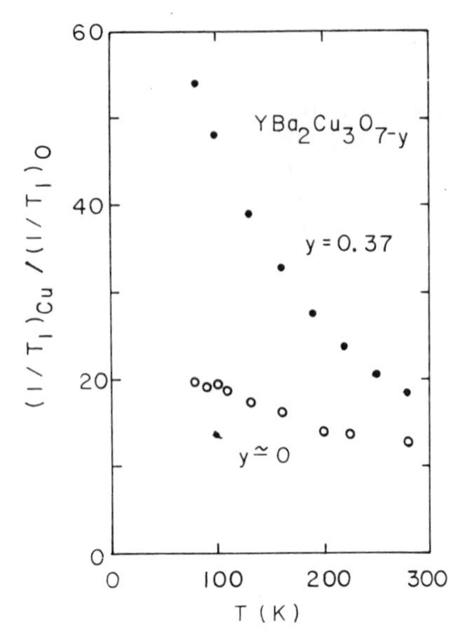

Fig. 7 The ratio of $(1/T_1)$ at Cu to $(1/T_1)$ at O is plotted against temperature for the y≃0 and the y=0.37 samples.

These relaxation behaviors in the y=0.37 materials are far less understood than the y≃0 material. It is interesting, however, so see to what extent the model of the AF correlation among Cu spins proposed to explain the data in the y≃0 material can be extended to the y=0.37 material. Since O relaxation is determined by spin fluctuations with q not close to $Q=(\pi,\pi)$, we expect the following relation,

$$\frac{1}{(T_1T)(O)} \propto \sum_{q\sim 0} \frac{\mathrm{Im}\,\chi(q,\omega_o)}{\omega_o} \sim \chi_s(T)/\Gamma \tag{12}$$

where Γ is the energy width at typical q not close to (π,π). Then the relation (11) requires that Γ is nearly T-independent. $1/(T_1T)$ at Cu, on the other hand, is dominated by spin fluctuations near $Q=(\pi,\pi)$. We assume that the staggered susceptibility $\chi(Q)$ is the product of the uncorrelated susceptibility $\chi_o(Q)$ and the enhancement factor $f(\xi)$ due to AF correlation in the spirit of the mean field expression, $\chi(Q)=\chi_o(Q)f(\xi)$. We further assume that $\chi_o(Q)$ has the same T-dependence as the uniform susceptibility χ_s. Then the ratio $1/(T_1T)$ (Cu) to $\chi_s(T)$, or almost equivalently, $(1/T_1)(Cu)/(1/T_1)(O)$ can be a measure of the AF enhancement factor $f(\xi)$. ($f(\xi) \propto \xi^2$ in the analysis of ref.33.) This ratio in the y=0.37 sample increases smoothly with decreasing temperature and we can hardly see an anomaly around 160K where $(1/T_1T)(Cu)$ has a peak. This suggest a possibility that the peak of $(1/T_1T)(Cu)$ is a combined result of the peculiar T-dependence of the spin susceptibility and the development of the AF correlation and may not necessarily require a spin gap.

In summary, we have described a picture of AF correlation among the Cu spins which gives a consistent account of the relaxation data in the y≃0 material and possibly in the y=-0.37 material. To obtain further understanding, the direct observation of the AF correlation in this system by neutron scattering experiment will be crucially important.

This work has been done in collaboration with P. C. Hammel, R. H. Heffner, A. P. Reyes, J. D. Thompson, Z. Fisk and K. C. Ott. The author appreciates stimulating discussion with D. Pines, T. M. Rice, H. Monien, D. S. Scalapino and A. J. Millis. Kind advice about Zr gettering method from W. D. Cooke and R. E. Walstedt is greatly acknowledged.

References

1. N. Nucker et al., Phys. Rev. B37, 5158 (1988).
2. A. Bianconi et al., Phys. Rev. B37, 7196 (1988).
3. P. S. List et al., J. Mag. Mat. 81, 151 (1989).
4. F. C. Zhang and T. M. Rice, Phys. Rev. B37, 3759 (1988).
5. V. J. Emery and G. Reiter, Phys. Rev. B38, 11938 (1988).
6. Y. Nakazawa and M. Ishikawa, Physica C158, 381 (1989).
7. D. C. Johnston, Phys. Rev. Lett. 62, 957 (1989).
8. Y. Endoh et al., Phys. Rev. B37, 7443 (1988).
9. G. Aeppli et al., Phys. Rev. Lett. 62, 2052 (1989).
10. R. J. Cava et al., Phys. Rev. B36, 5719 (1987).
11. M. Takigawa et al., Phys. Rev. B39, 300 (1989).
12. M. Takigawa, P. C. Hammel, R. H. Heffner and Z. Fish, Phys. Rev. B39, 7371 (1989).
13. M. Takigawa et al., Phys. Rev. Lett. 63, 1865 (1989).
14. M. Takigawa, P. C. Hammel, R. H. Heffner, Z. Fish, K. C. Ott and J. D. Thompson, Proceedings of the M^2-HTSC Conference, Stanford, 1989, to be published in Physica.
15. P. C. Hammel et al., Phys. Rev. Lett. 63, 1992 (1989).
16. S. E. Barrett, D. J. Durand, C. H. Pennington, C. P. Slichter, T. A. Friedman, J. P. Rice and D. M. Ginsberg, preprint.
17. F. Mila and T. M. Rice, Physica C157, 561 (1989).
18. T. Shimiu, K. Koga, H. Yasuoka T. Tsuda and T. Imai, Bulletin of Magnetic Resonance (Proceedings of the 10th ISMAR Meeting, Morzine, France, 1989) to be published.
19. R. F. Watson and A. J. Freeman, Hyperfine Interaction ed. A. J. Freeman and R. B. Frankel (Academic Press, NY, 1967) P. 53.
20 H. Monien, D. Pines and C. P. Slichter, preprint.
21. M. Horvatic et al., Physica C159, 689 (1989).
22. R. E. Walstedt, W. W. Warren, Jr., R. F. Bell, R. J. Cava, G. P. Espinosa, L. F. Schneemeyer and J. V. Waszczak, preprint.
23. H. Alloul, T. Ohno and D. Mendels, Phys. Rev. Lett. 63, 1700 (1989).
24. F. Mila and T. M. Rice, unpublished.
25. M. Oda, T. Ohguro, N. Yamada and M. Ido, J. Phys. Soc. Japan 58, 1137 (1989).
26. T. Moriya, J. Phys. Soc. Japan 18, 516 (1963).
27. T. Imai et al., J. Phys. Soc. Japan 57, 2280 (9188).
28. Y. Kitaoka, S. Hiramatsu, T. Kondo and K. Asayama, J. Phys. Soc. Japan 57, 30 (1988).
29. R. E. Walstedt, W. W. Warren, Jr., R. F. Bell and G. P. Espinosa, Phys. Rev. B40, 2572 (1989).
30. L. C. Hebel and C. P. Slichter, Phys. Rev. 113, 1504 (1959). L. C. Hebel, Phys. Rev. 116, 79 (1959).
31. D. R. Harshman et al., Phys. Rev. B239, 851 (1989).
32. Y. Kurada and C. M. Varma, preprint.
33. A. J. Millis, H. Monien and D. Pines, preprint; H. Monien, article is this volume.
34. D. L. Cox and M. Tree, preprint.
35. B. S. Shastry, Phys. Rev. Lett. 63, 1288 (1989).
36. N. Bulut, D. Hone, D. S. Scalapino and N. E. Bickers, preprint; N. Bulut, article in this volume.
37. H. Yasuoka, T. Imai and T. Shimizu, Strong Correlation and Superconductivity, ed. H. Fukuyama, S. Maekawa and A. P. Malozemoff (Springer-Verlag, 1989).
38. W. W. Warren, Jr. et al., Phys. Rev. Lett. 62, 1193 (1989).

^{17}O and ^{63}Cu NMR INVESTIGATION OF SPIN FLUCTUATIONS IN HIGH T_c SUPERCONDUCTING OXIDES

C. BERTHIER, Y. BERTHIER, P. BUTAUD, M. HORVATIC*,
Y. KITAOKA**, P. SEGRANSAN

Laboratoire de Spectrométrie Physique, Université J. Fourier Grenoble I
(LA08 CNRS), BP. 87, 38402 St Martin d'Hères, France
* Institute of Physics of the University, POB 304, 41001 Zagreb, Yugoslavia
** Department of Material Physics, Faculty of Engineering Science, Osaka University, Toyonaka, Osaka 560, Japan

INTRODUCTION

Since the discovery of the high T_c superconducting oxides, the electronic structure of these compounds in the normal state ($T > T_c$) is still a matter of controversy. It was reported early on by Emery [1] that the holes introduced in these materials by doping were primarily going into the O(2p) orbitals, leaving exactly one d-hole at the Cu site (Cu^{2+}) in the CuO_2 plane. Neutron inelastic scattering [2] and La NQR [3] in $La_{2-x}Sr_xCuO_4$ as well as Cu NMR spectroscopy [4-6] and neutron inelastic scattering[7,8] in $YBa_2Cu_3O_{6+x}$, support this scheme. In $YBa_2Cu_3O_7$ at least, however the temperature dependence of the Nuclear Spin Lattice Relaxation Rate (NSLRR)[9] and the Magnetic Hyperfine Shift (MHS) [10,11] of the Cu nuclei below T_c imply that these quasi-localized d-holes must be involved in the quasi-particles associated with superconductivity. Thus the nature of the coupling between these copper d-spins and the oxygen p-holes remains a central issue for the theoretical understanding of these high T_c materials [12-15]. ^{17}O NMR in these oxides is one of the most powerful tool to address this problem[16-23]. In the framework of the so-called t-J model [12,14], this coupling is so strong that a knowledge of the Cu(2) spin dynamical susceptibility $\chi_{Cu}(\vec{q}, \omega)$ should allow one to predict all NMR observables at O(2,3) and Y sites in the crystal. In contrast, two spin fluids models imply [15] or strongly suggest [13] the existence of a second spin susceptibility associated with the oxygen p-holes $\chi_h(\vec{q}, \omega)$ which at least above some temperature $T^* > T_c$ should behave differently from $\chi_{Cu}(\vec{q}, \omega)$.

In this paper we want to highlight two points: the first one is the dependence on the stoichiometry of the Cu(2) spin lattice relaxation rate and MHS tensor in $YBa_2Cu_3O_{7-\delta}$ which support the description of localized d-holes and point out the peculiar behaviour of $YBa_2Cu_3O_7$ as compared to its oxygen deficient parent compounds, whatever is their T_c value, 90 or 60 K. Secondly, we present the temperature dependence of the MHS tensor and the NSLRR of ^{17}O nuclei in $YBa_2Cu_3O_{6.65}$ (T_c = 60 K). The comparison of ^{17}O NMR data (this paper and ref. 18-23) with those obtained for ^{89}Y [24] shows that in this compound, the spins of the p-holes can have their own degree of freedom, at least above 120 K.

EXPERIMENTAL AND SAMPLES

All measurements were performed in a magnetic field H_o = 5.75 T. using pulsed NMR technique. As far as ^{63}Cu NMR is concerned, three samples have been investigated[6]: two "porous" single crystals[25] of composition $YBa_2Cu_3O_{6.9}$ (sample A) and $YBa_2Cu_3O_{6.75}$ (sample B) and an oriented powder sample $YBa_2Cu_3O_7$ (Sample C) This later was obtained by retreating under Oxygen crushed single crystals from the same batch as sample A. The NQR linewidth of sample C was found equal to 170 kHz.(full width at mid height FWMH) which is among the best values reported in the litterature. As discussed in Ref. 5-6, the distribution of oxygen vacancies in the CuO_2 plane was concluded to be homogeneous in sample A, whereas in sample B, a strong tendency to short range order has been observed. Details on the preparation of $YBa_2Cu_3O_{6.65}$ sample enriched with ^{17}O can be found in Ref. 18 and 23. Its oxygen composition was estimated from its quenching temperature (700°C)[26]. Details on powder orientation, the NMR technique and the site assignment can be found in Ref. 18.

RESULTS AND DISCUSSION

^{63}Cu NMR

We first would like to discuss our results on the Magnetic Hyperfine Shift tensor for the Cu(2) site. We have found its component along the c-axis K_{cc} to be independent of the temperature (above and below T_c) **and of the oxygen content**: K_{cc} = 1.28 ±0.03 % for samples A and C and K_{cc} = 1.31 ±0.05 % for sample B. [5,6] The absence of temperature dependence is a clear indication that this shift is purely orbital, due to some accidental cancellation of the spin contribution for this orientation of the magnetic field H_0.[27] Calculation of the orbital shift in a metallic band picture involve a summation over $\vec{k}$ of second order matrix elements of the type:

$$\frac{\langle\vec{k},n|L_1|\vec{k},n'\rangle\langle\vec{k},n'|L_2|\vec{k},n\rangle}{E_{\vec{k},n}-E_{\vec{k},n'}}\,(1-f_{\vec{k},n}).f_{\vec{k},n'} \tag{1}$$

between Bloch states of same $\vec{k}$ and of different band index.[28] In the present case, the $\vec{k}$,n states belong to the $d_{x^2-y^2}$ band, and the $\vec{k}$,n' to the other filled d-bands. This quantity should thus directly reflect the number of holes in the $d_{x^2-y^2}$ band. So the absence of dependence of K_{cc} on the oxygen composition is a strong evidence that the number of d-holes stays constant at least in the stoichiometry range δ =0 - 0.35 and finds its natural explanation in a description where the d-holes are quasi-localized (Cu^{2+}) in which case the orbital shift is an atomic quantity.[11,27]

Let us now discuss the Cu(2) NSLRR.[5,6] Results for samples A,B, and C are reported in Fig. 1. A sharp transition is observed at 90 K for sample C. The results are in excellent agreement with

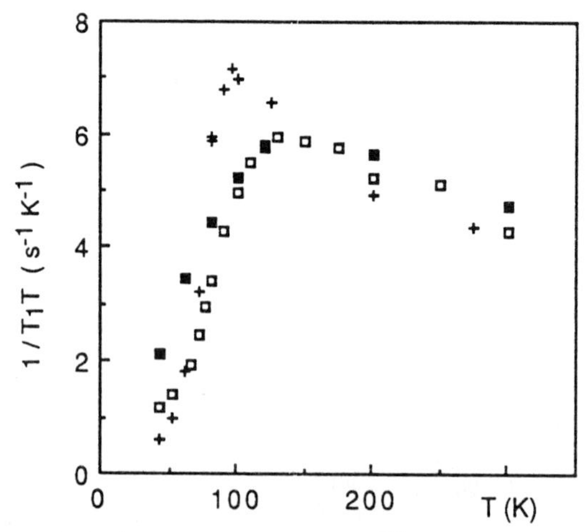

Fig. 1 Nuclear spin-lattice relaxation rate of $^{63}Cu(2)$ in $\delta = 0$ (+) oriented powder (sample C) and $\delta = 0.1$ (❑) and $\delta = 0.25$ (■) single crystals.

those of Hammel et al.[20] and quite close to the NQR data,[9] showing that there is no effect of the external magnetic field on the NSLRR at the transition but a small lowering of T_c. On the other hand, for samples A and B, a decrease of $(T_1T)^{-1}$ is observed starting around 120 K, which is well above T_c (90 and 60 K) for both samples. Such a decrease of $(T_1T)^{-1}$ was also observed in zero field by NQR in $YBa_2Cu_3O_{6.7}$ [29] and $YBa_2Cu_3O_{6.5}$ [30]. It is certainly related to the decrease observed for neutron inelastic scattering intensity at low energy in $YBa_2Cu_3O_{6.46}$ [7] and $La_{1.85}Sr_{0.15}CuO_4$ [31]; Although it is quite tempting to ascribe both phenomena to the opening of a gap for the AF fluctuations of the d-holes, this point has to be clarified in the future. We want to highlight that this decrease is not specific of oxygen compositions within the 60 K plateau, but is also present in the 90 K for oxygen composition far enough from $YBa_2Cu_3O_7$. ($\delta \approx 0.1$) It is thus likely that the peculiar behaviour of $YBa_2Cu_3O_7$ above T_c — i.e. a constant macroscopic susceptibility χ_m [32], which is reflected by constant MHS for O(2,3), [18,19] for Yttrium, [24] and for Cu(2) ($K_\perp$), [10,11] and an increase of $(T_1T)^{-1}$ between room temperature and T_c followed by a sharp decrease — is rather related to the absence of disorder introduced in the CuO_2 plane by oxygen vacancies than to the hole concentration. We shall further discuss the temperature dependence of the ^{63}Cu NSLRR and its origin after the presentation of the Oxygen data.

^{17}O NMR

^{17}O NMR in $YBa_2Cu_3O_7$ has shown that above T_c the MHS tensor $\overline{\overline{K}}$ [18,19] and the NSLRR [20] for the O(2,3) site stay constant (except a small anomaly between120 K and 90 K). These data thus cannot serve as a basis to discriminate between t-J or two-bands models. For this reason we have turned to the study of $YBa_2Cu_3O_{6.65}$, a composition which just falls in the middle of the concentration range corresponding to the 60 K plateau, and for which the macroscopic susceptibility χ_m presents a strong temperature dependence [32]

Following the position of the lines corresponding to the (1/2, -1/2) transition for the orientation H_0 // c-axis, between 50 K and 300 K, the temperature dependence of the MHS tensor along the c-axis K_{cc} was determined for the O(2,3) and the O(4) (bridging oxygen) sites. (Fig. 2.) (the narrow width of the lines in the whole temperature range is a very good indication of the homogeneity of our sample. The striking feature is the strong temperature dependence of K_{cc}[O(2, 3)] above T_c. It starts to decrease from room temperature down to T_c, whereas K_{cc}[O(4)] remains constant in the same temperature range Below T_c, K_{cc}[O(2,3)] sharply decreases to zero, indicating that in the decomposition of K_{cc} into its spin and orbital contribution, this latter is negligible. As shown in the inset $^{17}K_{cc}(T)$ versus the macroscopic susceptibility $\chi_m(T)$ [32] shows a linear dependence the origin of which will be discussed later. The difference in the temperature dependence of K_{cc}[O(2, 3)] and K_{cc}[O(4)] is similar to that observed in $La_{1.85}Sr_{0.15}CuO_4$,[21,22], where above T_c the value of $^{17}K_{cc}$[I] corresponding to the CuO_2 plane site is strongly temperature dependent and scales linearly with χ_m, while that of $^{17}K_{cc}$[II] (LaO plane) stays constant. This demonstrates that in the both types of compound the temperature dependence of χ_m finds its origin *within* the CuO_2 plane, and that the Cu(1) (in the chains) do not (or very weakly) contribute to it. This is also supported by the facts that the Yttrium MHS ^{89}K was found to scale linearly with χ_m for $x > 0.4$ in the system $YBa_2Cu_3O_{6+x}$,[24] as well as the value of ^{63}K[Cu(2), $H_0 \perp c$] in $YBa_2Cu_3O_{6.65}$ [33].

In order to obtain a full knowledge of the MHS tensor, we have investigated the temperature dependence between 360 K and 70 K of the (-1/2, 1/2) transition lineshape corresponding to the O(2, 3) sites for the orientation $H_o \perp$ c-axis. To extract K_{aa} and K_{bb}, experimental lineshapes were fitted to a computer simulation including an exact diagonalization of the total hyperfine Hamiltonian (quadrupolar and magnetic) for all field orientations in the a-b plane.[18] Typical spectra and corresponding computer simulations are shown in Fig. 3. Each $K_{\alpha\alpha}$ component can be decomposed into a spin and an orbital (Van Vleck) contribution

$$K_{\alpha\alpha} = K_{\alpha\alpha}^{spin} + K_{\alpha\alpha}^{orb} \qquad (2)$$

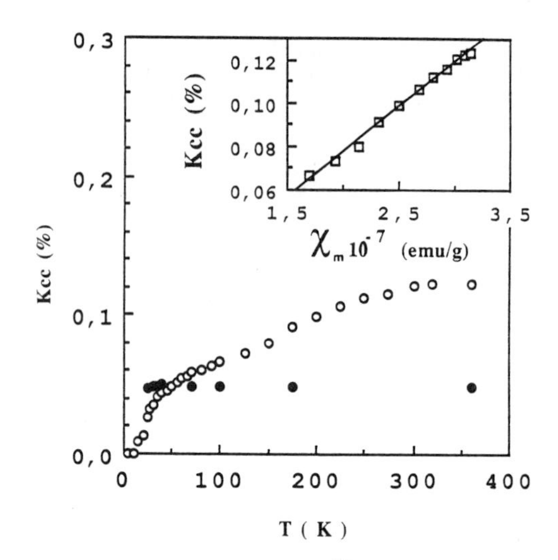

Fig. 2. Plot of the hyperfine magnetic shift of ^{17}O along the c-axis, versus temperature for the O(2,3) sites (open circle)) and for the O(4) site (solid circle). The insert shows the linear relation between K_{cc}[O(2,3)] and the macroscopic susceptibility (from ref 32).

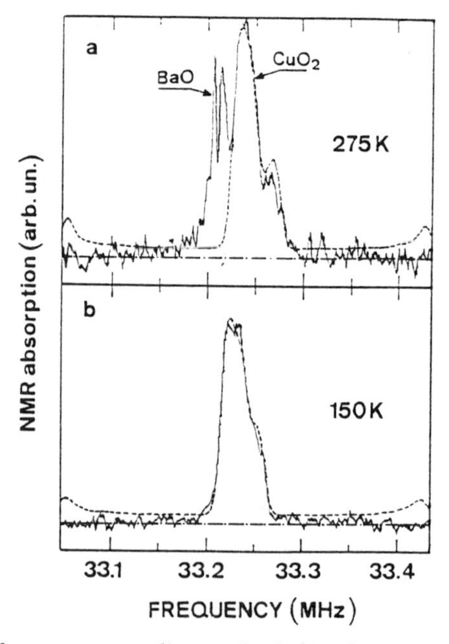

Fig. 3. Typical lineshape corresponding to the (1/2,-1/2) transition for the O(2,3) sites with $H_o \perp$ c-axis at 275K (fig. 3a) and at 150K (fig.3b). The dotted line is the computer simulation with an exact diagonalization of the total spin Hamiltonian.

One can thus define the experimental value of $K_{iso(ax)}$ as :

$$K_{iso} = \left(\sum_{\alpha} \frac{K_{\alpha\alpha}}{3} \right) = K_{iso}^{spin} + K_{iso}^{orb} \tag{3.a}$$

$$K_{ax} = \frac{2}{3}\left[K_{aa} - \frac{(K_{bb} + K_{cc})}{2} \right] = K_{ax}^{spin} + \frac{2}{3}\left[K_{aa}^{orb} - \frac{K_{bb}^{orb} + K_{cc}^{orb}}{2} \right] \tag{3.b}$$

(We assume in the following that the direction a corresponds to the Cu-O-Cu bond for the site under consideration).

K_{aa}, K_{bb}, K_{cc}, K_{iso}^{exp} and K_{ax}^{exp} are plotted in figure 4.

As mentioned in ref.18, the isotropic part of MHS tensor reflects the polarization of the O(2s) orbitals by the Cu(2) electronic spins due to their overlap with the $3d_{x^2-y^2}$ wave function[34]

$$K_{iso}^{spin} = 2f_s \frac{A^{2s}}{g\mu_B} \chi_{Cu} \tag{4}$$

where f_s is the fraction of unpaired spin on O(2s) orbitals, and χ_{Cu} is the local static spin susceptibility per Cu(2) atom. As far as the axial part of the tensor is concerned, it reflects the spin polarization of the O(2p) orbitals, which results on one hand from the covalence and overlap with the $3d_{x2-y2}$ Cu(2) wave function and on the other hand from the direct hyperfine coupling (dipolar and orbital) with an itinerant p-hole band if present in the electronic structure.

$$K_{ax}^{spin} = \frac{A^{2p}}{g\mu_B}\left(2f_p\chi_{Cu} + \chi_h\right) \tag{5}$$

$A^{2s}/2\mu_B$ and $A^{2p}/2\mu_B$ the local hyperfine coupling are respectively taken equal to 6000 kG/μ_B and 190 kG/μ_B respectively. We shall see later that the temperature dependence of K_{cc} mainly reflects that of K_{iso}^{spin}, thus from the inset of Fig.2 and equ.(4) one deduces $f_s = 0.6$ % and $\chi_{Cu} = 1.2\ 10^{-4}$ e.m.u./mole at 300K.

In order to descriminate between a single spin fluid ($\chi_h = 0$) or a two spin fluid model ($\chi_h \neq 0$), we need to consider $\alpha(T) = K_{ax}^{spin}(T) / K_{iso}^{spin}(T)$ which should be a constant $\alpha_o = f_p A^{2p} / f_s A^{2s}$ in the first case and temperature dependent in the second :

$$\alpha(T) = \frac{A^{2p}}{A^{2s}}\left[\frac{f_p + \chi_h/\chi_{Cu}}{f_s} \right] = \alpha_0 \left(1 + \frac{\chi_h(T)}{f_p \chi_{Cu}(T)} \right) \tag{6}$$

From equ. (3) we see that

$$\alpha(T) = \frac{K_{ax}^{exp}(T) - K_{ax}^{orb}}{K_{iso}^{exp}(T) - K_{iso}^{orb}} \tag{7}$$

and as $K_{iso}^{spin} \gg K_{iso}^{orb}$ the determination of α will depend crucially on the value of K_{ax}^{orb}.

From our measurements we can claim that K_{cc}^{orb} is smaller than 0.005 %. $K_{bb} - K_{cc}$ can be estimated from Fig. 4, to be 0.02 %, thus we conclude that K_{bb} is smaller than 0.03 %. However we did not measure K_{aa}^{orb} since the experimental lineshape of the (-1/2, 1/2) transition at very low temperature becomes too broad to allow a fit leading to reliable values of K_{aa} and K_{bb}.

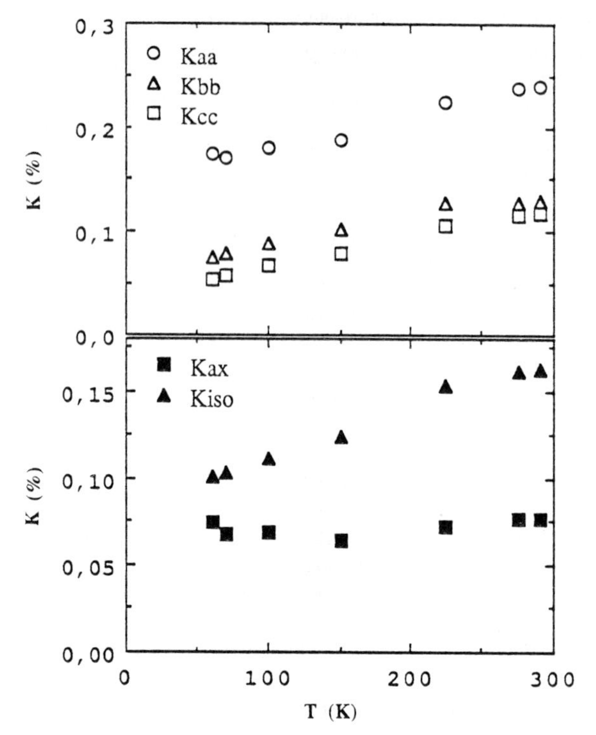

Fig. 4. The three components $K_{\alpha\alpha}$ of the magnetic hyperfine shift tensor $\overline{\overline{K}}$ for the O(2,3) sites are shown versus temperature in the upper part. The temperature dependence of the axial and isotropic part of $\overline{\overline{K}}$[O(2,3)] is shown below.

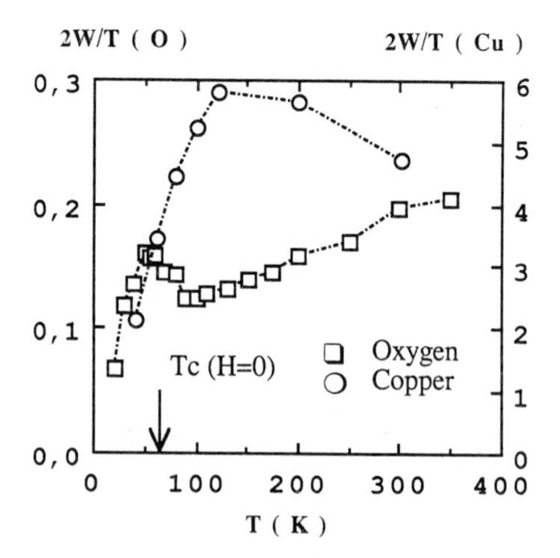

Fig. 5. Temperature dependence of the nuclear spin lattice relaxation rate of ^{17}O (left scale) in $YBa_2Cu_3O_{6,65}$ and of ^{63}Cu (right scale) in $YBa_2Cu_3O_{6,75}$ (see text).

In $YBa_2Cu_3O_7$ Takigawa et al [19] reported $K_{aa}^{orb} = 0.05 \pm 0.01$ %. This value should be considered as an upper limit in $YBa_2Cu_3O_{6.65}$ since the value of K^{orb} which reflects the Van Vleck susceptibility should decrease with a decreasing number of holes in the p band. Moreover, if the character of the p band at the Fermi level is p-σ along the a direction, the value of K_{aa}^{orb} should be zero from symmetry considerations.
Our results favour the existence of a constant contribution to K_{ax} due to a p-band with its own spin degree of freedom. However, for a set of values [0.05, 0.03, 0.01] for $\left[K_{aa}^{orb}, K_{bb}^{orb}, K_{cc}^{orb}\right]$, α could be constant within the error bars.
An alternative method to probe the existence of a second spin degree of freedom in the system is a comparison between the spin lattice relaxation rate of the ^{17}O and ^{89}Y for the same composition. We have measured the NSLRR for the site O(2,3) by studying the recovery of the nuclear magnetization of the (1/2, 3/2) transition. This technique allows a precise selection of one site in the crystal, contrary to measurements performed on the central line, which can pick up the contribution of several sites together. Experimental data were fitted to the theoretical expression given by Narath [35] for the (1/2, 3/2) transition (I = 5/2) in presence of a magnetic relaxation process. In order to check the validity of this procedure, we also measured at 200 K the recovery of the magnetization corresponding to the (3/2, 5/2) transition. The fit to the theoretical expression[35] gave the same value of T_1 within the experimental accuracy. The results are shown in Fig. 5 together with the NSLRR for the Cu(2) measured in single crystal of $YBa_2Cu_3O_{6.75}$ ($T_c = 60$ K). As we compare data obtained from samples with slightly different stoechiometry ^{17}O NSLRR in$YBa_2Cu_3O_{6.65}$ and ^{63}Cu NSLRR in $YBa_2Cu_3O_{6.75}$ [6], it is important to note that the large decrease of the ^{63}Cu NSLRR above T_c was also reported in the case of $YBa_2Cu_3O_{6.52}$ (see ref 30) . One of the striking features, which will be discussed later, is the hump in the temperature dependence of $^{17}(T_1T)^{-1}$, which starts increasing when the Cu(2) NSLRR $^{63}(T_1T)^{-1}$ starts decreasing, i.e. is well above T_c, which under an external field H = 5.7 Tesla should be lower than 50 K.
We come to the comparison between the ^{17}O and the ^{89}Y relaxation rates. Let us neglect for the moment the anisotropic part of the hyperfine coupling between the Cu(2) electronic spin $\vec{S}$ and the nuclei under consideration of atomic number μ (μ = 17, 89). Then $H = \sum_{<i,j>} {}^{\mu}\gamma_n \hbar \,.{}^{\mu}A\, {}^{\mu}\vec{I}_i . \vec{S}_j$
Within the t-J model, the relaxation rate for both nuclei can be expressed as a function of $\left\langle S_q^+(t) S_q^-(0) \right\rangle$ where $S_q^{\alpha} = \frac{1}{N} \sum_{\vec{R}} S^{\alpha}(\vec{R},t)\, e^{-i\vec{q}\vec{R}}$
For both nuclei one can write :

$$\frac{1}{T_1} \alpha \left({}^{\mu}\gamma_n\, {}^{\mu}A\right)^2 \sum_{\vec{q}} {}^{\mu}F_{\vec{q}} \int \left\langle S_q^+(t)\, S_q^-(0) \right\rangle e^{i\omega_n t} dt \tag{8}$$

where ${}^{\mu}F_{\vec{q}}$ is a geometrical form factor that depends on the position of the nuclei μ within the cluster of their first neighbour electronic spins. The importance of this form factor for the NSLRR of ligands in AF insulators was first noted by Marschall [36] and has been discussed by many authors [20,37-41] to explain the different behaviour of NSLRR on the Cu(2), O(2,3) and Y sites.

These form factors ${}^{17}F_{\vec{q}} = 2\,(1 + \cos q_{x(y)}a)$ (for a Cu-O-Cu bond along the x(y) axis , and ${}^{89}F_{\vec{q}} = 8(1 + \cos q_x a + \cos q_y a + \cos q_x a \cos q_y a)$ have the property of filtering the AF fluctuations around Q = (π/a, π/a) where

$$\frac{\chi''(\vec{q}, \omega_n)}{\omega_n} = \frac{1}{k_B T} \int \left\langle S_q^+(t) . S_q^-(0) \right\rangle e^{i\omega_n t}\, dt \tag{9}$$

is expected to be peaked from the AF structure of $YBa_2Cu_3O_6$. As a first approximation one can say that the Oxygen (Yttrium) nuclei probe 2 (8) independent channels of relaxation

$$\frac{1}{^{17}\left(T_1 T\right)} = 2\,C\left({}^{17}\gamma_n\,{}^{17}A\right)^2 \sum_{\vec{q}} \frac{\tilde{\chi}''\left(\vec{q},\omega_n\right)}{\omega_n}$$

$$\frac{1}{^{89}\left(T_1 T\right)} = 8\,C\left({}^{89}\gamma_n\,{}^{89}A\right)^2 \sum_{\vec{q}} \frac{\tilde{\chi}''\left(\vec{q},\omega_n\right)}{\omega_n} \qquad (10)$$

where $\tilde{\chi}''\left(\vec{q},\omega_n\right)$ is a filtered spectral density in which the contribution from the AF fluctuations has been removed and C is a constant. Note that we have taken the same $\tilde{\chi}''\left(\vec{q},\omega_n\right)$ for Oxygen and Yttrium nuclei, which we justify below

Millis, Monien and Pines [39] have expanded of $\frac{\tilde{\chi}''\left(\vec{q},\omega_n\right)}{\omega_n}$ around $Q = (\pi/a,\pi/a)$ within a RPA scheme and have calculated the relaxation rate for ^{17}O and ^{89}Y nuclei as a function of the correlation length ξ of the AF fluctuations and a parameter β which measures the strength of the AF fluctuations with respect to the background. Recent neutron results in the superconducting phase of $YBa_2Cu_3O_{6+x}$ for concentrations close to the AF phase boundary[7,8] have shown that ξ/a is within 2 and 1, and is likely to be close to unity for the stoichiometry we are dealing with.[7] In this latter case, ($\xi/a \approx 1$), Millis et al calculation shows that even for large value of β, the correction to the hypothesis of independent channels is fairly small.
Now the MHS is the coherent sum of the contribution of the Cu(2) in first neighbour position :

$$^{17}K_{iso} = 2\,\frac{^{17}A}{g\mu_B}\chi_{Cu}\left(\vec{q}=0,\omega=0\right) \quad ; \quad ^{89}K_{iso} = 8\,\frac{^{89}A}{g\mu_B}\chi_{Cu}\left(\vec{q}=0,\omega=0\right) \qquad (11)$$

Elimination of $^{\mu}A$ between equ.(10) and (11) leads to the result:

$$\frac{1}{^{17}\left(K_{iso}^2 T_1 T\right)} = \frac{C}{2}\,{}^{17}\gamma_n^2 g^2 \mu_B^2\, \frac{\sum_{\vec{q}} \tilde{\chi}''\left(\vec{q},\omega_n\right)}{\chi_{Cu}^2\left(\vec{q}=0,\omega=0\right)} = 4\left(\frac{^{17}\gamma_n}{^{89}\gamma_n}\right)^2 \frac{1}{^{89}\left(K_{iso}^2 T_1 T\right)} \qquad (12)$$

Although this result looks like a relationship between Korringa products, it is very important to notice that it is independent of any assumption on the temperature dependence of $K^2_{iso}T_1T$; Alloul et al [24] have claimed that for ^{89}Y, K^2T_1T could be considered as constant in $YBa_2Cu_3O_{6+x}$ ($0.4 < x < 1$) However, their published data can be fitted as well to KT_1T = const. with the same accuracy for all values of x.
Let us call

$$r(T) = \left(\frac{^{89}\gamma_n}{^{17}\gamma_n}\right)^2 \frac{^{89}\left(K_{iso}^2 T_1 T\right)_{exp}}{^{17}\left(K_{iso}^2 T_1 T\right)_{exp}} \qquad (13)$$

In the t-J model, there is only one source of NSLRR which in common both to ^{89}Y and to ^{17}O nuclei, and r (T) must be equal to 4 . Let us now suppose that an independent contribution to the ^{17}O relaxation comes from the O(2p) holes. Then

$$^{17}(T_1T)^{-1}_{exp} = {}^{17}(T_1T)^{-1}_{Cu} + {}^{17}(T_1T)^{-1}_{2p} \quad \text{and}$$

$$^{17}(T_1T)^{-1}_{2p} = \left[\frac{r(T) - 4}{r(T)}\right] {}^{17}(T_1T)^{-1}_{exp} \tag{14}$$

As the value of $^{17}(T_1T)^{-1}_{exp}$ is critically dependent on the origin K_o (orbital or chemical shift)taken for the Yttrium MHS, we have plotted in Figure 6 our values of $^{17}K_{iso}$, which is proportional to χ_{Cu}, versus ^{89}K. We find K_o = 280 ppm and, 220 ppm if we correct $^{17}K^{exp}_{iso}$ for a possible orbital contribution $^{17}K^{orb}_{iso}$ of 0.03 %, which is likely to be an overestimate. Finally in Figure 7 we compare oxygen $^{17}(1/T_1T)$ data and $4\left(\frac{^{17}\gamma_n}{^{89}\gamma_n}\right)^2 \frac{^{17}K^2_{iso}}{^{89}\left(K^2_{iso}T_1T\right)}$ which is the contribution of the Cu (2) electronic spin fluctuations to the O(2, 3) nuclei relaxation. $^{89}(K^2_{iso}T_1T)$ is calculated from Y data taken from Alloul et al [24] for the composition $YBa_2Cu_3O_{6.63}$. It is found that NSLRR $(T_1T)^{-1}$ of the O(2,3) nuclei can be decomposed into a temperature dependent part, with the same temperature dependence as for Yttrium, plus a constant part, which can only be attributed to the O(2p) conduction band. We thus claim that this is the signature of the existence of two independent spin fluids.For values of K_o ranging between 280 (Fig. 7) and 220 ppm, $(T_1T)_{2p}^{-1}$ falls in the range 0.075-0.05 $(sK)^{-1}$.
Obviously, one should discuss the case of $YBa_2Cu_3O_7$ along the same lines. Taking K_o = 300 ppm Alloul et al.[24] found r (T) = 7.6 by comparing their data with these of Hammel et al [20], which would lead to a fairly large value $(T_1T)_{2p}^{-1}$ compared to the value found above for $YBa_2Cu_3O_{6.65}$. However, with our estimation of $^{89}K_o$ (220< $^{89}K_o$ < 280 ppm), r(T) falls in the range 4.7-6.8 for $YBa_2Cu_3O_7$, which places $(T_1T)_{2p}^{-1}$ in a range 0.05-0.148 that is quite consistent with the values obtained in $YBa_2Cu_3O_{6.65}$
For a p band the relaxation rate can be decomposed into a dipolar and an orbital contribution both of which are proportional to the square of the density of states at the Fermi level $N_p(E_F)$. Obata[42] has shown that the orbital contribution is dominant and can be expressed as:

$$\frac{1}{(T_1T)} = \frac{32\,\pi}{\hbar}(\gamma_n\hbar)^2\, k_B\,[N(E_F)]^2 \left\langle\frac{\mu_B}{r^3}\right\rangle^2 \tag{15}$$

From this formula (which strictly speaking is valid only for cubic symmetry, but should give us the right order of magnitude), for $(T_1T)_{2p}^{-1}$ in the range 0.05-0.14 $s^{-1}.K^{-1}$ we obtain $N(E_F)$ = 0.12-0.21 states per eV and per spin direction, which leads to χ_h in the range 8.3 10^{-6} to 1.5 10^{-5} e.m.u./mole. In any cases this is an order of magnitude smaller than the susceptibility per Cu atom which we estimated above to be about 10^{-4} e.m.u./mole. We emphasize that in spite of the fact that the relaxation rate due to the O(2p) band is dominated by the orbital contribution, this relaxation is the signature of a second band with its own susceptibility χ_h decoupled from χ_{Cu} above 100 K.
Two objections can be made to the above discussion:
i) we have neglected the anisotropic part of the hyperfine tensor, which couples the Cu(2) electronic spins to the O(2, 3) nuclei. However, our analysis of the MHS data shows that this latter is small, most of the contribution to $^{17}K_{ax}$ being due either to an orbital shift or to a dipolar interaction with the O(2p) holes.
ii) This result is based under the comparison between r(T) and the value 4. This latter value is valid under the assumption that ξ/a is close to unity According to Millis et al. analysis,[39] it can be substantially increased for larger values of ξ, but such an hypothesis seems to be in desagreement with the neutron scattering data. [7,8] Moreover, their calculations are submitted to the constraints of a Fermi liquid picture, which has no firm NMR experimental basis at the moment, as discussed above.

Before concluding, we discuss the anomaly in ^{17}O NSLRR observed in the temperature range 100-50 K. First it must be underlined that it cannot be a Hebel-Slichter type anomaly[43] since it

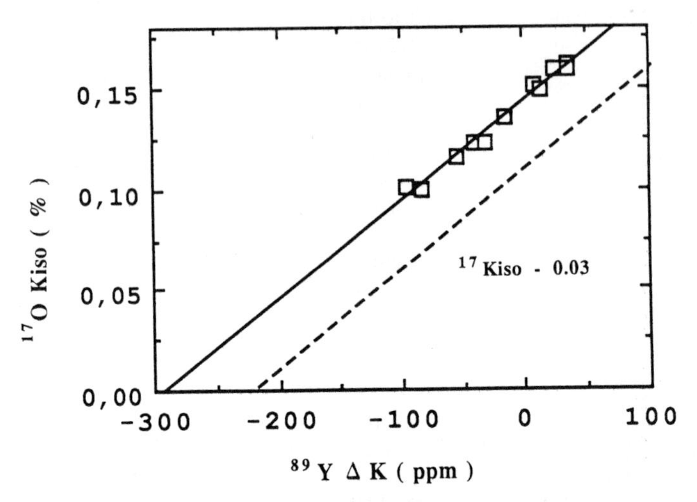

Fig. 6. The isotropic part of $K_{iso}(T)$ of the O(2,3) MHS versus $^{89}K(T)$ taken from Alloul et al data [24] with the temperature as an implicit parameter. (Some of the data have been interpolated). The origin of ^{89}K is found to range between 280 and 220 ppm depending on the value considered for $^{17}K_{iso}^{orb}$ (0 - 0.03 %).

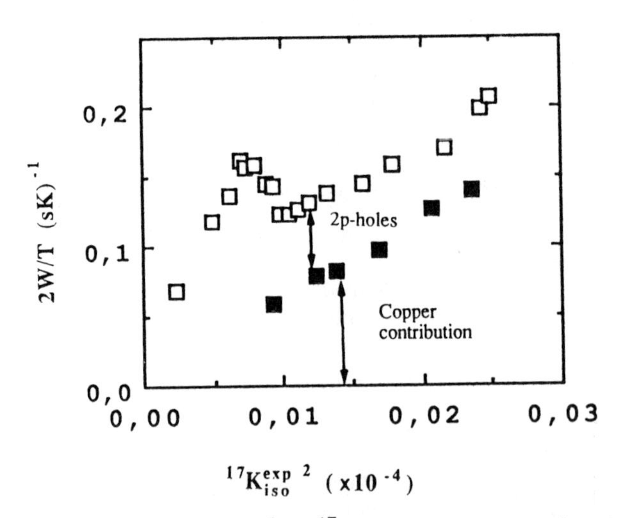

Fig. 7. Experimental values of $(T_1T)^{-1}$ for ^{17}O in the O(2,3) site as function of the square of the MHS isotropic part (open square). The solid squares represent the contribution attributed to the Copper, calculated from Y data (ref 22) as explained in the text. $^{17}K_{iso}^{exp}$ has not been corrected for any orbital contribution, and the origin of ^{89}K has been taken equal to K_o = 280 ppm.

starts well above T_c, which for $H_0 \neq 0$ is significantly lower than 60 K. An important point is to know whether the origin of the relaxation is purely magnetic or if some quadrupolar mechanism is involved. In order to clarify this point, we have measured the recovery of the magnetization corresponding to the (3/2, 5/2) transition at 70 K. Unlike the measurements performed at 200 K the value of $1/T_1$ extracted from the fit to the theoretical expression mentioned above was 30% larger than that obtained for the (1/2,3/2) transition. More generally , the data obtained above 100 K can be fitted with a very good agreement to the theoretical expression which account for a purely magnetic relaxation process.[35] Below this temperature, at the start of the anomaly , deviations occur which may be due to the onset of a quadrupolar mechanism. Thus the NSLRR anomaly could be due to some structural rearrangement in the CuO_2 plane which can be more or less pronounced depending on how the stoichiometry of the sample is close to a value corresponding to some superstructure of the oxygen vacancies. Recently, Takigawa et al have reported ^{17}O NMR data in a $YBa_2Cu_3O_{6,63}$ sample[44]. Their data strongly differ from ours below 120K, both for the MHS and for the NSLRR. In particular, no anomaly in the NSLRR was observed in their sample. The origin of these difference could be in the sample preparation.

CONCLUDING REMARKS

Let us now come back to the NSLRR of the Cu(2) nuclei. There has been several attempts to explain the relative magnitude and the different temperature dependence of the NSLRR of the Cu(2) nuclei on one hand and the ^{17}O and ^{89}Y nuclei on the other one, including two basic ingredients:
i) Antiferromagnetic fluctuations described either in the framework of some RPA susceptibility of itinerant electrons or by the excitations of a two-dimensional square lattice of 1/2 spins with Heisenberg coupling.
ii) Geometrical form factors for ^{17}O and ^{89}Y nuclei which filter these AF fluctuations around $\vec{q}=\vec{Q}_{AF}$ for these nuclei.

Most of these models refer to a Korringa behaviour for O(2,3)[20] and Yttrrium[24] sites, which is far from being established at the moment, the statistics of the data for ^{89}Y being too poor to settle this point. None of them accounts for the decrease of $(T_1T)^{-1}$ below $T \approx 120$ K >> T_c, except the one by Cox and Trees[41] who suggest that it could be due to a small anisotropy gap in the 2D Heisenberg excitation spectrum. Their model however does not take into account at this stage the interaction between the d and the p-holes. In a more phenomenological approach, Imai et al.[30] have suggested that the temperature dependence of the NSLRR of the Cu nuclei in the CuO_2 plane could be described in several high T_c superconducting oxides as $1/T_1 = a + bT$, where a accounts for AF fluctuations and b for some Korringa process where the fluctuations around $\vec{q}=\vec{Q}_{AF}$ have been removed. Such an explanation is not valid for most the compounds they consider, including $La_{1.85}Sr_{0.15}CuO_4$. and $YBa_2Cu_3O_{6.5}$ since in these compounds, ^{17}K and χ_m [21-23] are strongly temperature dependent above T_c. As a result, the term b becomes temperature dependent, and their decomposition fails.

Finally, we want to stress that any model attempting to explain the NSLRR in these oxides with excitations different from AF fluctuations [45] should account for the large enhancement at the Cu(2) site with respect to the O(2,3) and Y sites (filtering form factor) and a decrease of $(T_1T)^{-1}$ starting well above T_c except for $YBa_2Cu_3O_7$.

ACKNOWLEDGEMENTS

The Laboratoire de Spectrométrie Physique is "associé au Centre National de la Recherche Scientifique". This work was partly supported by the D.R.E.T.(Contract N° 89.34.045). We would like to thank H.Alloul and M.Takigawa for stimulating discussions.

REFERENCES

1 V. J. Emery, Phys. Rev. Lett. **58** (1987) 2794.
2 Y.Endoh, K. Yamada, R. J. Birgeneau, D. R. Gabbe, H. P. Jenssen, M. A. Kastner, C. J. Peters, P. J. Picone, T. R. Thurston, J. M. Tranquanda, G. Shirane, Y. Hidaka, M. Oda, Y. Enomoto, M. Suzuki, and T. Murakami, Phys. Rev. **B37**, (1988) 7443.

3 F. Borsa,M. Corti, T. Rega, A. Rigamonti, Novuo Cimento **11D**, 12 (1989); A. Rigamonti, F. Borsa,M. Corti, T. Rega and J. Zolio (preprint).
4 C.H. Pennington, D.J. Durand, C. P. Slichter, J.P. Rice, E. D. Bukowski, and D. M. Ginsberg, Phys. Rev. **B39** (1989) 274; ibid (1989) 2902.
5 M. Horvatić, P. Ségransan,C. Berthier, Y. Berthier, P. Butaud, J.Y. Henry, M. Couach, J. P. Chaminade, Phys. Rev. **B39** (1989) 7332;
6 M. Horvatić et al., Proceedings of M[2]SHTS Conference, Stanford USA (1989), (to be published in Physica C.) and to be published.
7 J. Rossat-Mignod, L. P. Regnault, M.J. Jurgens, C. Vettier, P. Burlet, J.Y. Henry, G. Lapertot, Proceedings of the International Conference of the Physics of Highly Correlated Electron Systems, Santa-Fe, Sept. 89, to be published and this Conference.
8 M. Tranquada, W.J.L. Buyers, H. Chou, T.E. Mason, M. Sato, S. Shamoto, and G. Shirane, preprint.
9 For a panel see Physica C **Vol 153-155.**
10 M. Takigawa, P. C. Hammel, R. H. Heffner, Z. Fisk, J. L. Smith, and R. B. Schwarz, Phys. Rev. **B39** (1989) 7371.
11 S.E. Barret, D.J. Durand, C.H. Pennington, C.P. Slichter, T.A. Friedman, J.P. Rice and D.M. Ginsberg, (preprint).
12 P.W. Anderson, in Frontiers and Borderlines in Many Particles Physics "Enrico Fermi",Course CIV eds R.A. Broglia and J.R. Schrieffer (North Holland, Amsterdam) (1988).
13 V. J. Emery and G. Reiter, Phys. Rev. **B38** (1989) 11938 .
14 F. C. Zhang and T. M Rice, Phys. Rev. **B37** (1988) 3759.
15 N. Andrei and P. Coleman, Phys. Rev. Lett., **62** (1989) 595, and references therein..
16 H. Bleier, P. Bernier, D. Jerome, J. M. Bassat, J.P. Coutures, B. Dubois, Ph. Odier, J. de Phys.(Paris) ,**49** (1988) 1825.
17 C. Coretsopoulos, H. C. Lee, E. Ramli, L. Reven, T.B. Rauchfuss and E. Oldfield, Phys. Rev. **B39**, 781 (1989);E. Oldfield, C. Coretsopoulos,S. Yang, L. Reven, H. C. Lee,J. Shore, O.H. Han and E. Ramli, Phys. Rev. **B40** (1989) 6832.
18 M. Horvatić, Y. Berthier, P. Butaud, Y. Kitaoka, P. Ségransan, C. Berthier, H. Katayama-Yoshida, Y. Okabe and T. Takahashi, Physica **C159** (1989) 689.
19 M. Takigawa, P. C. Hammel, R. H. Heffner, Z. Fisk, K. C. Ott, and J. D. Thompson, Phys. Rev. Lett. **63** (1989) 1865.
20 P. C. Hammel, M. Takigawa, R. H. Heffner, Z. Fisk and K. C. Ott Phys. Rev. Lett. **63** (1989) 1992 .
21 Y. Kitaoka, K. Ishida, F. Fujiwara, T. Kondo, K. Asayama, M. Horvatić, Y. Berthier, P. Butaud, P. Ségransan, C. Berthier, H. Katayama-Yoshido, Y. Okabe and T. Takahashi, in "Strong correlation and Superconductivity" eds H. Fukuyama, S. Maekawa and A.P. Malozemoff, Springer Verlag, Berlin (1989),p. 262.
22 Y. Kitaoka, Y. Berthier, P. Butaud, M. Horvatic, P. Ségransan, C. Berthier, H. Katayama-Yoshida, Y. Okabe and T. Takahashi, Proceedings of M[2]SHTS Conference, Stanford USA (1989), to be published in Physica C.
23 P. Butaud, M. Horvatić,Y. Berthier, P. Ségransan, Y. Kitaoka, C. Berthier and H. Katayama-Yoshida, to be published.
24 H. Alloul, T. Ohno, P. Mendels, Phys. Rev. Lett. **63** (1989) 1700.
25 J.Y. Henry, to be published.
26 K. Kishio, J. Shimoyama, T. Hasegawa, K. Kitazawa, K. Fueki and S. Tanaka, Jpn. J. Appl. Phys. 26, (1987) L125 .
27 F.Mila and T.M.Rice, Physica C157 (1989), 561.
28 J.Winter, Magnetic Resonance in Metals, Oxford Clarendon Press,(1971).
29 W.W.Warren Jr., R.E.Walstedt, G.F.Brennert, R.F.Bell, G.P.Espinosa and R.J.Cava Proceedings of M[2]SHTS Conference, Stanford USA (1989), to be published in Physica C.
30 T. Imai, T. Shimizu, H. Yasuoka, Y. Ueda, K. Yoshimura, K. Kosyge, Phys. Rev. B Rapid Communication to be published; H.Yasuoka, T.Imai and T.Shimizu in Ref. 19, p254.
31 G. Shirane, R.J. Birgeneau, Y. Endoh, P. Gehring, M.A. Kastner, K. Kitazawa, H. Kojima, I. Tanaka, T. R. Thurston, and K. Yamada, Phys. Rev. Lett. 63, 330, (1989).
32 S. Parkin, unpublished.
33 T. Shimizu, H. Yasuoka, T.Tsuda, K. Koga and Y. Ueda, Bulletin of Magnetic Resonance (1990) to be published.

34 R.G. Shulman and K. Knox, Phys. Rev. 119 (1960) 94.
35 A. Narath, Phys. Rev. 162, (1967) 320.
36 E.P. Marschall, Proceeding of the XVIth Congress AMPERE , Bucharest 1970, ed. I. Ursu, Publishing House of the Socialist Republic of Romania (1970),p485.
37 B. S. Shastry, Phys. Rev. Lett. 63 (1989) 1288.
38 N. Bulut, D. Hone, D.J. Scalapino, and N.E. Bickers, preprint UCSBTH-89-26 and 89-35.
39 A.J.Millis, H. Monien and D. Pines, preprint.
40 F. Mila and T.M. Rice, to be published in Phys. Rev.B
41 D.L.Cox and B.R.Trees (preprint).
42 Y. Obata, J. Phys. Soc. Jpn.18 (1963) 1020.
43 L. C. Hebel and C.P. Slichter,Phys. Rev. 113, (1959) 1504.
44 M. Takigawa et al, to be published in "Dynamics of Magnetic fluctuations in High Temperature Superconductors.",eds G. Reiter, P. Horsch, and G. Psaktakis, Plenum Press.
45 P.W.Anderson, in Ref.19, p. 2.

LOCAL HYPERFINE INTERACTIONS OF DELOCALIZED ELECTRON SPINS: ^{205}TL INVESTIGATIONS IN TL CONTAINING HIGH-T_c SUPERCONDUCTORS

M.Mehring, F.Hentsch and N.Winzek

2.Physikalisches Institut, Universität Stuttgart

Pfaffenwaldring 57, 7000 Stuttgart 80

Abstract

We report on the ^{205}Tl NMR (nuclear magnetic resonance) spectra of one-, two-, three-layer Tl containing compounds below and above the superconducting transition temperature T_c. Two different Tl nuclei were observed, namely those of the TlO layers (dominant line) and a Tl defect replacing Ca in the two- and three-layer compounds. Knight shifts, chemical shifts, Korringa type relaxation as well as highly correlated Cu spin dynamics was separately observed in the same compound.

1.0 Introduction

Since the discovery of high-temperature superconductors [1] numerous NMR investigations have appeared which report on the observation of spectral [2-10] and relaxational features [11-23] in these compounds. We restrict ourselves in this contribution to the discussion of the Tl containing high-T_c superconductors $Tl_2Ba_2CuO_6$ (T_c = 85K, one-layer), $Tl_2Ba_2CaCu_2O_8$ (T_c =110K, two layer) and $Tl_2Ba_2Ca_2Cu_3O_{10}$ (T_c = 125K, three-layer). Most of the observed features are, however, relevant also for the $YBa_2Cu_3O_7$ compound.

The corresponding ^{205}Tl NMR spectra are shown in Fig.1. Note that the "dominant line" which belongs to the Tl in the TlO layers shows strong shift anisotropy, which is shown to be correlated with the interlayer spacing in these compounds. The center of gravity (first moment) of all the TlO layer lines corresponds dominantly to an isotropic chemical shift which refers to Tl^{+3} ions (see marking in Fig.1). An additional <u>positive</u> Knight shift contribution is also observed which will be discussed in section 2.

Dynamics of Magnetic Fluctuations in High-Temperature Superconductors
Edited by G. Reiter *et al.*, Plenum Press, New York, 1991

The smallest line observed only in the Ca containing two- and three-layer compounds corresponds to a Tl defect which replaces Ca. This so-called "defect-line" is shifted strongly to lower frequencies beyond the Tl^{+} reference mark. It will be shown in section 2 that this is caused by a strong negative Knight shift.

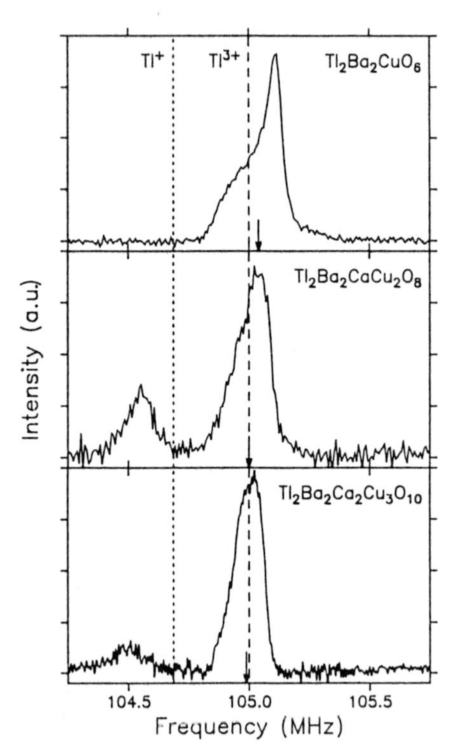

Figure 1. ^{205}Tl NMR spectra of the three different Tl containing high-T_c superconductors discussed here.

2.0 Chemical and Knight shifts

Two different types of NMR lineshifts are observed in conductors, namely "chemical shift" δ_c and Knight shift K_s. Both contribute to the total shift

$$\delta = \delta_c + K_s \qquad (1)$$

Both quantities are second rank tensors (3 x 3 matrix) whose principal axes are not necessarily co-linear and add up to the total shift tensor

(δ_{11}, δ_{22}, δ_{33}). We will restrict ourselves here to the isotropic part $\delta_{iso} = (\delta_{11}+\delta_{22}+\delta_{33})/3$. We believe that the separation made in Eq.(1) is appropriate, since even atoms, molecules and insulators show appreciable chemical shift, whereas an additional shift is observed in metals, namely the Knight shift K_s due to the spin susceptibility of the conduction electrons.

The electronspin dependent Knight shift K_s is therefore temperature dependent in the superconducting state. It should vanish for singulett superconductivity when T approaches zero. The total shift therefore becomes temperature dependent as

$$\delta(T) = \delta_c + K_s(T) + \delta_M(T) \quad (2)$$

where

$$\delta_M(T) = (1-N)\, M(T) \,/\, B_a \quad (3)$$

is the contribution due to the magnetic shielding M(T) at the applied magnetic field B_a (both quantities in SI units).

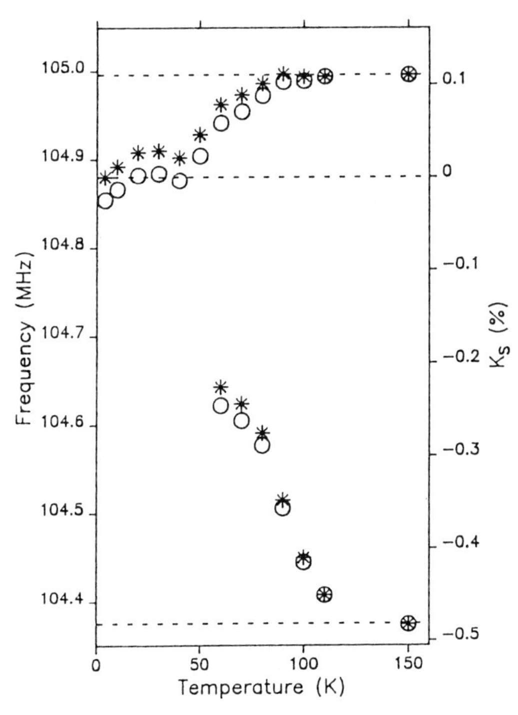

Figure 2. Temperature dependence of the lineshifts of the dominant line (a) (TlO layer) and of the "defect line" (b) (see text)

The demagnetizing factor ranges between $1/3 \leq N \leq 1$. We adopt N=0.6 as a powder average. Fig.2 shows the temperature dependence of both shifts namely the TlO layer line with shifts [21c]

$$T > T_c: \quad \delta_c = 1844 \text{ ppm} \qquad K_s = 1106 \text{ ppm}$$

and for the "defect line"

$$T > T_c: \quad \delta_c = 1844 \text{ ppm} \qquad K_s = -4815 \text{ ppm}$$

The shift reference is taken to be 104.88 MHz at 4.26 Tesla for Tl^{3+} ions. This analysis takes the temperature dependence of M(T) according to Eq.(3) into account. It is assumed that δ_c is temperature independent and K_s vanishes at T = 0.

We note that the positive Knight shift of the TlO layer nuclei arises from the direct overlap of the Tl orbitals with the bridging (pyramidal) oxygen which again overlaps with the Cu nuclei in the CuO_2 layer. The Knight shift of Tl in the TlO layer is therefore caused by superhyperfine interaction with the Cu^{2+} electron spins (d-holes). We point to section 3 where it is demonstrated that the relaxation (T_1) of these Tl nuclei correlates strongly with the temperature dependence of T_1 of the CuO_2 layer Cu nuclei

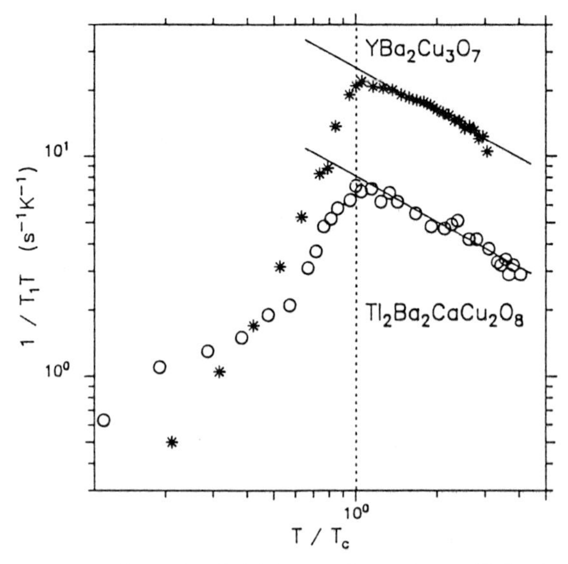

Figure 3. Comparison of spin-lattice-relaxation (T_1) in $Tl_2Ca_1Ba_2Cu_2O_8$ with ^{63}Cu relaxation in $YBa_2Cu_3O_7$ [21b].

weak temperature dependence shown in Fig.3 is caused by strong antiferromagnetic correlations. The filtering function [24-27] F(q) for the ^{205}TlO layer nuclei equals one. The spin lattice relaxation rate $1/T_1$ therefore reflects the integral of the dynamic susceptibility $\chi''(q,\omega)$ over the whole Brillouinzone.

In contrast, the ^{205}Tl spin lattice relaxation of the defect line (Ca position) shows completely different behaviour as is obvious from Fig.4. Here a Korringa like relaxation ($1/T_1 \propto K^2T$) is observed, completely distinct from the TlO layer ^{205}Tl relaxation. We point out that these distinct features are observed in the same compound, for the same types of nuclei being only in two different sites of the unit cell. The filtering function F(q) is in this case identical to the one for the Y nuclei [24-26] in $YBa_2Cu_3O_7$. It favours the $q \neq (\pi,\pi)$ dynamics of the electron spins, emphasizing the q = (0,0) components of the dynamic susceptibility $\chi''_{\perp}(q,\omega)$. This part of the susceptibility seems to resemble more that of a classical metal leading to the Korringa like relaxation.

At this point one could argue for either a two-spin fluid model, having two distinct spin entities, e.g. d -holes at the Cu^{2+} sites and p-holes at the oxygen sites or a one-spin-fluid model. The ^{205}Tl relaxation could be understood in the two-spin-fluid model by correlating the TlO layer nuclei with the Cu^{2+} hole spins and the Tl defect nuclei with the oxygen p-holes.
Their different relaxation behaviour would then point at highly correlated Cu^{2+} spins and less correlated more classical metallic like oxygen p-hole spins. The open question in this case is, however, why are the two entities not strongly coupled? On the other hand a one-spin-fluid model would do as well, since the relaxation of the different nuclei in the different places reflect different parts of the spin dynamics in q-space as it was discussed here. The open question remains which model can explain these complex features of $\chi''(q,\omega)$. Bulut et al. [25] have attempted such an approach based on the RPA treatment of an appropriate Hubbard model. Further information on this point may be found in the contribution by Bulut et al. [25] and Monien et al. [26] in these proceedings.

4.0 Vortex lattice

We finally mention that some interesting magnetic phenomena, namely the occurence of the vortex lattice in the superconducting state ($T<T_c$)

The strong negative Knight shift of the defect line resembles the behaviour of the ^{89}Y in $YBa_2Cu_3O_7$ (see reference 17b). Both nuclei occupy very similar positions between the CuO_2 layers. Their negative Knightshift points at an "indirect hyperfine coupling", i.e. without direct orbital overlap, but rather indirect overlap with the oxygen orbitals. This is in favour of σ-bonding in the CuO_2 layer.

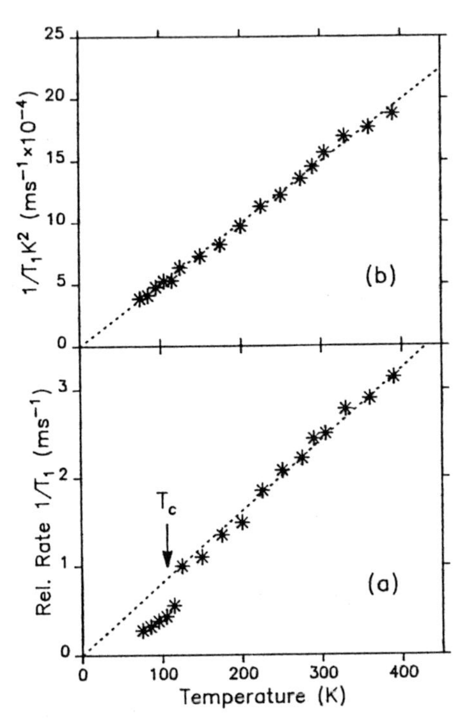

Figure 4. Korringa like relaxation of ^{205}Tl located at the Ca site in $Tl_2Ba_2Cu_1O_6$ (defect line, see text for discussion)

3.0 Relaxation

Spin lattice relaxation of ^{205}Tl nuclei in the three compounds discussed here shows very similar temperature dependence. Above T_c the relaxation rate $1/T_1$ is only slightly temperature dependent. In a plot $1/(T_1T)$ versus T (Fig.3) a $T^{-0.7}$ decrease is observed which parallels that of ^{63}Cu relaxation in the YBa_2CuO_3 compound for Cu in the CuO_2 layer. This supports the conjecture presented in section 2 that the Tl nuclei in the TlO layers are hyperfine coupled to the Cu^{2+} electron spins. THe ^{205}Tl nuclear relaxation of the "dominant line" (TlO layer) reflects directly the spin dynamics of the CuO_2 layer Cu^{2+} spins. Their can be observed by ^{205}Tl NMR. Since ^{205}Tl has spin 1/2 no quadrupolar broadening is present. We have oriented a powder sample of the one-layer compound ($Tl_2Ba_2CuO_6$) in the applied magnetic field (4.26 Tesla) and obtained a dominant orientation of the c-axis parallel to B_a.

In the superconducting state the field penetrates the sample in the form of flux lines (vortices) which form a regular triangular lattice. For a given average field $\bar{B}$ the relation

$$\bar{B} = \frac{2\,\phi}{\sqrt{3}\,d^2} \tag{4}$$

holds, where ϕ is the flux quantum (ϕ =2.0678 10^{-15} Tm^2). In our case ($\bar{B}$ = 4.26T) a vortex separation d=28.0 nm results.

The field inhomogenity due to the vortex lattice results in a characteristic asymmetric NMR lineshape as is observed in Fig.5. From the analysis of this asymmetric brodening a penetration depth of λ = 170nm results, very similar to the one observed for $YBa_2Cu_3O_7$. Further details can be found in references [27].

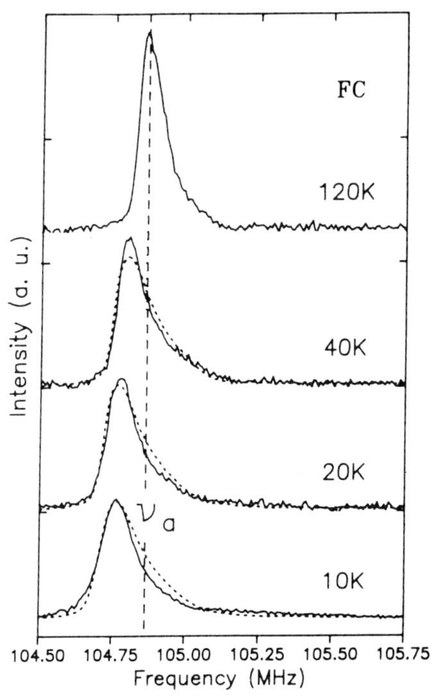

Figure 5. ^{205}Tl NMR spectra for $T>T_c$ (top) and $T<T_c$ in an aligned powder. Below T_c a triangular vortex lattice shows up in the characteristic asymmetric lineshape

Acknowledgements: We would like to acknowledge discussions with Hj.Mattausch, R.Kremer and A.Simon at the Max Planck Institute in Stuttgart. The Bundesminister für Forschung und Technologie (BMFT-Project-Nr. 13N5577) has given financial support.

References

1) J.G.Bednorz and K.A.Müller,Z.Phys. B64, 189 (1986)

2) (a) M.Mali, D.Brinkmann, L.Pauli, J.Roos, H.Zimmermann J.Hullinger, Physics lett. A124 (1987) 112
(b) M.Mali, J.Roos and D.Brinkmann, Physica C153 - 155 (1988)

3) H.Riesenmeier, Ch.Grabow, E.W.Scheidt, V.Müller and K.Lüders, Sol.State Commun. 64 (1987) 309

4) (a) W.W.Warren, R.E.Walstedt, G.F.Brennert, G.P.Espinosa and J.P.Rameika, Phys.Rev.Lett. 59 (1987) 1860
(b) R.E.Walsteddt, W.W.Warren, R.F.Bell, G.F.Brennert, G.P.Espinosa, R.J.Cava, L.F.Schneemeyer and J.V.Wascazak Phys. Rev. B 38 (1988) 9299

5) (a) Y.Kitaoka, S.Hiramatsu, K.Ishida, T.Kohara, Y.Oda, K.Amaya and K.Asayama, Physica 148B (1987) 298
(b) Y.Kitaoka, S.Hiramatsu, T.Kondo, Y.Oda and K.Asayama, J. Phys. Soc. Jpn. 57 (1988) 30

6) T.Imai, T.Shimizu, H.Yasuoka, Y. Ueda and K.Kosuge J.Phys.Soc.Jpn. 57 (1988) 2280

7) E.Lippmaa, E.Joon, I.Heinmaa, V.Miidel, A.Miller, R.Stern, I.Furo, L.Mihaly and P.Banki Physica C 153-155 (1988) 91

8) (a) M.Takigawa, P.C.Hammel, R.H.Heffner, Z.Fisk, J.L.Smith amd R.B.Schwarz, Phys. Rev. B 37 (1988) 7944
(b) M.Takigawa, P.C.Hammel, R.H.Heffner, Z.Fisk, J.L.Smith amd R.B.Schwarz, Phys. Rev. B 39 (1989) 300
(c) P.C.Hammel, M.Takigawa, R.H.Heffner, Z.Fisk and K.C.Ott Phys, Rev. Lett. 63 (1989), 1992

9) H.Lütgemeier, Physica C 153-155 (1988) 95

10) C.H.Pennigton, D.J.Durand, D.B.Zax, C.P.Slichter, J.P.Rice and D.M.Ginsberg, Phys. Rev. B37 (1988) 7944

11) I.Watanabe, K.Kumagai, Y.Nakamura, T.Kimura, Y.Nakamichi and H.Nakajima, J.of the Phys. Soc. Jpn 56 (1987) 3028

12) M.Horvatic, P.Segrasan, C.Berthier, P.Butaud, J.Y.Henry, M.Coach and J.P.Chaminade, Phys. Rev. B39 (1989), 7332

13) P.Wzietek, D.Köngeter, P.Auban, D.Jerome, J.P.Contures, B.Dubois Ph.Odier, Europhys. Lett. 8 (1989), 363

14) M.Horvatic, Y.Berthier, P.Butaud, Y.Kitaoka, P.Segrasan, C.Berthier, H.Katayama-Yoshida, Y.Okabe, T.Takahashi Physica C159 (1989), 689

15) E.Lippmaa, E.Joon, I.Heinmaa, A.Miller, R.Stern and S.Vija Paper presented at the International Seminar on High Temperature Superconductivity, Dubna (1989) USSR

16) C.Coretsopoulos, H.C.Lee, E.Ramli, L.Reven, T.B.Rauchfuss and E.Oldfield, Phys. Rev. B39 (1989), 781

17) (a) J.T.Markert, T.W.Noh, S.E.Russek, and R.M.Cotts, Solid state Commun. 63 (1987) 847
(b) H.Alloul, P.Mendels, G.Collin and P.Monod Phys.Rev.Lett. 61 (1988), 746
(c) H.Allouol, T.Ohno and P.Mendels, Phys.Rev.Lett. 63, (1989) 1700

18) (a) H.Seidel, F.Hentsch, M.Mehring, J.G.Bednorz and K.A.Müller, Europhys. Lett. 5 (1988) 647
(b) F.Hentsch, H.Seidel, M.Mehring, J.G.Bednorz and K.A.Müller Physica C 153-155, (1988) 727

19) H.Lütgemeier and M.W.Pieper, Solid State Commun. 64 (1987) 267

20) Y.Kitaoka, S.Hiramatsu, T.Kohara, K.Asayama, K.Oh-Ishi, M.Kikuchi and N.Koboyashai, J. J. of Appl. Phys. 26 (1987), 397

21) (a) F.Hentsch, N.Winzek, M.Mehring, Hj.Mattausch and A.Simon Physica C 158, 137 (1989)
(b) F.Hentsch, N.Winzek, M.Mehring, Hj.Mattausch and A.Simon to be published
(c) N.Winzek, F.Hentsch, U.Grosshans, M.Mehring, Hj.Mattausch and A.Simon (to be published)

22) K.Tompa, I.Bakonyi, P.Banki, I.Furo, S.Pekker, J.Vandlik, and L.Mihaly, Physica C 152 (1988), 486

23) K.Fujiwara, Y.Kitaoka, K.Asayama, H.Katayama-Yoshida, Y.Okabe, T.Takahashi, J. of the Phys. Soc of Jpn. 57 1988, 2893

24) F.Borsa and A.Rigamonti in Magnetic Resonance and Phase Transitions, p. 79, Academic Press (1979)

25) N.Bulut, D.Hone, D.J.Salapino and N.E.Bickers, Phys. Rev. B (in press)

26) H.Monien and D.Pines (to be published)

27) (a) M.Mehring, F.Hentsch, Hj. Mattausch and A.Simon Z.Phys B77, 355 (1989)
(b) M.Mehring, F.Hentsch, Hj. Mattausch and A.Simon (to be published)

WEAK COUPLING ANALYSIS OF SPIN FLUCTUATIONS IN LAYERED CUPRATES

Nejat Bulut

Department of Physics
University of California
Santa Barbara, CA 93106

INTRODUCTION

Layered cuprates have unusual properties both in the normal and the superconducting states[1–15]. It is believed that many of these have to do with the dynamics of low frequency spin fluctuations. For instance, in $YBa_2Cu_3O_7$, nuclear relaxation rate T_1^{-1} for O(2,3) nuclei exhibits a linear ("Korringa–like") temperature dependence, whereas the neighboring Cu(2) nuclei have an enhanced relaxation rate which deviates from Korringa behavior. However, in the superconducting state, both relaxation rates decrease sharply while keeping a nearly constant ratio for $(T_1^{-1})_{Cu}/(T_1^{-1})_O$. In addition, for both nuclei there are no Hebel–Slichter peaks. Inelastic neutron scattering also probes the dynamics of spin fluctuations. Measurements on $La_{2-x}Sr_xCuO_4$ show structure in temperature dependence for the spectral weight of low frequency spin fluctuations. A common feature of high T_c materials is the large and linear temperature dependence of their resistivity in the normal state. Hence, the contribution of spin fluctuations to the quasi–particle life time is of interest.

In this research our aim has been to see how well a description of these spin fluctuations can be provided by a weak coupling treatment of a single band model. We use a parameterized RPA form for the dynamic spin susceptibility $\chi(\mathbf{q},\omega)$ of a 2D Hubbard model. With this $\chi(\mathbf{q},\omega)$, in the normal state, we calculate nuclear relaxation rates for O(2,3) and Cu(2) nuclei, inelastic neutron scattering intensity and the quasi–particle life time τ due to spin fluctuations. After seeing a reasonable agreement with the experiment, we extend this weak coupling approach to the superconducting state by including the effects of a BCS type gap on $\chi(\mathbf{q},\omega)$. In the superconducting state our main objective is to make a distinction between the s and d–wave gap symmetries. For this purpose we present results for the Knight shift, nuclear relaxation rates and neutron scattering intensity obtained using s and d–wave gaps, and make comparisons with experiments.

We are using a very simple model, hence when we discuss NMR we will restrict ourselves to the $YBa_2Cu_3O_7$ material only. Many interesting properties observed in oxygen deficient $YBa_2Cu_3O_{7-\delta}$ require a more complete theory.

We start by studying the parameterized RPA susceptibility that we employ to calculate the physically observable quantities. Here we will try to justify our use of it by making a comparison with Quantum Monte Carlo (QMC) results[16] on an 8×8 lattice.

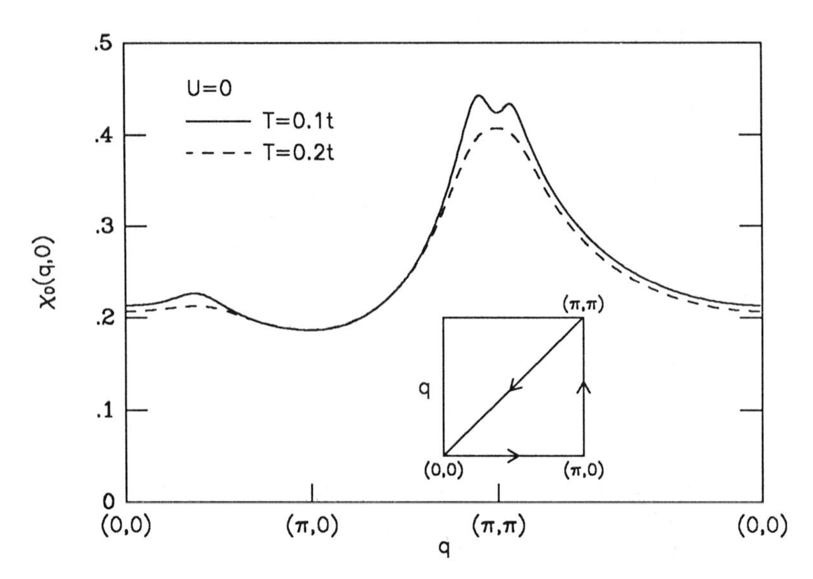

FIG. 1. Noninteracting susceptibility $\chi_0(\mathbf{q},0)$ as a function of $\mathbf{q}$ (see insert) for $n = 0.86$ at temperatures $T = 0.1t$ and $0.2t$.

RPA PARAMETERIZATION OF $\chi(\mathbf{q},\omega)$

Our starting point is the single–band Hubbard model on a square lattice

$$H = -t \sum_{<ij>,\sigma} (c^{\dagger}_{i\sigma}c_{j\sigma} + c^{\dagger}_{j\sigma}c_{i\sigma}) + U\sum_{i} n_{i\uparrow}n_{i\downarrow}, \tag{1}$$

which we use to describe the states formed from strongly hybridized planar Cu($3d_{x^2-y^2}$) and O($2p\sigma$) orbitals. Here $c_{i\sigma}$ ($c^{\dagger}_{i\sigma}$) is the fermion annihilation (creation) operator, t is the effective hopping matrix element and U is the onsite Coulomb interaction. An approximate form for $\chi(\mathbf{q},\omega)$ of the Hubbard model is given in RPA by

$$\chi(\mathbf{q},\omega) = \frac{\chi_0(\mathbf{q},\omega)}{1 - U\chi_0(\mathbf{q},\omega)}. \tag{2}$$

Here $\chi_0(\mathbf{q},\omega)$ is the noninteracting ($U = 0$) susceptibility

$$\chi_0(\mathbf{q},\omega) = \frac{1}{N}\sum_{\mathbf{q}} \frac{f(\varepsilon_{\mathbf{p}+\mathbf{q}}) - f(\varepsilon_{\mathbf{p}})}{\omega - (\varepsilon_{\mathbf{p}+\mathbf{q}} - \varepsilon_{\mathbf{p}}) + i0^+}, \tag{3}$$

with $\varepsilon_{\mathbf{p}} = -2t(\cos p_x + \cos p_y) - \mu$, where t is the hopping matrix element, $f(\varepsilon_{\mathbf{p}}) = (e^{\beta\varepsilon_{\mathbf{p}}} + 1)^{-1}$ the usual fermi factor, and μ the chemical potential.

Being an uncontrolled approximation, RPA gives a phase transition to an SDW state at a finite temperature for fillings above a critical filling n_c, which depends upon U/t. For an interaction strength of $U = 4t$, which corresponds to a moderate interaction strength of half the bandwidth, this critical filling is about 0.4, which is unphysically low. Hence, we set $U = 2t$, a weak coupling value in which case $n_c \simeq 0.865$, and use fillings less than n_c in order to stay in the paramagnetic regime of the RPA phase diagram[17] for the 2D Hubbard model. We use the filling n as a parameter and vary it to adjust the strength of antiferromagnetic (AF) fluctuations. In some sense, the parameterized values of the U and n represent an effective interaction and an effective filling that one would get, if one were to take into account the vertex and self–energy corrections that are not accounted for by RPA.

We first compare the noninteracting susceptibility with the RPA susceptibility. In Figs. 1 and 2, $\chi_0(\mathbf{q},0)$ and $\chi_{RPA}(\mathbf{q},0)$ are plotted as functions of $\mathbf{q}$ for a filling

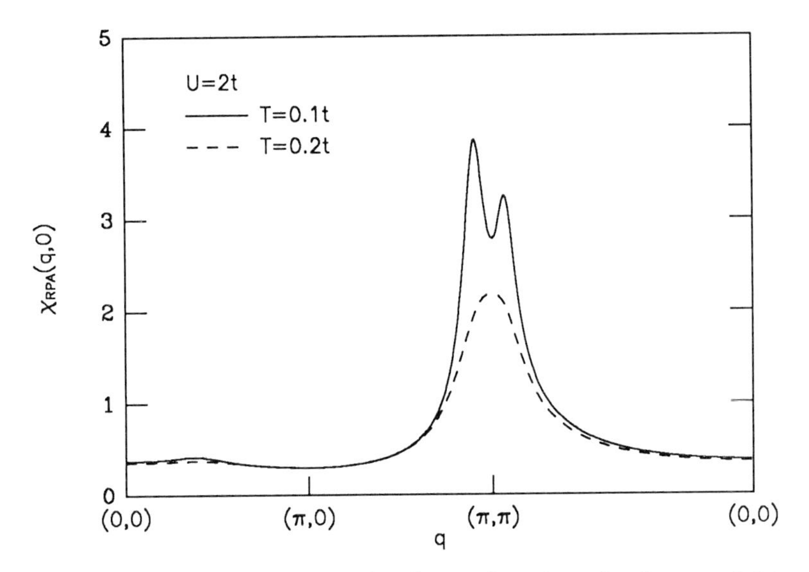

FIG. 2. RPA susceptibility $\chi_{RPA}(\mathbf{q}, 0)$ as a function of $\mathbf{q}$ for $n = 0.86$ and $U = 2t$ at temperatures $T = 0.1t$ and $0.2t$.

of $n = 0.86$ at temperatures $T = 0.1t$ and $0.2t$. Here the $\mathbf{q}$ values correspond to the path in the Brillouin zone shown in the insert of Fig. 1.

If we have a bandwidth $W = 8t$ of order an electron volt,* then $T = 0.1t$ corresponds to about $150K$. At low temperatures $\chi_0(\mathbf{q}, 0)$ is peaked at an incommensurate wave vector in the AF region of $\mathbf{q}$-space, since we are away from half filling. RPA enhances the AF part of $\chi(\mathbf{q}, 0)$ further.

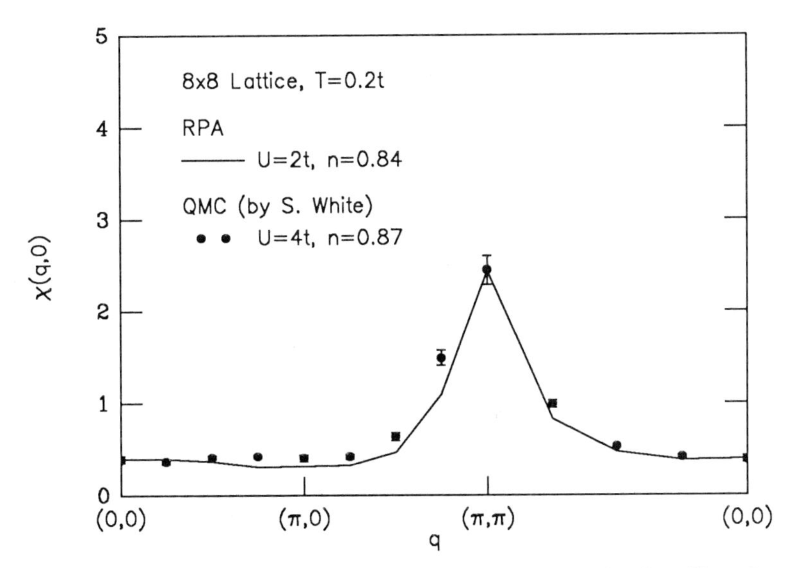

FIG. 3. Comparison of the RPA susceptibility, obtained using $U = 2t$ and $n = 0.84$, with the QMC susceptibility, obtained using $U = 4t$ and $n = 0.87$. Both calculations have been done on an 8×8 lattice at $T = 0.2t$.

* In this model t is an effective hopping which is renormalized by the strong correlations.

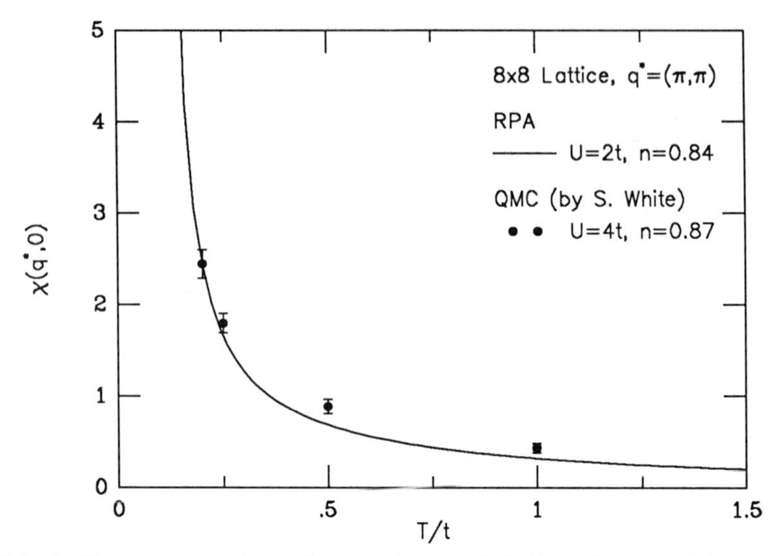

FIG. 4. Temperature dependences of $\chi_{RPA}(\mathbf{q}^*,0)$ and $\chi_{QMC}(\mathbf{q}^*,0)$ at $\mathbf{q}^* = (\pi,\pi)$ for the same parameters as in Fig. 3.

Next we make a comparison between parameterized RPA and QMC results obtained by S. White[16], both done on an 8×8 lattice. Fig. 3 shows $\chi(\mathbf{q},0)$ vs $\mathbf{q}$ obtained from QMC with $U = 4t$ and $n = 0.87$, and for RPA with $U = 2t$ and $n = 0.84$. A good agreement is obtained for both the $\mathbf{q} \sim 0$ and $\mathbf{q} \sim (\pi,\pi)$ regions of the Brillouin zone. The temperature dependence of $\chi(\mathbf{q}^*,0)$ for $\mathbf{q}^* = (\pi,\pi)$ is shown in Fig. 4.

These comparisons indicate that the parameterized RPA form captures some of the physics of the single band Hubbard model at intermediate couplings.

NUCLEAR RELAXATION RATE T_1^{-1}

The nuclear relaxation rate is given, with the appropriate form factors, by

$$T_1^{-1} = \frac{k_B T}{N} \sum_{\mathbf{q}} |A(\mathbf{q})|^2 \frac{\operatorname{Im}\chi(\mathbf{q},\omega)}{\omega}\bigg|_{\omega\to 0}. \tag{4}$$

For Cu(2) we use the Mila–Rice[18] hyperfine Hamiltonian, in which the Cu(2) nuclear spin has an onsite anisotropic and a transferred isotropic hyperfine coupling to the spins on the Cu(2) sites. The resulting form factor depends on the direction of the orienting magnetic field:

$$|A_{\mathrm{Cu}}(\mathbf{q})|^2 = \begin{cases} \left(\frac{1}{2}(A_{zz} - A_{xx}) - 4B\gamma_{\mathbf{q}}\right)^2 + \frac{1}{4}(A_{zz} - A_{xx})^2, & \mathbf{H} \parallel ab \\ (A_{xx} + 4B\gamma_{\mathbf{q}})^2, & \mathbf{H} \parallel c \end{cases}, \tag{5}$$

where $\gamma_{\mathbf{q}} = \frac{1}{2}(\cos q_x + \cos q_y)$. Here we take the onsite hyperfine couplings to be $A_{zz}/\gamma_n\hbar = -380\, kOe$ and $A_{xx}/\gamma_n\hbar = 80\, kOe$, and the transferred hyperfine coupling to be $B/\gamma_n\hbar = 85\, kOe$. For O(2,3) nuclei we assume only an isotropic transferred hyperfine coupling to the spins localized on the Cu sites. Hence, the oxygen form factor is

$$|A_{\mathrm{O}}(\mathbf{q})|^2 = \begin{cases} 4A_O^2 \cos^2(q_x/2) \\ 4A_O^2 \cos^2(q_y/2) \end{cases}, \tag{6}$$

where we take $A_O/\gamma_n\hbar = 110\, kOe$.

The resulting relaxation rates for Cu(2) with magnetic field in the ab plane, $(T_1^{-1})_{ab}$, and for O(2,3), $(T_1^{-1})_O$, are plotted in Fig. 5a–b as a function of temper-

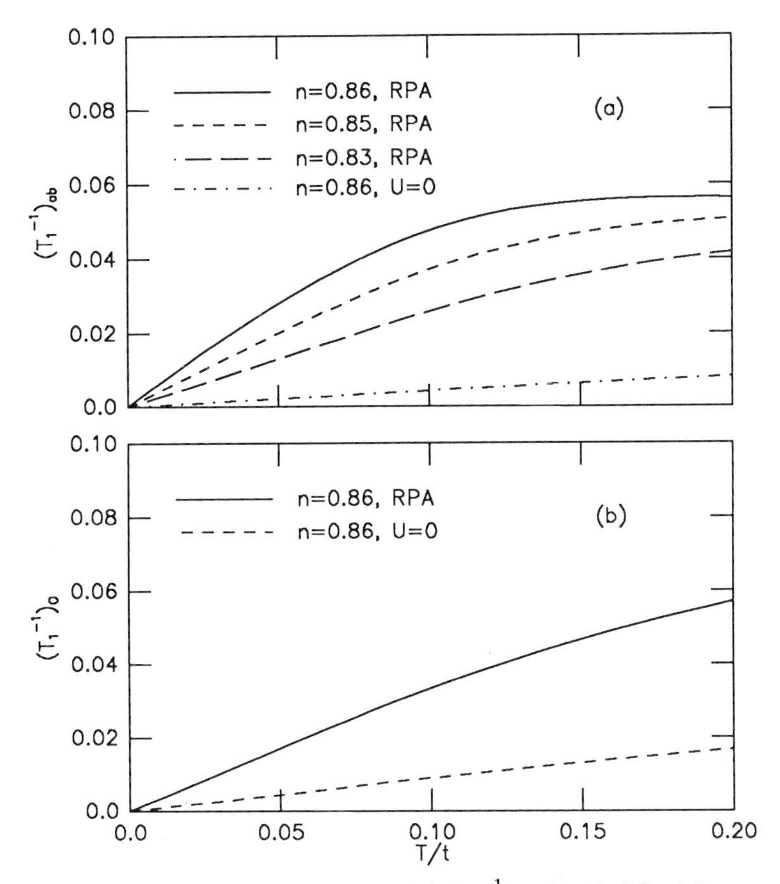

FIG. 5. The nuclear relaxation rates (a) $(T_1^{-1})_{ab}$ for Cu(2) with magnetic field in the ab–plane, in units of $\pi A_{zz}^2/\hbar t$, and (b) $(T_1^{-1})_O$ for O(2,3), in units of $\pi A_O^2/\hbar t$, as a function of temperature. Here $U = 2t$ and $n = 0.86$.

ature. In Fig. 5a we see the strong filling dependence of $(T_1^{-1})_{ab}$. O(2,3) relaxation rate follows an approximate enhanced Korringa law and does not depend so much on filling. The lowest curves are results for the noninteracting system and they obey a Korringa law for both Cu(2) and O(2,3) nuclei.

The relaxation rates for Cu(2) and O(2,3) nuclei are different because the Cu(2) form factor[2,7,19–23] is maximum in the AF region of $\mathbf{q}$–space whereas O(2,3) form factor vanishes at $\mathbf{q} = (\pi, \pi)$. Hence the Cu(2) and O(2,3) nuclei see different regions of $\mathbf{q}$–space. The non–Korringa behavior of T_1^{-1} for Cu(2) is due to the fact that the spectral weight of low frequency AF fluctuations first increases as we lower the temperature and then saturates below a characteristic temperature T^*. This is because away from half filling perfect nesting is not available. T^* is set by how far we are from perfect nesting and within RPA it is approximately $|\mu|/4$, where μ is the chemical potential.*

The slight deviation of T_1^{-1} for O(2,3) from linear T dependence is because away from half filling AF fluctuations peak at an incommensurate wave vector $\mathbf{q}^*$, and O(2,3) form factor vanishes only at $\mathbf{q} = (\pi, \pi)$. Hence, O(2,3) nuclei see some of the AF fluctuations.

A more direct comparison with experimental results is done in Fig. 6. Here in order to set the temperature scale we assumed a renormalized bandwidth of $1.2eV$.

* For $n = 0.86$ $\mu \sim -0.3t$.

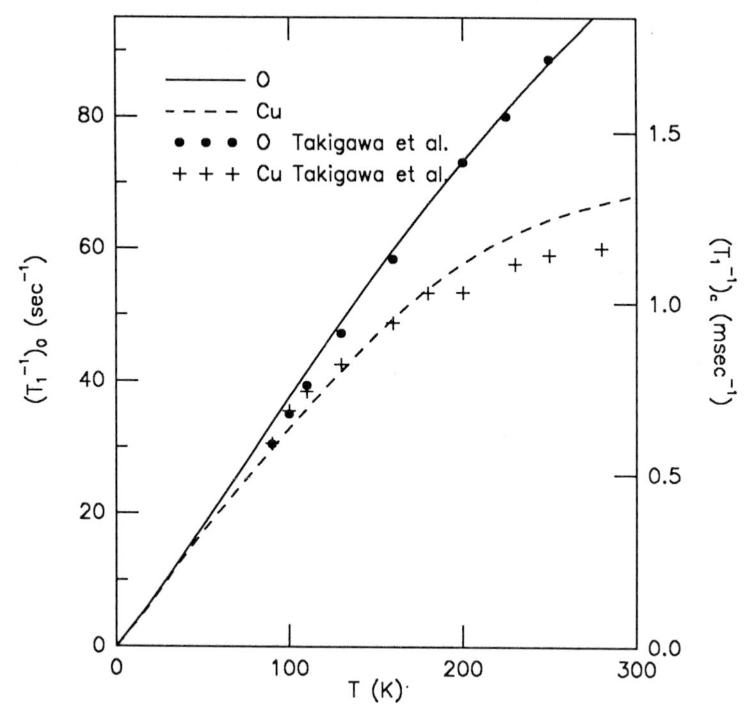

FIG. 6. A quantitative comparison with the experimental[7,13] relaxation rates with the orienting magnetic field parallel to c axis.

NEUTRON SCATTERING AND QUASI–PARTICLE LIFE TIME

Another probe of the magnetic fluctuations is neutron scattering. The inelastic neutron scattering cross section is proportional to $[n(\omega)+1]\ \mathrm{Im}\,\chi(\mathbf{q},\omega)$, where $n(\omega)$ is the Bose occupation number. Integrating this over $q = q_x = q_y$ from $\pi/2$ to $3\pi/2$ for various energy transfers ω gives the results for the integrated intensity $I(T)$ shown in Fig. 7.

The RPA results[21] appear similar to the features observed in neutron scattering experiments on $La_{2-x}Sr_xCuO_4$. The temperature dependence of $I(T)$ is a combined effect of the factor $n(\omega)+1$ and the spectral weight $\mathrm{Im}\,\chi(\mathbf{q},\omega)$ at AF wave vectors. It implies that as temperature increases the spectral weight of low frequency AF fluctuations starts to decrease above a characteristic temperature T^*, which is about $0.1t$ in this model.

Next we explore the magnitude and temperature dependence of the quasi–particle life time τ which arises from these spin fluctuations. The inverse lifetime of a quasi–particle of momentum $\mathbf{p}$ and energy ω, which is measured with respect to the chemical potential μ, is given by

$$\frac{1}{\tau(\mathbf{p},\omega)} = -2\left(1+e^{-\beta\omega}\right)^{-1} \mathrm{Im}\,\Sigma(\mathbf{p},\omega), \tag{7}$$

where $\Sigma(\mathbf{p},\omega)$ is the self energy of the quasi–particle due to interactions. To second order in U we have

$$\frac{1}{\tau(\mathbf{p},\omega)} = 2U^2\frac{1}{N}\sum_{\mathbf{q}}\left(1-f(\varepsilon_{\mathbf{p}-\mathbf{q}})\right)\frac{\mathrm{Im}\,\chi_0(\mathbf{q},\omega-\varepsilon_{\mathbf{p}-\mathbf{q}})}{1-e^{\beta(\omega-\varepsilon_{\mathbf{p}-\mathbf{q}})}}. \tag{8}$$

Within RPA, one replaces $U^2\chi_0(\mathbf{q},\nu)$ in Eq. (8) with

$$V_{eff}(\mathbf{q},\nu) = \frac{U^2\chi_0(\mathbf{q},\nu)}{1-U^2\chi_0^2(\mathbf{q},\nu)}\{1+U\chi_0(\mathbf{q},\nu)+U^2\chi_0^2(\mathbf{q},\nu)\}. \tag{9}$$

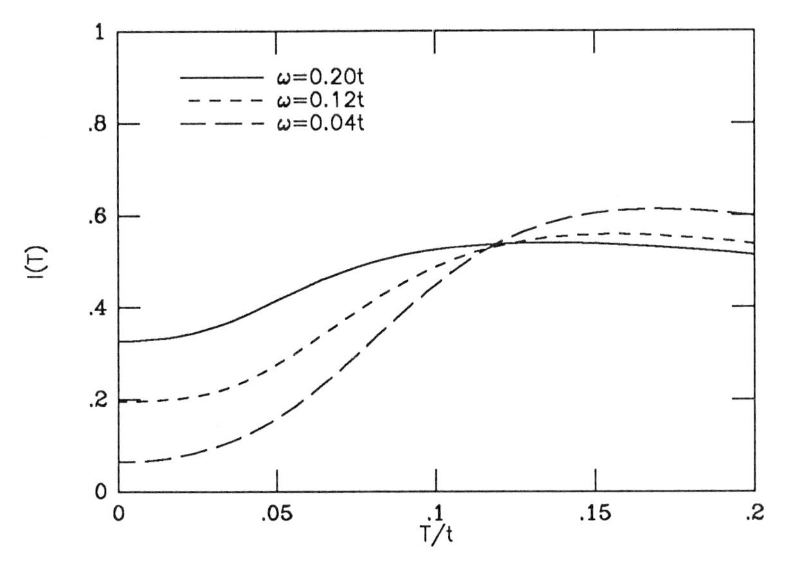

FIG. 7. Integrated neutron scattering intensity $I(T)$ vs T for different energy transfers ω.

Lee and Read[24] have analyzed the behavior of τ^{-1} for a tight binding model. They have shown that at half filling $1/\tau(\mathbf{p}_F) \sim T$ while away from half filling $1/\tau(\mathbf{p}_F)$ varies as T^2 upto a characteristic temperature set by the chemical potential and varies linearly above it. Here we numerically demonstrate their results and then extend it to RPA[25] using the parameters we have previously used to fit the NMR data. The long–dashed line in Fig. 8 is the second order result for $U = 2t$ and $n = 0.86$.

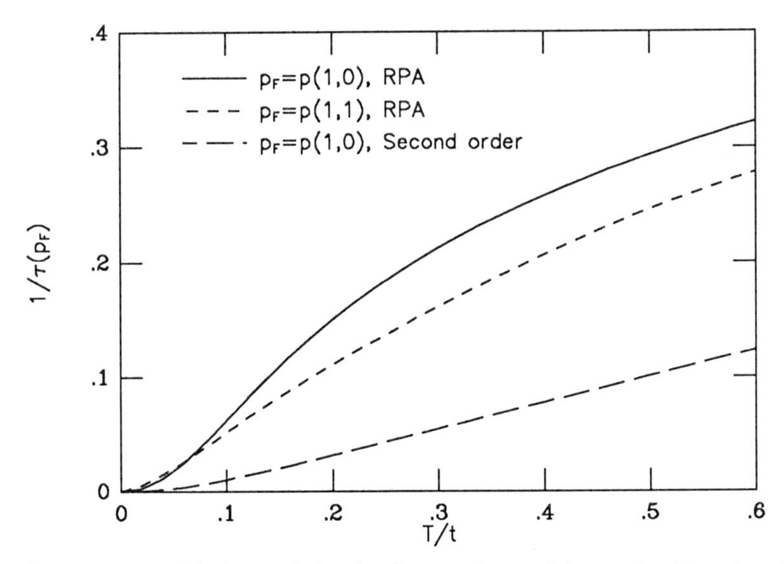

FIG. 8. Inverse lifetime $1/\tau(\mathbf{p}_F)$ of a quasi–particle on the Fermi surface calculated in second order (long–dashed line) and within RPA (full and short–dashed lines). $1/\tau(\mathbf{p}_F)$ has significant momentum dependence as shown here for $\mathbf{p}_F = p(1,0)$ and $p(1,1)$. The parameters are $U = 2t$ and $n = 0.86$.

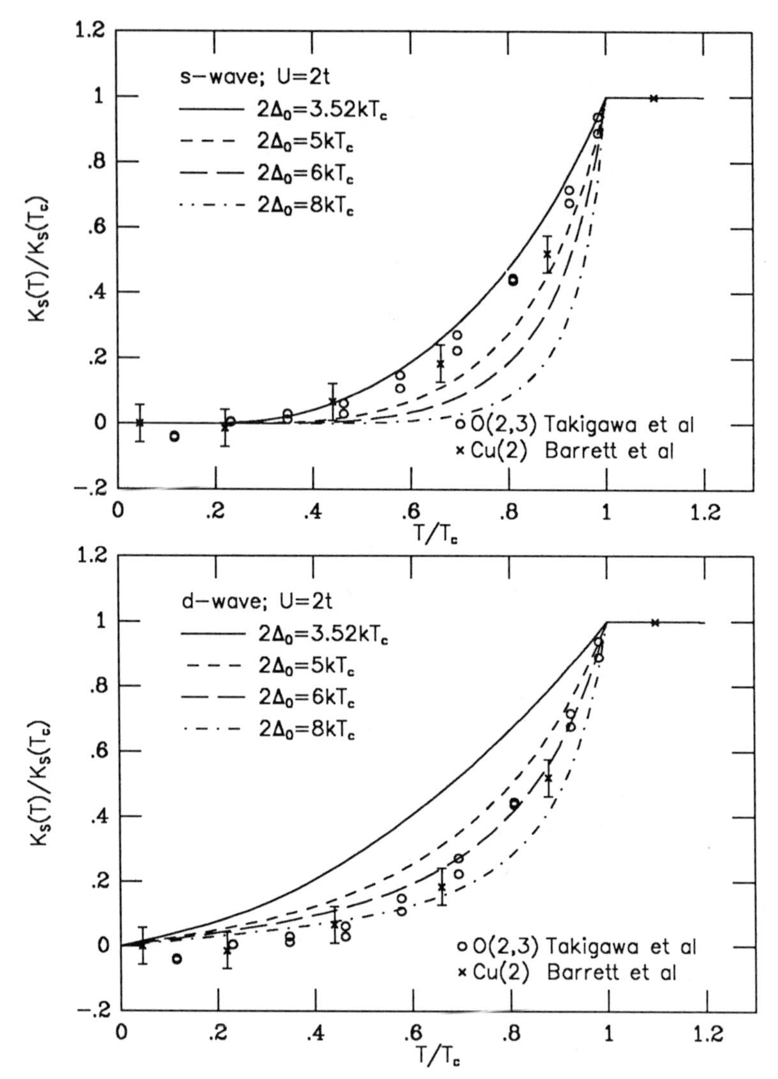

FIG. 9. Comparison of Knight shifts calculated within RPA using s and d–wave gap symmetries with the experimental[10,13] data.

As suggested by Lee and Read $1/\tau(\mathbf{p}_F) \sim T^2$ upto a characteristic temperature set by μ and $1/\tau(\mathbf{p}_F) \sim T$ above it. Within RPA, $1/\tau(\mathbf{p}_F)$ is enhanced and is proportional to T^2 upto $T^* \sim |\mu|/4$ and crosses over to linear T behavior above it. The quasi–particle damping rate, $1/\tau(\mathbf{p}_F)$, also depends on which point on the Fermi surface we put the quasi–particle. Even though there are deviations from linear T behavior, Fig. 8 shows that spin fluctuations can make a significant contribution to the quasi–particle life time.

SUPERCONDUCTING STATE

Here we extend the work on the normal state to the superconducting state by including the effects of a superconducting gap in computing the irreducible part of the susceptibility that enters the RPA expression[26,27]. In this simple model we treat the momentum dependence exactly, but do not take into account the self energy corrections due to strong spin fluctuations as Kuroda and Varma[28], and Coffey[29] did.

The BCS expression for $\chi_0(\mathbf{q},\omega)$ in the superconducting state is,

$$\chi_0(\mathbf{q},\omega) = \frac{1}{N}\sum_{\mathbf{p}}\{\frac{1}{2}\left[1+\frac{\varepsilon_{\mathbf{p+q}}\varepsilon_{\mathbf{p}}+\Delta_{\mathbf{p+q}}\Delta_{\mathbf{p}}}{E_{\mathbf{p+q}}E_{\mathbf{p}}}\right]\frac{f(E_{\mathbf{p+q}})-f(E_{\mathbf{p}})}{\omega-(E_{\mathbf{p+q}}-E_{\mathbf{p}})+i\Gamma}$$
$$+\frac{1}{4}\left[1-\frac{\varepsilon_{\mathbf{p+q}}}{E_{\mathbf{p+q}}}+\frac{\varepsilon_{\mathbf{p}}}{E_{\mathbf{p}}}-\frac{\varepsilon_{\mathbf{p+q}}\varepsilon_{\mathbf{p}}+\Delta_{\mathbf{p+q}}\Delta_{\mathbf{p}}}{E_{\mathbf{p+q}}E_{\mathbf{p}}}\right]\frac{1-f(E_{\mathbf{p+q}})-f(E_{\mathbf{p}})}{\omega+(E_{\mathbf{p+q}}+E_{\mathbf{p}})+i\Gamma}$$
$$+\frac{1}{4}\left[1+\frac{\varepsilon_{\mathbf{p+q}}}{E_{\mathbf{p+q}}}-\frac{\varepsilon_{\mathbf{p}}}{E_{\mathbf{p}}}-\frac{\varepsilon_{\mathbf{p+q}}\varepsilon_{\mathbf{p}}+\Delta_{\mathbf{p+q}}\Delta_{\mathbf{p}}}{E_{\mathbf{p+q}}E_{\mathbf{p}}}\right]\frac{f(E_{\mathbf{p+q}})+f(E_{\mathbf{p}})-1}{\omega-(E_{\mathbf{p+q}}+E_{\mathbf{p}})+i\Gamma}\} \tag{10}$$

This expression contains the usual coherence factors, the dispersion relation $E_{\mathbf{p}} = (\varepsilon_{\mathbf{p}}^2+\Delta_{\mathbf{p}}^2)^{1/2}$ and the gap $\Delta_{\mathbf{p}}$. For the gap we use an s-wave form, $\Delta_{\mathbf{p}} = \Delta_0(T)$, and a d-wave form, $\Delta_{\mathbf{p}} = \frac{\Delta_0(T)}{2}(\cos p_x - \cos p_y)$, where we take the BCS temperature dependence for $\Delta_0(T)$. We use a finite broadening Γ to control the logarithmic divergences in the calculations and treat $2\Delta_0/kT_c$ as a parameter. In a physical system the broadening will be due to the effects of spin fluctuations on the quasi–particle self energy.

Knight Shift

In this section we study the spin contribution to Knight shift, which is proportional to $\chi(\mathbf{q}\to 0,0)$. In Fig. 9, Knight shifts calculated using s and d–wave gaps are compared with the experimental results of Barrett *et al.*[10] and Takigawa *et al.*[13] Here the Stoner enhancement at $T=T_c$ is 1.76. For an s–wave gap the data is best fit with a small gap close to the usual BCS value $2\Delta_0 = 3.52kT_c$, which is surprising. As $T\to 0$, Knight shift for an s–wave gap decays exponentially in contrast with the linear T decay for a d–wave gap. For a d–wave gap the data is best fit with $2\Delta_0 = 6kT_c$. However, it is difficult to make a definite statement on the symmetry of the gap based on this comparison.

Nuclear Relaxation Rate T_1^{-1}

Before presenting results for Cu(2) and O(2,3) relaxation rates, we first study the relaxation rate for a noninteracting ($U=0$) system with an onsite hyperfine coupling. In Fig. 10 we show T_1^{-1} vs T given by

$$\left(T_1^{-1}\right)_{U=0} = \frac{T}{N}\sum_q \frac{\mathrm{Im}\,\chi_0(q,\omega)}{\omega}\bigg|_{\omega\to 0}. \tag{11}$$

For an s–wave gap T_1^{-1} has the well known Hebel–Slichter peak just below T_c, which is due to the piling up of states at the gap edge, and decays exponentially at low temperatures. The amplitude of the Hebel–Slichter peak is proportional to $\log(\Delta_0/\Gamma)$. A d–wave gap leads to a small peak below T_c, but this peak is nonsingular; it does not grow as Γ gets smaller. Due to the nodes of the gap on the Fermi surface, T_1^{-1} for a d–wave gap decays as T^3 at low temperatures.

Now we show results for O(2,3) and Cu(2) relaxation rates with $\mathbf{H}\parallel c$ using the full RPA susceptibility within BCS. In Fig. 11 T_1^{-1} for O(2,3) and Cu(2) are plotted for an s–wave gap.

For an s–wave symmetry, the absence of the Hebel–Slichter peak requires $2\Delta_0 = 8kT_c$ and a large broadening $\Gamma = 2.5T_c$.[28,29] In Fig. 12. we see that an RPA fit to experiments is possible using a d–wave gap symmetry with $2\Delta_0 = 8kT_c$ and $\Gamma\sim 1.5T_c$.

However, these results are to be taken into account cautiously, since in the present calculation the self–energy corrections due to spin fluctuations are not treated properly.

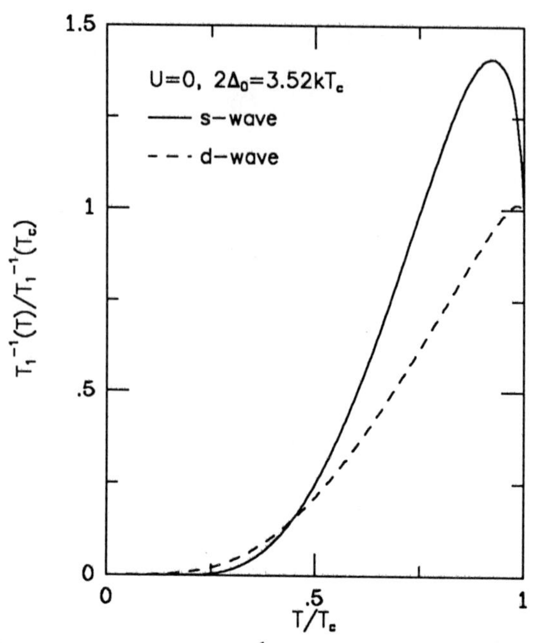

FIG. 10. Nuclear relaxation rate T_1^{-1} for a noninteracting system with an onsite hyperfine coupling using s and d–wave gap symmetries.

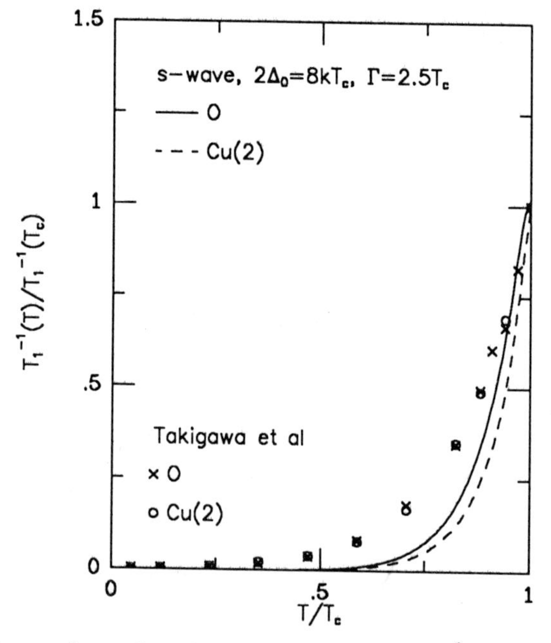

FIG. 11. Comparison of nuclear relaxation rate T_1^{-1} calculated within RPA using an s–wave gap symmetry with the experimental data[7,13].

Neutron Scattering

In Fig. 7 we have seen the integrated inelastic scattering intensity $I(T)$ vs T in the normal state. Here we examine the effect of the opening of a superconducting gap on $I(T)$.[26] Fig. 13 shows $I(T)$ vs T for s and d–wave gaps with $T_c = 0.04t$.

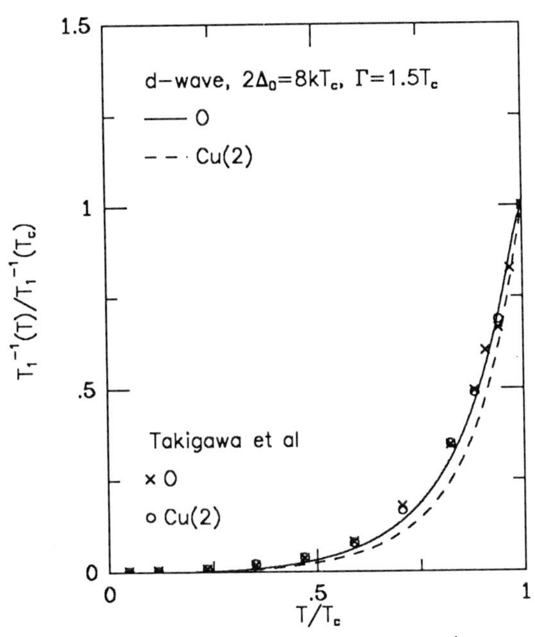

FIG. 12. Comparison of nuclear relaxation rate T_1^{-1} calculated within RPA using a d–wave gap symmetry with the experimental data[7,13].

An s–wave gap leads to a sharp decrease in $I(T)$ below T_c while a d–wave gap does not. In fact, for a d–wave gap $I(T)$ slightly increases in the superconducting state when $\omega > 2\Delta_0(T)$. This is due to the behavior of the coherence factors when the gap has a d–wave symmetry.

CONCLUSIONS

We have seen that a weak coupling treatment using an RPA parameterization of a single band Hubbard model can give a reasonable fit to the normal state NMR experiments. The susceptibility in this approximation compares well with the QMC susceptibility on an 8×8 lattice. Using the same fit parameters the temperature dependence of integrated neutron scattering intensity can be explained. Also spin fluctuations can give an enhanced, approximately linear contribution to the quasi–particle life time.

We extended this approach to the superconducting state by using the usual BCS expression for χ_0 and studied the distinction between s and d–wave symmetries. We found that a possible fit to the Knight shift data could be provided using a d–wave gap with a large Δ_0. An s–wave gap gave a surprisingly good fit but with a standard gap $2\Delta_0 = 3.52kT_c$. For nuclear relaxation rate, a d–wave gap provided the best fit to data but an s–wave gap also provided a reasonable fit with the Hebel–Slichter peak supressed by a large Δ_0 and Γ. Here we noted the need to properly treat the self–energy corrections due to spin fluctuations.[28,29]

ACKNOWLEDGMENTS

This paper is based on work done in collaboration with N.E. Bickers, D. Hone, H. Morawitz and D. Scalapino. I thank Barrett *et al.* and Takigawa *et al.* for allowing me to reproduce their data and S. White for allowing me to present his Monte Carlo results prior to publication. I thank D. Scalapino for a careful reading of the manuscript and many helpful comments. Partial support for this work was provided by the National Science Foundation under grant DMR86–15454. Numerical computations were performed at the San Diego Supercomputer Center.

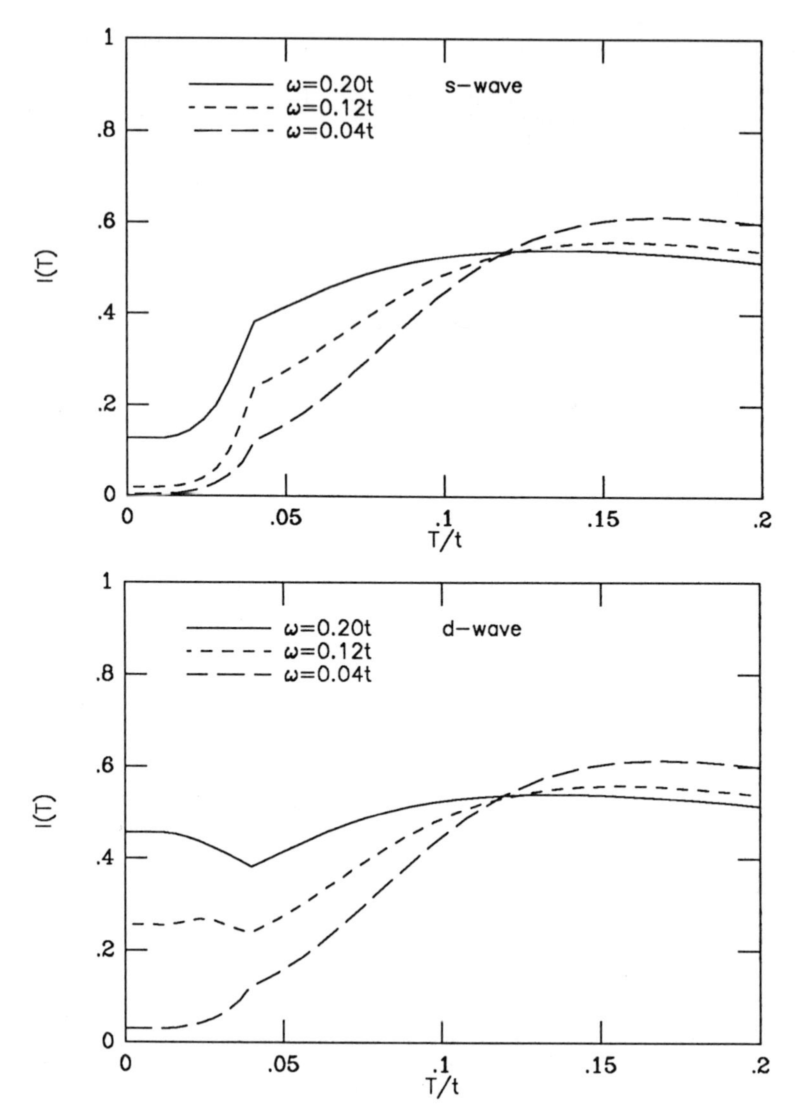

FIG. 13. Integrated neutron scattering intensity $I(T)$ vs T for different energy transfers ω using s and d–wave gap symmetries. Here $T_c = 0.04t$.

REFERENCES

1. For an extensive review of NMR experiments on $YBa_2Cu_3O_{7-\delta}$ see C.H. Pennington and C.P. Slichter,to appear in "Physical Properties of High Temperature Superconductors II" D.M. Ginsberg, (Ed.) World Scientific Publisheng Co., (1990); and Berthier et al., and M. Takigawa articles in this volume.

2. W.W. Warren, Jr., R.E. Walstedt, G.F. Brennert, G.P. Espinosa, and J.D. Remeika, Phys. Rev. Lett. 59:1860 (1987); R.E. Walstedt, W.W. Warren, Jr., R.F. Bell, G.F. Brennert, G.P. Espinosa, R.J. Cava, L.F. Schneemeyer, and J.V. Waszczak, Phys. Rev. B 38:9299 (1988).

3. T. Imai, T. Shimizu, H. Yasuoka, Y. Ueda, and K. Kosuge, J. Phys. Soc. Japan 57:2280 (1988).

4. C.H. Pennington, D.J. Durand, C.P. Slichter, J.P. Rice, E.D. Bukowski, and D.M. Ginsberg, Phys. Rev. B 39:2902 (1989).

5. M. Takigawa, P.C. Hammel, R.H. Heffner, and Z. Fisk, Phys. Rev. B 39:7371 (1989).
6. M. Takigawa, P.C. Hammel, R.H. Heffner, Z. Fisk, K.C. Ott, and J.D. Thompson, LANL preprint LA–UR–89–728, to be published.
7. P.C. Hammel, M. Takigawa, R.H. Heffner, Z. Fisk, and K.C. Ott, Phys. Rev. Lett. 63:1992 (1989).
8. R.E. Walstedt, W.W. Warren, Jr., R.F. Bell, and G.P. Espinosa, Phys. Rev. B 40:2572 (1989).
9. M. Horvatić, P. Ségransan, C. Berthier, Y. Berthier, P. Butaud, J.Y. Henry, M. Couach, J.P. Chaminade, Phys. Rev. B 39:1332 (1989).
10. S.E. Barrett, D.J. Durand, C.H. Pennington, C.P. Slichter, T.A. Friedmann, J.P. Rice, and D.M. Ginsberg, to be published in Phys. Rev. B.
11. M. Horvatić et al., Proceedings of the M^{2}SHTS Conference, Stanford USA (1989), (to be published in Physica C).
12. T. Imai, T. Shimizu, H. Yasuoka, Y. Ueda, K. Yoshimura, and K. Kosuge, ISSP, Tokyo, 1989, to be published in Phys. Rev. B.
13. M. Takigawa, P.C. Hammel, R.H. Heffner, Z. Fisk, K.C. Ott, and J.D. Thompson, Physica C 162-164, 853, (1989).
14. G. Shirane, R.J. Birgeneau, Y. Endoh, P. Gehring, M.A. Kastner, K. Kitazawa, H. Kojima, I. Tanaka, T.R. Thurston, and K. Yamada, Phys. Rev. Lett. 63:330 (1989).
15. J. Rossat–Mignod, L.P. Regnault, M.J. Jurgens, C. Vettier, P. Burlet, J.Y. Henry, G. Lapertot, Proceedings of the International Conference of the Physics of Highly Correlated Electron Systems, Santa–Fe, Sept. 89; and J. Rossat–Mignod article in this volume.
16. S. White, N.E. Bickers, D.J. Scalapino, unpublished.
17. H.J. Schulz, Phys. Rev. Lett. 64:1445 (1990).
18. F. Mila and T.M. Rice, Physica C, **157**, 561, 1989.
19. B.S. Shastry, Phys. Rev. Lett. 63:1288 (1989).
20. F. Mila and T.M. Rice, Phys. Rev. B 40:11382 (1989).
21. N. Bulut, D. Hone, D.J. Scalapino, and N.E. Bickers, Phys. Rev. B 41:1797 (1990); and preprint UCSBTH–89–35.
22. D. Cox and L. Trees, preprint.
23. A similar analysis of NMR experiments in the normal state has been done by A.J. Millis, H. Monien and D. Pines, to be published in Phys. Rev. B; and H. Monien article in this volume.
24. P.A. Lee and N. Read, Phys. Rev. Lett. 58:2691 (1987).
25. N. Bulut, D.J. Scalapino and H. Morawitz, preprint UCSBTH-90–14.
26. N. Bulut, D.J. Scalapino and N.E. Bickers, unpublished.
27. NMR in the superconducting state has also been studied by H. Monien and D. Pines, Rev. Mod. Phys.
28. Kuroda and Varma, preprint.
29. L. Coffey, Phys. Rev. Lett. 64:1071 (1990).

INFLUENCE OF THE ANTIFERROMAGNETIC FLUCTUATIONS ON THE NUCLEAR MAGNETIC RESONANCE IN THE CU-O HIGH TEMPERATURE SUPERCONDUCTORS

Hartmut Monien
Department of Physics
University of Illinois at Urbana - Champaign
1110 West Green Street
Urbana, 61801 IL

Abstract

We explore the influence of the antiferromagnetic correlations on the NMR experiments in the $YBa_2Cu_3O_7$ and $YBa_2Cu_3O_{6.63}$. We discuss several ways of determining the Cu and O hyperfine couplings. It is demonstrated that with an antiferromagnetic correlation length $\xi(T)$ which is growing with decreasing temperature it is possible to give a consistent account of the Knight shift and spin lattice relaxation time experiments in both materials.

Introduction

In the high temperature CuO superconductors the understanding of the magnetic properties plays a central role. The nuclear magnetic resonance experiments give a local probe for the magnetic properties of the low lying excitations which are important for superconductivity. We will demonstrate that the recent Knight shift and spin lattice relaxation time experiments by Takigawa [1] give evidence for a one band picture. This has very important implications for any theoretical approach for the high temperature superconductors. While the Knight shift measures the static long wavelength susceptibility $Re(\chi(q,\omega=0))$ the spin lattice relaxation time experiments measure the local spin dynamics at the nuclear site. To understand the nuclear magnetic resonance experiments we not only have to understand the spin correlations, i.e. $\chi(q,\omega)$, but also the coupling of the nuclear spin ***I*** to the electronic spin ***S***, i.e. the hyperfine Hamiltonian. We will first discuss the hyperfine Hamiltonian and then introduce our model for the spin - spin correlation function. We show that the with a model first introduced by Millis, Monien and Pines [2], hereafter referred to as MMP one can describe the Knight shift and nuclear magnetic relaxation time experiments in the metallic as well as in the strongly doped regime. Finally we will discuss the implications for the different CuO materials.

Dynamics of Magnetic Fluctuations in High-Temperature Superconductors
Edited by G. Reiter *et al.*, Plenum Press, New York, 1991

The hyperfine Hamiltonian and the spin-spin correlation function

Here we will explore the implication of a one band model, i.e. we will assume that there is only one spin degree of freedom **S** per unit cell which is responsible for the relaxation of the Cu and O nuclear spin as well. This spin **S** which mainly resides on the Cu site may interact with the nucleus at the same site (direct hyperfine coupling) or at a different site (transferred hyperfine coupling). We will assume that the Cu nuclear spin ^{63}I has a direct as well as a transferred hyperfine coupling. We will further assume that the O as well as the Y nucleus have only a negligible small direct hyperfine coupling and that the NMR experiments on these nuclei are determined by a transferred hyperfine coupling only. Now we can write down the hyperfine Hamiltonian.

$$H_{hf} = \sum_{i\alpha} {}^{63}I_{i\alpha} A_{\alpha\alpha} S_{i\alpha} + B \sum_{<ij>\alpha} {}^{63}I_{i\alpha} S_{j\alpha}$$

$$+ C \sum_{<ij>\alpha} {}^{17}I_{i\alpha} S_{j\alpha} + D \sum_{<ij>\alpha} {}^{89}I_{i\alpha} S_{j\alpha} \tag{1}$$

Here $A_{\alpha\alpha}$ is the direct hyperfine coupling tensor, containing the core polarization and the anisotropic dipolar coupling and spin - orbit coupling, B is the transferred hyperfine coupling first proposed by Mila and Rice [3]. C is the oxygen and D is the the Yttrium transferred hyperfine coupling. The symbol <ij> denotes neighboring sites i and j.

Having specified the hyperfine Hamiltonian, we have to describe the correlation of the spins. Most of the microscopic theories of the CuO high temperature superconductors do not allow to evaluate the spin-spin correlation function for very small ω and arbitrary $\mathbf{q}$ explicitly. Bulut et al. [4] calculated the dynamic structure factor from a 3 band Hubbard model in the RPA. Here we use a phenomenological model for the spin-spin correlation function $\chi(\mathbf{q}, \omega)$ first proposed by Millis, Monien and Pines [2]. According to their model the spin-spin correlation function consists of two parts. A long wavelength part which is very much Fermi liquid like and a second part which contains the effect of the antiferromagnetic correlations and is peaked around the antiferromagnetic wavevector $\mathbf{Q} = (\pi / a, \pi / a)$.

$$\chi(\mathbf{q}, \omega) = \chi_{FL}(\mathbf{q}, \omega) + \chi_{AF}(\mathbf{q}, \omega) \tag{2}$$

We denote the Fermi liquid part at $\mathbf{q}$=0 with $\chi_{FL}(\mathbf{q}=0,\omega) = \bar{\chi}$ We will make the assumption that we describe the imaginary part of the Fermi liquid like part of the spin-spin correlation function with

$$\chi_{FL}(\mathbf{q}, \omega) = \pi \frac{\omega}{\Gamma} \bar{\chi}, \tag{2a}$$

independent of the wavevector $\mathbf{q}$. The antiferromagnetic part can be motivated by examining the spin-spin correlation function of a strongly antiferromagnetically correlated Fermi liquid. Taking a random phase approximation for the correlation function and expanding the spin - spin correlation function for the noninteracting system we obtain:

$$\chi_{AF}(\mathbf{q}, \omega) = \frac{\chi_0(\mathbf{Q},0)\ (\ \xi\ /\ \xi_0\ )^2}{1 + \xi^2\ q^2 - i\ (\ \omega\ /\ \omega_{SF}\ )}, \tag{2b}$$

where we have introduced the temperature dependent antiferromagnetic correlation length $\xi(T)$ and the energy scale ω_{SF} typical for the spin fluctuations. The susceptibility $\chi_0(\mathbf{Q},0)$ is related to the quasiparticle like contribution $\bar{\chi}$ by $\chi_0(\mathbf{Q},0) = \bar{\chi}\ (\ \xi\ /\ \xi_0\ )^2$. The wavevector $\mathbf{q}$ is measured from the antiferromagnetic wavevector $\mathbf{Q} = (\pi\ /\ a, \pi\ /\ a)$. The total susceptibility at zero wavevector can be expressed in terms of the quasiparticle contribution:

$$\chi = \bar{\chi}\ (\ 1 + \frac{\sqrt{\beta}}{2\pi^2}\ ) \tag{3}$$

For a typical value of $\beta = 3.0$ the quasiparticle contribution dominates the static susceptibility. The contribution of the antiferromagnetic part is only 17% of the total static susceptibility. The spin fluctuation energy ω_{SF} can be related to the energy scale of the noninteracting system Γ by:

$$\omega_{SF} = \Gamma \left(\frac{\xi_0}{\xi}\right)^2. \tag{4}$$

As we shall see later on the spin fluctuation energy scale ω_{SF} is proportional to the temperature T for temperatures larger than a few 10 K. This is exactly the same behavior as observed for the density-density response function in the Raman scattering experiments. The latter led Varma et al. [5] to propose that the imaginary part of the density density response function behaves like ω / T in the low energy regime. Our phenomenological model is an expansion of the spin-spin correlation function around small ω so that we do not attempt to extrapolate to very large ω as one encounters for example in the neutron scattering experiments. For further details of the MMP model we refer the reader to the paper by Millis, Monien and Pines [2].

Determination of the hyperfine couplings

In a one band model of the elementary excitations of the CuO high temperature superconductors all Knight shifts are determined by the same static susceptibility $\chi_0(T)$. The difference in the magnitude would just arise from the different hyperfine couplings of the various nuclear spins.

$$^{63}K_{\perp}(T) = \frac{A_{\perp} + 4\,B}{^{63}\gamma_n \gamma_e \hbar^2}\ \chi_0(T) \qquad (5.a)$$

$$^{63}K_{\|}(T) = \frac{A_{\|} + 4\,B}{^{63}\gamma_n \gamma_e \hbar^2}\ \chi_0(T) \qquad (5.b)$$

$$^{17}K_{iso}(T) = \frac{2\,C}{^{17}\gamma_n \gamma_e \hbar^2}\ \chi_0(T) \qquad (5c)$$

$$^{89}K(T) = \frac{8\,D}{^{89}\gamma_n \gamma_e \hbar^2}\ \chi_0(T) \qquad (5.d)$$

For the 90 K superconductor $YBa_2Cu_3O_7$ the static susceptibility is temperature independent so that there is no way of proving this hypothesis In the 60 K material the Cu and O Knight shifts in the normal state are temperature dependent . Indeed Takigawa [1] finds for the $YBa_2Cu_3O_{6.63}$ material that the Cu and the O Knight shift are proportional to each other in the normal state.

The Cu Knight shift in the c - direction is very small so that in a one component picture we have to assume that the direct hyperfine coupling $A_{\|}$ and the transferred hyperfine coupling B nearly cancel each other. Therefore we take $A_{\|} = -4B$. The resonance experiments in the antiferromagnetic state by Yasuoka et al. [6] allow to put another constraint on the relation between the direct and transferred hyperfine coupling. In an antiferromagnetic background the resonance frequency is given by:

$$\left(\frac{\mu}{\mu_{eff}}\right) \frac{|A_{\perp} - 4\,B|}{^{63}\gamma_n \gamma_e \hbar} = 160 \text{ KOe} \qquad (6)$$

From neutron scattering [7] we know that the effective magnetic moment is about $\mu_{eff} = 0.6\ \mu_B$. Therefore the Cu hyperfine couplings are known apart from an overall factor. The oxygen isotropic hyperfine coupling C can be inferred from the ratio of the Cu(2) Knight shift for a field applied in the a or b direction to the O(2,3) isotropic Knight shift. The anisotropy of the Cu relaxation rate gives another constraint relating the Cu hyperfine couplings $A_{\|}$, $A_{\perp}$, and B. Monien, Pines and Slichter [8] analyzed the anisotropy of the relaxation rates for the case of completely uncorrelated spins and strongly antiferromagnetically spins in the 90K material. The following set of hyperfine couplings is consistent with the Knight shift results in the 60K and 90K materials and the analysis of Monien, Pines and Slichter:

$$A_{\|} = -167 \,..\, -172 \text{ kOe}/\mu_B \qquad (7a)$$

$$A_{\perp} = 33 \,..\, 39 \text{ kOe}/\mu_B \qquad (7b)$$

$$B = 42 \,..\, 43 \text{ kOe}/\mu_B \qquad (7c)$$

The oxygen was determined by Millis, Monien and Pines to be

$$C = \quad 70 \,..\, 74 \text{ kOe} / \mu_B \qquad (7d)$$

which is in good agreement with ratio of the Cu(2) to the O(2,3) Knight shift in the 60K and 90K materials, see e.g. [9] and [10]. Having specified the hyperfine Hamiltonian we can now analyze the spin lattice relaxation time experiments.

Spin lattice relaxation rates

The spin lattice relaxation rate is determined by the imaginary part of the spin - spin correlation function at the nuclear site. If the hyperfine Hamiltonian, as in our case contains an on site as well as a nearest neighbor coupling term the hyperfine couplings are **q** dependent. The relaxation rate is given by:

$$\frac{1}{T_1 T} \sim \sum_{\mathbf{q}} |A(\mathbf{q})|^2 \, S(\mathbf{q},\omega) \, , \qquad (8)$$

where $S(\mathbf{q},\omega)$ is the dynamic structure factor which in the low frequency regime is simply given by:

$$S(\mathbf{q},\omega) = \frac{T}{\omega} \, \mathrm{Im} [\, \chi(\mathbf{q},\omega) \,] \qquad (9)$$

We will use our ansatz for the spin-spin correlation function Eq. (2). With the hyperfine Hamiltonian (1) the various relaxation rates are given by (e.g. [2]):

$$^{63}W_{||} = \frac{3}{4} \frac{1}{\mu_B{}^2 \hbar} \sum_{\mathbf{q}} [A_\perp - 2\,B\,(\cos(q_x a) + \cos(q_y a))]^2 \, S(\mathbf{q},\omega) \qquad (10a)$$

$$^{63}W_{\perp} = \frac{3}{8} \frac{1}{\mu_B{}^2 \hbar} \sum_{\mathbf{q}} [A_{||} - 2\,B\,(\cos(q_x a) + \cos(q_y a))]^2 \, S(\mathbf{q},\omega) + {}^{63}W_{||} \qquad (10b)$$

$$^{17}W = \frac{3}{4} \frac{1}{\mu_B{}^2 \hbar} \sum_{\mathbf{q}} [\, 2\, C^2 \, (1 - \cos(q_x a)) \,] \, S(\mathbf{q},\omega) \qquad (10c)$$

For the MMP model the **q** averages can be carried out analytically. It is useful to introduce the moments of the structure factor $S(\mathbf{q},\omega)$:

$$S_0 = \left(\frac{a}{2\pi}\right)^2 \int d^2q \; S(\mathbf{q},\omega) \qquad (11a)$$

$$S_1 = \left(\frac{a}{2\pi}\right)^2 \int d^2q \; [1 - \tfrac{1}{2}(\cos(q_x a) + \cos(q_y a))] \; S(\mathbf{q},\omega) \qquad (11b)$$

$$S_2 = \left(\frac{a}{2\pi}\right)^2 \int d^2q \; \left(\frac{4}{5}\right) \; [1 - \tfrac{1}{2}(\cos(q_x a) + \cos(q_y a))]^2 S(\mathbf{q},\omega) \qquad (11c)$$

These integrals depend on the detailed form of the spin - spin correlation function. The moment S_0 is dominated by the antiferromagnetic correlation length ξ, $S_0 \sim (\xi / a)^2$, whereas S_1 depends only weakly on ξ via $\ln(\xi / a)$. S_2 does not depend on the coherence length at all. For the detailed expressions of the moments we refer the reader to [2]. We can now express the spin lattice relaxation times for the different nuclei in terms of the moments of the structure factor.

$$^{63}W_{||} = \frac{3}{4} \frac{1}{\mu_B{}^2\hbar} [(A_\perp - 4B)^2 S_0 + 8B (A_\perp - 4B) S_1 + 20 B^2 S_2] \tag{12a}$$

$$^{63}W_\perp = \frac{3}{4} \frac{1}{\mu_B{}^2\hbar} \{ (A_\perp - 4B)^2 S_0 + 8B (A_\perp - 4B) S_1 + 20 B^2 S_2 + (A_\perp - 4B)^2 S_0 + 8B (A_\perp - 4B) S_1 + 20 B^2 S_2 \} \tag{12b}$$

$$^{17}W = \frac{3}{2} \frac{1}{\mu_B{}^2\hbar} C^2 S_1 \tag{12c}$$

The copper relaxation rate is dominated by S_0 and therefore proportional to $(\xi / a)^2$, whereas the oxygen relaxation rate is nearly independent of ξ. The difference in the temperature dependence of the copper and the oxygen relaxation rates is coming from this difference.

Comparison with experiment

In the MMP model the copper relaxation rate is strongly influenced by the antiferromagnetic correlations. Since the copper relaxation rate $^{63}(1 / T_1)$ is nearly independent of the temperature, the factor T from the Bose - factor in the structure factor has to be canceled by the build up of the antiferromagnetic correlations. Since the copper relaxation rate is determined by the moment $S_0 \sim \xi^2$ we make an ansatz for the correlation length which describes the buildup of the antiferromagnetic correlations as one lowers the temperature:

$$\frac{\xi(T)}{a} \sim \frac{\text{const.}}{\sqrt{T + T_x}}, \tag{13}$$

where T_x is a typical energy scale on which the coherence length varies and the proportionality constant determines the enhancement of the copper relaxation rate $1 / T_1$. This model for the temperature dependence of the coherence length implies that the relaxation time of the Cu(2) behaves like $^{63}T_1 \sim T + \text{const.}$.We find very good agreement with the experimental results in $YBa_2Cu_3O_7$ for the Cu relaxation rate (see e.g. [11], [12]) and reasonable good agreement as shown in Fig. 1 , where we have used the mean field expression for the temperature variation of the coherence length.

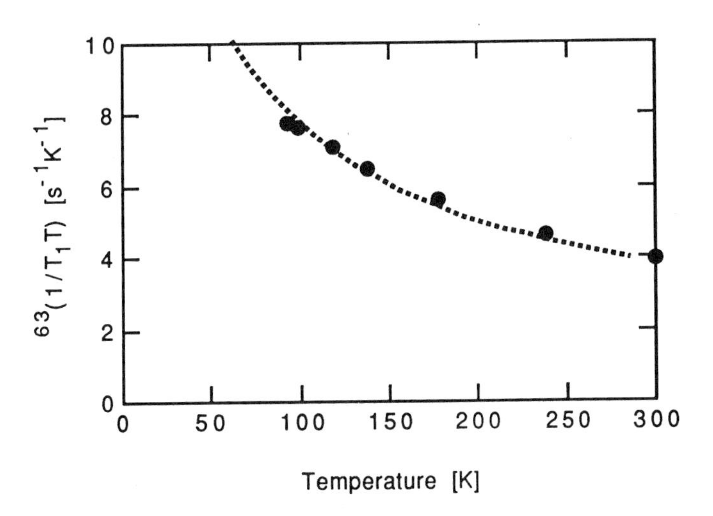

Figure 1. The nuclear magnetic relaxation rate of the copper nucleus as a function of temperature. The dots mark the experimental points by Barrett et al. [12]. The dashed line gives the fit with the MMP theory with a T_X = 61K and a correlation length ξ of 2.5 *a* at 100K and a Γ =0.5eV.

The only justification for choosing the particular form for the coherence length is the comparison with experiment. Now we can proceed and calculate the oxygen relaxation rate. Again we find reasonable good agreement with the oxygen spin lattice relaxation rate as measured by Hammel et al. [13]

The strength of the antiferromagnetic contribution is determined by $(a/\xi)^4$ which was determined to be 10.0. Now let us turn to the $YBa_2Cu_3O_{6.63}$ material.The static susceptibility which determines the Knight shift becomes temperature dependent. As shown by M. Takigawa [1] the Cu and the O Knight shift are proportional to each other with a temperature independent

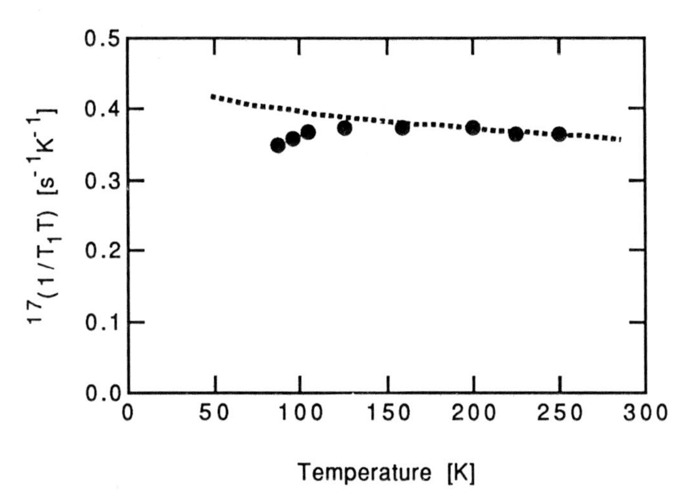

Figure 2. The nuclear magnetic relaxation rate of the copper nucleus as a function of temperature. The dots mark the experimental points by [13]. The dashed line gives the fit with the same parameters as above.

proportionality factor. This shows that in the long wavelength limit there is only one spin per unit cell. The other interesting observation is that the oxygen spin lattice relaxation rate which is Korringa like in the 90K material is proportional to the Knight shift in the 60K material. We do not know the reason why the static susceptibility is temperature dependent. One might suspect some sort of very low energy spin gap. This led us to propose that the antiferromagnetic part of the spin spin correlation function, Eq. 2, scales like $\chi_0(T)$. This is quite different from the Fermi liquid behavior where the spin lattice relaxation time is proportional to the Knight shift squared. We have repeated a similar analysis like MMP with the additional assumption that there is an additional temperature dependence in the structure factor.

We use the same hyperfine Hamiltonian as for the $YBa_2Cu_3O_7$ material, with the same hyperfine couplings, and again the simple form for the temperature dependence of $\xi(T)$, Eq. 13. The copper relaxation rate is determined by two competing temperature dependent quantities. The first being the antiferromagnetic correlation length $\xi(T)$ which is growing with decreasing temperature. The other is the static susceptibility $\chi_0(T)$ which decrease sharply if when one lowers the temperature. At lower temperature the static susceptibility decreases much faster than the correlation length increases. At higher temperatures the static susceptibility becomes saturates at 280K whereas the antiferromagnetic correlation length is still decreasing with increasing temperature. The copper relaxation rate reflects therefore the interplay between two

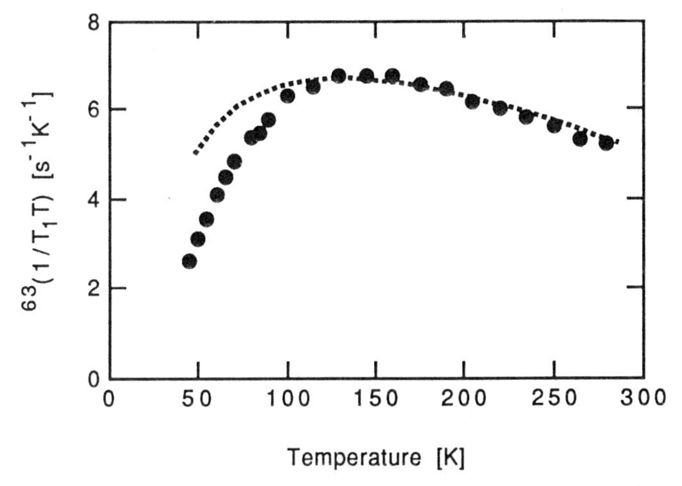

Figure 3. The nuclear magnetic relaxation rate of the copper nucleus as a function of temperature. The dots mark the experimental points by Takigawa. [10]. The dashed line gives the fit with the MMP theory with a $T_X = 31K$ and a correlation length of 4.0 *a* at 100K and a $\Gamma = 0.5eV$.

temperature dependent physical quantities. We do not know if the temperature dependence of the correlation length and the static susceptibility are related to each other. We show the fit to the experimental data of Takigawa [1] for the Cu(2) spin lattice relaxation time in Fig. 3 and the resulting fit with the same parameters for the oxygen relaxation spin lattice relaxation rate in Fig. 4.

The agreement with the measured copper relaxation rate is quite remarkable. It is interesting to note that after taking out the **q** independent prefactor $\overline{\chi}(T)$ the Cu relaxation rates for the $YBa_2Cu_3O_7$ and the $YBa_2Cu_3O_{6.63}$ material are very similar. The next test is the oxygen relaxation rate. Since the oxygen spin lattice relaxation rate depends only weakly (via a logarithm) on ξ, the relaxation rate should be proportional to the static susceptibility $\bar{\chi}(T)$.

As we can see in Fig. 2 this is indeed the case. The measurement of the oxygen relaxation rate gave the hint that the spin lattice relaxation does not depend on a density of states squared so that it would be proportional to the Knight shift but that it is proportional to a density of states or the Knight shift.

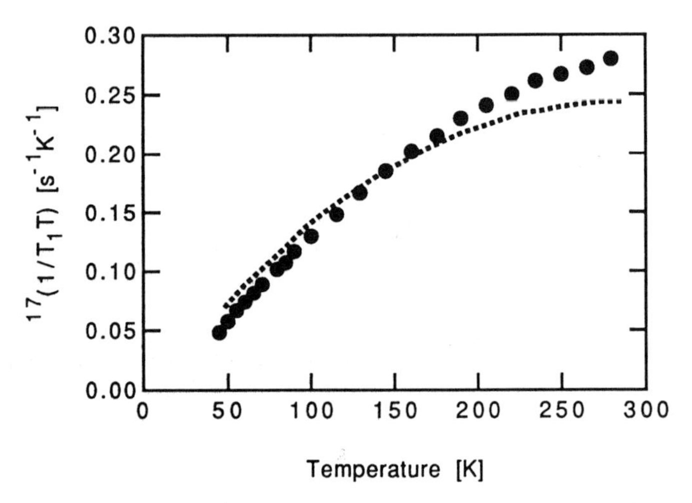

Figure 2. The nuclear magnetic relaxation rate of the copper nucleus as a function of temperature. The dots mark the experimental points by [13]. The dashed line gives the fit with the same parameters as above.

Conclusions

The first interesting remark one can make is that the antiferromagnetic correlation length increases in going from the 90K to the 60K material. The second is that the temperature scale T_X is decreased in the 60K material. It is surprising that the hyperfine Hamiltonian Eq. 1 is describing both the 90K and the 60K material with the same hyperfine coupling indicating that the material has some very ionic features.

The second remark is that the relaxation rate shows a very interesting behavior : It does not go like the Knight shift squared but is directly proportional to the Knight shift. This comes about because in addition to the energy scale of the spin fluctuations of the noninteracting system, Γ, there is a second energy scale for the spin fluctuations in the interacting system ω_{SF} which goes approximately like the temperature. This behavior is in sharp contrast to a Fermi liquid where the only relevant energy scale is E_f.

It is not *a priori* clear how the model spin-spin correlation function, Eq. 2, can be extended to higher energies. The small energy scale ω_{SF} indicates that the structure of the spin-spin correlation function in **q** space, which is responsible for the difference in the copper and oxygen relaxation rate, might be completely wiped out when one is coming to energies which are used in the neutron scattering experiments. One could only speculate how the spin-spin correlation function, Eq. 2, could be modified to describe the neutron scattering experiments.

The temperature dependence of the spin fluctuation energy ω_{SF} resembles the energy dependence, postulated by Varma et al. [5], of the energy scale in the imaginary part of the selfenergy . This is indicating that the spin and the charge degrees of freedom have very similar behavior and, which is more important, have the same energy scale. The exchange of the antiferromagnetic spin fluctuation could also reduce the energy range in which the imaginary part of the self energy goes like $(\varepsilon - \varepsilon_f)^2$ as required by Fermi liquid theory.

Finally we remark that the tendency toward antiferromagnetism is much stronger in the $YBa_2Cu_3O_{6.63}$ than it is in the $YBa_2Cu_3O_7$ material. The correlation length which can be determined from experiment by comparing the oxygen and copper relaxation rate is larger by a factor of two at 100K and the antiferromagnetic correlation length is increasing much sharper with decreasing temperature for the $YBa_2Cu_3O_{6.63}$ material. The temperature scale T_x is decreased by nearly a factor of two which is a further indication that one is coming closer to an antiferromagnetic transition.

Acknowledgements

I would like to thank my collaborators A. J. Millis and D. Pines. It is a pleasure to acknowledge many useful discussions with M. Takigawa, P. C. Hammel, R. H. Heffner and C. P. Slichter, C. Pennington, D. Durand and S. Barrett. This work was supported in part by the Science and Technology center for Superconductivity at the University of Illinois.

References

1. M. Takigawa, this volume
2. A. J. Millis, H. Monien and D. Pines, Phys. Rev. B in print
3. F Mila and T. M. Rice, Physica **C157**, 561 (1989)
4. N. Bulut, D. Hone, D. S. Scalapino, and N. E. Bickers, unpublished
5. C. M. Varma, P. B. Littlewood, S. Schmitt-Rink, E. Abrahams, and A. E. Ruckenstein, Phys. Rev. Lett, 1996 (1989)

6. H. Yasuoka, T. Shimizu, Y. Ueda and K. Kosuge, J. Phys. Soc. Jpn. **57**, 2659 (1988)
7. J. M. Tranquada, D. E. Cox, W. Kunnmann, H. Mouden, G. Shirane, M. Suenga, P. Zallikar, D. Vaknin, S. K. Sinha, M. S. Alvarer, A. Jacobsen, and D. C. Johnston, Phys. Rev. Lett **60**, 156 (1988)
8. H. Monien, D. Pines, C. P. Slichter, unpublished
9. C. H. Pennington, D. J. Durand, C. P. Slichter, J. P. Rice, E. D. Bukowski and D. M. Ginsberg, Phys. Rev. B **39**, 2902 (1989)
10. M. Takigawa, P. C. Hammel, R. H. Heffner, Z. Fisk, Phys. Rev. B **39**, 7371 (1989)
11. R. E. Walstedt, W. W. Warren, Jr., R. F. Bell, and G. P. Espinoza, Phys. Rev. B **40**, 2572 (1989)
12. S. E. Barrett, D. J. Durand, C. H. Pennington, C. P. Slichter , T. A. Friedmann, J. P. Rice, and D. M. Ginsberg, unpublished
13. P. C. Hammel, M. Takigawa, R. H. Heffner, Z. Fisk, K. C. Ott, Phys. Rev. Lett. **63**, 1992 (1989)

Microscopic Models for Spin Dynamics in the CuO_2-planes with application to NMR

T. M. Rice
Theoretische Physik
ETH-Hönggerberg
8093 Zürich, Switzerland

Although the general form of the electronic structure of the CuO_2-planes on the larger energy scale ($\sim 1\ eV$) is generally agreed upon, the reduced model on the relevant energy scale for superconductivity ($\sim 10^{-1}\ eV$) continues to be debated. The key issue is usually represented in terms of the adequacy of a one-band model to describe this low energy region. Such a model was proposed at the outset of the high-T_c problem by Anderson [1] and derived from a more general starting model with $Cu\,d$-states and $O\,p$-states by Zhang and Rice [2]. In this derivation only the leading $dp\sigma$ hybridization and Coulomb repulsion on Cu sites were kept and terms such as the direct $p-p$ hybridization between O orbitals or Coulomb repulsion on the O sites were ignored. These latter terms are not small and could result in important changes in the model. Recently, these questions have been examined in some detail by a number of groups[3],[4] who solved for the eigenstates, wavefunctions and response functions of small clusters, starting from the full model with all Cu- and O-orbitals and interactions. As an example, Hybertsen, Stechel, Schlüter and Jennison [4] exactly diagonalized small clusters up to $Cu_5\,O_{16}$ and found the low lying energy spectrum and wavefunctions. For the case of Cu^{2+}-valence the low energy spectrum has spin degrees of freedom only and is well described by an antiferromagnetic Heisenberg model with nearest neighbor (n.n.) coupling only. The strength J is also in good agreement with experimental values. Introducing an extra hole or electron on the cluster allows charge degrees of freedom at low energy. These in turn are well represented by assuming that the extra hole (or electron) is tightly bound to a Cu^{2+}-spin to form a spin singlet which however can hop with n.n. and n.n.n. matrix elements t and t' respectively. Hybertsen et al. find relatively small values for the ratio $|t'/t|$ ($\approx 1/6$). However Eskes, Tjeng and Sawatzky [5] report larger value for this ratio. The one-band $t-J$ model with a possible n.n.n. t'-correction is then the most plausible reduced model to describe the low energy electronic structure of the doped CuO_2-planes.

In the $t-J$ model one starts from a localized electronic spin model in contrast to an itinerant band model. This in turns influences the form of hyperfine coupling that will exist between the nuclear and electronic spins. In particular if one examines the coupling to the Cu-nuclei, then the anisotropy in this coupling is quite different. Mila and Rice [6] starting from the localized spin model examined the form of this coupling including both on-site and n.n. transferred hyperfine coupling. Only by including the latter is it possible to explain the different crystalline anisotropies in the Knight shift and relaxation rate of the planar Cu-nuclei in $YBa_2Cu_3O_7$ [7]-[9]. The O- and Y-nuclear spins couple to the electronic spins through the O-orbitals. Therefore these couplings are sensitive to the form of the extra holes, namely do these extra holes introduce extra electronic spin degrees of freedom or are they bound in local singlets so that, as the $t-J$ model predicts, only Cu^{2+}-spin degrees of freedom remain. Mila and Rice [10] have argued that the scaling of the Y-Knight shift and relaxation rate with total susceptibility found recently by Alloul and coworkers [11] in the series $YBa_2Cu_3O_{7-x}$, shows that Cu^{2+}-spins which dominate the susceptibility, are also determining the Y-nuclear response. This scaling is hard to reconcile with models which have extra electronic spin degrees of freedom on the O-orbitals.

References

(1) P.W. Anderson, Science 235, 1196 (1987) and in Proc. Int. School E. Fermi, Course CIV, eds. R.A. Broglia and J.R. Schrieffer, p. 1, 1987.

(2) F.C. Zhang and T.M. Rice, Phys. Rev. B 37, 3759 (1988).

(3) A. Ramsak and P. Prelovsek, Phys. Rev. B 40, 2234 (1989).
W.H. Stephan, W. von der Linden and P. Horsch, Phys. Rev. B 39, 2924 (1989).
M. Ogata and H. Shiba, J. Phys. Soc. Japan 58, 2836 (1989).
C.H. Chen, H.B. Schüttler and A.J. Fedro, Phys. Rev. B 41, 2581 (1990).

(4) M.S. Hybertsen, E.B. Stechel, M. Schlüter and D.R. Jennison, preprint.

(5) H. Eskes, L.H. Tjeng and G.A. Sawatzky, Springer Ser. in Mater. Science Vol. II, 20 (1989).

(6) F. Mila and T.M. Rice, Physica C 157, 561 (1989).

(7) M. Takigawa, P.C. Hammel, R.H. Heffner and Z. Fisk, Phys. Rev. 39, 7371 (1989).
M. Takigawa, P.C. Hammel, R.H. Heffner, Z. Fisk, J.L. Smith and R. Schwarz, Phys. Rev. B 39, 300 (1989).

(8) R.E. Walstedt, W.W. Warren, Jr., R.F. Bell, G.F. Brennert, G.P. Espinosa, R.J. Cava, L.F. Schneemayer and J.V. Waszczak, Phys. Rev. B 38, 9299 (1988).
(9) C.H. Pennington, D.J. Durand, C.P. Slichter, J.P. Rice, E.D. Bukowski and D.M. Ginsberg, Phys. Rev. B 39, 2902 (1989).
(10) F. Mila and T.M. Rice, Phys. Rev. B 40, 11382 (1989).
(11) H. Alloul, T. Ohno and P. Mendels, Phys. Rev. Lett. 63, 1700 (1989).

Recent Topics of μSR Studies on High-T_c Systems

Y.J. Uemura[1], G.M. Luke[1], B.J. Sternlieb[1], L.P. Le[1], J.H. Brewer[2], R. Kadono[2], R.F. Kiefl[2], S.R. Kreitzman[2], T.M. Riseman[2], C.L. Seaman[3], J.J. Neumeier[3], Y. Dalichaouch[3], M.B. Maple[3], G. Saito[4], H. Yamochi[4]

1. *Department of Physics, Columbia University, New York City, New York 10027*
2. *TRIUMF and Department of Physics, University of British Columbia Vancouver, B.C., Canada*
3. *Department of Physics, University of California, San Diego, La Jolla California 92093*
4. *Institute for Solid State Physics, University of Tokyo, Roppongi, Minato-Ku, Tokyo 106, Japan*

Recent progress of our muon spin relaxation studies on high-T_c systems is outlined. The topics include the phase diagrams of the hole-doped and electron-doped systems, search for possible effects of anyons, and the study of the relation between T_c and superconducting carrier density (based on the μSR measurements of penetration depth) in the 214, 123, 2212, 2223, as well as Pr-doped 123 systems. The combined neutron scattering and μSR results on a single crystal of $(La_{1.94}Sr_{0.06})CuO_4$ demonstrate the slowing down of dynamic spin fluctuations in a wide temperature range above the freezing temperature, which resemble the results in canonical spin glasses CuMn or AuFe. We also present the μSR results of the penetration depth in an organic superconductor $(BEDT-TTF)_2Cu(NCS)_2$ and in a Uranium compound U_6Fe.

Introduction

History of the muon spin relaxation (μSR) measurements dates back to the late 1950's and 60's following the discovery of parity violation. The application of μSR to condensed matter physics has steadily developed[1] in the 70's and early 80's; mainly in the study of magnetic properties of ferro- or antiferromagnets and spin glasses. Extensive μSR studies on high-T_c systems[2-4] have increased the recognition of this technique as one of the very powerful experimental methods in the study of magnetism and superconductivity.

In this paper, we would like to provide a review of our recent μSR studies on high-T_c systems[3], focussing on several selected topics. The measurements on static magnetic order by zero-field μSR have enabled us to determine the magnetic phase diagrams of both hole-doped and electron-doped systems (topic 1), as well as to search for the possible effect of "anyons" (topic 2). Based on μSR measurements of the magnetic field penetration depth, interesting relations have been found between T_c and the superconducting carrier density (topic 3). Combined results of our neutron scattering and μSR measurements on a single crystal specimen of $(La,Sr)CuO_4$ elucidate the spin dynamics of the specimen in the compositional region where "spin glass" like properties have been predicted (topic 4). We also describe the ongoing effort to extend the μSR measurements to cover other exotic superconductors, i.e., heavy-fermion and organic superconductors (topic 5). Although we could not include it in this paper, we should also mention that μSR is very useful in studying the flux pinning and depinning phenomena (see ref. 5).

Dynamics of Magnetic Fluctuations in High-Temperature Superconductors
Edited by G. Reiter *et al.*, Plenum Press, New York, 1991

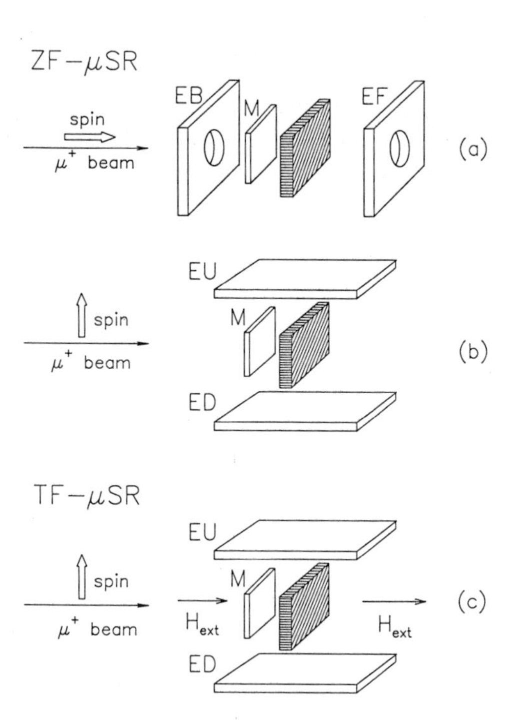

Fig. 1 Schematic views of experimental configurations of zero-field μSR (ZF-μSR) measurements (a) and (b); and of transverse-field μSR (TF-μSR) (c). In the case of (b) and (c), the spin polarization of incident muon is changed perpendicular to the beam direction with the use of a Wien Filter. Shaded areas denote specimens, M the muon defining counter, and EU, ED, EB and EF the positron detection counters.

Experimental Aspects

Details of technical aspects of the muon spin relaxation experiments have been described in earlier literature[6]. Figure 1 illustrates the experimental configurations of the Zero-Field μSR (ZF-μSR) and Transverse-Field μSR measurements at TRIUMF (Vancouver). All the μSR studies described here have been obtained at TRIUMF using the low energy (4.1 MeV) surface muon beam. The incident positive muon is identified by the counter M, stopped within about 200 mg/cm^2 of the specimen, and usually resides at an interstitial site in the crystal until it decays into a positron with the mean lifetime of 2.2 μsec. Two sets of positron counters (EU and ED or EF and EB) detect the muon decay positron which is emitted preferentially towards the muon spin direction. More than 10^6 events of such an individual muon decay are accumulated one by one to produce exponential life-time histograms. The time evolution of the muon spin direction and polarization in the specimen can be determined from the asymmetry of the positron histograms, (EU-ED)/(EU+ED), after correcting for the solid angle factors.

Originally, the muon spin is polarized along its flight direction when the μ^+ is created by the decay of pions. μSR measurements can be performed with this initial spin direction, as illustrated in Fig. 1(a). It is, however, also possible to rotate the incident muon spin direction to be perpendicular to the beam direction, as shown in Fig. 1(b), by using an electro-magnetic Wien Filter. Then, the ZF-μSR measurements can be performed with two different experimental configurations. This feature is very helpful, especially for the study of static magnetic order using single crystals, since one can accurately determine the direction of local fields with respect to the crystal axes. For example, we have detected[7] the changes of the Cu spin direction at $T \sim 65$ K and 35 K in the antiferromagnet Nd_2CuO_4 ($T_N \sim 250K$), as well as the static magnetic order in $Sr_2Cl_2CuO_2$ below $T_N = 260K$ (ref. 8), by performing the ZF-μSR measurements on single-crystal specimens using the two configurations.

For the TF-μSR measurements with a high transverse external magnetic field H_{ext}, the configuration in Fig. 1(c) is used. The bending of the incident beam due to the Lorentz force can be avoided in this configuration, and the incident beam can be focussed on the specimen. In the μSR measurements of the magnetic field penetration depth of type-II superconductors, the configuration in Fig. 1(c) has another advantage. With the plate-shape thin specimens, the measurements can be made with the demagnetizing factor N = 1. The magnetic field H, magnetization M and the magnetic induction B of type-II superconductors are related[9] as $H = B - 4\pi M$ and $H = H_{ext} - 4\pi NM$. Therefore the average magnetic induction B can be kept equal to H_{ext}, even below T_c, in this configuration. This feature is helpful in avoiding the complications due to the demagnetizing field. The effect of flux trapping can also be minimized in field-cooled measurements with this configuration, since the flux vortices do not have to move macroscopic distance below T_c as long as B does not change above and below T_c.

Topic 1. Magnetic Phase Diagrams

With its superb sensitivity to static magnetic order, μSR has been extensively applied to determine magnetic phase diagrams of the high-T_c systems[10–14]. Following the discovery[15] of the electron doped superconductors $(Nd_{2-x}Ce_x)CuO_4$, where Nd^{3+} is substituted by Ce^{4+}, we have performed[14] detailed ZF-μSR measurements on this system using the reduced specimens with $x = 0.0 \sim 0.20$. Figure 2 compares the magnetic and superconducting phase diagrams of the electron-doped system $(Nd_{2-x}Ce_x)CuO_4$ and the hole doped counterpart $(La_{2-x}Sr_x)CuO_4$. The magnetic ordering temperature was determined by μSR.

In both systems, the doping of carriers into the parent undoped antiferromagnetic compounds first causes the reduction of Néel temperature, followed by the destruction of the magnetic order. The onset of superconductivity is seen around the composition at which the static magnetic order disappears. There is no clear signature that magnetic order and superconductivity coexists; it is shown[14,16] by μSR that good superconducting specimens of both hole-doped and electron-doped high-T_c cuprate superconductors do not show static magnetic order. This indicates that the onset of superconductivity is tied to the destruction of static magnetic order, rather than being a separate uncorrelated transition. Therefore, the phase diagrams in Fig. 2 strongly suggest an important role of magnetic interactions in the mechanisms of superconductivity.

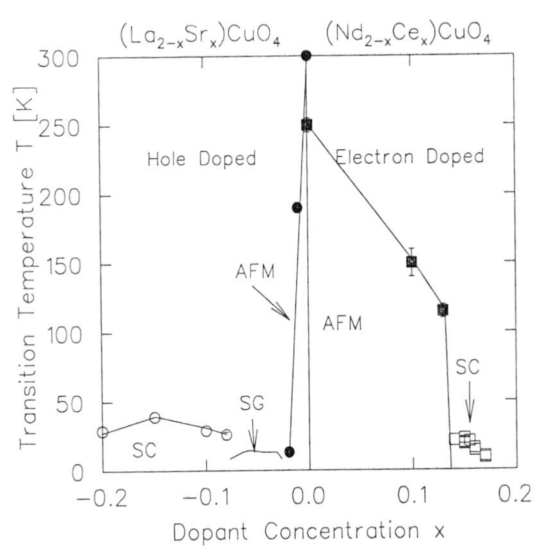

Fig. 2 Magnetic and superconducting phase diagrams of the electron-doped system $(Nd_{2-x}Ce_x)CuO_4$ and the hole-doped system $(La_{2-x}Sr_x)CuO_4$. The magnetic ordering temperature is determined by μSR and the superconducting T_c by bulk measurements. AFM denotes the antiferromagnetic phase, SC the superconducting phase, while SG indicates possible spin glass state.

Recently, we have performed[20] more detailed studies of the temperature and orientation dependence of the zero-field muon spin relaxation rate in $YBa_2Cu_3O_7$ and $Bi_2Sr_2CaCu_2O_8$. There is no change of the width of the static random local field at T_c and no significant orientation dependence was observed. So far, we have not found any clear signature of B_{dia} in the μSR experiments.

The unlikely but possible cases for us to fail in detecting B_{dia} are:

(1) μ^+ happens to rest at a site where the effect of B_{dia} cancels due to symmetry (such as, the muon site located just in the center of the adjacent CuO_2 planes which have opposite directions of B_{dia}).

(2) B_{dia} appears only on the surface or at grain boundaries.

(3) B_{dia} is not static, and a kind of motional narrowing effect (caused by the motion of anyons with alternating directions of b or B_{dia}) eliminates the static averaged field.

(4) B_{dia} appears not only below T_c but perhaps at some temperature higher than T_c.

(5) B_{dia} is smaller than $\sim 0.2\ G$.

Further experiments using single crystal specimens are currently underway.

Topic 3. T_c versus Carrier Density

When an external magnetic field H_{ext} ($H_{c1} < H_{ext} < H_{c2}$) is applied to type-II superconductors, it penetrates into the specimen by forming flux vortices. The local magnetic field near the vortex core is somewhat larger than that at places distant from the vortices; the local field has a distribution with a width ΔH. This width is determined by the penetration depth λ and by the structure of the flux vortex lattice. Pincus et al. [21] showed that ΔH is nearly independent of H_{ext} and $\Delta H \propto 1/\lambda^2$ in a wide range of H_{ext}.

In μSR measurements, one detects more than 10^6 individual muon decay events, each of which reflects the local field at an interstitial site of the crystal where the μ^+ particle rests. Thus, μSR samples the local field distribution within the specimen. In μSR measurements of the penetration depth λ, H_{ext} is applied perpendicular to the initial muon spin direction, and the muon spin precession around H_{ext} is observed. The width ΔH of the local field in the vortex state causes the dephasing of the muon spin precession; the relaxation rate σ of the Gaussian envelope $exp(-\sigma^2 t^2/2)$ is proportional to ΔH. Therefore, by measuring the relaxation rate σ, one can determine the penetration depth λ through the relation $\sigma \propto \Delta H \propto 1/\lambda^2$.

The coherence lengths ξ of cuprate high-T_c superconductors are very short; typically about 10 $\sim$ 20 Å within the CuO_2 planes. On the other hand, the normal state resistivity, measured[22] on single crystals of $YBa_2Cu_3O_y$ with $y = 6.5 \sim 7.0$, indicates that the mean free path l of charged carriers on the CuO_2 plane is longer than 100 Å at low temperatures $T \ll T_c$. In such a "clean limit" ($\xi/l \ll 1$), one can expect $1/\lambda^2 = (4\pi n_s e^2/m^* c^2)$ where n_s is the density of superconducting carriers and m^* the effective mass. Then the values of n_s/m^* can be determined in μSR since $\sigma \propto \Delta H \propto 1/\lambda^2 \propto n_s/m^*$.

We have performed TF-μSR measurements on more than 30 different specimens of cuprate high-T_c superconductors[23,24]. Figure 4 shows a plot of T_c versus $\sigma(T \rightarrow 0) \propto n_s/m^*$, both determined by μSR on these systems. With increasing carrier density, T_c initially increases, then saturates, and finally is suppressed in the heavily doped region. This tendency can be seen universally[23] in the single layer $La_{2-x}Sr_xCuO_4$ (214), double layer $YBa_2Cu_3O_y$ (123), as well as the triple layer $Bi_2Sr_2Ca_2Cu_3O_{10}$ (2223) and similar systems. Moreover, the initial increase of T_c with increasing n_s/m^* follows a straight line shared by the 214, 123, and 2223 systems.

Although it is impossible to separate the effect of n_s from that of m^* in the μSR measurement alone, it is quite likely that the results in Fig. 4 predominantly reflect the differences of n_s among various specimens. For a typical value of $m^* = 5\ m_e$, $\sigma = 1 \mu sec^{-1}$ corresponds to a carrier density of $n_s \sim 2 \times 10^{21} cm^{-3}$. Therefore, the results in Fig. 4 are generally consistent with the estimates of the carrier density based on the stoichiometry and valency. The μSR measurement, however, has an advantage since it reflects the actual superconducting carrier density.

There are, however, some differences between the hole-doped and electron-doped systems. Electron doping appears far less effective in destroying static magnetic order: the static order disappears around $x = 0.06$ for the hole-doped material while surviving up to as high as $x = 0.14$ in the electron-doped system. (Hall effect measurements[17] confirmed that the nominal concentration x represents the number of carriers actually doped in the CuO_2 plane for the case of the two systems in Fig. 2.) In an electron-doped system, the charge carriers are located primarily at the Cu site, changing Cu^{2+} to Cu^{1+} which acts as a non-magnetic Zn atom. This corresponds to the "magnetic dilution" of the antiferromagnetic CuO_2 plane. In contrast, the doped holes are mainly located at the oxygen atom, changing O^{2-} to the magnetic O^{1-}. The resulting unpaired spin at the oxygen site would mediate an effective ferromagnetic interaction between the adjacent Cu moments, frustrating the original antiferromagnetism. It is known that the dilution is much less effective than frustration in destroying the magnetic long range order. Then, the difference in the area of the antiferromagnetic phases in Fig. 2 can be understood as reflecting the different roles of holes and electrons in modifying the superexchange interaction between the Cu moments.

Topic 2. Search for Anyons

Various theoretical models have been proposed to explain the condensation mechanism of high-T_c superconductors. One of them, recently advocated by Laughlin, Halperin, Wilczek and others[18], is based on a concept of charged particles in 2-dimensional systems obeying fractional statistics. Those particles are referred to as " anyons ", since they could have a symmetry which is intermediate between that of fermions and bosons. According to the theory of Laughlin et al., each anyon experiences a fictitious field b, of the order of 10^6 Gauss, as a many-body force from the other anyons. This fictitious field creates Landau levels (even in the absence of external fields), opens an energy gap, leading to superconductivity.

Halperin et al.[18] suggested a few possible ways to test the existence of anyons in experiments. One of them is based on orbital diamagnetism, like Landau diamagnetism, which should exist since anyons are moving in the fictitious magnetic field. This diamagnetic field B_{dia} is expected to be several orders of magnitude smaller than the fictitious field b, but may still be observable experimentally. The superb sensitivity of zero-field μSR to small static magnetic fields ($\sim 1G$) is then very helpful in the search for the effect of B_{dia}. In conjunction with earlier TF-μSR studies of penetration depths, ZF-μSR measurements have been performed to make sure that there is no effect from static magnetic order of Cu spins[19]. Figure 3 shows an example of such ZF-μSR data on superconducting $(Tl_{0.5}Pb_{0.5})Sr_2CaCu_2O_7$ (T_c = 75 K). There is no difference between the zero-field relaxation function observed just above T_c and at $T \ll T_c$. The observed slow depolarization of muon spins corresponds to the static random local fields of 2.0 G. This immediately gives an upper limit for B_{dia}. Moreover, the magnitude of this random field is consistent with the nuclear dipolar fields expected at the muon site. Thus, there is no signature of the effect of B_{dia} in Fig. 3.

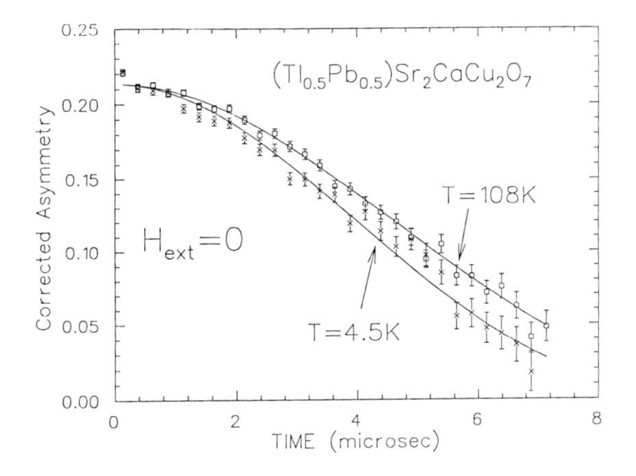

Fig. 3 Zero-field muon spin relaxation function observed in a superconductor $(Tl_{0.5}Pb_{0.5})Sr_2CaCu_2O_7$ ($T_c = 75K$) at $T = 4.5$ K and 108 K. The observed width of the random local fields corresponds to about 2 G. There is no significant difference between results at the two different temperatures (systematic uncertainty of the experiment is larger than the apparent slight difference between the results).

This advantage is illustrated in the $YBa_2Cu_3O_y$ system. There is a "plateau" at $T_c = 60$ K in the plot of T_c versus oxygen concentration y. This led some scientists to speculate that the 123 system with $T_c = 60$ K has some special "electronic phase". In a plot of T_c versus *measured* superconducting carrier density, such as Fig. 4, however, no anomaly is present at $T_c = 60$ K. This is because the number of carriers on the CuO_2 plane remains nearly unchanged with increasing y in the "60K plateau" region of $6.6 < y < 6.75$, as seen by μSR[23] and Hall effect[25] measurements. The extra holes with increasing y in this region are located therefore in some other part of the crystal. Thus, T_c increases smoothly as a function of the real carrier density n_s on the CuO_2 plane. This demonstrates that the $T_c = 60$ K plateau is merely an artifact of the crystal chemistry.

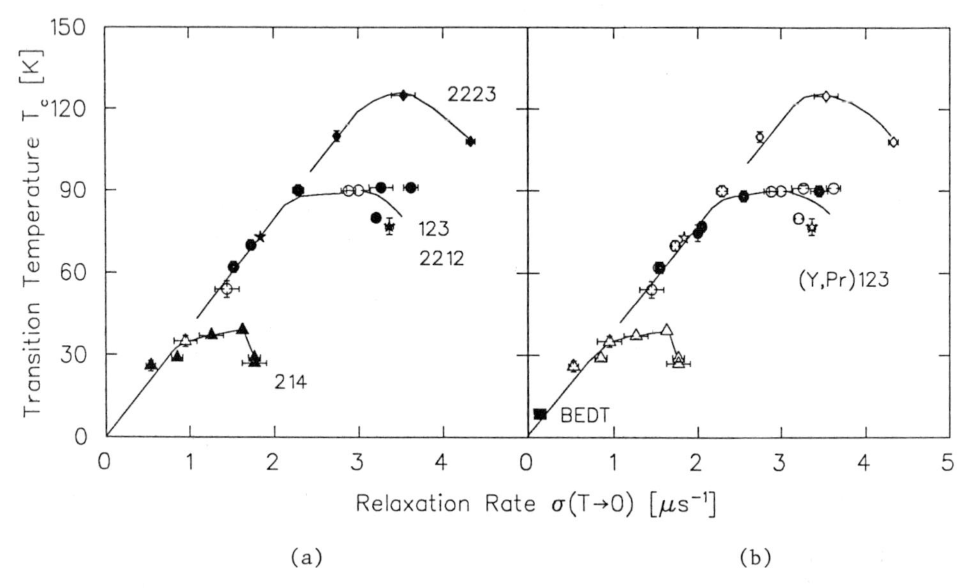

Fig. 4 *The superconducting transition temperature T_c of various high-T_c superconductors plotted versus the low temperature values of the muon spin relaxation rate $\sigma(T \to 0)$. Both T_c and σ have been determined by μSR measurements. In the "clean limit" $\xi/l \ll 1$, $\sigma \propto 1/\lambda^2 \propto n_s/m^*$. The left figure (a) represent points from Fig. 2 of ref. 23 (see the figure caption in ref. 23 for details), where the points for the 123 system are obtained with the oxygen depleted specimens $YBa_2Cu_3O_y$. The closed circles in the right figure (b) represent $(Y_{1-x}Pr_x)Ba_2Cu_3O_7$ with $x = 0.3$ to 0.05, and the closed square in (b) near the origin represents $(BEDT-TTF)_2Cu(NCS)_2$.*

We confirmed further aspect of this by measuring $\sigma(T \to 0)$ for several specimens of $(Y_{1-x}Pr_x)Ba_2Cu_3O_7$ (ref. 24). In the Pr doped 123 system, the carrier density is changed by the substitution of Pr^{4+} for Y^{3+}, avoiding complications from the crystallographic ordering of oxygen. As shown in Fig . 4(b), the μSR results on the Pr doped 123 system give points in the T_c versus σ plot which agree very well with those of oxygen depleted 123 system. This is a further evidence that the transition temperature is determined by the carrier density in the CuO_2 plane, being insensitive to the method by which n_s is controlled. We should also note here that the carrier density n_s can be regarded as either the 3-dimensional or 2-dimensional density. Since the average distance between the CuO_2 planes is 6 ± 1 Å for all the different systems shown in Fig. 4, they share the approximately same conversion factor between the 2-d and 3-d densities.

We have to wait further theoretical development to fully understand the underlying mechanism which would give the results in Fig. 4. We can, however, note the following points:

(1) The BCS theory gives the transition temperature as $T_c = \hbar\omega_B exp(-1/VN(0))$ in the weak coupling limit, where $\hbar\omega_B$ is the energy scale of the boson which mediates the pairing (for conventional superconductors, it is the Debye frequency) and N(0) the density of states at the Fermi level. The assumption here is that the Fermi energy ϵ_F is much larger than $\hbar\omega_B$. In the above formula of T_c, we can not expect a large dependence of T_c on the carrier density, since $\hbar\omega_B$ does not depend on n_s, and N(0) in the 2-d noninteracting electron system is constant. Therefore, Fig. 4 encourages the development of theories different from the weak-coupling BCS theory with phonon as the mediating boson.

(2) The linear relation between T_c and n_s/m^* can be expected[25] if $\hbar\omega_B$ is larger than ϵ_F. In this case, the pre-exponential factor of the above equation of T_c will be ϵ_F instead of $h\omega_D$: ϵ_F of 2-d electron gas is proportional to n_s/m^*.

(3) The linear relation between T_c and n_s/m^* can also be expected in theories based on Bose-Einstein condensation[26]. The short coherence length of high-T_c superconductors motivates the development of this type of theory. Although the B-E condensation does not occur in 2-dimensional systems, slight three dimensional features could lead to a reasonable values of T_c comparable to those of the existing systems.

(4) The relation $T_c \propto n_s/m^*$ can also be explained in "anyon" theory[18].

(5) The saturation and suppression of T_c in the heavily doped region has been predicted by theories based on spin frustrations due to the hole at oxygen site, such as proposed by Aharony et al.[27]

Topic 4. The Spin Glass Region

In the magnetic and superconducting phase diagrams of the cuprate high-T_c systems, one finds non-superconducting and insulating parent compounds which exhibit antiferromagnetic order in the lightly-doped region and the superconducting phase without static magnetic order in the region with large hole concentrations. With increasing doping of carriers into the parent antiferromagnetic compounds, the Néel temperature is sharply reduced, and the spatial spin correlation of the ordered Cu moments become increasingly random. Around the compositional region between the antiferromagnetic and superconducting phases, it has been discussed that the magnetic properties may resemble those of canonical spin glasses such as CuMn or AuFe. Experimental information was, however, often inconclusive, due to possible spreads in the sample stoichiometry and to the limited time window of each probe for dynamical phenomena.

Recently we have performed[28] combined μSR and neutron scattering studies on a large single crystal $La_{1.94}Sr_{0.06}CuO_4$ (NTT # 10) which has a composition lying in this possible "spin glass" region. In neutron scattering, the dynamic slowing down of the spin fluctuations can be detected as an increase in the scattering intensity within the quasielastic energy window. We selected the quasielastic magnetic scattering intensity with an energy window of $\Delta E \sim 0.5meV$; the signal reflects spin fluctuations with frequencies lower than a cut-off frequency $\nu_c = \Delta E/\hbar \sim 10^{12}$ Hz. The scattering intensity was spread around the 2-dimensional "Bragg Rod" in reciprocal space. As shown in Fig. 5, the quasielastic intensity at the "Bragg Rod" increases with decreasing temperature below $T \sim 30$ K and saturates around $T \sim 20K$, indicating that dynamic spin fluctuations become slower than ν_c below $T = 20$ K.

We have also performed Zero and Longitudinal-Field μSR measurements on the same single crystal. As shown in Fig. 6, the zero-field relaxation functions below $T \sim 6K$ are characteristic of static random local fields. The time window for the "static" field in μSR in determined by $t > 1/\omega$ where ω is the muon spin precession frequency around the instantaneous value of the local field. In the present case, we are detecting the spin freezing with the time window of $t \sim 10^{-8}sec$; the freezing temperature T_g is around 6 K. The relaxation function observed at $T = 3.9$ K shows quick damping without significant oscillations, characteristic of widely random local fields and a highly disordered spatial spin structure[29]. At $T > 7$K, the muon spin relaxation rate is reduced via the "exchange narrowing" effect; in the language of magnetic resonance, due to dynamical spin fluctuations.

These combined neutron and μSR results demonstrate that: (1) the spin fluctuations slow down gradually in a wide temperature region above T_g, from $T = 20$ K with the characteristic time of $t \sim 10^{-12} sec$ to $T \sim 6$K with $t \sim 10^{-8} sec$; and (2) that the spatial spin structure is very random without 3-dimensional long range order. These features are important characteristics of spin glass systems, as have been demonstrated in the canonical spin glasses CuMn and AuFe (see ref. 29 - 31). Thus, the present results provide strong support to the picture that there is indeed a "spin glass" state in the border region of the antiferromagnetic and superconducting phases.

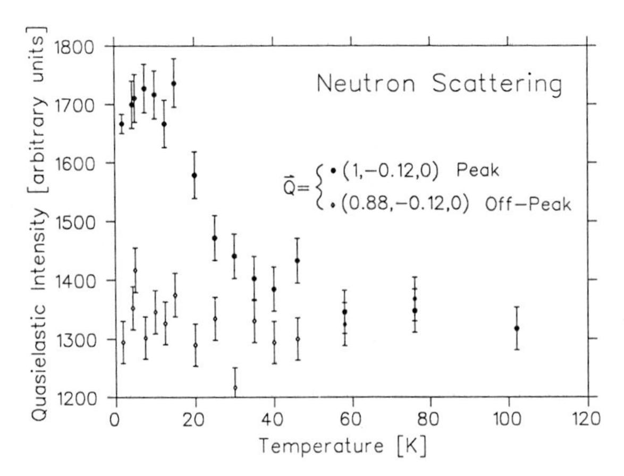

Fig. 5 Temperature dependence of the quasielastic neutron scattering intensity from a single crystal (NTT # 10) $(La_{1.94}Sr_{0.06})CuO_4$, measured at the "Bragg rod" (peak) in reciprocal space with the energy resolution of $\Delta E = 0.5 meV$ HWHM. The intensity at the off-peak position is also shown.

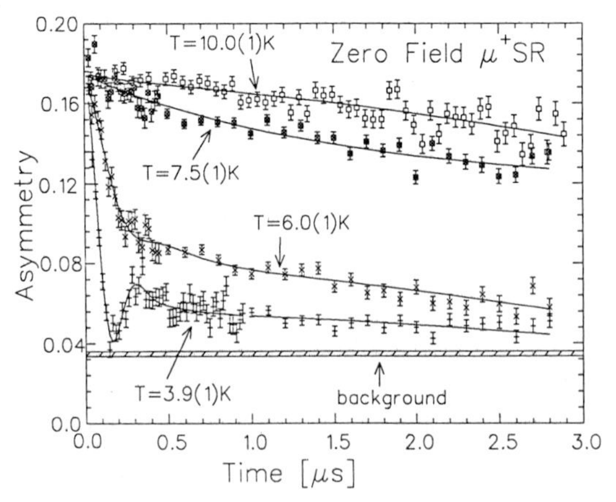

Fig. 6 Zero-field muon spin relaxation functions observed in NTT #10 crystal of $(La_{1.94}Sr_{0.06})CuO_4$. The increase of the relaxation rate reflects the slowing down of Cu spin fluctuations as the temperature approaches the freezing temperature $T_g \sim 6K$ from above. (see ref. 28 for details).

Topic 5. Extension to Organic and Heavy-Fermion Superconductors

The μSR measurements have also been applied to study static magnetic order and the penetration depth of other exotic superconducting materials: heavy-fermion and organic systems. In the heavy-fermion superconductors UPt_3 (ref. 32), $CeCu_{2.1}Si_2$ (ref. 33) and $(U,Th)Be_{13}$ (ref. 34), μSR measurements have revealed the static magnetic order with very small average ordered moments which coexists with superconductivity. Static order has not been detected in ZF-μSR measurements of pure UBe_{13}.

In order to measure the magnetic-field penetration depth λ in the superconducting state, we have performed TF-μSR measurements on the heavy fermion systems UBe_{13} (T_c = 1.0 K) and U_6Fe (T_c = 3.9 K) (see ref. 35). In UBe_{13}, we could not detect any significant change of the relaxation rate σ above and below T_c, which indicates that the penetration depth at $T = 0$ is longer than $\sim$ 8000 Å in this material. This is reasonable in view of the heavy effective mass m^* (indicated by $\gamma \sim 1600 mJ/mole\ deg^2$), as $\sigma \propto 1/\lambda^2 \propto n_s/m^*$. Then, we tried U_6Fe which has much lighter effective mass[35] ($\gamma \sim 25 mJ/mole\ deg^2$) than UBe_{13}. The temperature dependence of σ in U_6Fe, shown in Fig. 7, indicates that there are no anomalous zeros in the energy gap of this systems; suggesting s-wave superconducting pairing. The absolute value of $\sigma(T \rightarrow 0)$, which corresponds to $n_s/m^* = 2.7 \times 10^{20} cm^{-3}/m_e$ if the clean limit is assumed. We are now working on the measurements of URu_2Si_2.

Recently, we have extended the study to the organic superconductor $(BEDT-TTF)_2Cu(NCS)_2$. This system has the highest T_c (10.8 K in zero-field) among other organic superconductors[36]. The transport and H_{c2} measurements[36] indicate highly 2-dimensional electronic properties, as in the case in the high-T_c systems. Figure 8 shows our current data on the temperature dependence of the muon spin relaxation rate σ measured with single crystal specimens with the transverse external field H_{ext} = 3.1 kG applied perpendicular to the conductive plane (b-c plane). It should be noted that T_c is reduced to about 8 K in the field. At the moment, we do not have enough statistical accuracy of the data to

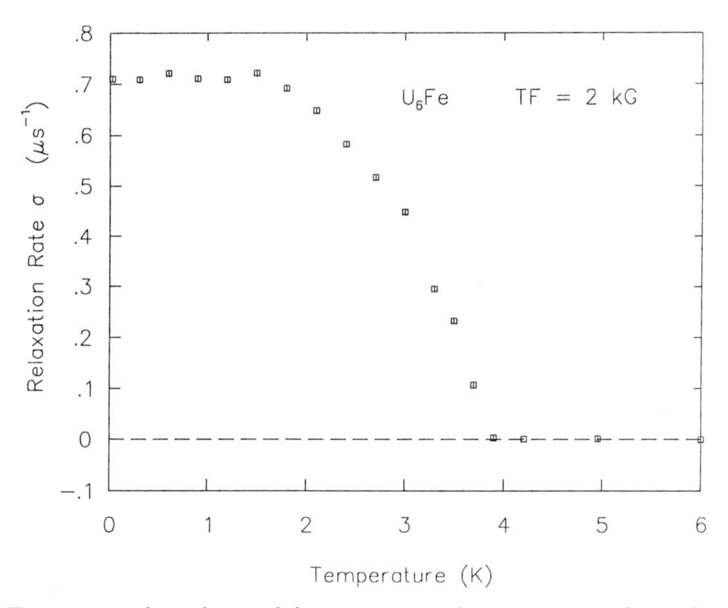

Fig. 7 Temperature dependence of the muon spin relaxation rate σ observed in a Uranium compound U_6Fe with the transverse external field of 2 kG. The flat temperature variation at low temperatures suggests s-wave pairing.

clearly distinguish the symmetry of superconducting pairing (more data in the low temperature region are required). After correction for the background relaxation contributions, the effect of the penetration depth at $T \rightarrow 0$ is found to be $\sigma \sim 0.14\mu sec^{-1}$, which corresponds to a ground state penetration depth of $\lambda \sim 7000$ Å and $n_s/m^* = 5.4 \times 10^{19} cm^{-3}/m_e$ in the clean limit. In the case of $(BEDT-TTF)_2Cu(NCS)_2$, it is likely that one carrier exists per molecule, each of which having a very large volume (844 Å^3 per molecule): the nominal carrier density is $n = 1.2 \times 10^{21} cm^{-3}$. Then the observed small relaxation rate (long penetration depth) can be understood to result partly from the low carrier density. Additional possible causes for the reduction of σ in the organic system include: (1) m^* larger than the bare mass m_e, (2) n_s smaller than n, and/or (3) deviation from the clean limit. It is also interesting to note that the results of σ and T_c for this organic system $(BEDT-TTF)_2Cu(NCS)_2$ gives a point on the σ versus T_c plot of Fig. 4 near the straight line shared by the cuprate high-T_c superconductors, as shown in Fig. 4(b). Further detailed μSR measurements on various organic superconductors are now underway.

Acknowledgement

We would like to thank J.F. Carolan, W.N. Hardy, J.R. Kempton, P. Mulhern, X. Li, B.X. Yang, H. Zhou, W.J. Kossler, X.Y. Yu, C.E. Stronach, A.W. Sleight, M.A. Subramanian, J. Gopalakrishnan, S. Uchida, H. Takagi, Gang Xiao, C.L. Chien, B.W. Statt, Y. Hidaka, T. Murakami, R.J. Birgeneau, T. Thurston, P. Gehring, K. Yamada, and many other scientists for collaboration on the μSR and neutron experiments described in this paper; V. Emery, R. Friedberg, B. Halperin, T.D. Lee, G. Shirane, F. Wilczek and M.K. Wu for stimulating discussions; NSF (DMR-89-13784, DMR-87-21455), USDOE (DE-AC02-CH00016, DE-FG03-86ER-45230), David and Lucile Packard Foundation, and NSERC of Canada for financial support.

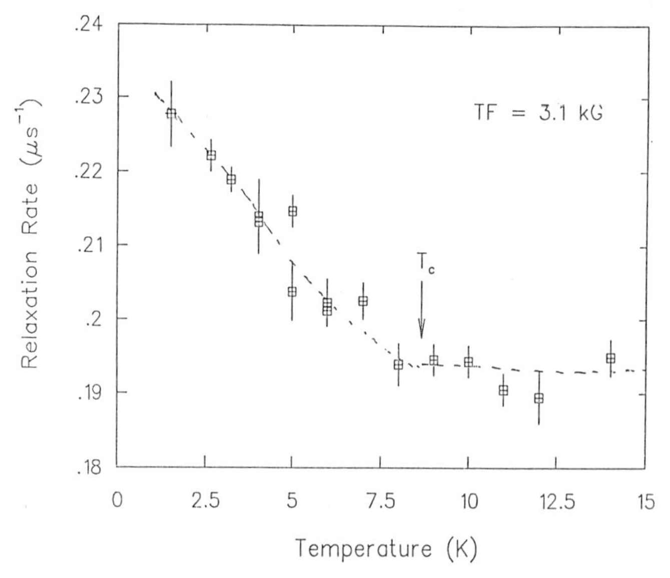

Fig. 8 *Temperature dependence of the muon spin relaxation rate σ observed in single crystal specimens of an organic superconductor $(BEDT-TTF)_2Cu(NCS)_2$ with the external field $H_{ext} = 3.1kG$ applied perpendicular to the conductive b-c plane. The low temperature relaxation rate ($\sigma = 0.14\mu sec^{-1}$ after correction for background relaxation) corresponds to the penetration depth of $\lambda \sim 7000$ Å . The broken line is guide to the eye.*

References

1. For historical development of μSR, see proceedings of four previous international conferences, Hyperfine Interactions 6 (1979); 8 (1981); 17 – 19 (1984); 31 (1986).
2. In addition to the present paper, there are three other μSR papers presented at this meeting by E.J. Ansaldo, J.I. Budnick, and R. De Renzi.
3. For earlier review papers of the Columbia-UBC collaboration studies at TRIUMF, see Y.J. Uemura *et al.*, Physica C162-164, 857 (1989); Y.J. Uemura *et al.*, J. Phys. (Paris) Colloq 49, C8-2087 (1988).
4. H. Keller, IBM J. Res. Develop. 33, 314 (1989).
5. B.J. Sternlieb *et al.*, Physics C162-164, 679 (1989).
6. See, for example, A. Schenck, *Muon Spin Rotation Spectroscopy*, Hilger, Bristol (1985).
7. G.M. Luke *et al*, unpublished.
8. L.P. Le *et al.*, unpublished.
9. J.A. Cape and J.M. Zimmerman, Phys. Rev. 153, 416 (1967).
10. Y.J. Uemura *et al.*, Phys. Rev. Lett. 59, 1045 (1987).
11. J.I. Budnick *et al.* , Europhys. Lett. 5, 647 (1988); D.R. Harshman *et al.* Phys. Rev. B38, 852 (1988).
12. J.H. Brewer *et al.* Phys. Rev. Lett. 60, 1073 (1988); N. Nishida *et al.*, J. Phys. Soc. Japan 57, 599 (1988).
13. B.J. Sternlieb *et al.*, Phys. Rev. B40, 11320 (1989).
14. G.M. Luke *et al.*, Nature 338, 49 (1989); Physica C162-164, 825 (1989).
15. Y. Tokura *et al.*, Nature 337, 345 (1989).
16. R.F. Kiefl *et al.*, Phys. Rev. Lett. 63, 2136 (1989).
17. H. Takagi *et al.*, Phys. Rev. B40, 2251 (1989); Physica C162-164, 1677 (1989).
18. R. Laughlin, Phys. Rev. Lett 60, 2677 (1988); Science 242, 525 (1988); B.I. Harperin *et al.*, Phys. Rev. B40, 8726 (1989).
19. The earliest reference on this can be found in W.J. Kossler *et al.*, Phys. Rev. B35, 7133 (1987).
20. R.F. Kiefl *et al.*, submitted to Phys. Rev. Lett.
21. P. Pincus *et al.*, Phys. Lett. 13, 31 (1964).
22. H. Takagi and S. Uchida, unpublished data.
23. Y.J. Uemura *et al.*, Phys. Rev. Lett. 62, 2317 (1989).
24. C.L. Seaman *et al.*, unpublished.
25. Z.Z. Wang *et al.* Phys. Rev. B36, 7222 (1987).
26. T.D. Lee, in *Festschrift in Honor of Luigi Radicati*, in press (1990).
27. A. Aharony *et al.*, Phys. Rev. Lett. 60, 1330 (1988).
28. B.J. Sternlieb *et al.*, submitted to Phys. Rev. B.
29. Y.J. Uemura *et al.*, Phys. Rev. B31, 546 (1985).
30. F. Mezei and A.P. Murani, J. Magn. Magn. Matrs. 14, 211 (1979).
31. Y.J. Uemura *et al.*, J. Appl. Phys. 57, 3401 (1985).
32. D.W. Cooke *et al.*, Hyperfine Interactions 31, 425 (1986).
33. Y.J. Uemura *et al.*, Phys. Rev. B39, 4726 (1989).
34. R.H. Heffner *et al.*, Phys. Rev. B40, 806 (1989).
35. Z. Fisk *et al.*, Physica C153-155, 1728 (1988).
36. K. Oshima *et al.*, Physica C153-155, 1148 (1988); for a review of organic superconductors, see T. Ishiguro, Physica C153-155, 1055 (1988).

RECENT RESULTS IN THE APPLICATION OF μSR TO THE STUDY OF MAGNETIC PROPERTIES OF HIGH-T_c OXIDES[1]

E. J. ANSALDO

Department of Physics
University of Saskatchewan
Saskatoon, Canada S7N0K3

ABSTRACT

An overview is given of recent muon spin rotation-relaxation (μSR) measurements of magnetic penetration depths, λ(T) (with emphasis on the Bi-Sr-Ca-Cu-O system) and magnetic ordering in its interplay with superconductivity for $La_{2-x}Sr_xCuO_{4-y}$ and $YBa_2Cu_3O_x$. Available results for the dependence of T_c on carrier concentration in Bi-Sr-Ca-Cu-O samples of different stoichiometries depart from the general trends found – and posited to be a universal property of the CuO_2 layers – in work on other high-T_c families. More work is required to separate extrinsic factors (vortex lattice morphology, phase inhomogeneities, defect structure) from intrinsic (mechanisms, electronic and crystalline structure) effects, before the significance of empirical correlations *vis a vis* theoretical explanations can be assessed. Results on magnetic ordering and fluctuations are reviewed in the context of their possible coexistence with superconductivity. It is found that existing μSR data, although yielding tantalizing results, do not provide a complete and unambiguous picture of the interaction between magnetic and superconducting order parameters.

INTRODUCTION

The μSR (muon spin rotation-relaxation) technique has proven of unique value in the determination of flux line lattice (FLL) properties and magnetic ordering (especially for slowly fluctuating electronic moments) in oxide superconductors and their precursors. Basically this is because the positive muon in interstitial lattice sites (close to oxygen ions) is a truly microscopic probe of internal field distributions in the oxides. Such fields are due to the FLL in the

[1] Results presented here obtained at TRIUMF (Canadian meson facility, Vancouver, B. C.) in collaboration with D. R. Harshman and G. Aeppli (AT&T Bell Labs.), and T.M Riseman and D. Williams (University of British Columbia), on samples prepared by the groups of B. Battlog,R.J. Cava and G. Espinosa (AT&T Bell Labs), and N. D. Spencer (Grace Corp.)

mixed state of type II superconductors and/or the effective internal fields due to electronic moments in magnetically ordered substances. In the latter case, the measurements can take place in zero applied field (ZF) or in longitudinal (quenching, LF) applied field so that the origin of the field distribution (e.g. antiferromagnetism *vs.* spin glass), and its dynamics may be assessed within the limits set by the probe (magnitudes to *ca.* 1T, fluctuation rates down to 10^{-12} s.). The technique is uniquely suited to the determination of coexistence with great sensitivity, but unfortunately without the site discrimination of NMR. Here we summarize (published) results on cuprate oxides obtained mostly on samples prepared at AT&T in experiments carried out using the μSR facility at the TRIUMF cyclotron, in the context of extensive similar research carried out by other groups, and with emphasis on the unsolved questions requiring more complete data sets than currently available.

FLL IN THE VORTEX STATE

As discussed in previous papers on $YBa_2Cu_3O_x$ and $La_{1.85}Sr_{0.15}CuO_4$ [1-6] and recent similar work on two-layer Bi-2212[7,8] and a one-layer Bi-2111 ($Bi_2(Sr,Ca)_2CuO_{6.15}$) sample,[9] the penetration depth λ is determined (assuming that the muons are randomly distributed in the vortex lattice) from the relaxation of the muon polarization in an external field (TF)such that the separation between the vortices is smaller than λ. The μ spin relaxation rate $\Lambda = 1/T_2$ is then independent of applied field and given by the second moment $M_2 = <|\Delta H|^2>^{1/2}$, of the microscopic field distribution. It is usual to fit the data for powder samples to a Gaussian relaxation function $e^{-(t/T_2)^2}$. Such approximation is inappropriate for single crystals, as discussed in the only single crystal experiment published so far (see discussion in References 5 and 10). The penetration depth is related to the relaxation time T_2 by,[10] $\lambda = \sqrt{0.043\phi_o\gamma_\mu T_2}$, where ϕ_o is the magnetic flux quantum and γ_μ the muon's gyromagnetic ratio. In the simple London picture the relaxation rate is then $\Lambda \propto n_s/m^*$, where n_s and m^* are the superconducting carrier density and effective mass, respectively. For large anisotropy, the directionally averaged λ^{powder} so obtained is dominated by the in-plane (or "hard") penetration depth, given by $\lambda_{ab}= \lambda^{powder}/1.23$.[12] The temperature dependence of λ may in principle be compared to theoretical predictions, i.e. strong *vs* weak coupling, dirty *vs* clean limit.[8] In practice, however, the comparison is doubtful because of extrinsic effects such as T-dependent pinning strength, FLL thermal motion[11], different defect structures for samples of similar composition, *etc.*, i.e. heuristic factors affecting the topology of the FLL as a function of temperature. The above considerations are illustrated in Figures 1 and 2, for the $YBa_2Cu_3O_x$ samples of Reference 6 and for a (BiPb)-2223 sample,[13] respectively. The latter also shows the sensitivity of μSR to flux pinning effects, observed in zero field cooled measurements. For Bi-Sr-Ca-Cu-O , unlike the $YBa_2Cu_3O_x$ case, the reversible effects extend well below T_c [14]. Such pinning-depinning ("melting") effects are the first important difference between the two systems, but have not been investigated in detail for the BiPb-2223 or Tl-2223 systems so far. Results for the temperature dependence of λ below 20K, however, are reproducible and always show only a small variation of with temperature, compatible with conventional (nodeless order parameter) s-wave pairing in all cases. It should be noted here that the power law T-dependence of the μSR T_2 has been observed recently for the polar heavy electron case, UPt_3.[15]

The μSR results have thus been useful mainly to study the low temperature behaviour and magnitude of the penetration depth, as illustrated by Figures 1 and 2, and to establish an empirical T_c *vs* $\lambda(0)$ behaviour, i.e. the dependence of T_c on the ratio n/m^* in the simple isotropic one band clean limit London picture. A comparison of our results with other (published) data on $La_{2-x}Sr_xCuO_{4-y}$, $YBa_2Cu_3O_x$, and (Bi,Pb,Tl)-Sr-Ca-Cu-O samples,[1-9] and including the BiPb-2223 case of Figure 2 is given in Figure 3. In the $YBa_2Cu_3O_x$ system

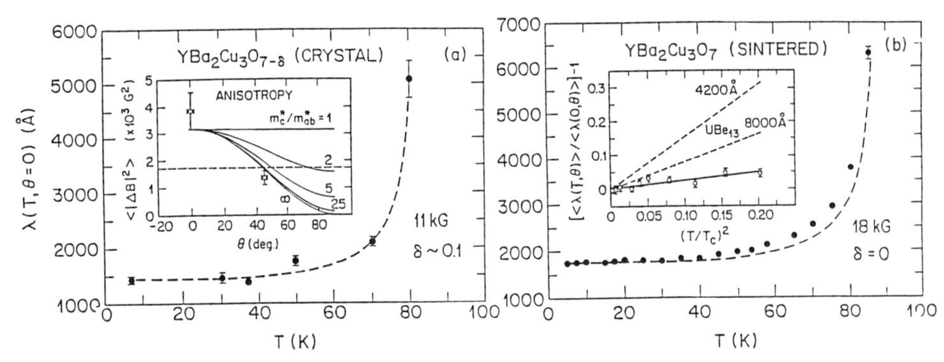

FIGURE 1. Temperature dependence of λ for $YBa_2Cu_3O_x$, crystal and sintered samples. The insets show determination of the effective mass anisotropy from the crystal data and a comparison with power law behaviour for the p-wave heavy electron case UBe_{13}, respectively. The dashed lines are fits of the data to the two fluid relation $\lambda(T) = \lambda(0)[1-(T/T_c)^4]^{-1/2}$ yielding a hard penetration depth $\lambda_{ab}(0)$=1415 Å . (From Reference 5.) Oxygen deficient samples show a greater departure from the two fluid temperature dependence.

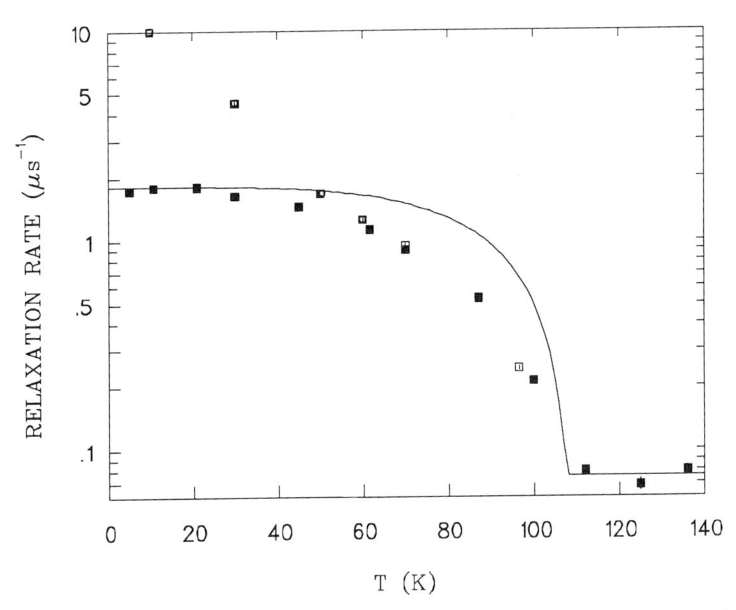

FIGURE 2. Temperature dependence of the relaxation rate for a sintered sample of $Bi_{1.8}Pb_{0.2}Sr_2Ca_2Cu_3O_{10}$. Field cooled data (filled symbols) compared to the two-fluid model (solid line) yield $\lambda_{ab}(0)$=1720Å. The open symbol data points were obtained after zero field cooling, and show that the onset of reversibility (FLL mobility) is at T=70 K for this sample in the 0.4 T applied field. The relaxation above T_c is due to interaction with nuclear moments.

T_c first increases directly proportional to $\Lambda(0)$ as x increases from 6.65, but then saturates for x≥6.85, even as $\Lambda(0)$ continues to increase. The linear trend resumes with the 3-layer $(BiPb)_2Sr_2Ca_2Cu_3O_x$ cases to saturate again for Tl-2223 samples.

In contrast to $YBa_2Cu_3O_x$, as is evident in Figure 3, changing stoichiometry in the Bi-Sr-Ca-Cu-O system has a very different effect. In this case T_c appears to decrease slightly with increasing carrier concentration, eventually crossing the $YBa_2Cu_3O_x$ line, and the measured n_s/m^* values are generally lower than for $YBa_2Cu_3O_x$. The comparatively lower value of the measured penetration depth is in agreement with a high field magnetization measurement,[16] and in accord with the observation of superconducting fluctuations around T_c for Bi-Sr-Ca-Cu-O samples.[17] Notice also that $\Lambda(0)$ for the Bi-2111 sample of Reference 9 corresponds to its one-layer nature, but its T_c is similar to values for the bilayer samples, blurring the distinction between one- and two-layer cases. The T_c *vs* $\Lambda(0)$ trend is in qualitative agreement with the results of Morris *et al.* and Tallon *et al.*,[18] who showed that T_c decreases with increasing carrier concentration, as the oxygen content is changed in a controlled fashion, in a manner similar to the other systems in the "heavy doping" limit. There is a dramatic difference between the Bi-2212 case and the Tl-2212 and 2223 samples obtained by Pb/Tl substitutions. For example, a sample of $Tl_{0.5}Pb_{0.5}Sr_2CaCu_2O_7$ is shown in Reference 7 to have $\Lambda(0) \sim 2.4\mu s^{-1}$, about 2.5 times higher than for Bi-2212 with a similar T_c of ~74K.[7]

MAGNETIC ORDERING

The 3D magnetic ordering of La_2CuO_{4-y} and $La_{2-x}Sr_xCuO_{4-y}$ was elucidated early by μSR in conjunction with neutron scattering experiments.[19,20] It should be noted that the muon, as a local probe, responds to short range ordering (and is sensitive to small moments) thus complementing staggered magnetization neutron scattering measurements. Figure 4 illustrates the main result of our study of sintered $La_{2-x}Sr_xCuO_{4-y}$ samples[20] with $x < 0.06$.

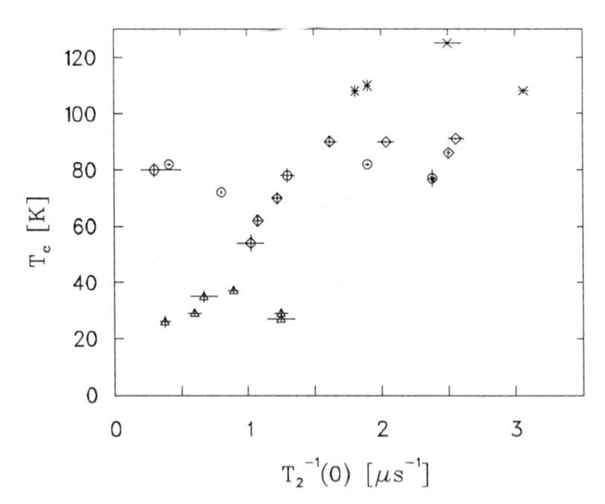

FIGURE 3. Plot of T_c *vs* extrapolated (T→0) μSR relaxation rate $\Lambda(0)$ for $La_{2-x}Sr_xCuO_{4-y}$ (triangles), $YBa_2Cu_3O_x$ (diamonds), Bi-Sr-Ca-Cu-O (circles), and (Bi-Pb-Tl)-2223 (asterisks)[1–9]. Notice that, except for x =0.15 and 0.20, samples of the $La_{2-x}Sr_xCuO_{4-y}$ system as well as $YBa_2Cu_3O_x$ samples with x below *ca.* 0.5 also display static magnetic ordering effects (*vide infra*). The 60K "plateau" region has not been investigated in detail. Data were obtained by the different authors in a variety of fields, from 0.1 to 1.8 T. See also Luke *et al.*, these proceedings.

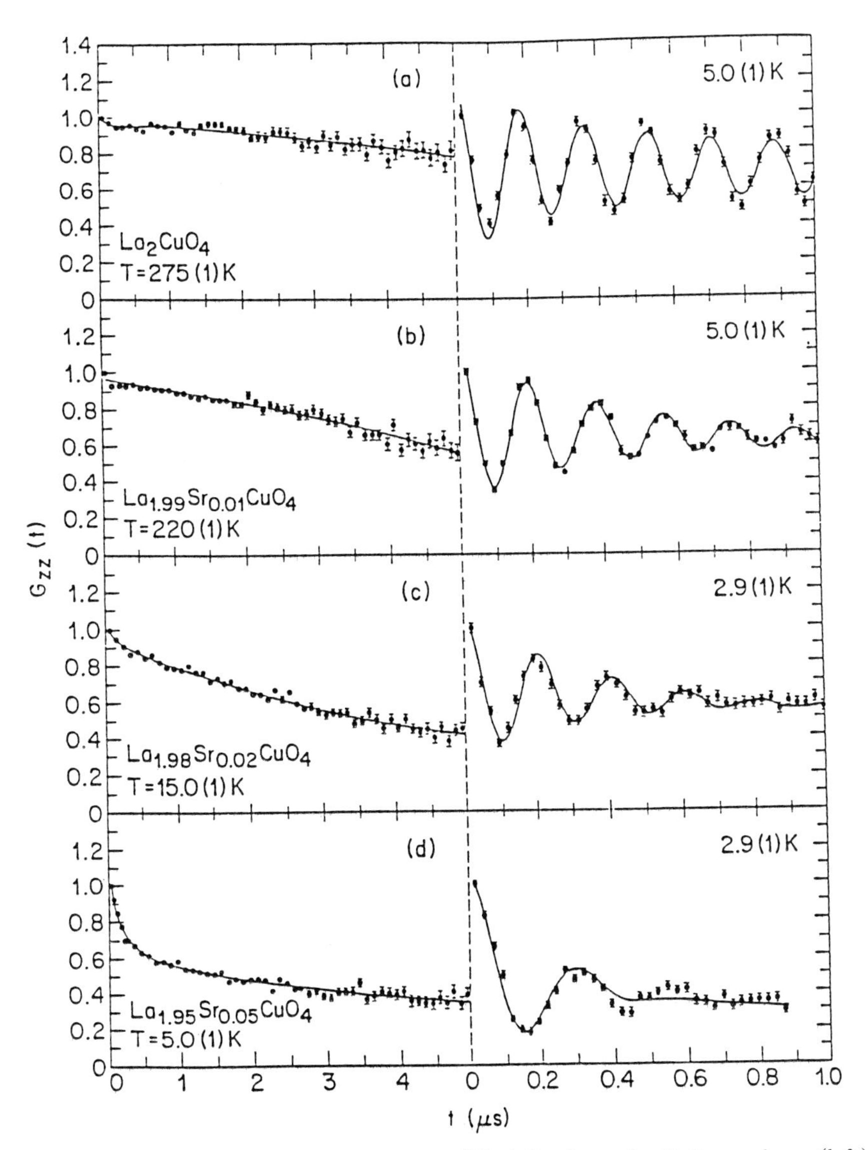

Figure 4. Example of ZF relaxation functions (G_{zz}) for $La_{2-x}Sr_xCuO_{4-y}$ above (left) and below (right) the "freezing" temperature. This shows the increase of the relaxation of the oscillatory signal due to inhomogeous broadening as the order becomes short range (evolving into spin glass type of magnetic order) as the Sr concentration is increased. Dynamical effects on a fraction of the muons above the transition result in the long time tail of G_{zz}. (From Reference 19.)

The data show that random freezing of fixed moments supplants the Nèel order as doping is increased, as probed by muons in interstitial low symmetry locations. Evidence was also found for the frustration effects that produce ferromagnetically aligned in-plane pairs of copper spins, for a fraction of the muons residing in high-symmetry sites with zero net field when $x = 0$. In such case dynamic effects due to fluctuations result in a long-time component of the relaxation function, which is not quenched in longitudinal fields below 0.2T, and which should be searched for on careful measurements (high satistics, low backgrounds, tests with longitudinal fields) in all samples.

Effects of the static ordering were observed in later work for higher values of the Sr concentration, in the superconducting range, which has been interpreted variously as indication of the intrinsic connection between the magnetic ordering and the superconductivity[21] or as evidence (reinforced by Meissner fraction reduction effects) for more prosaic effects of electronic inhomogeneities due to phase-miscibilty problems.[22] Those experiments were not sensitive enough to detect the contribution of a possible fraction of the muons experiencing dynamic fields or located in a superconducting paramagnetic regions. Later evidence, from ZF experiments below 90 mK, is that the static magnetic freezing is totally absent for $x = 0.15$[23], in agreement with the TF data of Reference 22.

Similar ordering phenomena, i.e. the appearance of Nèel ordering evolving into random, glassy magnetic order have also been detected in $YBa_2Cu_3O_x$ for x varying from 6.0 to 6.5, with definite spin glass ZF and LF relaxation functions around $x = 6.45$.[24] In this case two or more muon sites are apparent in the data. No sign of magnetic ordering was found for fully oxygenated $YBa_2Cu_3O_x$ ($x = 7.0$) at 20 mK.[23] (see Figure 5.)

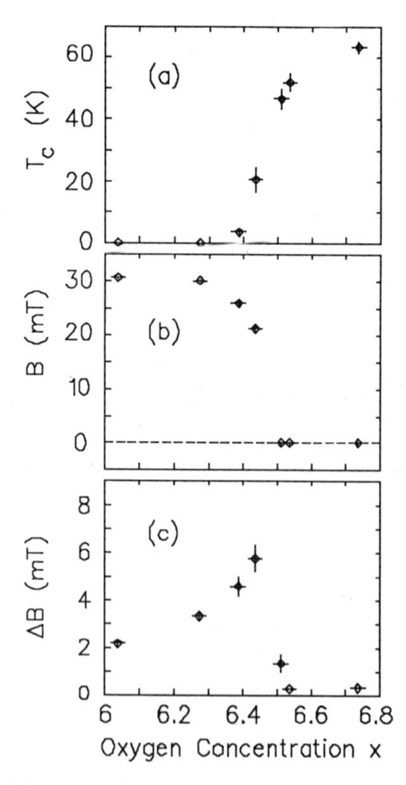

Figure 5. Evolution of the critical temperature (a), average field (b), and width (c, line broadening) of the effective field distribution at the main muon site in $YBa_2Cu_3O_x$ (From Kiefl *et al.*, Reference 23.)

MIXED CASES AND COEXISTENCE

Recent experiments relate to the possible coexistence of static magnetic order or spin freezing with superconductivity. As discussed above, and simply stated, ZF-μSR experiments in cuprate oxides exhibit signals that are consistent with quasi-static magnetism in samples that are also superconducting. For example, $La_{2-x}Sr_xCuO_{4-y}$ samples with $0.08 < x < 0.15$[21] and $YBa_2Cu_3O_x$ with $6.4 < x < 6.5$, as shown in Figure 5.[23,24] In the $La_{2-x}Sr_xCuO_{4-y}$ system the effects appearing as coexistence have been attributed to subtle effects of the electronic inhomogeneity, i.e. samples with composition away from the optimum $x = 0.15$ consist in reality of a mixture of phases, as evidenced also by their reduced Meissner fraction.[22] It should be noted, however, that this is not a case of a simple separation into two distinct phases. Such case can be determined fairly unambiguosly by a combination of ZF and TF measurements, as done for $La_2CuO_{4+\delta}$.[25] In that case the two distinct phases are clearly stoichiometric La_2CuO_4 and a oxygen-rich supercoducting phase (60% by volume with a 30% Meissner fraction). By contrast, in the superconducting region of $La_{2-x}Sr_xCuO_{4-y}$, separation of the μSR signal into components has not been carried out in detail, and may indeed not be possible at all if instead of a segregated phase the concentration inhomogeneities are on a microscopic scale and result in both glassy and dynamical ordering effects, as for samples in the spin glass non superconducting region. From the more complete set of measurements of Brewer *et al.*[24] on oxygen deficient $YBa_2Cu_3O_x$, it appears as if a fraction of the samples lose their superconductivity as the temperature is lowered (for x in the region where ZF spin glass- like relaxation functions develop), as if the onset of static magnetic order actually quenches the superconductivity in a fraction of the sample volume which itself is a function of temperature below T_c. It is interesting to speculate whether the coupling of magnetic and superconducting order parameters is intrinsic to the oxide materials in terms of the disorder associated with doping (i.e. a function of hole concentration) in the sense that the electronic (composition) inhomogeneities giving rise to measured effects are of spatial extent comparable to interlayer spacings, or to the coherence length and spin correlation lengths found in neutron scattering (interhole distances). There may be pair-breaking effects due to the ferromagnetic coupling or magnetic fluctuations, as well as superconducting proximity effects.[26] It may be rather a question of semantics whether such coexistence is considered "intrinsic", but it may be necessary at a fundamental level, for in some theories phase separations between hole rich regions may arise naturally (Reference 26 and Emery *et al.*, these proceedings.)

CONCLUSIONS

A remarkable correlation has been found between T_c and the ratio of (superconducting) carrier concentration to effective mass for a selected group of sintered copper oxide high-T_c samples, but with substantial departure for one- and two-layers Bi-2111 and Bi-2212 samples.[7] Such effects should be studied in detail, since the Bi-oxide samples have shown other differences, such as flux pinning and superconductive fluctuation behaviour, in addition to absence of magnetic ordering in the undoped compounds. In addition, other one-layer families also obtain that display high T_c's comparable to the two-layer cases and that are variable over a wide range, but which have not been studied by μSR so far (or results have not been published.) For the $La_{2-x}Sr_xCuO_{4-y}$ and $YBa_2Cu_3O_x$ families the dropoff in T_c as the carrier concentration is reduced may be related to the onset of microscopic phase separation and the coupling between the order parameters in that region of composition, rather than an intrinsic feature of the superconducting coupling alone. A common feature remaining is the broad maximum in T_c in such a diagram, whose similarities and differences between families may have a bearing on the detailed similarities and differences in electronic structure (density of states, interlayer coupling, *etc.*) It should also be remarked that no detailed heuristic comparison has been carried out so far between the FLL obtained for single crystals and the bulk of the measurements on sintered samples, except for the $YBa_2Cu_3O_x$ case of Reference 5 (which is affected by background subtraction problems), and on aligned powder samples.[27] Single crystal μSR experiments should clarify the important questions of how well the actual field distribution obtained compares to the ideal Abrikosov lattice and the effects of anisotropy on the temperature dependence of the penetration depth measured, in addition to pinning behaviour of the FLL. As for magnetic ordering effects, a dichotomy exists between μSR and neutron scattering experiments (for summaries see the articles by Rossat-Mignod, Shirane, and Tranquada in these proceedings): most of the μSR results outlined above were obtained on sintered samples, while the neutron scattering experiments showing magnetic fluctuations are carried out on larger size crystals, which may be rather more inhomogeneous in composition than the sintered samples. As discussed above, it is not clear at present which measured effects are a result purely of sample dependent inhomogeneities, indeed to what extent the inhomogeneities are intrinsic and essential. In any case, the interesting magnetic fluctuations detected by neutron scattering and NMR[28] are not apparent in the μSR data, although so far no careful detailed measurements were carried out to search for subtle dynamical effects in the μSR data. The most obvious conclusion is that the fluctuations are too rapid to be detectable by the muons, but the fact remains that both types of experiments were carried out on substantially different samples. Current and planned experiments by μSR on cuprate oxides are designed to address these, and the above discussed FLL questions in greater detail.

REFERENCES

1. G. Aeppli *et al.*, *Phys. Rev.* **B35**, 7129 (1987).
2. F. N. Gygax *et al.*, *Europhys. Lett.* **4**, 473 (1987).
3. D. R. Harshman *et al.*, *Phys. Rev.* **B36**, 2386 (1987).
4. W. J. Kossler *et al.*, *Phys. Rev.* **B35**, 7133 (1987).
5. D. R. Harshman *et al.*, *Phys. Rev.* **B39**, 851 (1989).
6. Y. J. Uemura *et al.*, *Phys. Rev. Lett.***62**, 2317 (1989), and references therein.
7. E. J. Ansaldo *et al.*, *M^2S-HTSC, Physica C*, in press, (1989).
8. R. Lichti *et al.*, *J. Appl. Phys.* **54**, 2361 (1989).
9. P. Birrer *et al.*, *Physica C* **158**, 230 (1989).
10. E. H. Brandt, *Phys. Rev.* **B37**, 2349 (1988).
11. P. L. Gammel *et al.*, *Phys. Rev. Lett.* **61**, 1666 (1988).
12. W. Barford and J. Gunn, *Physica C* **153-155**, 691 (1988).
13. N. D. Spencer *et al.*, *J.J.A.P.* **28**, L1564, (1989).
14. B. Pumpin *et al.*, *Z. Phys. B72*, 175 (1988),
    and B. Sternlieb *et al.*, *M^2S-HTSC, Physica C* in press, (1989).
15. C. Broholm *et al.*, *March APS meeting, B.A.P.S.* to be published, (1990).
16. S. Mitra *et al.*, *Phys. Rev.* **B40**, 2674, (1989).
17. W. Schnelle *et al.*, *Physica C* **161**, 123, (1989).
18. D. E. Morris *et al.*, *Phys. Rev.* **B39**, 6612 (1989),
    and J. L. Tallon *et al.*, *Physica C* **158**, 247 (1989).
19. Y. J. Uemura *et al.*, *Physica C* **153-155**, 768 (1988).
20. D. R. Harshman *et al.*, *Phys. Rev.* **B39**, 851 (1989).
21. A. Weidinger *et al.*, *Phys. Rev. Lett.* **62**, 102 (1989).
22. D. R. Harshman *et al.*, *Phys. Rev. Lett.* **63**, 1187 (1989).
23. R. Kiefl *et al.*, *Phys. Rev. Lett.* **63**, 2136 (1989).
24. J. H. Brewer *et al.*, *Phys. Rev. Lett* **60**, 1073 (1989),
    and *M^2S-HTSC Physica C* in press (1989).
25. E. J. Ansaldo *et al. Phys. Rev.* **B40**, 2555 (1989).
26. R. S. Markiewicz and B. G. Giessen, *Physica* C160, 497 (1989), and
    R. Dupree *et al.*, *Phys. Rev. Lett.* **63**, 688 (1989).
27. T. M. Riseman *et al. Physica C***153-155**, (1988), and
    M^2S-HTSC Physica C in press (1989).
28. P. C. Hammel *et al.*, *Phys. Rev. Lett.* **63**, 1992 (1989).

ON THE PHASE DIAGRAM OF BISMUTH BASED SUPERCONDUCTORS

R. De Renzi, G. Guidi, C. Bucci, R. Tedeschi

Dipartimento di Fisica
Università di Parma, I-43100 Parma, Italy

and G. Calestani

Istituto di Strutturistica Chimica
Università di Parma, I-43100 Parma, Italy

Introduction

μSR has proven an ideal tool to probe magnetic ordering, thanks to the fact that muons can be implanted in virtually any compound and that they can most often be considered passive probes of the intrinsic behaviour of their hosts. In the study which we present here we have addressed the issue of characterizing the basic magnetic properties of the 2-2-1-2 materials of the family of the $Bi_2Sr_2CaCu_2O_8$ superconductor. The primary aim of this work is to ascertain the analogies between this and the other generations of high T_c materials (in particular $La_{2-x}Sr_xCuO_4$ and $YBa_2Cu_3O_{7-y}$).

The first studies on this topic[1,2,3] already recognized the existence of an antiferromagnetic end member also for the 2-2-1-2 family, obtained by substituting a trivalent cation, like Y, for the divalent Ca in the original chemical formula. This establishied the universality of the characteristic phase diagram of these cuprous perovskites, in which a true antiferromagnetic order of the copper spins seems to be antagonistic with respect to the superconducting properties of the materials, although strong magnetic correlations seem to survive well inside the superconducting phases. In this line a determination of the phase diagram of the 2-2-1-2 compounds was required.

A natural estension of this approach has been the attempt to identify an analog of the so-called electron superconductors[4], $Nd_{2-x}Ce_xCuO_4$, which we pursued by looking for a similarly doped compound, namely $Bi_2Sr_2Nd_{1-x}Ce_xCu_2O_{8+\delta}$. Although

Dynamics of Magnetic Fluctuations in High-Temperature Superconductors
Edited by G. Reiter *et al.*, Plenum Press, New York, 1991

a superconducting phase could not be reached, indications are found of a disruption of the antiferromagnetic order which is brough about by Ce doping.

Another topic addressed in this paper is that of the temperature dependence of the spontaneous magnetisation for these materials, which are generally believed to be well described by a two dimensional Heisenberg model. Contraddictory evidence can be extracted from different experimental techniques. The peculiar shape of the magnetisation curves obtained from μSR experiments is discussed in this context.

In the first section of this paper a brief account is given on the sample preparation and characterisation. An outline of the μSR zero field measurement is presented in the second section, together with a description of the μSR results for a representative sample. The origin of the local magnetic field experienced by the muons is briefly discussed therein. The third section is devoted to the temperature dependence of the spontaneous magnetisation and to the phase diagram for the 2-2-1-2 materials, as obtained from our experiments.

The samples

Fine powders of $Bi_2Sr_2Y_{1-x}Ca_xCu_2O_{8+\delta}$ were prepared by the normal method described elsewhere[5]. Their characterisation was already discussed in reference 1, where the dependence of the c lattice parameter on x is taken as indicating selective substitution Y at the Ca site. Here we present results on a number of samples with $0.0 < x < 0.35$. The mixture of divalent and trivalent cations indirectly controls the concentration of holes in the CuO_2 planes.

Electrons could be similarly forced onto these planes by a suitable tetravalent cation substitution. In this line Ce doping was attempted directly on $Bi_2Sr_2YCu_2$-$O_{8+\delta}$. However the 2-2-1-2 phase could not be obtained even for quite low dopant concentrations, indicating that Ce catalyzes the formation of a different crystal structure[6], as is revealed by X-ray powder diffraction. A more favourable situation was found for the composition $Bi_2Sr_2NdCu_2O_{8+\delta}$, in which a restricted range of Ce concentrations could be forced into the Nd planes. Nominal compositions of the μSR samples were $Bi_2Sr_2Nd_{1-x}Ce_xCu_2O_{8+\delta}$, with $x = 0.0$, 0.1, 0.2 and 0.3, although, unlike the (Y,Ca) case, the X-ray diffraction patterns for these samples indicate the presence of impurity phases, namely a very small fraction of $Bi_2Sr_2CuO_6$ and minute traces of CeO_2, both non magnetic. Therefore the precise Ce content of each specimen is not known. However we have consistent indication that the actual concentration is proportional to x.

The μSR measurements

The experiments were performed on the ISIS μSR facility[7], at the Rutherford Appleton Laboratory, using the Dizital equipment.

In a Muon Spin Rotation experiment positive muons are implanted in a sample, where they come to rest with no appreciable loss of their initial spin polarisation, *P(t=0)*. The time evolution of this quantity, *P(t)*, can be measured experimentally thanks to parity violation in the muon decay,

$$\mu^+ \rightarrow e^+ + \nu_\mu + \nu_e$$

which results in a distribution of the emitted positrons peaked in the direction of the muon spin. As a consequence the rate of counts in a given positron detector, forming an angle θ with the direction of the initial polarisation of the muons, is given by:

$$N(t) = N_\circ e^{-t/\tau_\mu}(1 + \mathcal{A}\cos\theta P(t))$$

where $\tau_\mu = 2197$ ns is the muon lifetime, $\mathcal{A}$ is an asymmetry parameter and $N_\circ$ a normalisation constant.

In a zero field μSR experiment the time evolution of the muon spin is determined uniquely by the local field at the site of implantation. If this field has a well defined value, as is the case for a magnetically ordered material, the muon Larmor precession around it causes the polarisation to oscillate,

$$P(t) = G(t)cos(2\pi\gamma_\mu \mathrm{B}_\mu t)$$

where $\gamma_\mu = 135.5$ MHz/Tesla is the muon magnetogyric ratio, B_μ is the local field and $G(t)$ is the spin relaxation function; in this case $P(t)$ can be easily obtained from two identical detectors, a forward one, placed at $\theta = 0°$, with rate N_F, and a backward one, at $\theta = 180°$, with rate N_B, by taking the ratio:

$$\frac{N_F - N_B}{N_F + N_B} = AP(t)$$

If more than a stopping site is present $P(t)$ will result in a superposition of different oscillating terms.

The powder nature of our samples does not complicate too much this simple picture. The precession frequency, proportional to $|\mathbf{B}_\mu|$, does not depend on the orientation of the different cristallites, which determines only the apertures of the precession cones. The powder average of these cones will just result in the appearence of a non precessing term, due to the components of $\mathbf{B}_\mu$ parallel to the initial muon polarisation, and in a reduction of the oscillation amplitude.

Figure 1 shows several polarization plots, obtained for a sample of $Bi_2Sr_2Y_{1-x}$-$Ca_xCu_2O_{8+\delta}$ with $x = 0.1$ at different temperatures, together with their best fit. The presence of oscillations indicates the existence of ordered magnetic moments in the

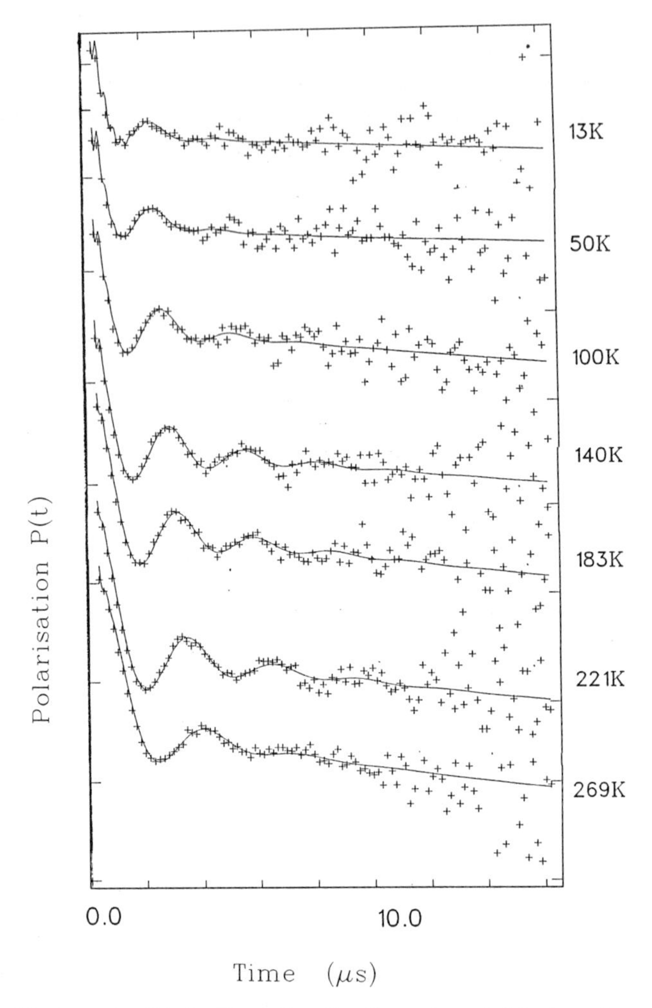

Figure 1 - *μSR data for* $Bi_2Sr_2Y_{1-x}Ca_xCu_2O_{8+\delta}$ $(x = 0.1)$ *at various temperatures. The solid line is a three component fit with a high frequency oscillation, a low frequency one and a non oscillating term.*

material. In this representative case three components of $P(t)$ are fitted at all temperatures, although they are not always easily seen by eye, as their relative amplitudes and decay rates are changing with temperature. This pattern has been already discussed in previuos works[1,2,3], where the occurrence of antiferromagnetic ordering in the material where Ca is fully substituted by Y was first reported.

We shall summarize here a few findings which are common to all the materials of this family. Two muon stopping sites are clearly identified by these measurements, and more[2] are present whose signal fall outside the passband of our spectrometer[1]. The abundance of stopping sites could well be expected in view of the large unit cell of $Bi_2Sr_2CaCu_2O_8$, where many interstitials can accomodate the positive muon. The value of the local fields B_μ, extrapolated to T = 0 K, are of the order of 3 and 30 mTesla, respectively, for the two observed components. The assignment of these values to a site in the lattice and the origin of B_μ are two strictly connected issues which must be tackled in order to correlate the experimental observation to the spontaneous magnetisation.

The first consideration which is due is that muons are known[8] to be at rest up to room temperature in these perovskites, so that their diffusional dynamics is not interfering with the magnetic interactions which they probe.

The dominant contribution to B_μ is the dipolar field from the ordered Cu moments. This field can be calculated site by site on the basis of the known crystal structure, if one assumes a magnetic structure (see inset on figure 2) and a copper magnetic moment consistent with experimental results from the other copper perovskites.

A selection of the values of B_μ, calculated along a few lines which run parallel to the c axis, is plotted in figure 2. These lines cross probable muon sites in the lattice (where empty space, the electrostatic potential and the vicinity to an oxygen ion are the three criteria adopted to search for such sites). A number of possible 30 mTesla sites are thus identified close to the oxygen belonging to the SrO layers – including a position (0.5,0.0,0.625), referred to the tetragonal pseudo-cell of constants a = 3.6 Å, b = 3.6 Å, c = 30.9 Å, whose analog was already considered a likely candidate for $YBa_2Cu_3O_{7-y}$[9]. A value of B_μ close to 3 mTesla is found for at least two sites, (0.25,0.25,0.7) and (0.5,0.0,0.7), both within the BiO bilayers.

In what follows we shall concentrate mostly on the lower field component. The determination of its muon site must be considered only indicative, as the level of approximation in this calculation is quite rough (the copper moment - $\mu_{Cu} = 0.5\mu_B$ - is only inferred from that of similar compounds). Furthermore a well known modulation of the orthorombic structure[5] induces particularly large displacements of the Bi atoms and of their oxygens; the muon is probably forming a muoxyl bond with one of these oxygens and its precise position will depend on the fine details of this modulation.

However the main point in our dipolar calculation is to show that the 3 mTesla site is very likely far away from the copper planes, so that we are allowed to neglect transferred hyperfine couplings to the copper electronic moments. In this instance, and as long as the magnetic order has correlation lengths in excess of 10-15 Å and correlation times larger than a few μs, the measured value of $B_\mu(T)$ is directly proportional to the spontaneous staggered magnetisation, $M_s(T)$.

Figure 3 shows the temperature dependence of B_μ for the x = 0.1 case of figure 1. Similar plots can be obtained for other compositions. The other parameters of the data fit besides B_μ, namely the muon asymmetry, proportional to the fraction A of the total implanted muons precessing around the lower field, and the decay rate λ of the oscillation, are plotted in the inset.

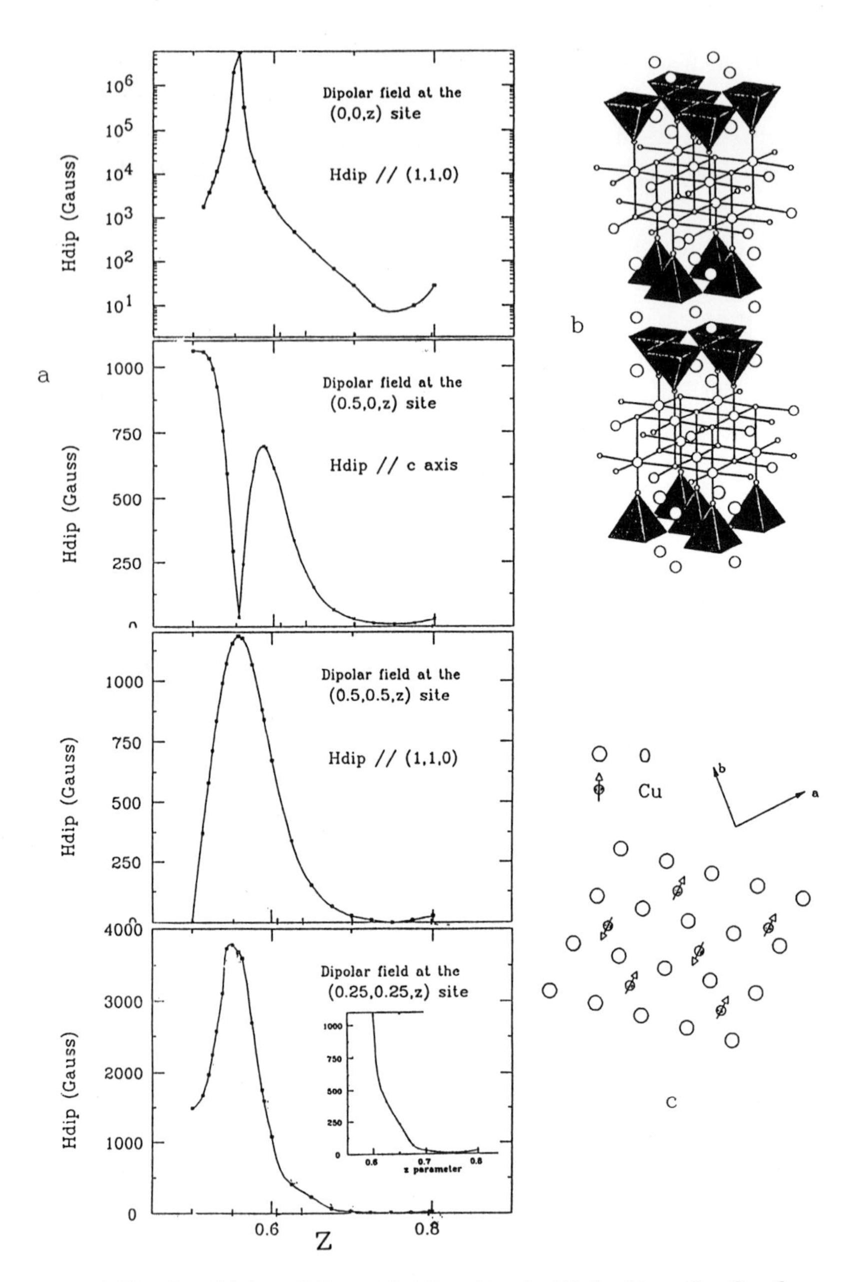

Figure 2 - *a) Dipolar field at different lattice sites in* $Bi_2Sr_2Y_{1-x}Ca_xCu_2O_{8+\delta}$, *along lines parallel to the c axis. b) The lattice. c) The assumed ordering of* $1/2\mu_B$ *copper moments in the* CuO_2 *planes.*

Problems could arise from the expected strong dependence of the Nèel temperature T_N on composition. If we had very inhomogeneous samples we would be measuring some average of $M_s(T)$ over different Ca concentrations. In this case, however, we would expect to see a gradual reduction of A with increasing temperature, proportional to the volume fraction of the sample already in the paramagnetic phase (the non precessing component should correspondingly increase). We take the fact that A

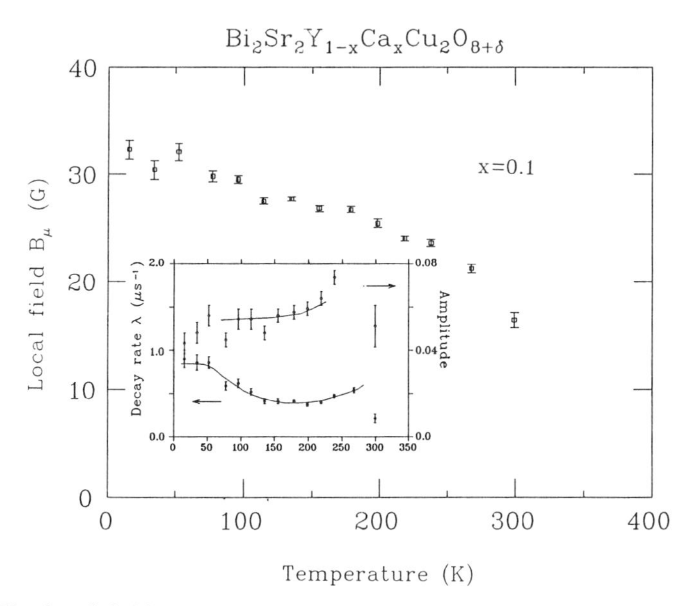

Figure 3 - *The local field at the low field muon site, extracted from the fit of figure 1. This quantity is proportional to $M_s(T)$. The inset shows the amplitude and decay rate of the fitted oscillation.*

is non decreasing up to room temperature to indicate that our samples are relativaly homogeneous in composition.

While the increase of λ above 250 K is probably related to the incipient muon diffusion, its jump below 50 K might have to do with three-dimensional disordering of the copper planes, induced by freezing of the hole gas, which has been invoked also for the interpretation of neutron scattering data [10]. The disorder in the stacking of the CuO_2 bilayers, which shows up as a reduction of the magnetic peak amplitudes in neutron diffraction, increases the mean square value of the field at the muon site, thus determining a faster relaxation. No significant variation is observed in the average field value within the same temperature interval, although it must be noted that here the faster damping of the wave increases considerably the error bars on the fitted field value. The proposed explanation of the low temperature rise in λ is, for the moment, only speculative; on the other hand in the intermediate temperature range, where λ is constant, the assumption that $B_\mu(T)$ follows the behaviour of the three dimensional magnetisation remains valid.

The spontaneous magnetisation and the phase diagram

Figure 4 shows the results for various compositions (x = 0, 0.02, 0.05, 0.1, and 0.2), all plotted on the same scale. Two other samples with larger Ca concentrations

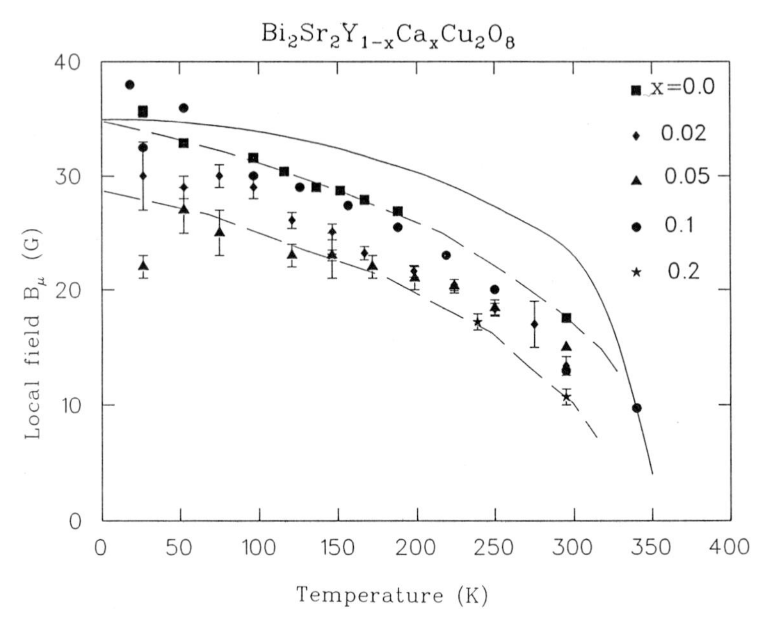

Figure 4 - *Summary of $B_\mu(T)$ for different (Y,Ca) compositions. The solid line is the mean field curve, while the broken lines are just guiides to the eye.*

showed no sign of ordering down to 20 K. All the data of figure 4 display a very similar temperature dependence, indicating that the behaviour of $M_s(T)$ does not vary in the explored interval of x. The shape of this function seems however significantly different from that of the mean field theory, given by the solid line in figure 4, a fact which is also evident from other μSR experiment on end members of the superconductiong families, in particular on La_2CuO_4[11]. A large reduction of M_s already at low temperatures is in fact expected for magnetic systems with strong two dimensional charactcr[12] like ours, where the intraplanar isotropic exchange coupling constant, which makes the 2-D Heisenberg model applicable, is estimated to be four to five orders of magnitude smaller than any other coupling, like the interplanar exchange or possible anisotropies[13].

Both $B_\mu(0)$ and T_N can be derived by extrapolating our data to zero and high temperature, respectively. Both quantities suffer from large systematic errors, due to the fact that the extrapolation depends on the shape that one assumes for the magnetisation curve. However fittings performed with the same curve for all data ensure much smaller relative errors. The first quantity, proportional to $M_s(0)$, shows only rather small variations up to $x = 0.2$, a fact which was already noted in the $YBa_2Cu_3O_{7-y}$ family. Also $T_N(x)$ is almost constant in this range and it comes to an abrupt fall around $x = 0.25$. This result is summarized in the left side of the magnetic phase diagram of figure 5.

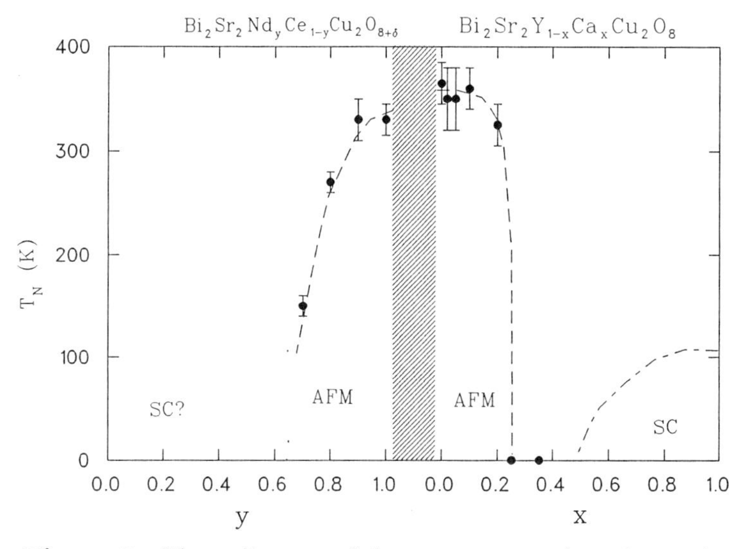

Figure 5 - *Phase diagram of the two compounds under study.*

The abscissae of the phase diagram deserve a further comment. In figure 5 they represent Ca concentration, although plotting T_N against the hole density in the copper planes would be probably more appropriate. Unfortunately this latter quantity is not directly accessible. The disadvantages of substituting it with chemical compositions were already pointed out in the case of $YBa_2Cu_3O_{7-y}$, mentioned above, where the formation of CuOCu dimers interferes with the direct doping of holes into the Cuo_2 planes. The case is even more complicated for these 2-2-1-2 compounds, where the doping mechanism probably involves an interplay between the valence of the cation acting as a spacer within the Cuo_2 bilayer and the valence of Bi, responsible for the superstructure of these compounds.

The extension of this work towards tetravalent cation dopings started from the measurement of $Bi_2Sr_2NdCu_2O_{8+\delta}$, which demonstrated that Nd is a substituent equivalent to Y. The muon local field value was measured also for materials of the type $Bi_2Sr_2Nd_{1-x}Ce_xCu_2O_{8+\delta}$; the curves relative to the $x = 0.0$ and $x = 0.2$ concentrations of this compound are plotted together in figure 6. A reduction of T_N with increasing x is quite evident from these data. Although the homogeneity of these samples is certainly not perfect, the nature of the impurity phases, already discussed in the second section, and the monotonic trend displayed by the data themselves, which can be seen in the right half of the phase diagram of figure 5, indicates that we are definitely observing the intrinsic effect of Ce doping. Whether this mechanism implies

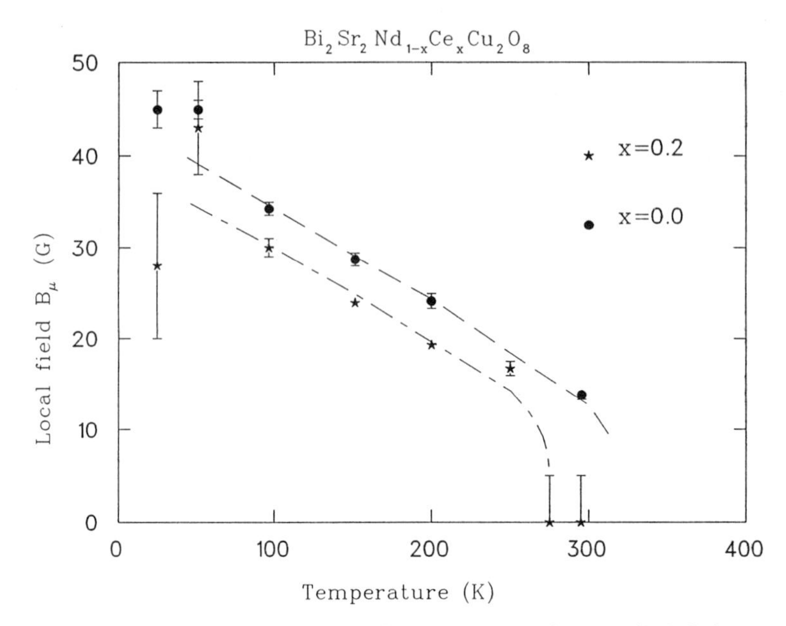

Figure 6 - *Local field value at the low field muon site for two (Nd,Ce) compositions.*

electrons coming into play is still an open question. We can barely note that, if this is the case, the distruction of antiferromagnetic ordering seems less abrupt on the right side of the phase diagram of figure 5 than it is on the left side, a fact which was already observed for $Nd_{2-x}Ce_xCuO_4$.

In conclusion we believe that the common magnetic phenomenology of all superconducting cuprous perovskites, which probably includes the existence of an electron doped superconducting phase (an hypothesis which is supported by evidence from the present paper), is a key feature to the understanding of high T_c superconductivity. However more information must still be gathered on the similarities and on the peculiarities of the different materials before a complete picture can be drawn.

Aknowledgements We wish to thank the staff of the ISIS μSR facility, and in particular Dr C. A. Scott for their support during the experiments. We are also grateful to M. G. Francesconi for her help in sample preparation and in their X-ray characterisation.

References

[1] - R. De Renzi et al. - Phys. Lett. A 135 (1989) 132.

- R. De Renzi, et al. - Proc. Symp. on Electronic Properties of High T_c Superconductors, ed. A. Bianconi, Pergamon, Oxford (1989), p.141.

- R. De Renzi et al. - to appear on Physica C.

[2] J.H. Brewer et al. - Phys. Rev. Lett. 60 (1988) 1073.
[3] n. Nishida et al. - J. Phys Soc. Japan 57 (1988) 597.
[4] Y. Tokura et al. - Nature 337 (1989) 345.
[5] G. Calestani et al. - Physica C, in press.
[6] C. Paracchini et al. - submitted to Physica C.
[7] G.H. Eaton et al. - Nucl. Instrum. Methods, A269 (1988) 483.
[8] C. Bucci et al. - Phys. Lett. A 127 (1988) 115.
[9] A. Schenck et al. private communication.
[10] J. Rossat-Mignod, this workshop
[11] J. Budnick et al. - Europhys. Lett. 5 (1988) 651.
[12] E. Rastelli and A. Tassi, private communication
[13] G. Shirane, this workshop

RAMAN SCATTERING FROM SPIN FLUCTUATIONS IN THE CUPRATES

K. B. Lyons, P. A. Fleury, R. R. P. Singh, P. E. Sulewski

AT&T Bell Laboratories, Murray Hill, NJ 07974

Introduction

Following the discovery of the new class of oxide superconductors in 1987[1] some workers rapidly suggested a possible link between the magnetic properties of the systems and their superconducting properties.[2] The cuprate systems are unique in that they exhibit a square planar lattice of copper spins, with spin ½, coupled by superexchange via the oxygen atoms. Although the earliest of these ideas have since been shown to be inapplicable to the specific systems involved, it is still true that a detailed understanding of the superconducting mechanism has remained elusive, and much interest is centered on the possibility that magnetic phenomena may lie at the heart of the pairing mechanism. In this context, it has become important to understand the magnetic behavior of both the parent and superconducting systems, in hopes of shedding light on the mechanism of the superconductivity. As these studies have progressed, it has become clear that the magnetic system itself, described by a Heisenberg 2D antiferromagnetic interaction for the insulators, is not amenable to description in terms of previous theoretical results. The first portion of this paper is devoted to a report of these deficiencies, especially as they are reflected in the spin dynamics seen through inelastic light scattering, and the manner in which they may be resolved. This problem has taken on interest in its own right, independent of its relation to superconductivity. As we shall show, the conventional spin wave theory fails to describe the short wavelength dynamics for these systems, and must be supplanted by a treatment which takes explicit account of quantum fluctuations in the ground state.

The second portion of this paper reports on similar light scattering measurements in doped systems, with emphasis on the influence that carriers have on the magnetic fluctuation spectrum. In metallic systems we find that the spin fluctuations are still observed, manifesting the presence of localized spins on some short time scale. The spectrum is strongly modified, and shifted to lower frequencies, as might be expected due to the randomizing effect of the carriers. Nevertheless, the fact that localized spins are preserved in these systems increases the likelihood that they might be involved in the superconductivity mechanism itself. Indeed, even in some of the superconducting materials themselves we find clear evidence of the spin fluctuations.[3] We regard this evidence as a crucial piece of the puzzle of the high temperature superconductors, which must be addressed by any valid theoretical approach.

Light Scattering from Spin Fluctuations

It is well known by now[4] that the Cu spins in the cuprate CuO_2 planes order into an antiferromagnetic structure due to the oxygen superexchange interaction, with a 3D ordering temperature which ranges from 500 K down, depending on details of stoichiometry and composition. In this ordered state, the insulators are isomorphic to the situation encountered two decades ago, in K_2NiF_4. The difference lies in the spin magnitude, which is unity for Ni and ½ for Cu, in these respective cases. As we shall see, this difference causes a profound difference in the applicability of the conventional spin wave theory.

Dynamics of Magnetic Fluctuations in High-Temperature Superconductors
Edited by G. Reiter *et al.*, Plenum Press, New York, 1991

In both the cuprates and K_2NiF_4 the spin interactions may be expressed in terms of the Heisenberg Hamiltonian:

$$\mathbf{H}=\sum_{<ij>} JS_iS_j \qquad (1)$$

where the sum is carried out over all distinct nearest neighbor pairs in the lattice, and J is the microscopic superexchange integral. A conventional spin-wave solution to (1), assuming that the ground state is represented by strict Neel order, yields a spin-wave spectrum

$$E_k^2=(SJZ+g\mu H_A)^2-(SJZ\gamma_k)^2 \qquad (2)$$

where $\gamma_k=½(\cos k_x a+\cos k_y a)$, $Z=4$ is the number of nearest neighbors, $S=½$, and J is the exchange interaction constant. Here k is the wave vector of the excitation. This result was first obtained by Parkinson,[5] who derived results for general S, and applied them to K_2NiF_4, with good success.

Therefore, it was natural to expect that a similar treatment would work in the cuprates when early neutron scattering investigations[4] demonstrated the presence of antiferromagnetism and showed that the spin wave velocity was very high in the cuprate insulators, exhibiting strongly 2D character. Due to the energy limitation on the neutron scattering experiment, though, it was not possible to measure the spin wave dispersion with any degree of accuracy. Hence, the value of J could not be determined. Nor could the validity of (2) be ascertained.

A different probe of the spin dynamics is provided by inelastic light scattering. In the case of K_2NiF_4,[5] the dominant term of the interaction Hamiltonian is just

$$\mathbf{H}_R=\sum_{<ij>} (\vec{E}_i\cdot\vec{\sigma}_{ij})(\vec{E}_o\cdot\vec{\sigma}_{ij})\vec{S}_i\cdot\vec{S}_j \qquad (3)$$

where $\vec{E}_o$ and $\vec{E}_s$ represent the incident and scattered magnetic field, and $\vec{\sigma_{ij}}$ represents the vector connecting the nearest neighbor sites i and j. The sum is again carried out over all distinct nearest neighbor pairs $<ij>$. This interaction Hamiltonian, which preserves total spin, does not contribute to single magnon scattering, but provides a very strong intensity to the two-magnon spectrum. The latter scattering process consists of the production of two magnons at equal and opposite wave vectors $\vec{k}$, which thus sum to zero to preserve momentum conservation in the light scattering process. If the dispersion curve (2) is assumed, then the sum may be carried out analytically, providing a spectral lineshape for comparison with the data. This lineshape deviates significantly from the two-magnon joint density of states, which, for spin ½, would exhibit a peak and cutoff at its zone boundary value of $4J$. This deviation is caused by the proximity of the magnons produced, which leads to a strong attractive interaction between them, with a concomitant decrease in the energy of the process. In fact, for spin ½ the theory of Parkinson predicts a peak near $2.7J$. The corresponding values for unit spin are $8J$ for the zone boundary energy of two magnons and $6.8J$ for the peak in the scattered spectrum.

In fact, since the light scattering process is averaged over the entire Brillouin zone, in a known fashion, it is easily seen that the excitation may be viewed as a local spin pair flip, to a good approximation. In this picture, the energy involved, from a naive bond counting argument, would be $3J$ for spin ½, in near agreement with the calculation of Parkinson. Thus, we see that the long range order of the system is in fact rather immaterial to the inelastic light scattering process, since the short wavelength magnons dominate the scattering process. It is essentially an inversion of a pair of neighboring spins, which then have direct interactions with only six other neighbors. If order is present on this very short length scale, we expect to see a magnetic component in the spectrum, albeit with its shape modified somewhat by the single-spin dynamics in a non-ordered system.

The success of this picture for unit spin is striking, as may be seen in Fig. 1.[5] The predicted line width in this case is about 8% of the energy where the spectral maximum is located, in excellent agreement with the spectrum observed. The theory and data are practically indistinguishable, save for the noise on the experimental spectrum. Moreover, the anticipated selection rules, which predict scattering only in a B_{1g} symmetry, were verified. A later investigation[6] failed to obtain polarization data, apparently due to poor sample quality.

Spin Fluctuation Scattering in Insulating Cuprates

The result is strikingly different for the insulating cuprates, as illustrated for the simplest of these systems, La_2CuO_4, in Fig. 2. In this case, the predicted width is about 11% of the energy at the spectral maximum, or some $0.3J$. In fact, as shown in the figure, the observed fractional width is near 38%. Moreover, if we identify the spectral maximum with the energy $2.7J$, the resulting value of J=990 cm^{-1} predicts a cutoff in the spectrum below 4000 cm^{-1}, according to the theory based upon two-magnon scattering. The observed spectrum deviates strikingly from this expectation as well. Moreover, spectral components are observed in A_{1g} and B_{2g} geometries. Initially,[7] it was plausible to attribute these deviations

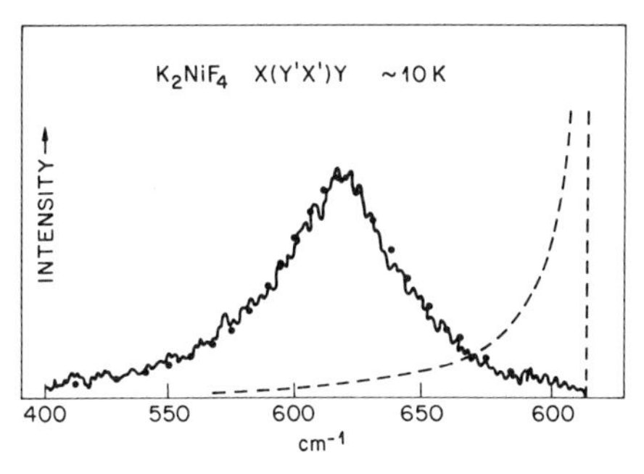

Fig. 1. Spin fluctuation spectrum of K_2NiF_4. The solid line is the observed spectrum, while the dots represent the theoretical calculation according to Parkinson.[5] The dashed line shows the joint density of states, with no allowance for magnon interaction.

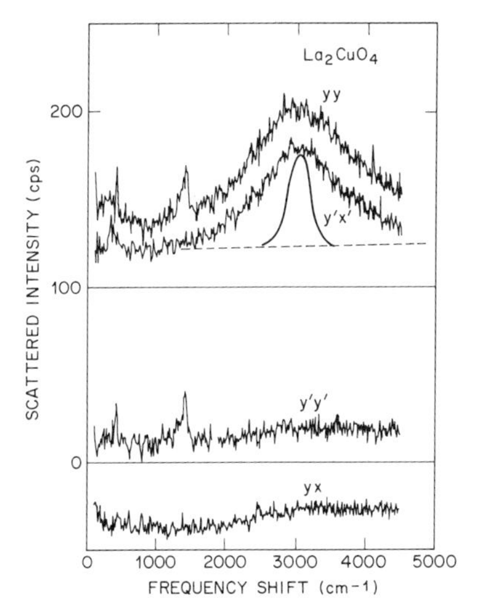

Fig. 2. Spin fluctuation spectra in La_2CuO_4 for several geometries. The smooth solid line is the calculated spectral line shape for the B_{1g} component, against which it is displayed. The phonons at low frequency are not well resolved here, and the A_{1g} feature near 1470 cm^{-1} may be either a two-phonon or a defect-induced one-magnon feature.

from the theoretical expectations to extrinsic origins, such as impurities or non-magnetic scattering. However, studies in many different systems, including systems with different out-of-plane structure,[8] and with deliberate doping[9] gave evidence that perhaps *all* of these observations in fact were intrinsic to the planar CuO_2 system. Given this fact, it appeared necessary to revisit the model behind the interpretation of the spectra, and to ascertain whether shortcomings of that model might be responsible for the deviations observed.

In fact, there is reason to suspect that the description based upon spin wave excitations from a Neel state might be inapplicable for spin ½. A spin fluctuation in a spin ½ system represents a complete *flip* of a pair of spins, not just a tilting as is true in systems of higher spin. Thus, the influence on the magnetic order is greater, and one could expect a greater impact on the spin dynamics. Mathematically speaking, the spin wave descriptions represents an expansion in $1/2S$, which is naturally questionable for $S=½$. It is therefore iimperative that spin fluctuations be taken into account.

The usual way of taking into account the quantum fluctuations in the ground state leads to a simple renormalization of the effective exchange integral[10] to give $J'=Z_cJ$. The expressions obtained for all magnetic phenomena then remain the same, save that J is replaced with J'. This procedure assumes that the relevant quantities, such as Z_c, do not take on any wave vector dependence. The fact that spin wave theory works so well for $S=1$ is evidence that such dependences are small. The fluctuation correction is not small, being some 12% for the $S=1$ case, but the resulting expressions describe the spectra well, as detailed above. The correction for $S=½$ is only somewhat larger, $Z_c\cong1.18$, but the spectral *shape* in this case bears little resemblance to that predicted by the model. Hence, the extraction of a value of J by simple identification of the peak position must remain rather suspect.

A procedure recently developed for the calculation of matrix elements in the Heisenberg AFM ground state by expansions about the Ising limit, which is precisely calculable,[11] enables us to address the quantum fluctuation problem in more detail. If the fluctuations are included in the ground state description, then the fluctuation-dissipation theorem tells us that the spectral moments may be calculated from appropriate matrix elements of this full ground state. To see this explicitly, assume that the Raman interaction Hamiltonian H_R is known. We can then express the scattered spectrum as

$$I(\omega)=\sum_i \delta(\omega-(E_i-E_o)) \,|<0\,|\,H_R\,|\,i>|^2 \quad , \tag{4}$$

where E sub i is the ith eigenenergy of (1) and $|0>$ is the ground state. As an example, the first spectral moment may then be written as

$$\rho_1=\int \omega I(\omega)d\omega/I_T=<0\,|\,H_R[H,H_R]\,|\,0>/I_T$$

where I_T is the integral of (4) (the zeroeth spectral moment). Higher moments may be written as matrix elements of operators of successively higher complexity.

This method proceeds then by dividing the Hamiltonian (1) into two pieces

$$H=J_z\sum_{<ij>} S_i^z S_j^z + J_{xy}\sum_{<ij>}(S_i^x S_j^x + S_i^y S_j^y) \quad ,$$

where the Heisenberg model corresponds to $J=J_{xy}=J_z$. We expand the matrix elements of interest in the quantity $x=J_{xy}/J_z$, and extrapolate the series so obtained to estimate the value for the Heisenberg limit (x=1). Each term in such a series may be calculated exactly by making an exhaustive calculation of all spin clusters, imbedded in a Neel state, which contribute to the specified order in x.

From the first few moments of the spectrum, we can obtain the peak position (actually the center of gravity of the spectrum, given by the first moment), the width (actually the second cumulant), and the skewness (given by the first three moments). From the first moment, we extract the value of J, which is the only parameter in the problem. Once this value is chosen, the other comparisons become parameter free. We find that the width and skewness of the B_{1g} peak are described quantitatively by the results of the calculation, as shown in Table I.

TABLE I. Comparison of Measured and Calculated Spectral Moments, La_2CuO_4

symmetry	ρ_1 (cm^{-1}) theor	exp	M_2/ρ_1 theor	exp	M_3/ρ_1 theor	exp
B_{1g}	3680	3680 J≡1030 cm^{-1}	0.23	0.27	0.28	0.25
A_{1g}	3600	3700	0.34	0.38		
B_{2g}	4100	4500	0.28	0.27		

Moreover, we find that the presence of spin fluctuations in the ground state will activate a new scattering process, corresponding to the flip of two *diagonal next-nearest neighbors*. The Raman Hamiltonian corresponding to this process is the same as (3), but with σ_{ij} now replaced by σ'_{ij} which lies diagonally across the Cu-O square. In the absence of quantum fluctuations, such a process is not allowed, since it does not conserve spin. However, when pairs of spins are flipped by the quantum fluctuations in the ground state, it becomes allowed. Moreover, it is clear that there are two distinct types of such process which will be allowed by the existence of a single spin pair flip in the lattice: one that results in a final state with a nearest neighbor pair flipped and one that results in two isolated spins flipped. Clearly the latter, which gives the only contribution in B_{2g} geometry, will have the highest energy. Both will contribute in A_{1g} geometry. Thus, this single term in the Raman Hamiltonian provides four additional quantities to be checked with the experimental spectra: the positions and widths of the A_{1g} and B_{1g} features. These comparisons are all *parameter free*.

We have utilized a CCD camera, interfaced to a Spex Triplemate spectrometer, to obtain high quality to high frequency shift, in order to study the entire spectral profile of the observed components. The spectra show clearly that *all three* components represent real scattering, and that fluorescence contributions are small. The high quality of the spectra obtained is shown by the typical runs in Fig. 3. Moreover, if we extract the spectral moment values, as shown in Table I, for the A_{1g} and B_{2g} components, we obtain excellent agreement with the values predicted by calculation. We conclude, therefore, that *all* of the scattering we see about roughly 1800 cm^{-1} is magnetic in origin, and that the account of quantum fluctuations in the ground state is crucial for a correct description of the spin dynamics. This agreement gives us good confidence in the value of J extracted by this analysis, namely J=1030±50 cm^{-1}, which predicts a spin-wave slope of 0.83±0.04 eV-Å near zone center. Recent neutron measurements to much higher energy, although very difficult, have recently been performed, yielding good agreement with this value, namely a slope of 0.85±0.1 eV-Å.

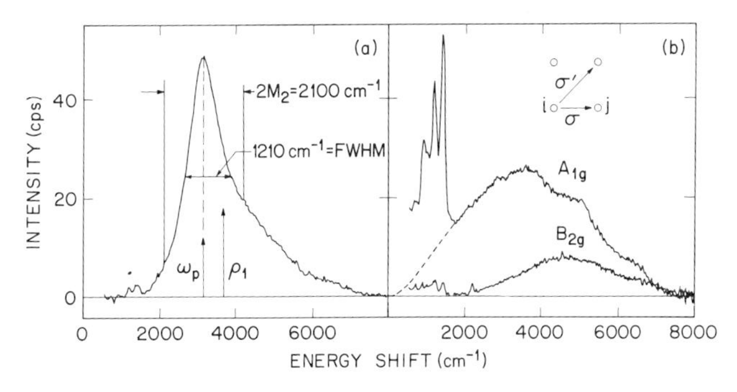

Fig. 3. Spectra obtained by use of a CCD camera in order to improve spectral quality and to extend the range of the spectrum to 8000 cm^{-1} (1 eV). All three spectral components are shown, B_{1g} in the left half of the figure and A_{1g} and B_{2g} on the right.

It should be commented that the strong intensity associated with the A_{1g} and B_{2g} components is rather unexpected, since the spin coupling across the diagonal of the basal plane square should be small. In fact, it apparently is small in the ground state of the system, since we need to introduce no higher order couplings in the Hamiltonian of the system, (1), in order to describe the spectra. By contrast, the high intensity seen necessitates that it be not small, but in fact quite large, in the excited intermediate state associated with the Raman scattering. Since the Raman scattering exhibits strong resonance[12] it is plausible that such a situation could exist. Indeed, calculations of the charge transfer modes in the Cu-O planes suggest that the lowest state may be the A_{2u}, which corresponds to transfer of an electron from the Cu ion on the corner of a square onto the four oxygens of that square, as sketched in Fig. 3. This intermediate state will communicate equally well with all four copper ions around the square, and thus would be consistent with a roughly equivalent intensity for the B_{1g} and A_{1g} processes. It is possible, then, that further study of the resonance excitation profiles for the various spectral components may yield information about the charge transfer states in the planes.

When the same formalism is used to describe the spin fluctuation spectra of other related compounds, a similar success is achieved. For example, Fig. 4 shows 27 spectra obtained for three different geometries with three different laser frequencies in three different compounds. Tables II, and III summarize the comparison of the spectral moments with the calculation results. As may be seen, the comparison is excellent in virtually all cases. The value of J changes somewhat between the compounds, in correlation with the lattice parameter; but, for each compound, once J is fixed using the first moment of the B_{1g} spectrum, the other spectral moments and cumulants are obtained correctly.

Spin Fluctuations in Doped Materials

When carriers are added to the material by doping, the spin fluctuation spectrum changes. For small carrier concentration x, less than that required to produce a metallic state, the changes are small. When the material becomes metallic, and the carriers become highly mobile, the effect is greatly magnified. For example, Fig. 5 shows spectra obtained on Sr-doped La_2CuO_4 for several different concentrations, spanning the insulator-metal transition. As is evident there, a small change in x which traverses this transition causes a very large change in the observed spectrum. The magnetic component is still present, as shown by the differences among the different symmetry components, but it is greatly broadened, even extending to higher energy shift.

Thus, there are two effects associated with the carriers. One is a simple randomization effect, that is not influenced by the carrier dynamics, and tends to move the spectrum to lower energy. This result is expected, since the presence of excess holes removes spins or bonds, depending on whether the holes reside on the copper or oxygen sublattices. Furthermore, the randomization inherent in the presence of these holes is expected to broaden the spectrum, as observed. There is a second effect, though, influenced by the carrier dynamics.

This dynamic effect may be seen directly in a system where the carriers freeze out on cooling. $La_{2-x}Sr_xNiO_4$ is such a system, for x=0.2. In this case, the resistivity increases several orders of magnitude below 50 K, manifesting the freezing of the carriers. As this change occurs, the spin fluctuation spectrum undergoes the modification shown in Fig. 6b. The spectrum in the stoichiometric pure material is shown in Fig. 6a. Clearly, the presence of excess holes, even when they are relatively stationary as at 10 K, causes a very large renormalization of the spin fluctuation energy. When they become mobile, a further change occurs, as may be seen in the spectrum at 300 K, where the spin fluctuation component appears as a broad wing without any well defined peak.

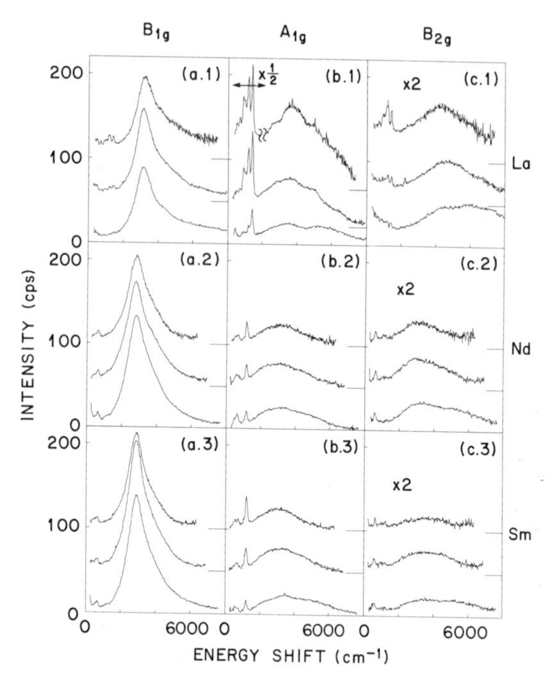

Fig. 4. Each column of three figures refers to the specified geometry, while each row refers to the specified cation R in R_2CuO_4. Spectra are shown in each figure, top to bottom, for three laser frequencies: 5145Å, 4880Å, and 4579Å.

A different behavior is observed when the magnetic system is randomized by isoelectronic substitution of Ni at the Cu site. Spectra of Ni-doped material show that substitution of up to 3-4% Ni causes virtually no change in the observed spectrum for the $Ba_2YCu_3O_{6.0}$ system.

Spin Fluctuations in Superconducting Material

In the $Ba_2YCu_3O_{6+x}$ system, it is possible to vary the hole concentration continuously from the insulator (x=0), where the spectrum is quite similar to that of La_2CuO_4 into the metallic phase, where superconducting behavior up to 92 K is observed for x=1.0. A sequence of these spectra is shown in Fig. 7 for different values of x.

It is apparent that the spin fluctuation feature is still quite evident in the 60 K superconducting phase. Even in a superconductor near the highest doping achievable, there appears to be a remnant of the spin fluctuation scattering, which in this case becomes a gradually decaying background extending to rather high energy shift. We note that in the latter case it is difficult to be sure that the scattering is magnetic in origin, since it is conceivable that electronic scattering could be present in the B_{1g} geometry. The fact that there is a marked difference between the B_{1g} and B_{2g}, though, argues for the magnetic interpretation. We may conclude that localized spins exist in the region of the sample probed by the light, at all concentrations x.

Although this conclusion is inescapable, it is more difficult to make a definitive statement about the material in the bulk of the sample, since the $Ba_2YCu_3O_{6+x}$ material is known to be rather unstable, and might very well lose some oxygen near the surface. The penetration depth of the light is 500-1000Å, hence any process which influences a significant portion of this layer could modify the observed spectra.

In an effort to address this issue, we have focused on the 60 K material, since it is chemically the more stable of the superconductors in the $Ba_2YCu_3O_{6+x}$ system. Moreover, the plateau in T_c as a function of x at T_c=60 K suggests an insensitivity of the superconducting properties to oxygen loss which should make our experiments with this material less susceptible to artifacts due to surface modification. We have used samples which yield very high IR reflectivity, and, moreover, show evidence of a gap in the IR reflectivity, quite close to the value anticipated for weak-coupling BCS theory ($2\Delta=3.5kT_c$). In this case, we can employ two different scattering techniques in one experiment to reveal the simultaneous presence of superconducting material and spin fluctuations in our material. Working at high resolution, and with an iodine cell to remove the light scattered elastically by the surface, we observe the spectra shown in Fig. 8a. The spectra are all shown here with a spectrum at 90 K subtracted. Clearly there is a feature which develops in the gap region which reflects the quasiparticle spectrum of the superconductor. The gradual nature of the gap "edge" precludes any extraction of a true gap on the basis of these spectra, but it clearly indicates the presence of superconducting material in our scattering volume.

TABLE II. Values of the Cumulants of the B_{1g} Spectra

Sample	λ_L (Å)	ω_p (cm^{-1})	M_1 (cm^{-1})	M_2 (cm^{-1})	M_3 (cm^{-1})	M_2/M_1
	5145	3210 ± 10	3760 ± 50	850 ± 50	840 ± 100	0.23
La_2CuO_4	4880	3130	3650	940	830	0.26
	4579	3133	3700	1050	990	0.28
	5145	2806	3014	722	460	0.24
Nd_2CuO_4	4880	2760	3110	860	675	0.28
	4579	2800	3240	910	780	0.28
	5145	2858	3044	693	460	0.23
Sm_2CuO_4	4880	2859	3167	794	633	0.25
	4579	2897	3330	920	820	0.28
Theory			3.6 J	0.8 J	1.0 J	0.23

TABLE III. Cumulants and Intensities of A_{1g} and B_{2g}, Ratioed to B_{1g}

		A_{1g}		B_{2g}		I_Γ	
Sample	λ_L (Å)	M_1/J	M_2/M_1	M_1/J	M_2/M_1	B_{1g}/A_{1g}	B_{2g}/A_{1g}
	5145	3.7	0.38	4.1	0.22	0.39	0.13
La_2CuO_4	4880	3.6	0.39	4.7	0.25	0.84	0.24
	4579	4.2	0.39	5.1	0.23	1.5	0.37
	5145	3.7	0.34	3.9	0.21	2.6	0.27
Nd_2CuO_4	4880	3.8	0.38	4.0	0.25	2.3	0.31
	4579	3.9	0.37	4.6	0.30	2.8	0.45
	5145	3.6	0.33	4.1	0.23	2.6	0.15
Sm_2CuO_4	4880	3.9	0.35	4.1	0.24	2.8	0.24
	4579	4.2	0.35	4.6	0.29	3.1	0.27
Theory		3.5	0.34	3.9	0.28		~0.4

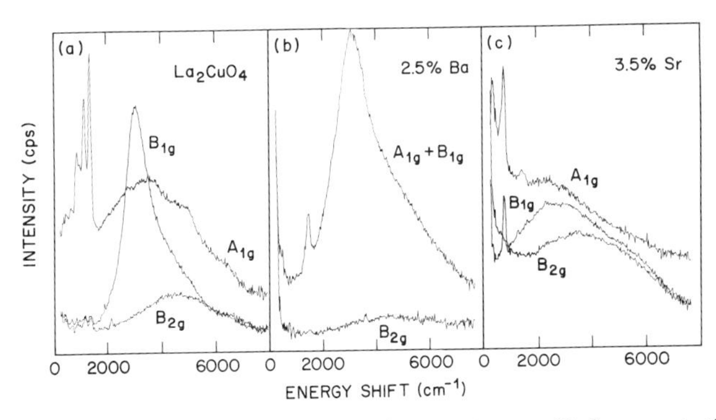

Fig. 5. B_{1g} spectra obtained in $M_{2-x}Sr_xCuO_4$ for the specified concentrations x, for M = La,Ba.

In the same experiment, we can work at lower resolution to study the high frequency scattering in the spin fluctuation region. These spectra are shown in Fig. 8b. We note a very clear difference between the B_{1g} and B_{2g} components, with a shape to that difference which changes only slightly on traversing T_c.

From these two observations, we conclude that localized spins coexist in our sample with superconductivity. It is also clear from the comparison of the spectrum with that of the insulator (Fig. 8b) that the observed spectrum cannot be described as a superposition of a spectrum with and without the spin fluctuation component of the insulator. Hence we must conclude either (i) that the spins exist in the superconducting regions or (ii) that the non-superconducting (AFM) regions are small enough to have their local spin dynamics strongly modified from the bulk material. In either case, the localized spins must exist in intimate contact with the superconducting material: either as a single phase containing both or as a two-phase material, with the two-phase structure occurring on a very small length scale.

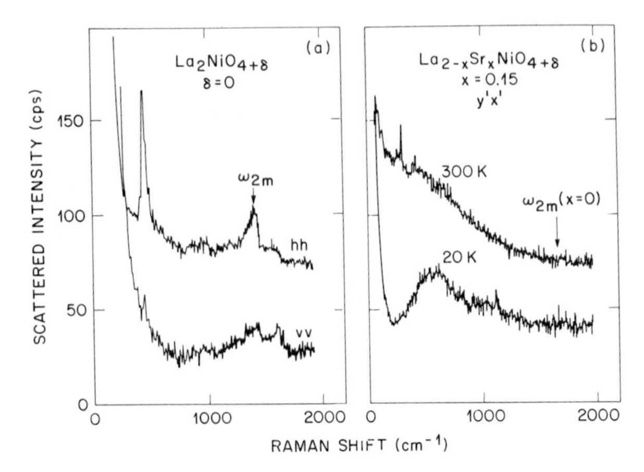

Fig. 6. Spin fluctuation spectra obtained in stoichiometric La_2NiO_4 (sample protected comletely from oxygen exposure after anneal, part a) and in Sr-doped material, part b, at two temperatures. The *xx* and *yy* spectra for the former are obtained on a single-domain region of the sample surface.

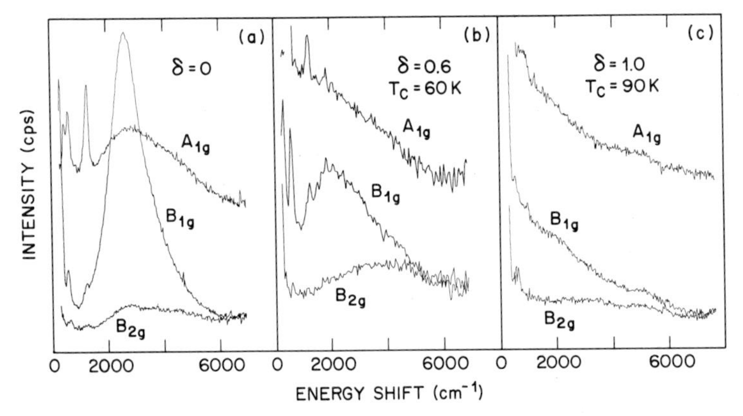

Fig. 7. Spectra obtained in the three relevant geometries for x=0 (insulating phase), x=0.6 (60 K superconductor), and x=0.9 (89 K superconductor). The three components are shown on the same intensity scale for each material.

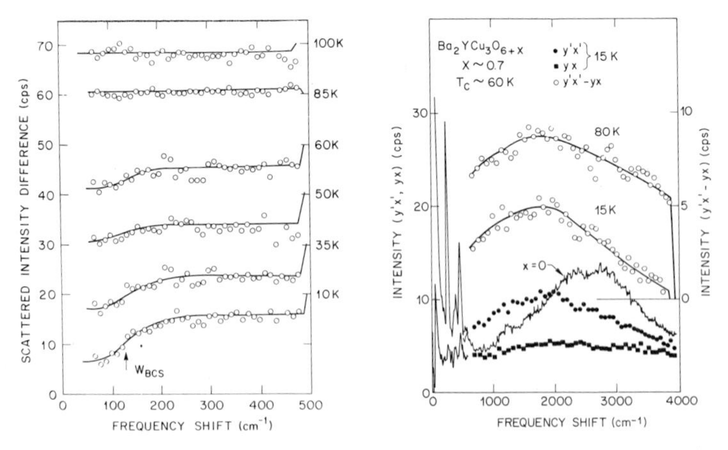

Fig. 8. Part (a) shows spectra obtained at high resolution in the low frequency region. Part (b) shows the spin fluctuation spectral region, with the spectrum of the insulating material (x=0) superimposed as the dashed line. Spectra are shown for the B_{1g} and B_{2g} geometries at 15 K. Subtracted spectra, B_{1g}–B_{2g}, are shown against the scale at the right, for 15 K and 80 K.

Conclusions

Light scattering spectra have provided a means for accurate measurement of the superexchange constant J in cuprate insulators. Proper analysis of these spectra demonstrates that the spin dynamics are strongly modified by quantum fluctuations in the ground state. The analysis yields quantitative agreement with the spectral moments observed. The value of J so obtained is found to correlate with lattice parameter when the values are compared among different compounds in the cuprate family.

Experiments on doped samples show that doping with carriers yields a substantial modification of the spin fluctuation spectrum, while isoelectronic doping with other magnetic ions leads to no observable effect. There are two effects, one due to the randomizing effect of stationary carriers on the spin system, and the other due to carrier motion. In some cases, these two effects are separable, and may be amenable to theoretical analysis in the future.

Finally, we have demonstrated that localized spins persist in the superconducting phases of $Ba_2YCu_3O_{6+x}$. The most unequivocal evidence for this is in the case of the material at x=0.6, with T_c=60 K. A substantial spin fluctuation peak is observed in this case, while experiments on the same sample and on the same temperature run show the presence of a superconducting gap in the Raman spectrum. We thus conclude that localized spins exist in the superconducting material at least on a time scale as large as our inverse line width (which translates in this case to about 2 fs). We cannot draw direct implications as to the role of spin interactions in the mechanism of the superconductivity. However, it is clear that the spin dynamics pose a theoretical puzzle which must be addressed by any theory that describes the cuprate systems in sufficient detail to explain the superconductivity. At the very least, it is clear that the carriers and the spins are strongly coupled, as seen by the changes brought about in the spectrum by carrier mobility, and thus some role for the spins in mediating the superconductivity is indeed likely.

Acknowledgements

The authors wish to acknowledge the help of numerous colleagues in the work summarized in this paper. Samples were provided by D. Buttrey, A. S. Cooper, G. Espinosa, J. P. Remeika, L. F. Schneemeyer, and J. V. Waczcak. Numerous helpful discussions of the results have involved L. Cooper, P. B. Littlewood, G. A. Thomas, and C. M. Varma.

References

1. K. Kishio, K. Kitazawa, et al, Chem. Lett. 429 (1987).
2. P. W. Anderson *et al*, Phys. Rev. Lett. **58**, 2790 (1987).
3. K. B. Lyons, P. A. Fleury, L. F. Schneemeyer and J. V. Waszczak, Phys. Rev. Lett. **60**, 732 (1988).
4. G. Shirane, Y. Endoh, R. J. Birgeneau, M. A. Kastner, Y. Hidaka, M. Oda, M. Suzuki, and T. Murakami, Phys. Rev. Lett. **59**, 1613 (1987).
5. P. A. Fleury and H. J. Guggenheim, Phys. Rev. Lett. **24**, 1346 (1970).
6. S. R. Chinn, H. J. Zeiger, and J. R. O'Conor, Phys. Rev. B **3**, 1709 (1971).
7. K. B. Lyons, P. A. Fleury, J. P. Remeika, A. S. Cooper, T. J. Negran, Phys. Rev. B **37**, 2353 (1987).
8. P. E. Sulewski, P. A. Fleury, et al, Phys. Rev. B **41**, 225 (1990).
9. K. B. Lyons and P. A. Fleury, J. Appl. Phys. **64**, 6075 (1988).
10. Keffer, Handbuch der Phys, Vol. XVIII
11. R. R. P. Singh, Phys. Rev. B **39**, 9760 (1989).
12. I. Ohana, D. Heiman, M. S. Dresselhuas, and P. J. Picone, Phys. Rev. B **40**, 2225 (1989).

ELECTRONIC STRUCTURE OF $Bi_2Sr_2CaCu_2O_8$ SINGLE CRYSTALS AT THE FERMI LEVEL

R. Manzke, G. Mante, S. Harm, R.Claessen, T. Buslaps, and J. Fink*

Institut für Experimentalphysik, Universität Kiel
D-2300 Kiel 1, FRG
*Kernforschungszentrum Karlsruhe, Institut für Nukleare Festkörperphysik, D-7500 Karlsruhe, FRG

ABSTRACT

Angle-resolved photoemission has been performed on $Bi_2Sr_2CaCu_2O_8$ single crystals with high energy and angle resolution to study for $T>T_c$ the dispersion of the electronic states close to the Fermi level E_F. Resonant photoemission has indicated that the states at E_F are localized in the Cu-O planes and essentially of O(1) 2p character. The experimental band structure displays only weak dispersions as compared to theoretical results due to important correlation effects. From the Fermi level crossing of the weakly dispersing O2p band along the ΓX direction the existence of a Fermi surface can be proven. The shape is close to the result of band structure calculations and the unoccupied part becomes visible up to the Brillouin zone boundary in angle-resolved inverse photoemission spectra. Below T_c, the opening of the superconductivity gap in the Fermi surface can be observed along ΓX, yielding a value of $2\Delta/k_BT_c \sim 8$. This indicates that $Bi_2Sr_2CaCu_2O_8$ is a strong-coupling superconductor.

INTRODUCTION

For any microscopic model of high-T_c superconductivity a detailed knowledge of the electronic structure of the perovskite cuprates is of crucial importance. Photoelectron spectroscopy, in particular in its angle-resolving mode, is a suitable tool to obtain information about the character, energy location, and dispersion of the electronic states.

Due to the surface-sensitivity of photoemission high-quality sample surfaces are required, which are chemically stable in ultra-high vacuum conditions, and, for angle-resolved measurements, maintain crystalline order after preparation. Among the high-T_c superconductors discovered so far these requirements are met only by $Bi_2Sr_2CaCu_2O_8$, from which large single crystals can be grown, which are easily cleavable perpendicular to the c-axis.

In this contribution we discuss the nature and dispersion of the electronic states close to the Fermi level. Along the ΓX direction the dispersion has been measured by angle-resolved photoemission (ARPES) and inverse photoemission (ARIPES), for the ΓM direction only by ARPES. These results will be compared with predictions from band structure calculations performed within the local density approximation. Moreover, because the superconductivity gap of the high T_c compounds has become accessible by the energy resolution of photoemission, we have been able to determine the size of the gap directly.

EXPERIMENTAL

The photoemission data presented here have been obtained with synchrotron radiation at the Hamburger Synchrotronstrahlungslabor (HASYLAB) and a hemispherical electron energy analyzer applying an energy resolution as low as 60 meV and an angular resolution of $\pm 0.5^{\circ}$. The data on the empty electronic states were measured by angle-resolved *inverse* photoemission (ARIPES) employing a band-pass photon detector operating at 9.9 eV photon energy with an overall energy resolution of 640 meV.

The preparation of the $Bi_2Sr_2CaCu_2O_8$ single crystals is described elsewhere.[1] From ac magnetic susceptibility a T_c of 83K is determined. The crystal consists dominantly of the n=2 phase. Several samples prepared from the same crystal were introduced into the spectrometer chamber and cleaved after moderate backing (about 100°C for 10 hours) at a base pressure of 10^{-10} mbar. The surface structure was controlled by LEED yielding for the surface lattice and the superstructure along the b^* direction identical values as for the bulk.[2] In addition, we found from low temperature LEED experiments that the pattern remains unchanged down to about 20K.

For an energy reference the Fermi edge of an in situ evaporated gold film was used. The good chemical stability of the $Bi_2Sr_2CaCu_2O_8$ surfaces is demonstrated by the fact that we did not observe any time dependence of the spectra as may be due to, e.g., oxygen loss or contamination.

RESULTS AND DISCUSSION

Early photoemission spectra[3] of $Bi_2Sr_2CaCu_2O_8$ already displayed a resonant enhancement of intensity near the Fermi level at a photon energy of 18 eV. By utilizing constant-initial-state (CIS) spectroscopy (in which the initial state energy remains fixed, while photon and final state energy are swept simultaneously) we have studied this resonance in more detail[4]. At normal emission it was found to be most pronounced for initial state energies around 1.5 eV binding energy. The resonance occurs close to the O 2s-O 2p absorption edge, and can be interpreted by a super-Coster-Kronig decay of the excited core hole ($h\nu + O(2s^2 2p^5) \rightarrow O(2s^1 2p^6) \rightarrow O(2s^2 2p^4) + e^-$) which couples resonantly to the direct photoemission. Since it is known from electron energy loss data[5] that the unoccupied O 2p-states close above E_F involved in this process have the symmetry expected for the states in the Cu-O planes, we may conclude that also the filled states near the Fermi level are localized at the O(1)-site, i.e. also in the Cu-O planes.

In order to map both the occupied and unoccupied band structure parallel to the ab-plane we have performed angle-resolved photoemission and inverse photoemission[2,6], respectively, along the high-symmetry directions ΓX and ΓM of the Brillouin zone ($k_{\parallel}$-spectra). The ARPES spectra show clear structure down to about 6 eV binding energy due to Cu 3d and O 2p derived valence band states. Though it is rather difficult to distinguish individual bands, as they are predicted by recent band structure calculations (e.g. Ref. 7), as an overall result the experimental bands seem to display much less dispersion than the calculated ones. A more sophisticated approach to the overall valence band structure by using the polarization dependence of the spectra and applying numerical fits has been carried out in Ref. 8. We rather focus on the fine structure of the electronic states close to the Fermi level.

Figure 1 shows highly resolved (ΔE=60meV) ARPES spectra taken along the ΓX and ΓM direction of the Brillouin zone given in the insert. The photoemission spectra clearly show a peak at around 0° emission angle (Γ-point) dispersing upwards in energy for higher angles. At about $\vartheta=7.5°$ the turning point of the emission onset is located exactly at the independently determined Fermi level, while the width of the onset can be well described by a Fermi-Dirac distribution convoluted with the experimental resolution. Thus, the observed dispersion of the peak can be interpreted due to a crossing of a band through the Fermi energy. Correspondingly, the ARIPES spectra (see Ref. 2) display an increase of intensity at E_F between 30° and 40° incidence angle, which shows the occurence of the band above the Fermi level. It should be realized that, according to the discussion above, this band must be dominantly of O(1) $2p_{x,y}$ -character.

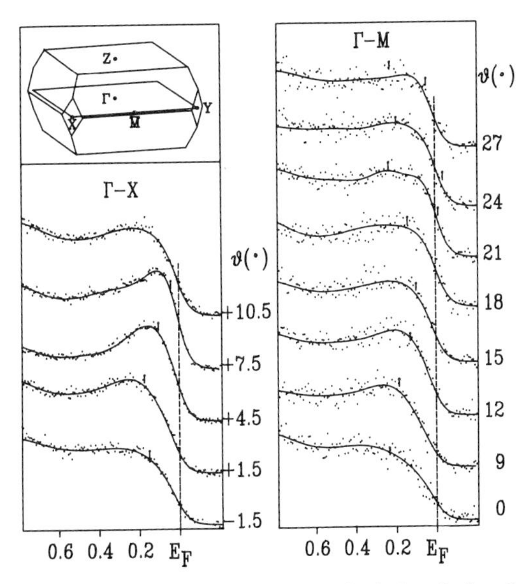

Fig. 1. Angle-resolved photoemission spectra of $Bi_2Sr_2CaCu_2O_8$ along the ΓX (left panel) and ΓM direction (right panel) taken with 18 eV photon energy at 300K. The lines result from a fit of the original spectra by Gaussians, the maxima close to E_F are indicated. The insert shows the Brillouin zone.

Along the ΓM direction this band disperses also upwards to E_F, but the dispersion is much weaker than in ΓX. However, a Fermi edge is clearly visible for 21° and larger emission angles. In particular from the 21° spectrum, a second emission maximum is evident very close to E_F. In ΓM these states are responsible for the Fermi edge and seem to push the O(1) 2p states below E_F.

In Fig. 2 the experimental band structure for both, ΓX and Γ$\overline{M}$ is compared to the calculation of Krakauer and Pickett[7] using the local density approximation (LDA). Along the ΓX direction we find the experimental conduction band intersecting E_F at about the same point in reciprocal space as predicted by the theory. A similar result has been published by C. Olson et al.[9] Along Γ$\overline{M}$ this band remains below E_F, while an additional band forms an electron pocket around the $\overline{M}$-point. The experimental band width of 0.5 eV is drastically reduced compared to about 3eV

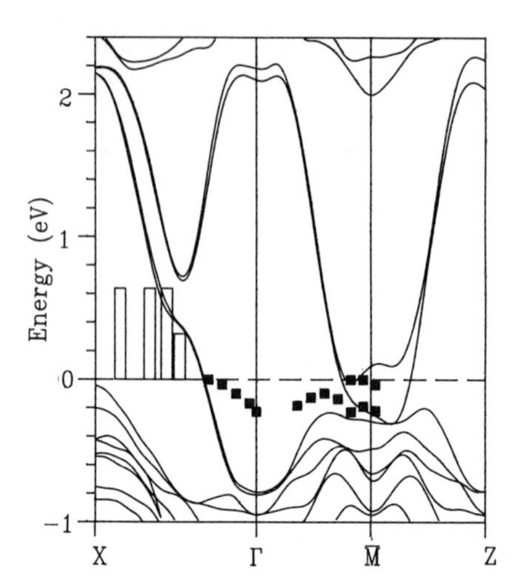

Fig. 2. Experimental band structure of $Bi_2Sr_2CaCu_2O_8$ close to the Fermi level compared to the LDA calculation of Ref. 7. The small full squares represent occupied states from ARPES and the large open ones empty states from ARIPES (from Ref. 2).

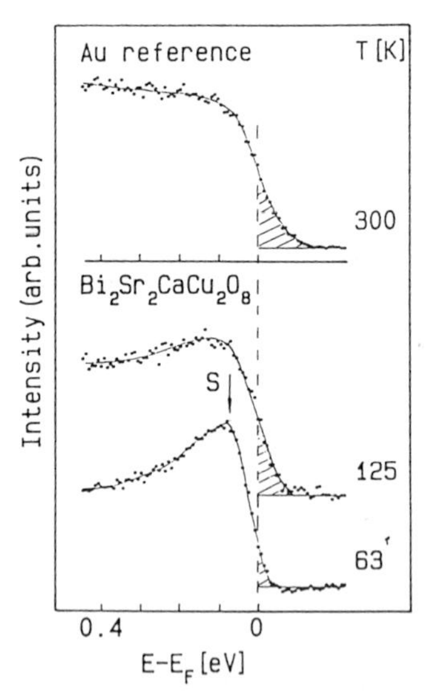

Fig. 3. High resolution ARPES spectra of $Bi_2Sr_2CaCu_2O_8$ at 9° emission angle along ΓX above and below T_c. Note the enhancement S and the shift of spectral weight to higher binding energy below T_c. For comparison a spectrum of evaporated gold taken at identical conditions is also shown.

in the LDA calculation. This discrepancy is possibly due to strong correlation effects, which cannot be properly taken into account in the LDA scheme. However, the experimental data confirm the existence and, in particular, the shape of the Fermi surface as calculated.

In order to show that it is the O(1) derived conduction band observed along ΓX which is responsible for superconductivity, temperature-dependent measurements have been performed[10], the results of which are shown in Fig. 3. Spectra have been taken at an emission angle along the ΓX direction, where for $T>T_c$ a clear Fermi-Dirac cutoff can be seen, corresponding to a point in k-space on the Fermi surface. For $T<T_c$ a shift of intensity towards higher binding energy is found, resulting in an enhanced intensity maximum S. This peak S can be interpreted as the singularity of the quasiparticle density of states in the superconducting phase, broadened by the experimental resolution, as one would expect it from a BCS-type superconductor. The shift of the turning point of the emission onset to higher binding energy then yields the superconductivity gap of $2\Delta=60$meV, which is in excelent agreement with other photoemission and tunnelling studies (see Ref. 11). This gap translates into a ratio of $2\Delta/k_BT_c \sim 8$, about twice as large as the BCS value, indicating a strong-coupling mechanism in this high-T_c compound.

ACKNOWLEDGEMENTS

The authors thank Z.X. Zhao for providing the high-quality single crystals. This work is supported by the 'Bundesministerium für Forschung und Technologie', Fed. Rep. Germany (project 401 AAI).

REFERENCES

1. Z.X. Zhao, Physica C 153, 1144 (1988)
2. R. Claessen, R. Manzke, H. Carstensen, B. Burandt, T. Buslaps, M. Skibowski, and J. Fink, Phys. Rev. B 39, 7316 (1989)
3. T. Takahashi, H. Matsuyama, H. Katayama-Yoshida, Y. Okabe, S. Hosoya, K. Seki, H. Fujimoto, M. Sato, and H. Inokuchi, Phys. Rev. B 39, 6636 (1989)
4. R. Manzke, T. Buslaps, R. Claessen, G. Mante, and Z.X. Zhao, Solid State Commun. 70, 67 (1989)
5. N. Nücker, H. Romberg, X.X. Xi, J. Fink, B. Gegenheimer, and Z.X. Zhao, Phys. Rev. B 39, 6619 (1989)
6. R. Manzke, T. Buslaps, R. Claessen, M. Skibowski, J. Fink, VUV 9, Hawaii 1989, to be published in Physica Scripta
7. H. Krakauer and W.E. Pickett, Phys. Rev. Lett. 60, 1665 (1988)
8. R. Böttner, N. Schroeder, E. Dietz, U. Gerhardt, W. Aßmus, and J. Kowalewski, submitted to Phys. Rev. B
9. C.G. Olsen R. Liu, D.W. Lynch, B.W. Veal, Y.C. Chang, P.Z. Jiang, J.Z. Liu, A.P. Paulikas, A.J. Arko, and R.S. List, VUV 9, Hawaii 1989, to be published in Physica Scripta
10. R. Manzke, T. Buslaps, R. Claessen, and J. Fink, Europhys. Lett. 9, 477 (1989)
11. G. Margaritondo, D.L. Huber, and C.G. Olson, Science 246, 770 (1989), and references therein

CALCULATION OF PHOTOEMISSION SPECTRA FOR THE t-J MODEL AND THE EXTENDED HUBBARD MODEL

Walter Stephan and Peter Horsch

Max-Planck-Institut für Festkörperforschung

D-7000 Stuttgart 80, Federal Republic of Germany

INTRODUCTION

There are a wide variety of experimental probes of condensed systems (XPS, UPS, BIS, EELS, etc.) [1,2] for which a theoretical description may be formulated simply in terms of a single-particle (-hole) propagator. In the particular case of transition metal oxides, these measured excitation spectra cannot be understood fully using conventionally calculated band structures. These discrepancies may however be understood to arise from correlation effects which are not properly treated in single particle band structure calculations. These features may most simply and successfully be handled within an impurity model [3] , where a single metal ion coordinated by the appropriate ligands models the crystal. This approach has the advantage of being simple enough to allow for a realistic treatment of many atomic orbitals with crystal-field effects etc. On the other hand, there are interesting questions concerning for example quasiparticle dispersion, i.e. width of bands, which cannot readily be considered within the impurity approach. The exact diagonalization of finite clusters with periodic boundary conditions presented here may be considered as a step from the impurity problem toward the crystal.

We discuss here results obtained for two models presently under consideration for the electronic structure of high-temperature superconductors (HTSC). For the Emery model [4] (extended Hubbard model for the CuO_2-planes) photoemission and inverse photoemission spectra have been calculated for the insulating (undoped) as well as for hole and electron doped cases. High-energy satellites and new low-lying excitations in the charge- transfer energy gap appear as a consequence of the strong Coulomb correlations on the Cu-sites. These low-energy states appear to be responsible for conductivity and superconductivity in hole-doped HTSC's. The atomic character of these states and their spin-correlations is examined.

We proceed with the discussion of the spectral function for a single hole moving in a Heisenberg antiferromagnet, i.e the t-J model. This model was proposed by Anderson [5] and by Zhang and Rice [6] as a generic model for the copper oxide superconductors. It is widely agreed upon that the magnetic properties of the undoped system are well described by the isotropic spin-1/2 Heisenberg model, viz. the atomic limit of the 1-band Hubbard model at half-filling. This is a direct consequence of the strong correlations on the Cu-sites, that is the stabilization of the Cu d^9-configuration.

Whether the 3-band model reduces in the hole-doped case to the t-J model through the formation of local singlets as argued by Zhang and Rice remains a matter of controversy [7,8,9]. The fact that doping destroys long-range antiferromagnetic order at rather low concentration of holes (3% in $La_{2-x}Sr_xCuO_4$) is a direct indication of the strong coupling between the holes and the Cu-spins. We address here the reverse problem, namely the effect of the spin-system on the nature of the carriers.

As central result it is found that the spectral function for a single hole in 2D is characterized by a quasiparticle (QP) peak at the low-energy side of an otherwise incoherent spectrum [10,11]. The coherent motion of the QP is described by the dispersion relation $E_{\vec{k}} \sim \frac{J}{2}(\cos k_x + cos k_y)^2$ [11]. The propagation for realistic values for t and J is mainly due to the action of the spin-flip term in the Heisenberg operator, leading to an effective second nearest neighbor hopping of the hole. The one-dimensional case is different in the sense that quasiparticle-like states are absent. [12]

This contribution is organized as follows. After a description of the numerical approach to the calculation of spectral functions, we consider the photoemission and inverse photoemission spectra obtained for the 3-band model. Particular emphasis will be put on the characterization of the low-lying excitations in this model. The next topic is the spectral function of a hole in the t-J model. Finally we conclude by comparing the low-energy physics of the Emery model and the excitation spectrum of the t-J model.

CALCULATION OF THE SPECTRAL FUNCTION

The method used to calculate the spectral function is an application of the recursion method of Haydock, Heine and Kelly [13] based on the Lanczos algorithm, which has previously been used in a many-body context by Gunnarsson and Schönhammer [14], and others [15,16,11]. The spectral density of the single-hole excitations may be defined as

$$A_{k\sigma}(\omega) = \frac{1}{N_c} \sum_m |\langle\psi_m(N-1)|a_{k\sigma}|\psi_0(N)\rangle|^2 \delta(\omega - E_0(N) + E_m(N-1)). \quad (1)$$

Here the operator $a_{k\sigma}$ annihilates a particle with momentum k and spin σ, $|\psi_0(N)\rangle$ is the ground state eigenfunction of an N-particle system, and $|\psi_m(N-1)\rangle$ is an eigenstate of the (N-1)-particle system, which have energies $E_0(N)$ and $E_m(N-1)$, respectively. An analogous definition applies for the particle spectral function. Naturally momentum conservation restricts the possible final states $|\psi_m(N-1)\rangle$ which may contribute to the spectral function. The hole spectral function may be rewritten as imaginary part of the single-particle Greens function:

$$A_{k\sigma}(\omega) = \frac{1}{\pi} Im\langle u_0(N-1)|\frac{1}{E+H-i\delta}|u_0(N-1)\rangle, \quad (2)$$

where $|u_0(N-1)\rangle = a_{k\sigma}|\psi_0(N)\rangle$, and $E = \omega - E_0(N)$ is the excitation energy measured from the ground state energy of the N-particle system, and $\delta \to 0^+$. In practice a finite value of δ leads to a useful smoothing of the spectra, which would otherwise consist of a set of spikes due to the relatively small number of k-points used.

The desired expectation value is readily calculated iteratively by using the Lanczos algorithm [17,18] to generate a basis in which the Hamiltonian is tri-diagonal. To begin with the N-particle ground state eigenvector $|\psi_0(N)\rangle$ is calculated using the standard Lanczos technique. Then a particle is annihilated by the operator a_k which may depend on the experiment which is being mimicked when there is more than a single site per unit cell. This new (N-1) -particle state is then expressed in the appropriate basis, and the problem of calculating the propagator has been reduced to the evaluation of the expectation value of the matrix $(H+z)^{-1}$ in this state $|u_0(N-1)\rangle$. The desired expectation value may be found by renewed application of the Lanczos algorithm, with the starting vector given by $|u_0(N-1)\rangle$. After M Lanczos iterations, an M dimensional tridiagonal representation of the Hamiltonian is generated. The coefficients of this matrix may then be used in a continued-fraction expansion [13] of the inverse $(H+z)^{-1}$, or equivalently this expectation value may be expressed in terms of the eigenvalues and eigenvectors of the (small) M-dimensional tridiagonal matrix. The latter approach is to be preferred if frequency integrals over the density of states are desired.

SINGLE-PARTICLE EXCITATIONS OF EXTENDED HUBBARD MODELS

A widely accepted model for the CuO_2-planes is the 3-band model [19,4], which includes both orbitals on Cu $(d_{x^2-y^2})$ and on O (p_σ). Coulomb interactions are characterized by 3 terms, Hubbard repulsions $U_d(U_p)$ for two holes on Cu(O)-sites, respectively, and a Coulomb repulsion $V=U_{pd}$ between nearest neighbor Cu-O pairs. The model, which is also known as Emery's model, is able to describe a variety of situations, e.g. insulators with Mott-Hubbard or charge-transfer gaps.

$$H=\sum_{i,\sigma}\epsilon_i n_{i,\sigma}+\sum_{i,j,\sigma}t_{i,j}a^+_{i,\sigma}a_{j,\sigma}+U_i\sum_{i,j}n_{i,\uparrow}n_{i,\downarrow}+\frac{V}{2}\sum_{i,j,\sigma,\sigma'}n_{i,\sigma}n_{j,\sigma'}. \tag{3}$$

The $a^+_{i,\sigma}$ are creation operators for holes in copper $3d_{x^2-y^2}$ or oxygen $2p_x(2p_y)$ orbitals, respectively, and $n_{i,\sigma}$ are the corresponding particle number operators. We employ the hole picture, i.e. the vacuum state with no holes corresponds to Cu^+ and O^{2-}, i.e. all 3d and 2p states occupied. The insulating (undoped) ground state has one hole per CuO_2-unit which resides predominantly on Cu, as $\epsilon=\epsilon_p-\epsilon_d>0$. We put $\epsilon_d=-\epsilon_p$. A strong hybridisation $t_{pd}(t_{pd}\leq\epsilon)$ is characteristic for the copper oxides, i.e. leading to a large covalent splitting of the bands. The stabilization of the Cu d^9-configuration is a direct consequence of the strong correlations on Cu $(U_d>\epsilon)$. This interaction also forces additional holes essentially on the O-sites, i.e. the model describes a charge-transfer insulator. In the reverse case if $\epsilon>U_d>t_{pd}$ the model describes a Mott-Hubbard insulator. We may also include the nearest neighbor repulsion $V=U_{pd}$ [20] and a finite hopping matrix element t_{pp} between neighboring oxygen sites which are expected to be significant. Parameter sets pertinent to the copper oxides have been derived from spectroscopic data [7] and calculated on the basis of the constrained density functional approach [21].

The practical limit for exact diagonalization is at present reached with a cluster of only 4 CuO_2 units, even with the restriction of the model to only a single orbital per site. This is a rather severe limitation because the discrete nature of the spectra of finite systems may cause difficulties in the interpretation of the calculated spectra. However through the use of modified periodic boundary conditions [22,23] this problem is at least partially overcome. This consists of the inclusion of complex phase factors $\exp(i\phi)$ when a particle crosses the boundary of the cluster and enters again at the other side. Through this device it is possible by varying the angle ϕ to alter the set of allowed single particle momenta continuously. The accompanying energy shifts result in a filling in of the calculated spectra.

We begin with a qualitative discussion of the spectral function and of the main changes expected from correlation effects on the large energy scale. Figure 1(a) gives a sketch of the density of states for the 3-band model. For the undoped system the antibonding band is half-filled, that is the bandstructure is that of a metal. If we include the large Coulomb correlation on Cu this band is expected to split, forming a lower and upper Hubbard band (Fig. 1b). These 'bands' correspond to transitions $d^9 \rightarrow d^8$ and $d^9 \rightarrow d^{10}$, respectively. The d^8 final state (2 holes on Cu) is shifted by an energy $\sim U_d$. Such a spectrum is characteristic for charge-transfer insulators [3,24] with a CT-gap E_{CT}.

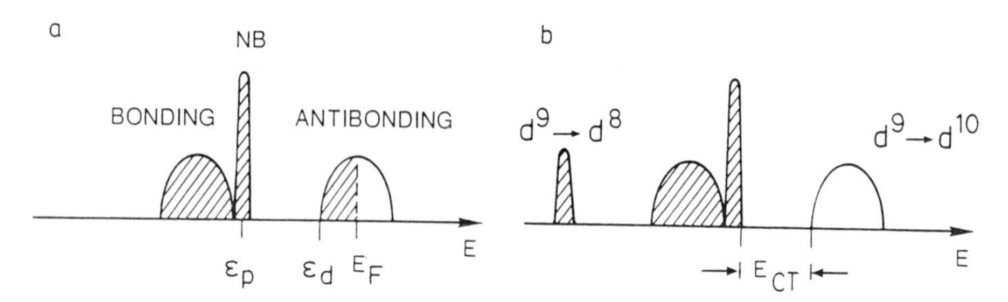

Fig. 1. (a) Sketch of density of states for the 3-band model. The antibonding band is half-filled in the undoped case (d^9p^6). The nonbonding band (NB) is dispersionless for $t_{pp} = 0$. (b) Sketch of Photoemission (shaded) and inverse PES in the presence of strong correlations on Cu (large U_d). The antibonding band is split into a lower and upper Hubbard band as a result of correlations. In the case $U_d \gg \epsilon$ shown here the lowest particle-hole excitations are across the charge transfer gap E_{CT}.

Figure 2(a,b) shows typical photoemission and inverse photoemission spectra (PES) starting from the undoped ground state. The Cu and O related spectral weights were calculated by adding or removing an electron on a single atom per unit cell (Cu, O_1, or O_2) in the form of a Bloch state. We present here the k-integrated spectra which correspond to the local creation at a given atom. An example where a_k was chosen to correspond to an antibonding wave function may be found in References 11 and 16 for comparison. The parameters chosen are $\epsilon = 2.0$, $U_d = 6.0$, $U_p = 3.0$, $U_{pd} = 0.0$, with $t = 1.0$ taken as the unit of energy. Hence all energy scales must be multiplied by $t_{pd} \sim 1.5eV$ to obtain energies in eV. This parameter set is representative for the values found in the constrained-density-functional approach [21] . The existence of a gap between the top of Fig. 2(a) and the bottom of Fig. 2(b) is consistent with an insulating ground state at half-filling, as expected. The structures in the range from E=-3 to -1 in PES (2a) derive from the nonbonding and bonding bands, respectively. A sharp and well separated $d^9 \rightarrow d^8$ satellite appears, however, only for larger values for U_d [25] . For the parameter set of Fig. 2 a more complex 3-peak structure (E=-9 to -5) results due to a mixing with other final states, e.g. $d^{10}p^4$. This is a signature of the fluctuations in the ground state between d^9p^6 and $d^{10}p^5$ ($< n_d > \sim 0.7$ holes).

The most interesting feature are the low-lying excitations at $E \sim 0.5$. which have no correspondence in the bandstructure, and appear as result of correlations [11] . This also implies an important modification of the picture as sketched in Fig. (1b). We will discuss these states below.

The most notable change in spectra (2c) and (2d) of the hole-doped system is the appearance of states in the inverse PES within the energy region corresponding to the gap at half-filling shown in Figs. 2(a) and (b). The emergence of these states upon doping is actually seen e.g. in O 1s absorption spectra for various copper oxides [24,26] . The existence of a pseudogap between these states and the upper Hubbard band sets a lower limit for ϵ. The 'oxygen related' gap states are, however, strongly hybridized and have considerable Cu-character. Evidence for such a strong hybridization of the

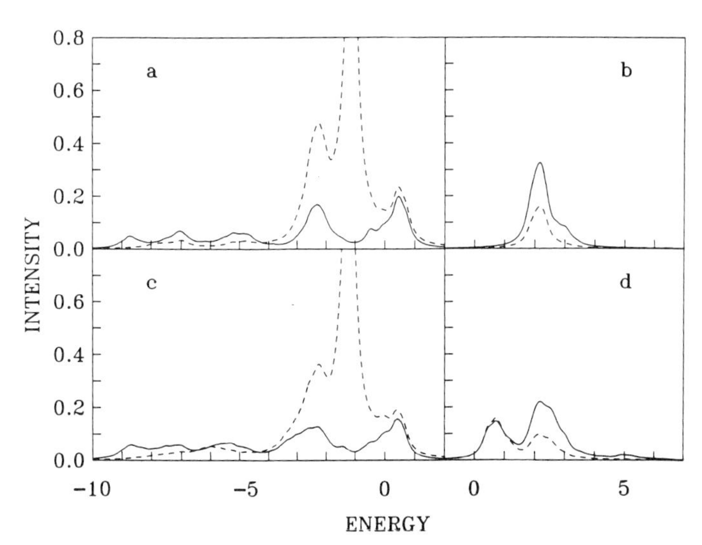

Fig. 2. Photoemission spectra for the parameters $\epsilon = 2.0$, $U_d = 6.0$, $U_p = 3.0$, $U_{pd} = 0.0$, with $t = 1.0$ taken as the unit of energy. Cu-(solid) and O-spectra (dashed lines) are given seperately. Figures 2(a,b) show PES and inverse PES, respectively, for the undoped initial state. Figs. 2(c,d) represent a repetition of the same calculations starting from an initial state with one extra hole, which implies a doping concentration of 25%.

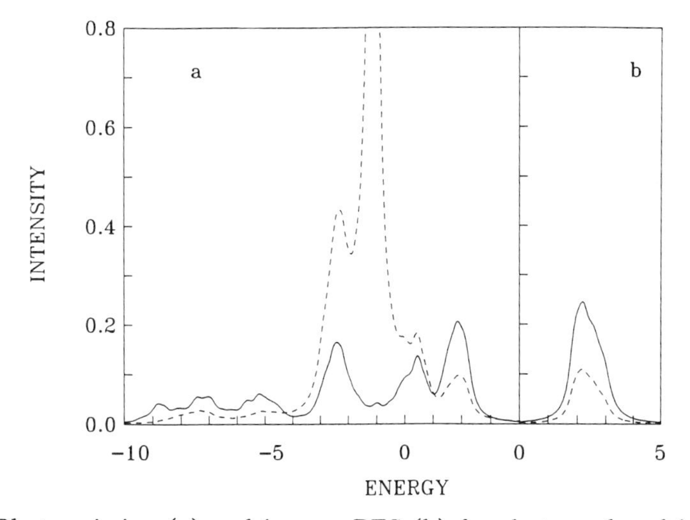

Fig. 3. Photoemission (a) and inverse PES (b) for electron doped initial state. Cu- and O- spectra are represented by solid and dashed lines, respectively. Parameters as for Fig.2. The initial state contains one extra electron and the corresponding doping concentration is 25%.

states close to the Fermi level has been found in recent photoemission experiments of the $Bi_2Sr_2CaCu_2O_8$ materials [27] . It is evident that the upper Hubbard band has also some oxygen character, because in the ground state $< n_d >\sim 0.7$.

In the electron doped case [28] (Fig. 3) the chemical potential moves into the upper Hubbard band. The characteristic feature are new states in the PES arising from the Hubbard band at about $E \sim 2$. The considerable width of this peak may be attributed to the correlated nature of carriers in this band. This suggests the association of the low-energy peak observed in $Nd_{2-x}Ce_xCuO_4$ (XAS) with the upper Hubbard band. This could explain its only marginal concentration dependence. However there are also other explanations. [24] Moreover there is still no clear experimental evidence for a related increase of Cu^+, which might indicate the necessity to extend the model.

LOW-LYING SINGLE PARTICLE EXCITATIONS

A particularly interesting feature in Fig. 2 is as already emphasized the appearance of the low-lying single-hole excitations at energy $E \sim 0.5$. In order to further characterize these states we calculate the local singlet and triplet correlation functions

$$C_s = \sum_i < \psi_{si}^+ \psi_{si} >, \quad C_t = \sum_i < \psi_{ti}^+ \psi_{ti} >, \quad \psi_i = \frac{1}{\sqrt{2}}(d_{i\uparrow} P_{i\downarrow} \mp d_{i\downarrow} P_{i\uparrow}), \tag{4}$$

where the $-(+)$ sign refers to the local singlet (triplet) operator. The operator P^+ creates a hole in a symmetric linear combination of oxygen orbitals around a Cu site, i.e. the O-orbital has the same symmetry as $d_{x^2-y^2}$.

For the lowest final state in PES we find a large singlet correlation C_s and almost vanishing triplet correlation function (Fig. 4), that is a strong antiferromagnetic correlation between an additional oxygen hole and its Cu-neighbors. Similar results for C_s have been obtained by finite-temperature Monte-Carlo calculations [29] . These results support arguments given by Zhang and Rice [6] and also by Eskes and Sawatzky [7] . They argued that the singlet state

$$\psi_s^+ |0 >= \frac{1}{\sqrt{2}}(P_\downarrow^+ d_\uparrow^+ - P_\uparrow^+ d_\downarrow^+)|0 > \tag{5}$$

formed by two holes in a CuO_4-cluster should also describe the low-lying excitations of the CuO_2-planes. The singlet-triplet splitting for $U_d \gg \epsilon \gg t$ is given by

$$E_{st} = -8\,(\frac{t_{pd}^2}{U_d - \epsilon} + \frac{t_{pd}^2}{\epsilon}). \tag{6}$$

Equation (5) may also be interpreted as Heitler-London limit of a covalent bond formed out of the $d_{x^2-y^2}$ and the corresponding symmetric oxygen orbital [25] . The formation of local singlets gives a natural explanation of the linear increase of C_s as function of the number of added holes as observed in Fig.4 .

Doping introduces holes into these low-lying states. This leads to a strong suppression of the antiferromagnetic order between Cu-sites, which may be measured e.g. by the spin-spin correlation function between nearest neighbor Cu-sites

$$C_1 = \frac{1}{N} \sum_{i,\vec{\delta}} < S_{Cu}^z(\vec{r}_i + \vec{\delta}) S_{Cu}^z(\vec{r}_i) >, \tag{7}$$

$$C_2 = \frac{1}{N} \sum_{i,\vec{\delta}} < S_{Cu}^z(\vec{r}_i + \vec{\delta})\; n_O(\vec{r}_i + \vec{\delta}/2)\; S_{Cu}^z(\vec{r}_i) > . \tag{8}$$

Here $\vec{\delta}$ connects neighboring Cu-sites. The second CF, C_2, is particularly interesting because it measures the spin-correlation function between two Cu-neighbors only when an oxygen hole sits in between.

With one additional hole, corresponding to a doping concentration of 25%, C_1 is strongly reduced, yet still antiferromagnetic. C_2 on the other hand turns out ferromagnetic [16], that is the oxygen hole tends to align its neighboring Cu-spins ferromagnetically when it moves through the crystal. We stress that the correlation between the spins on O and on Cu is antiferromagnetic, according to the singlet correlation function. This has similarity with the frustration model of Aharony [30], albeit the size of the local ferromagnetic correlation is much smaller than in Aharony's model [16], which does not allow for both hybridisation and the motion of the carriers.

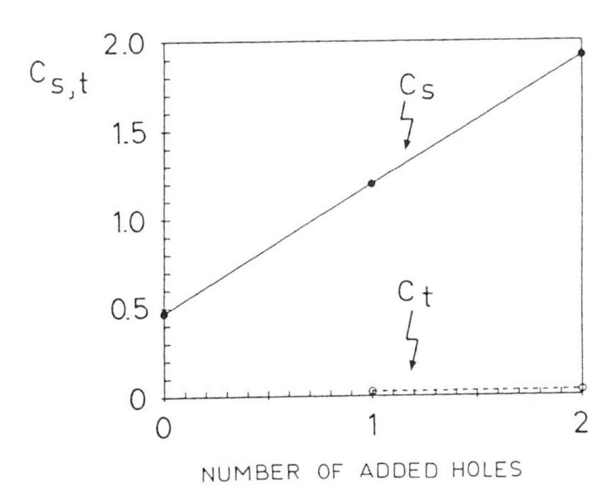

Fig. 4. Local singlet (C_s) and triplet (C_t) correlation functions in the ground state of the Emery model with 0,1, and 2 additional holes. Parameter set as for Fig. 2.

EFFECTIVE HAMILTONIAN FOR THE STRONG CORRELATION LIMIT: t-J MODEL

It was emphasized by Anderson that the essence of the low-energy physics of the CuO_2-planes is contained in the single-band Hubbard model. The t-J-model is derived from the Hubbard Hamiltonian in the limit $t/U \ll 1$. By canonical transformation doubly occupied configurations are eliminated [31,32,33] leading to the following Hamiltonian up to order t^2/U

$$H_{tJ^*} = -t\sum_{i,j,\sigma} c^+_{i,\sigma}c_{j,\sigma} + \frac{t^2}{U}\sum_{j,\sigma}\sum_{\delta,\delta'}(c^+_{j+\delta,\sigma}c^+_{j,\bar{\sigma}}c_{j,\bar{\sigma}}c_{j+\delta',\sigma} - c^+_{j+\delta,\sigma}c^+_{j,\bar{\sigma}}c_{j,\sigma}c_{j+\delta',\bar{\sigma}}), \quad (9)$$

where $j+\delta$ and $j+\delta'$ are nearest neighbors of j, respectively. It is understood that creation and annihilation operators are restricted from here on to the space without double occupancy. The t^2/U-terms describe a hop from $j+\delta$ to $j+\delta'$ via a virtual intermediate state with a double occupancy at j.

The t-J-Hamiltonian with $J = 4t^2/U$ follows from the 2-site contributions ($\delta = \delta'$) after rewriting in terms of spin and number operators.

$$H_{tJ} = -t\sum_{i,j} c^+_{i,\sigma}c_{j,\sigma} + J\sum_{\langle i,j\rangle}\left(\vec{S}_i\cdot\vec{S}_j - \frac{n_i n_j}{4}\right). \quad (10)$$

The 3-site terms ($\delta \neq \delta'$) in Eq.(9) only contribute in the presence of a hole. These terms are frequently neglected, which is a valid approximation close to half-filling when one is interested in e.g. the ground state energy. The corrections are of the order $\delta_h J$ with δ_h being the concentration of holes. For our problem, i.e. the propagation of a hole, it is not at all obvious that they can be neglected. We have investigated this question by comparing the excitation energies of the low-energy states for the t-J^* (9) and the t-J model (10) with exact results for the 1-band Hubbard model in 1D and 2D [10]. The inclusion of the 3-site terms improves the agreement with the Hubbard model significantly. However, as the total dispersion of the low-energy states changes only by a small amount when omitting these terms, we conclude that they are not crucial for the understanding of the propagation of carriers.

Zhang and Rice [6] argued that the low lying excitation spectrum of the Emery model [4] can be mapped on the t-J-model. This correspondence is considered valid in the limit that U_{Cu} and the level spacing $\epsilon = \epsilon_p - \epsilon_d$ between Cu($d_{x^2-y^2}$) and O(p_σ) orbitals are large compared to the hybridisation t_{pd}. In this limit the hopping matrix element t describing the motion of a singlet [6] and the superexchange J are given by

$$t \sim \frac{t_{pd}^2}{\epsilon}, \quad J = \frac{4t_{pd}^4}{\epsilon^2}\left(\frac{1}{U_d} + \frac{1}{\epsilon}\right). \tag{11}$$

We note that the upper and lower Hubbard band of the effective single-band Hubbard model correspond to structures in the spectra of the Emery model around $E \sim 0.5$ (Fig. 2(a)) and $E \sim 2$ (Fig. 2(b)) , respectively. The effective U is of the order of the charge- transfer gap. We will return below to the question of the equivalence of the low-lying excitations of the Emery model and those of the t-J model. In the following we discuss the spectral function for the t-J model in 2D. For the one-dimensional problem see Reference 12.

SPECTRAL FUNCTION FOR THE t-J MODEL IN TWO DIMENSIONS

The spectral function for a single hole moving in a classical Neel-state (J=0), that is moving via the restricted hopping only has been worked out by Brinkman and Rice (BR) [34]. The single particle Greens function for this case

$$G(\omega) = \frac{1/\omega}{1 - \frac{z}{z-1}\left[\frac{1}{2} - \sqrt{\frac{1}{4} - (z-1)\frac{t^2}{\omega^2}}\right]} \tag{12}$$

describes a k-independent continuum, i.e. the Greens function in real space is given by $G_{ij}(\omega) = \delta_{ij} G(\omega)$. In 2D the number of nearest neighbors is z=4. The density of states $A(\omega)$ defined by (12) is in good agreement with the numerical result for J=0, when the Neel-state is taken as initial state in (1) [10]. The lowest spin-1/2 states lie at $\omega_0 \approx -3.45\,t$ for N=16, which is very close to the band edge $\omega_0 = -2\sqrt{3}\,t$ in the retraceable path approximation of BR. As emphasized by BR there is in addition a band tail extending to -4t. The final states in this tail have high spin and their spectral weight is small [10].

The $S = 1/2$-states become lowest when the antiferromagnetic coupling between the spins exceeds a critical value. Exact diagonalization studies showed the existence of a critical value of J_c for finite systems [35,36] above which the fully spin-polarized state, which is stable according to Nagaoka's theorem [37] in the limit $J = 0$, becomes unstable. For 16 sites the ferromagnetic state becomes unstable for $J_c^{min} \approx 0.06$, while above $J_c^{max} \approx 0.075$, the ground state already has spin $S = 1/2$. As an estimate for the thermodynamic limit one finds $J_c \approx 0.005$ when neglecting $H_J^{\perp}$. [38] That is, in the physically interesting regime the low-energy states have low spin (S=1/2), and are not of the Nagaoka type.

A different picture emerges if one considers the spectral function of a hole moving in a spin fluctuating background ψ_0 for finite J [11,10]. A typical spectrum for $A(\vec{k},\omega)$ at $\vec{k}=(\frac{\pi}{2},\frac{\pi}{2})$ is shown in Fig. 5 for $N=16$. The form of this spectrum with a peak at the low-energy side of a continuum of width 7t is similar to spectra at other $\vec{k}$-values for $N=16$ [10] and $N=18$ [11], except for those at the special points $\vec{k}=(0,0)$ and (π,π).

The momentum integrated spectral function starting from the singlet ground state is shown in Fig. 6. The set of peaks, which arise from different $\vec{k}$-values at the bottom of the continuum, appear well separated from the incoherent part of the spectrum. The separation is of the order of 2J and might be related to a density of states effect of the spin excitations as suggested by Kane et al. [39] The depletion of $A(\omega)$ in the interval $0 \leq \omega \leq 2t$ is already present for $J \to 0$ and is a signature of the singlet ground state. It is not seen in calculations starting from the classical Neel-state. The peak at $\omega \approx 1$ stems from the spectrum at $\vec{k}=(\pi,\pi)$.

The momentum-dependence of the QP-peak is well described [11,10] by the dispersion

$$E_{\vec{k}} \sim \frac{J}{2}(\cos k_x + \cos k_y)^2 \tag{13}$$

after subtracting a k-independent energy. The form of this dispersion $E_{\vec{k}}$ is related to an effective next-nearest neighbor hopping. The lowest energies are found on the Fermi surface for noninteracting fermions, i.e., $\vec{k}=(\frac{\pi}{2},\frac{\pi}{2})$ and $(\pi,0)$. The precise location of the absolute minimum is, however, not possible due to an accidental degeneracy at these points in the case N=16. A more accurate scaling of the QP bandwidth W on J is given by $W(J)=-0.23+2.66\,J^{0.91}$ for 16 sites and J between 0.04 and 0.5. The negative intercept indicates two conflicting mechanisms: (a) the coherent motion via spin-flip and (b) higher order processes in the hopping [40]. For larger values for J the spin-flip process is dominant. The "reversed" dispersion is actually observed for $J \leq 0.06$, as well as in the pure Ising case. Since a reliable scaling to the thermodynamic limit is lacking, we cannot decide on the basis of our data whether the bandwidth is linear in J or obeys a sub-linear form as proposed by Gros and Johnson. [41]

Our finding of a well defined peak at the bottom of the continuum confirms the dominant pole approximation in the work of Kane, Lee and Read [39] in their extension of the approach of Schmitt-Rink et al [42]. Surprisingly their perturbative treatment about the Ising limit, i.e., considering the anisotropy $\alpha = J^{\perp}/J^{z}$ as small parameter, gives the proper width for the quasiparticle band at the isotropy point ($\alpha=1$). The emergence of the QP may be understood as a consequence of the string energy associated with the motion of a hole in 2D, which results in a ladder type spectrum for $J^{\perp}$=0. Inclusion of spin-waves leads to a smearing of these structures except for the low-energy excitation which remains undamped. We emphasize that the spectral weight of the quasiparticle peak is large over the whole Brillouin zone except at the special points $\vec{k}=(0,0)$ and (π,π) where other states dominate the spectrum.

We finally mention that the features of Fig. 5 in particular the well separated quasiparticle peak have been confirmed also by other groups [43]. Similar spectra are also obtained if we perform the same calculations for the XY- and the Ising-Hamiltonian [10].

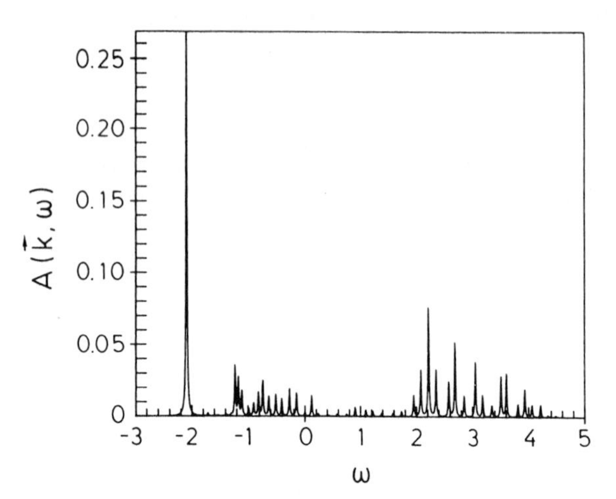

Fig. 5. Spectral function $A(\vec{k},\omega)$ at $\vec{k} = (\frac{\pi}{2}, \frac{\pi}{2})$ for a single hole inserted into the exact singlet ground state of the Heisenberg antiferromagnet in 2D ($N = 4 \times 4$ and J/t=0.2).

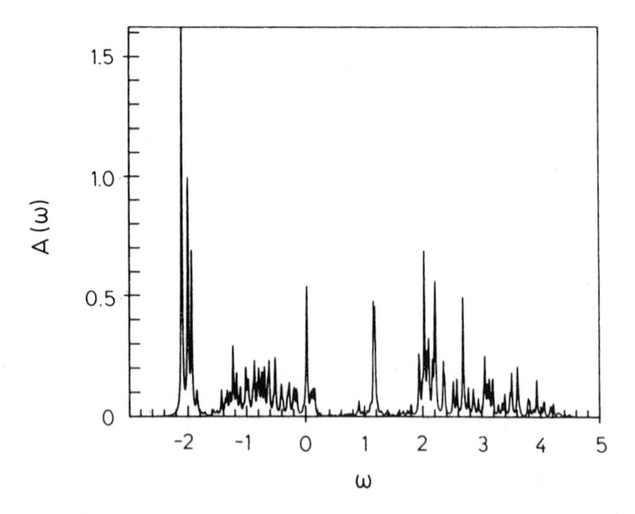

Fig. 6. Momentum integrated spectral function (density of states). Same parameters as in Figure 5.

EQUIVALENCE OF LOW-LYING EXCITATIONS OF THE EMERY MODEL AND THE t-J MODEL

Numerical studies are still limited to supercells with $4CuO_2$-units which corresponds to a 4-site t-J model. [44] We have identified the low-lying 1-hole excitations in the Emery model according to their singlet and triplet character, Eq.(4). We find that the singlet structures in the Emery model lie significantly below the triplet excitations for $\epsilon > 4$. Similar trends for the different $\vec{k}$-values are found in both models. To be more specific, for e.g. $\vec{k} = (\pi, 0)$ there are only two peaks in the t-J model for $N = 2 \times 2$. These peaks are separated by an energy of 7t, this is by coincidence precisely the width of the continuum for the larger system ($N = 4 \times 4$) in the t-J model. The corresponding low-energy structures in the Emery model have singlet character and are below the 'triplet' states. For smaller Cu-O splitting the upper state merges with triplet states and looses its singlet character, whereas the state at low energy retains its local singlet character. This makes it plausible to expect the quasiparticles found in the t-J model also for the Emery model in the realistic parameter regime. We finally mention that finite t_{pp} does not destroy this result. Actually the singlet is further stabilized, which has also been reported by others [45] .

SUMMARY

The most notable structures in the photoemission spectra of the Emery model are besides the satellites at high energy the appearance of low-energy states below the band-like states (Zhang-Rice singlet). Occupation of these states leads to the frustration of antiferromagnetic order between Cu-sites. In the hole-doped systems these states appear in the charge-transfer gap and are seen in inverse PES. Remnants of the charge-transfer gap are expected to remain as a pseudo-gap for sufficiently large p-d level separation. The spectra calculated for realistic parameters also show that the states in the CT-gap are strongly hybridized, i.e., additional holes do not exclusively reside on oxygen. We finally mention that the energetic position of the satellites calculated for parameters taken from the constrained-density-functional approach are in good agreement with experiment.

New puzzles are posed, however, by the recently found electron doped superconductors with CuO_2-planes [28] , e.g.$Nd_{2-x}Ce_xCuO_4$. In the framework of the Emery model these electrons would go into the upper Hubbard band corresponding to the formation of Cu^+. The chemical potential is pinned at the edge of the upper Hubbard band. The low-energy states found in inverse photoemission (XAS) in these compounds [24] could be taken as confirmation of this picture. Yet these states have also been interpreted differently [24] , and furthermore there is still no clear experimental evidence for the substantial increase of Cu^+. This might indicate the necessity to extend the model.

The motion of a carrier (hole) strongly interacting with the spins of a Heisenberg antiferromagnet was studied in detail in the case of the t-J model. Our central finding is a coherent quasiparticle band with dispersion $E_{\vec{k}} \sim \frac{J}{2}(\cos k_x + cosk_y)^2$ and $S = 1/2$, which emerges at the low-energy side of a broad incoherent spectrum of width 7t. Studies for finite doping have not been performed yet. However, it is tempting to speculate that this might lead to a Fermi- liquid like picture at finite doping concentration.

ACKNOWLEDGEMENTS

We are particularly grateful to M. Ziegler, K.v.Szczepanski, W. von der Linden who have contributed to this work, and to P. Fulde, J. Fink, T.A. Kaplan, A. Muramatsu, A.M. Oles, P. Prelovsek, and J. Zaanen for many stimulating discussions.

REFERENCES

1. J. C. Fuggle, J. Fink, and N. Nücker, Int. J. Mod. Phys. B1,1185 (1988) and references therein.
2. R. Manzke, this volume.
3. J. Zaanen, G.A. Sawatzky, and J.W. Allen, Phys.Rev.Lett. 55,418 (1985), J. Zaanen, C. Westra, and G. A. Sawatzky, Phys.Rev. B33,8060 (1986).
4. V. J. Emery, Phys. Rev. Lett. 58,2794 (1987).
5. P. W. Anderson, Science 235,1196 (1987), Proceedings of the International School of Physics 'Enrico Fermi', July 1987, (North Holland, Amsterdam,1989), and Physics Reports 184, 195 (1989).
6. F. C. Zhang and T. M. Rice, Phys.Rev. B37,3759 (1988).
7. H. Eskes and G.A. Sawatzky, Phys. Rev. Lett. 61, 1415 (1988).
8. V.J. Emery and G. Reiter, Phys. Rev. B38, 11938 (1988).
9. F.C. Zhang, Phys. Rev. B39, 7375 (1989).
10. K.J. von Szczepanski, P. Horsch, W. Stephan, and M. Ziegler, Phys. Rev. B41, Febr. (1990).
11. P. Horsch, W.H. Stephan, K. v. Szczepanski, M. Ziegler, and W.v.d. Linden, Physica C 162-164, 783 (1989).
12. M. Ziegler and P. Horsch, this volume.
13. R. Haydock, V. Heine, and M. J. Kelly in Solid State Physics, Vol.35, edited by H. Ehrenreich, F. Seitz, and D. Turnbull (Academic, New York, 1980).
14. O. Gunnarsson and K. Schönhammer, Phys. Rev. B31, 4815 (1985).
15. C.A. Balseiro, A.G. Rojo, E.R. Gagliano, and B. Alascio, Phys. Rev. B38, 9315 (1988).
16. P. Horsch and W.H. Stephan, in 'Interacting Electrons in Reduced Dimensions', ed. by D. Baeriswyl and D. Campbell (Plenum,New York,1989).
17. C. Lanczos, J. Res. Natl. Bur. Stand. 45, 255 (1950).
18. J. Borysowicz, T.A. Kaplan, and P. Horsch, Phys. Rev. B31, 1590 (1985).
19. C.M. Varma, S. Schmitt-Rink, and E. Abrahams, Solid State Commun. 62, 681 (1987).
20. W.H. Stephan, W. von der Linden, and P. Horsch, Phys. Rev. B39, 2924 (1989) and Int. J. Mod. Phys. B1, 1005 (1988).
21. M. S. Hybertsen, M. Schlüter and N. E. Christensen, Phys. Rev. B 39, 9028 (1989); A.K. McMahan, R.M. Martin, and S. Satpathy, Phys. Rev. B38, 6650 (1988); E.B. Stechel and D.R. Jennison, Phys. Rev. B40, 6919 (1989).
22. R. Jullien and R. M. Martin, Phys. Rev. B26,6173 (1982).
23. A. M. Oles, G. Treglia, D. Spanjard, and R. Jullien, Phys. Rev. B32,2167 (1985).
24. J. Fink et al. in 'Earlier and recent aspects of Superconductivity', ed. K.A. Müller and G. Bednorz; Springer Series of Solid-State Sciences.
25. P. Horsch, Helv. Phys. Acta 63, xxx (1990).
26. F. J. Himpsel et al., Phys. Rev. B38,11946 (1988).
27. R.S. List et al., Physica C 159, 439 (1989).
28. Y. Tokura, H. Takagi, and S. Uchida, Nature 337, 345 (1989).
29. G. Dopf, A. Muramatsu, and W. Hanke, to be published.

30. A. Aharony, R. J. Birgeneau, and M. A. Kastner, Int.J.Mod.Phys. B1,649 (1988).
31. L.N. Bulaevskii, E.L. Nagaev, and D.L. Khomskii, Sov. Phys. JETP 27, 836 (1968).
32. J.E. Hirsch, Phys. Rev. Lett. 54, 1317 (1985).
33. J. Zaanen and A.M. Oles, Phys. Rev. B37, 9423 (1988).
34. W.F. Brinkman and T.M. Rice, Phys. Rev. B2, 796 (1970).
35. J. Bonca, P. Prelovsek, and I. Sega, Phys. Rev. B 39, 7074 (1989).
36. D. Poilblanc, Phys. Rev. B 39, 140 (1989) and W. von der Linden, private communication.
37. Y. Nagaoka, Phys. Rev. 147, 392 (1966).
38. B.I. Shraiman and E.D. Siggia, Phys. Rev. Lett. 60, 740 (1988).
39. C.L. Kane, P.A. Lee, and N. Read, Phys. Rev. B 39, 6880 (1989).
40. S.A. Trugman, Phys. Rev. B 37, 1597 (1988).
41. C. Gros and M.D. Johnson, Phys. Rev. B40, 9423 (1989).
42. S. Schmitt-Rink, C.M. Varma, and A.E. Ruckenstein, Phys. Rev. Lett. 60, 2793 (1988).
43. E. Dagotto et al. (preprint), S. Trugman (preprint).
44. A. Ramsak and P. Prelovsek, Phys. Rev. B40, 2239 (1989).
45. H. Eskes, G.A. Sawatzky, and L.F. Feiner, Physica C160, 424 (1989).

MICROWAVE ABSORPTION OF SUPERCONDUCTORS IN LOW MAGNETIC FIELDS

K.W. Blazey

IBM Research Division
Zurich Research Laboratory
8803 Rüschlikon, Switzerland

The results of modulated microwave absorption experiments on ceramics, thin films and single crystals of $YBa_2Cu_3O_{7-\delta}$ are described with emphasis on the different behavior of the various forms. All show microwave losses due to the presence of Josephson junctions which are a handicap for their potential application.

INTRODUCTION

On cooling a superconductor through its transition temperature the direct current resistivity drops sharply to zero. Resistivity measurements at different frequencies do not all show this same dramatic effect. At low frequencies the surface impedance is also found to drop but not to zero. Conversely, at higher frequencies as in the far infra-red there is an increase of absorption on cooling through T_c. Measurements at microwave frequencies show results like the low frequency probe. The losses, which decrease on cooling, at these frequencies are due to unpaired carriers which freeze out as the temperature is lowered. However even at the lowest temperatures the losses never disappear completely and there is always a residual surface impedance.

In the high-T_c copper oxide superconductors,[1] that were first made in ceramic form, the grains are weakly coupled by intergranular Josephson junctions, which dominate the magnetic properties[2] and contribute a large part of this residual surface impedance. Even in thin films and single crystals, the effect of junctions occurring naturally at the surface generates losses and limit the lowest surface impedance attainable. Due to the weak coupling and hence lower critical field, fluxons are created in the junctions at very low fields, much less than 1 Oe. This has the effect of giving rise to considerable microwave absorption by viscous fluxon motion[3] already in small applied magnetic fields.

Many of the recent microwave investigations of the high-T_c superconductors have been carried out in commercial electron spin resonance spectrometers. These instruments measure the derivative of a resonance absorption by applying a modulation field parallel to the external field and detecting the microwave absorption synchronously with it by lock-in techniques. The microwave absorption of a superconductor is neither resonant nor reversible and because of this latter property their modulated microwave absorption is not simply the derivative of the direct absorption. However the high sensitivity of these instruments has helped to distinguish the absorption processes contributing to the surface impedance.

Dynamics of Magnetic Fluctuations in High-Temperature Superconductors
Edited by G. Reiter *et al.*, Plenum Press, New York, 1991

The modulated microwave absorption of ceramic or powdered cuprate superconductors is illustrated in Fig. 1 and has been widely discussed in the literature.[4-11] The signal is quite rich in fluctuations, some of which may be removed by screening, showing they are due to environmental magnetic noise.[12] The main properties of the spectrum of Fig. 1 are an open hysteresis loop rotated slightly with respect to the axes. The loop is widest for small modulation fields and closes, first at high fields, as the modulation amplitude is increased. This behavior has its origin in the superconducting critical state that exists at the surface.[13] Two processes contribute to the modulated absorption signal, one due to the modulation of the fluxon density which is independent of the direction of the external field sweep and another due to the surface critical current.[14] The latter component will of course change sign with change of direction of the external field sweep and thus cause a change of sign in that signal component. This is the origin of the open hysteresis loops at low modulation fields. It has already been mentioned that fluxons entering the Josephson junctions network will contribute to the microwave losses by viscous fluxon motion. The linear change of the fluxon density with modulation field is apparent in the variation of the modulated microwave absorption with larger modulation amplitude shown in Fig. 2. The hump in Fig. 2 at small modulation amplitudes is due to the nonlinear critical component that saturates at larger modulation fields. This has been attributed to Josephson junction decoupling[15] or fluxon pinning / depinning[13] over a cycle of the modulation field. Both processes would give a nonlinear component for modulation fields smaller than the critical field

$$H_{c1}{}^{*} = \frac{4\pi}{c} \lambda J_c$$

where λ is a fluxon relaxation distance and J_c the critical current for depinning.

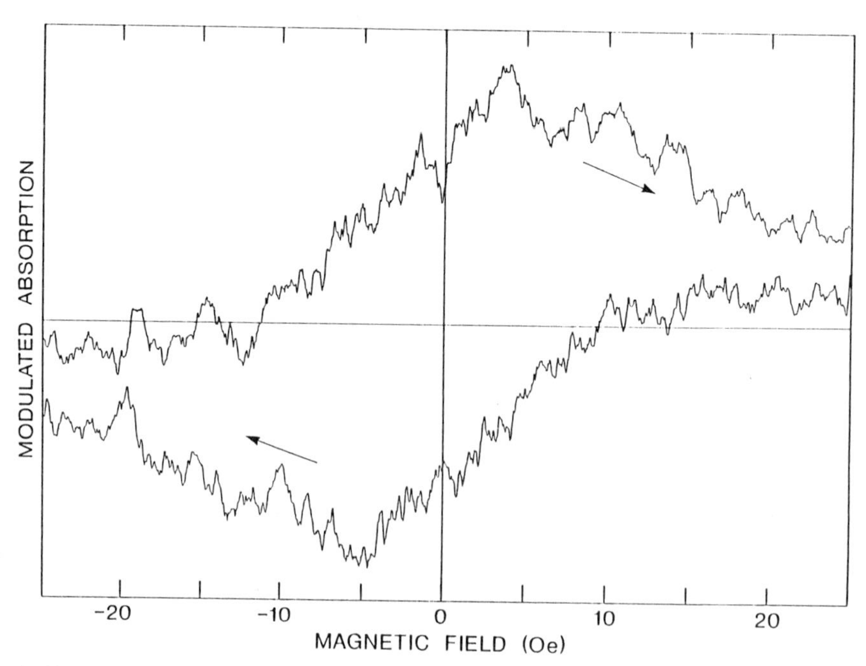

Fig. 1. Modulated microwave absorption of a $YBa_2Cu_3O_{7-\delta}$ ceramic at 67 K. The microwave power was 1 μW and the modulation field amplitude 0.1 Oe.

The same kind of hysteretic absorption due to granularity at the surface is also seen in conventional superconductors Nb[13] and $PbMo_6S_8$[16] where the surfaces are inhomogeneous. Grain boundaries in Nb polycrystals or thin films oxidize giving rise to Josephson junctions, and the sulphide is prepared as a hot sintered compressed powder and hence resembles the cuprate oxide ceramics.

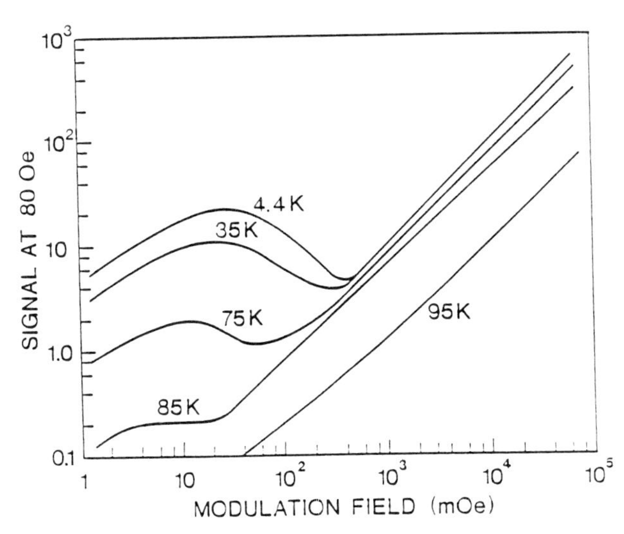

Fig. 2. Variation of the modulated microwave absorption of a $YBa_2Cu_3O_{7-\delta}$ ceramic at 80 Oe with the modulation field amplitude for different temperatures.

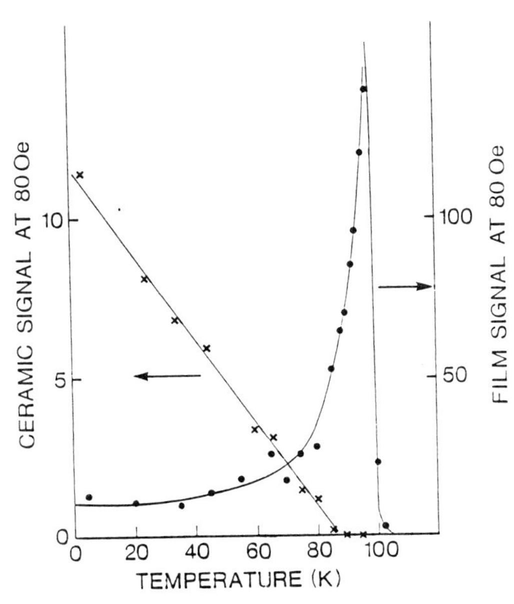

Fig. 3. Temperature dependence of the modulated microwave absorption in arbitrary units at 80 Oe for a $YBa_2Cu_3O_{7-\delta}$ ceramic with a modulation amplitude of 5 mOe and a thin film with 100 mOe modulation amplitude.

THIN FILMS

Because of their potential importance for applications the surface impedance of thin films has been widely studied[17-19] and also discussed[20] in terms of the weakly coupled granular model. This is borne out further by the modulated microwave absorption spectrum which is very similar to Fig. 1 when the external field is applied perpendicular to the film.[21] However, the temperature dependence of the signals is not the same for both as shown in Fig. 3. Whereas those due to the ceramic decrease on approaching T_c, those due to the thin film show a maximum similar to the imaginary part of the susceptibility. The weak intergranular Josephson junctions have been taken as the source of the absorption in the ceramics and this is still true in the thin films at low temperatures even though there are probably fewer weak junctions in the latter. At higher temperatures, near T_c, the intragranular fluxons would appear to dominate the modulated microwave absorption of the thin films as they do the susceptibility. The ceramics are dominated more by the weak junctions.

SINGLE CRYSTALS

The microwave absorption of single crystals[22,23] bears more resemblance to that of the thin films than the ceramics just described when large modulations are applied. This again is consistent with the presence of fewer weak junctions but there are still many twin domain boundaries present in $YBa_2Cu_3O_{7-\delta}$ crystals that may behave as Josephson junctions.[24]

Single crystal measurements at low microwave powers and small modulation fields revealed another source of microwave losses in the form of a regular line spectrum as shown in Fig. 4.[25] Such lines can be seen superimposed on the signals obtained from ceramics but due to the many grains and incorporated junctions they are not always clearly resolved as in the single crystal spectra.[5,11,26] However

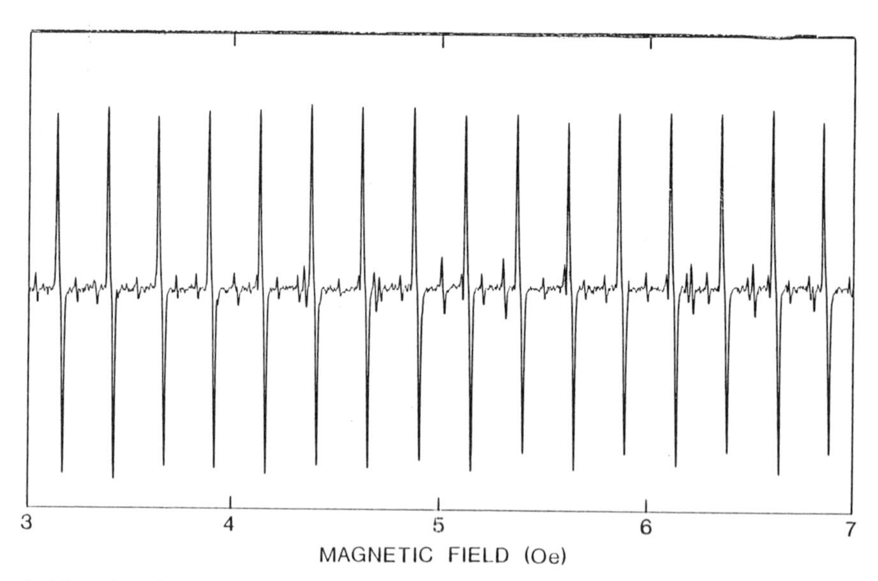

Fig. 4. Modulated microwave absorption of $YBa_2Cu_3O_{7-\delta}$ single crystal at 4.4 K showing the regular absorption line series due to a naturally occurring SQUID structure in the crystal. The incident microwave power is 3 μW and the modulation amplitude 10 mOe.

they are to be widely found, not only in the cuprate oxides, but also in conventional superconductors such as Nb[27] and $PbMo_6S_8$.[25] Such a periodicity indicates their origin to be related with fluxons appearing in a SQUID structure but the microscopic form of interferometer remains uncertain.

The spectra are highly anisotropic with respect to the external field and are determined by

$$H_0 \cos \phi = \pm (p + \tfrac{1}{2}) \Delta H$$

where p is an integer, ϕ the angle between the external field H_0 and a $\langle 110 \rangle$ direction and ΔH the minimum line separation which is given by

$$\Delta H = \frac{\phi_0}{S}$$

where ϕ_0 is the flux quantum and S the active area of cross section intersecting the magnetic field. Since the active area is in a $\langle 110 \rangle$ the twin boundaries are probably involved and fluxons are created at the top and bottom end of a junction in the c-direction. Striations associated with these twin boundaries are visible through the microscope with 1 μm separation. This separation and the crystal thickness yield an area S which fits the observed line periodicity. Apparently only a few junctions which absorb microwaves exist in the crystals shortly after growth and these change with time, possibly due to surface decomposition which occurs on exposure to air at extended planar defects.[28]

Close to zero field the line absorption is absent below a certain threshold microwave power. Above threshold the lines broaden directly proportional to the excess microwave field in the cavity over the threshold value according to

$$\Delta H = H_1 - H_{c1J}$$

where H_1 is the microwave field, H_{c1J} the junction critical field and δH the measured linewidth as shown in Fig. 5. Thus absorption begins when the microwave field is large enough to induce the critical current across the junction when fluxon creation and annihilation occurs. As the critical current of a Josephson junction decreases with increasing temperature the threshold for microwave absorption also begins at lower powers. Also evident in Fig. 5 is additional structure within the absorption linewidth in the higher power absorption spectra. This secondary fluxon excitation extrapolates back to the same threshold power as the primary absorption process.[29] Another facet of the SQUID behavior is the field dependence of the threshold microwave power which shows oscillations with approximately 50 Oe periodicity.[29]

As already mentioned the absorption line interval, ΔH, is related to the cross section, S, of one of the $\langle 110 \rangle$ oriented twin boundary related striations. This area is the product of the crystal thickness, t, and the junction width, w, including the penetration depth λ_L into the bulk on either side of the junction

$$S = t(w + 2\lambda_L) = \frac{\phi_0}{\Delta H} \quad .$$

The observation of different spectra with different field spacings of ΔH shows there is some variation in the paths connecting active regions of the junctions at the surfaces. One spectra observed over the entire temperature range up to T_c has

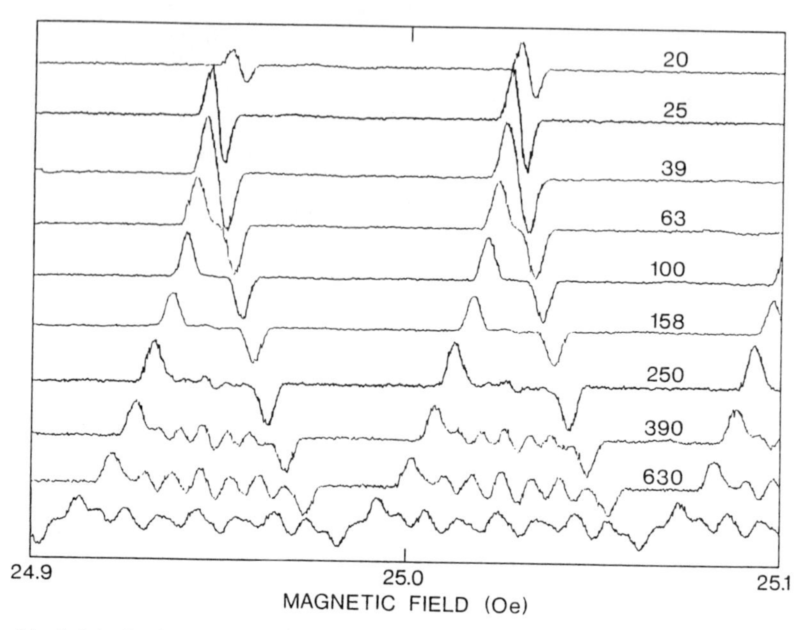

Fig. 5. Modulated microwave absorption of a $YBa_2Cu_3O_{7-\delta}$ single crystal at 4.4 K with 5 mOe modulation and the incident microwave power in μW indicated on each spectrum.

a period of $\Delta H = 82$ mOe at low temperatures indicating a larger effective cross section, S, than that originally observed to correspond to the striations seen on the crystal surface. As the crystal remained 50 μm thick this is attributed to another junction configuration.

Any temperature variation of ΔH may be reasonably attributed to the temperature dependence of the penetration depth. The temperature variation of the area S obtained by substituting the BCS temperature dependence of λ_L near T_c in the above equation gives

$$S = tw + \frac{\sqrt{2}\, t\, \lambda(0)}{\left(1 - \frac{T}{T_c}\right)^{\frac{1}{2}}} .$$

This equation was found to be obeyed over the entire temperature range and not just near T_c which is 92 K for this crystal.[30]

From the gradient and intercept of this BCS equation a value of the penetration depth $\lambda_L(0) = 0.42$ μm for screening currents perpendicular to the a,b-plane was determined and $w = 4.5$ μm which is exceptionally wide for a Josephson junction, suggesting that this structure may be due to microcracking[31] or an accumulation of twin boundaries. Due to the uncertainty in the active thickness of the crystal, the value of $\lambda_L(0)$ should be considered a lower limit. Nevertheless its value is in reasonable agreement with that estimated from other single crystal determinations. Analysis of low field magnetization[32] and μSR[33] measurements on single crystal $YBa_2Cu_3O_{7-\delta}$ finds $\lambda_L(0) = 0.14$ μm for screening currents in the a,b-plane, i.e. one-third the same quantity in the orthogonal direction. These latter experiments do not yield directly $\lambda_L(0)$ for screening currents perpendicular to the a,b-plane. But low field magnetization[34] does yield the two corresponding values of the lower critical

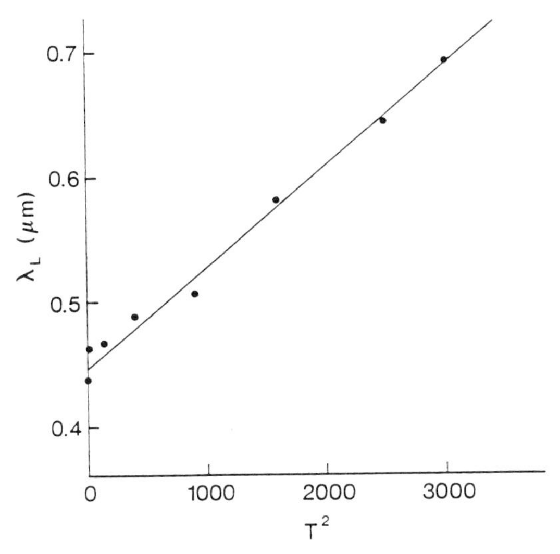

Fig. 6. Variation of the penetration depth for screening currents perpendicular to the *a,b*-plane with T^2.

field H_{c1}, as 180 mOe for $H \perp c$ and 530 mOe for $H \| c$, which should scale as the inverse of the penetration depth. The ratio of these two values is 2.9 which is in good agreement with the ratio of the two penetration depths. The same anisotropy factor of 3 has been found for H_{c2} as well.[35]

The temperature dependence of the penetration depth for screening currents in the *a,b*-plane is flat on approaching $T = 0$.[32,33,36] This is not found for the orthogonal penetration depth determined from the line separation which shows an easily measurable shift on warming above helium temperatures. The low temperature variation follows a T^2 temperature dependence as shown in Fig. 6. This result is similar to that found in the heavy fermion system UBe_{13}[37] and in $YBa_2Cu_3O_{7-\delta}$ powders.[38] A possible cause of this quadratic temperature dependence is the anisotropy of the superconducting gap as inferred from infrared reflectivity measurements[39] and found in the model calculations of Schneider and Frick.[40]

SUMMARY

Microwave absorption measurements are extremely sensitive to the presence of Josephson junctions at the sample surface. Several loss mechanisms are possible due to these junctions as well as those predicted from a two fluid model in the bulk and identification of the dominant source of loss is not always obvious.

REFERENCES

1. J.G. Bednorz and K.A. Müller, *Z. Phys. B* **64**, 189 (1986).
2. J.R. Clem, *Physica C* **153-155**, 50 (1988).
3. A.R. Strnad, C.F. Hempstead and Y.B. Kim, *Phys. Rev. Lett.* **13**, 794 (1964).
4. R. Durny, J. Hautala, S. Ducharme, B. Lee, O.G. Symko, P.C. Taylor, D.J. Zhang and J.A. Xu , *Phys. Rev. B* **36**, 2361 (1987).
5. J. Stankowski, P.K. Kahol, N.S. Dalal and J.S. Moodera, *Phys. Rev. B* **36**, 7126 (1987).

6. K.W. Blazey, K.A. Müller, J.G. Bednorz, W. Berlinger, G. Amoretti, E. Buluggiu, A. Vera and F.C. Matacotta, *Phys. Rev. B* **36**, 7241 (1987).
7. K. Khachaturyan, E.R. Weber, P. Tejedor, A.M. Stacy and A.M. Portis, *Phys. Rev. B* **36**, 8309 (1987).
8. S. V. Bhat, P. Ganguly, T.V. Ramakrishnan and C.N.R. Rao, *J. Phys. C* **20**, L559 (1987).
9. A. Dulcic, B. Leontic, M. Peric and B. Rakvin, *Europhys. Lett.* **4**, 1493 (1987).
10. A.I. Tsapin, S.V. Stepanov and L.A. Blumenfeld, *Phys. Lett. A* **132**, 373 (1988).
11. S. Tyagi, M. Barsoum and K.V. Rao, *J. Phys. C* **21**, L827 (1988).
12. S. Tyagi and M. Barsoum, *Supercond. Sci. Technol.* **1**, 20 (1988).
13. K.W. Blazey, A.M. Portis and J.G. Bednorz, *Solid State Commun.* **65**, 1153 (1988).
14. M. Pozek, A. Dulcic and B. Rakvin, *Solid State Commun.* **70**, 889 (1989).
15. A. Dulcic, B. Rakvin and M. Pozek. Preprint.
16. A.M. Portis, K.W. Blazey, C. Rossel and M. Decroux, *Physica C* **153-155**, 633 (1988).
17. J.S. Martens, J.B. Beyer and D.S. Ginley, *Appl. Phys. Lett.* **52**, 1822 (1988).
18. J.P. Carini, A.M. Awasthi, W. Beyermann, G. Grüner, T. Hylton, K. Char, M.R. Beasley and A. Kapitulnik, *Phys. Rev. B* **37**, 9726 (1988).
19. N. Klein, G. Müller, H. Piel, B. Roas, L. Schultz, U. Klein and M. Peiniger, *Appl. Phys. Lett.* **54**, 757 (1989).
20. T.L. Hylton, A. Kapitulnik, M.R. Beasley, J.P. Carini, L. Drabeck and G. Grüner, *Appl. Phys. Lett.* **53**, 1343 (1988).
21. K.W. Blazey and A. Höhler, submitted to Solid State Commun.
22. S.H. Glarum, J.H. Marshall and L.F. Schneemayer, *Phys. Rev. B* **37**, 7491 (1988).
23. A. Dulcic, R.H. Crepeau and J.H. Freed, *Phys. Rev. B* **38**, 5002 (1988).
24. G. Deutscher and K.A. Müller, *Phys. Rev. Lett.* **59**, 1745 (1987).
25. K.W. Blazey, A.M. Portis, K.A. Müller and F.H. Holtzberg, *Europhys. Lett.* **6**, 457 (1988).
26. V.F. Masterov, A.I. Egorov, N.P. Gerasimov, S.V. Kozyrev, I.L. Likholit, I.G. Savel'ev, A.V. Federov and K.F. Shtel'makh, *JETP Lett.* **46**, 365 (1987).
27. K.W. Blazey, A.M. Portis and F.H. Holtzberg, *Physica C* **157**, 16 (1989).
28. H.W. Zandbergen, R. Gronsky and G. Thomas, *Phys. Stat. Sol. (a)* **105**, 207 (1988).
29. K.W. Blazey and F.H. Holtzberg, *IBM J. Res. Develop.* **33**, 324 (1989).
30. K.W. Blazey, *Physica Scripta* **T29**, 92 (1989).
31. K.L. Keester, R.M. Housley and D.B. Marshall, *J. Crystal Growth* **91**, 295 (1988).
32. L. Krusin-Elbaum, R.L. Greene, F.H. Holtzberg, A.P. Malozemoff and Y. Yeshurun, *Phys. Rev. Lett.* **62**, 217 (1982).
33. D.R. Harshman, L.F. Schneemeyer, J.V. Waszczak, G. Aeppli, R.J. Cava, B. Batlogg, L.W. Rupp, E.J. Ansaldo and D.Ll. Williams, *Phys. Rev. B* **39**, 851 (1989).
34. L. Krusin-Elbaum, A.P. Malozemoff, Y. Yeshrun, D.C. Cronemeyer and F.H. Holtzberg, *Phys. Rev. B* **39**, 2936 (1989).
35. K. Nakao, N. Miura, K. Tatsuhara, H. Takeya and H. Takei, *Phys. Rev. Lett.* **63**, 97 (1989).
36. A.T. Fiory, A.F. Hebard, P.M. Mankiewich and R.E. Howard, *Phys. Rev. Lett.* **61**, 1419 (1988).
37. D. Eingel, P.J. Hirschfeld, F. Gross, B.S. Chandrasekar, K. Andres, H.R. Ott, J. Beuers, Z. Fisk and J.L. Smith, *Phys. Rev. Lett.* **56**, 2513 (1986).
38. J.R. Cooper, C.T. Chu, L.W. Zhou, B. Dunn and G. Grüner, *Phys. Rev. B* **37**, 638 (1988).
39. R.T. Collins, Z. Schlesinger, F. Holtzberg and C. Field, *Phys. Rev. Lett.* **63**, 422 (1989).
40. T. Schneider, these proceedings.

Magnetic Properties of a Granular Superconductor

R. Hetzel [a] and T. Schneider [b]

a) Institute for Theoretical Physics, University of Heidelberg
Philosophenweg 19, 6900 Heidelberg, West Germany
b) IBM Research Division, Zurich Research Laboratory
Rüschlikon, Switzerland

ABSTRACT

We investigate the magnetic properties of the disordered and frustrated XY-model to model weakly coupled superconducting grains in a magnetic field. Extending the mean field approximation to include the induced field, we treat the magnetic field self-consistently and derive the internal field distribution. Furthermore we calculate numerically the critical line in the (H,T)-phase diagram, providing evidence for the experimentally observed $H^{2/3}$ behavior. Moreover we study the nonequilibrium properties in terms of the relaxation of the excess magnetization following a field jump. Our numerical results exhibit the aging effect previously found in spin glasses and in high temperature superconductors.

INTRODUCTION

The high-temperature superconductors share several properties with granular superconductors.[1-4] The broad resistive transitions are highly sensitive to weak magnetic fields, and the relaxation of their magnetization is very slow and exhibits a pronounced hysteresis effect. The susceptibilty of a field-cooled sample differs significantly from the zero-field-cooled case. This behavior is well known in spin glasses and gave rise to the superconducting glass model.[1] It was suggested that these properties originate from the small coherence length in the oxides,[2] which favors a network of weakly coupled grains in the ceramic materials.

Here we investigate a model widely used to describe networks of Josephson junctions in granular superconductors in an external magnetic field.[5] Starting from the mean field approximation, we treat the magnetic field self-consistently and derive the distribution of internal fields numerically. Furthermore we calculate the critical line in the (H,T)-phase diagram. We will then turn to the nonequilibrium properties. Using the Monte Carlo technique, we study the decay of the excess magnetization numerically. In particular we will determine its functional form and demonstrate its dependence on the waiting time, thus providing numerical evidence for the aging phenomena[6] – which has only been discovered recently in high-temperature superconductors[7] – for the first time.

Dynamics of Magnetic Fluctuations in High-Temperature Superconductors
Edited by G. Reiter *et al.*, Plenum Press, New York, 1991

THE MODEL

Consider a network of N superconducting grains which are connected by Josephson junctions:[5,8] we assume that each of these grains can be characterized by a single superconducting order parameter $\psi_j = |\psi_j| \exp(i\varphi_j)$, where φ_j is the phase. The energy of a single junction connecting grains i and j is given by $E_{ij} = J_{ij}(1 - \cos(\varphi_j - \varphi_i))$, where J_{ij} is the coupling energy. To a first approximation we take the intergrain energies, J_{ij}, to be constant and independent of the particular junction, i.e. $J_{ij} = J$. Neglecting any charging effects, the Hamiltonian of the full network in the presence of a magnetic field then reads[5]

$$\mathscr{H} = - \sum_{(i,j)} J \cos(\varphi_j - \varphi_i - A_{ij}) \ . \tag{1}$$

The sum extends over nearest neighbor pairs only. The bond variable

$$A_{ij} = \frac{2\pi}{\phi_0} \int_i^j A \cdot d\ell \tag{2}$$

accounts for the magnetic field $h = \nabla \times A$, where A is the vector potential and ϕ_0 is the elementary flux quantum. In this Hamiltonian it is implicitly assumed that the magnetic penetration depth is large compared to the grain size (so that the coupling energy is independent of the shapes of the individual grains) and even as large as the sample dimensions (so that the inner junctions experience the magnetic field as well). Within the limit of an infinite London penetration depth, the vector potential corresponds precisely to the applied field.

It should be noted that the Hamiltonian is known from statistical mechanics as is the frustrated XY-model.[9] This is a classical 2-dimensional spin system where the gauge field acts as a source of frustration to phase alignment.

SELF-CONSISTENT MEAN FIELD APPROXIMATION

To construct the Ginzburg-Landau functional, we rewrite the Hamiltonian (1) in the form

$$- \beta\mathscr{H} = \frac{1}{2} \mathbf{v}^{+}\mathbf{P}\mathbf{v} - K_N \ , \tag{3}$$

as originally introduced by Choi and Doniach.[10] Here $\mathbf{v}$ is a spin vector with components $v_j = \exp(i\varphi_j)$. $\mathbf{P}$ is a positive definite Hermitian matrix with the elements

$$P_{ij} = K_{ij} \exp(-iA_{ij}) + \delta_{ij} \sum_{\ell} K_{i\ell} \ , \tag{4}$$

$$K_{ij} = \begin{cases} \beta J & \text{for nearest neighbors} \\ 0 & \text{otherwise,} \end{cases}$$

and

$$K_N = \frac{1}{2} \sum_{(i,j)} K_{ij}$$

is a constant. The Hubbard-Stratanovich transformation yields

$$Z = \mathrm{Tr}\, \exp(-\beta \mathcal{H}) = \int \Pi\, dz_j\, dz_j^* \exp(-F) \ , \tag{5}$$

with the free energy

$$F = \frac{1}{2} \left\{ \mathbf{z}^{+} \mathbf{P}^{-1} \mathbf{z} - \frac{1}{2} \sum_i |z_i|^2 + \frac{1}{32} \sum |z_i|^4 + O\left(|z_i|^6\right) \right\} . \tag{6}$$

Neglecting quartic and higher-order terms, steepest descent results in the eigenvalue problem

$$\mathbf{Pz} = E\mathbf{z} \ . \tag{7}$$

The largest eigenvalue E_{max} will then determine the mean field transition temperature T_c via the condition

$$\frac{1}{E_{max}} - \frac{1}{2} = 0 \ , \tag{8}$$

where E_{max} is a function of T and H.

As the temperature is lowered below T_c, the linearized free energy is no longer appropriate, and screening must be taken into account. The free energy has to be supplemented by an additional term, i.e.

$$\frac{\beta V}{4\pi N} \sum_{i=1}^{N} h_i^2 \ ,$$

where the h_i terms are the local magnetic fields. In the presence of an external magnetic field $\mathbf{H}$ = $(0,0,H)$ the energy to be minimized is the Gibbs free energy. Introducing the induced fields $h_{si} = h_i - H$, the Gibbs free energy

$$G = F - \frac{\beta V}{4\pi N} \sum_{i=1}^{N} h_i H$$

reduces, for small induced fields, to

$$G = \frac{1}{2N} \sum_{i=1}^{N} \left\{ \left(\frac{1}{E(H)} - \frac{1}{2} - \frac{h_{si}}{E^2(H)} \frac{dE}{dH} \right) c^2 |\psi_i|^2 + \frac{1}{32} c^4 |\psi_i|^4 + \frac{\beta V}{4\pi} \left(h_{si}^2 - H^2\right) \right\} \tag{9}$$

where $z_i = c\psi_i$. Invoking stationarity, we obtain an expression for the variance of the induced field, i.e.

$$\sigma^2 = < h_s^2 > - < h_s >^2 = \left(4\pi M^2\right) (\beta_A - 1) \ . \tag{10}$$

β_A denotes the Abrikosov parameter which is given by

$$\beta_A = \frac{<|\psi|^4>}{<|\psi|^2>^2}$$

and $<...>$ is the average over sites. Thus the magnetization is proportional to the variance of the distribution of the induced fields, a quantity which is directly related to the μSR line width. Recent μSR experiments on $YBa_2Cu_3O_{7-\delta}$ confirm our findings that $\sigma \sim M$.[11]

CRITICAL LINE T_c *(H)*

To derive quantitative predictions, we next consider ordered and disordered arrays of grains. In the ordered case, we assume that the grains are placed on the lattice points of a 2-dimensional square lattice with lattice constant *a*. Adopting the Landau gauge $A_{ij} = Hx \cdot \hat{y}$, A_{ij} takes the form

$$A_{ij} = \frac{2\pi}{\phi_0} H \frac{x_i + x_j}{2} (y_j - y_i) \ . \tag{11}$$

Using the Bloch ansatz $z_i = z_{nm\ell} = ue^{i(\nu m + \omega \ell)}$, the eigenvalue problem (7) reduces to Harper's equation

$$eu_n = 2u_n \cos\omega + 2u_n \cos(2\pi fn - \nu) + u_{n-1} + u_{n+1} \tag{12}$$

where $f = Ha^2/\phi_0$. This tight-binding problem was extensively studied by Hofstadter.[12] We evaluated the largest eigenvalue E_{max} numerically; our results for the critical line $T_c(H)$ are depicted in Fig. 1. The periodicity of one flux quantum per square is clearly seen. Experiments on high-temperature superconductors revealed evidence for[1]

$$T_c(0) - T_c(H) \sim H^{2/3} \tag{13}$$

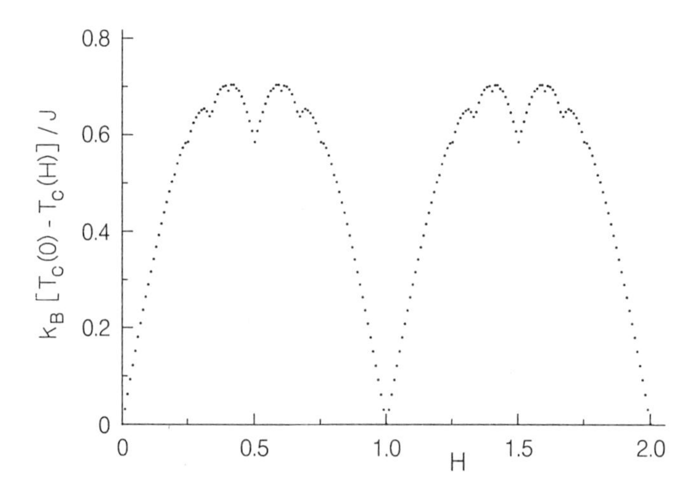

Fig. 1. Mean-field transition temperature T_c versus *H* for grains placed on a two-dimensional square lattice. From Ref. 8.

for intermediate field strength. The corresponding plot for our data is shown in Fig. 2. In support of (13) we identify linear behavior within a temperature range of $0.1 < k_B[T_c(0) - T_c(H)] < 0.5$.

Next we allow the locations of the grains to undergo random fluctuations about the grid points of the reference square lattice, which is certainly a more appropriate model for granular superconductivity than an ordered array. To explore the effects of disorder, we consider the simplest case, allowing the x-coordinate of the grains to fluctuate randomly in a small interval, Δ, around each lattice point. The eigenvalue problem (12) is slightly modified. As expected, the periodicity of Fig. 1 is removed but the relationship (13) still holds for intermediate fields, as shown in Fig. 3. Considering two-dimensional disorder, we have also demonstrated that it has essentially the same effect on the critical line as one-dimensional disorder.[13]

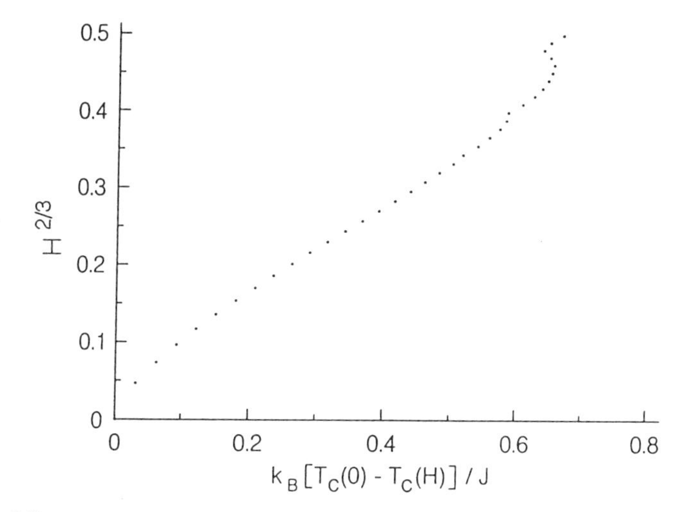

Fig. 2. $H^{2/3}$ versus Reduced Temperature $k_B[T_c(0) - T_c(H)]/J$ for the data shown in Fig. 1. From Ref. 8.

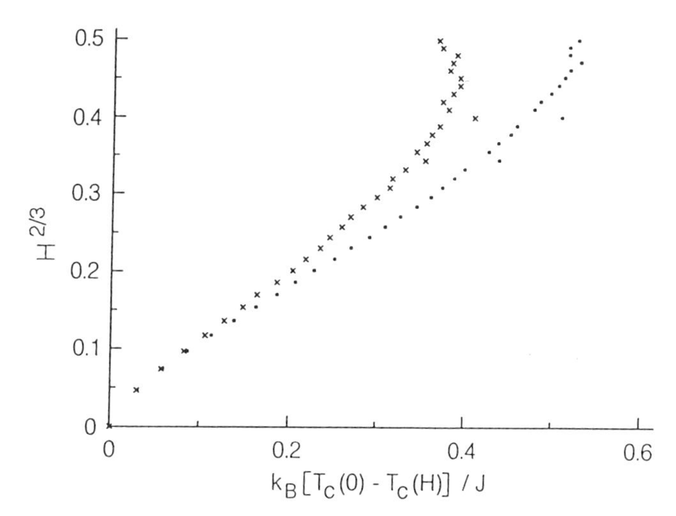

Fig. 3. $H^{2/3}$ versus Reduced Temperature $k_B[T_c(0) - T_c(H)]/J$ for bond disorder along the x-axis, $\Delta = 1/4$ (■) and $\Delta = 1/2$ (**x**). From Ref. 8.

It should be added that below the transition temperature $T_c(H)$ the array is phase locked and trapped in a globally superconducting state, exhibiting irreversibility in its intergranular magnetic properties. Above the transition temperature however, the network is phase decoupled and resistive.

To clarify the nature of the phase transition further, we also performed Monte Carlo calculations.[14] In the case of the unfrustrated XY-model where $H=0$, the phase transition is known to be of Kosterlitz-Thouless type.[15] In the fully frustrated but ordered model, where $H=1/2$, there are indications of both Kosterlitz-Thouless and Ising behavior.[9,16] In the presence of disorder, however, the Ising behavior is known to disappear.[17] Using the standard Monte Carlo algorithm, we computed various thermodynamic quantities, including the specific heat, the susceptibility and the helicity modulus as a function of temperature *and* system size. The Kosterlitz-Thouless theory predicts a universal jump in the helicity modulus at T_c and a finite specific heat below T_c.[15] While disorder suppresses the universal jump of the helicity modulus, the susceptibility nevertheless diverges in the low temperature phase. Our results provide strong evidence that the disordered and frustrated XY-model undergoes a nonuniversal Kosterlitz-Thouless transition.[14]

AGING

The phenomena of aging was discovered in 1983 by Lundgren in spin glasses.[6] It has since then received an increasing amount of interest.[18] Only recently, a related effect — the magnetic memory effect — was found in high-temperature superconductors.[7] Aging manifests itself in the relaxation of excess magnetization in terms of the dependence on the waiting time which has elapsed since the preparation of the sample.

In a typical relaxation experiment one rapidly cools the sample in a small magnetic field to a temperature below the transition temperature. After waiting for some time, one switches off the field and observes the relaxation of the magnetization which is called the thermoremanent magnetization. Alternatively one may cool down the sample in zero field, and then let some time elapse before switching on the field. One then follows the relaxation of the magnetization towards its equilibrium value; it is now denoted as zero-field-cooled magnetization. Experimentally one finds an extremely slow relaxation in both spin glasses[6] and superconductors.[1]

For temperatures just below T_c and small magnetic fields, the magnetization in high-temperature superconductors decays logarithmically with time, up to a measuring time equal to the waiting time. Then it starts to deviate from the logarithmic behavior. This echolike response to a field jump is called the magnetic memory effect.[7]

Applying the frustrated XY-model, we have performed a systematic numerical study of the time decay of the magnetization following a field jump. The magnetization, as derived from the supercurrents circulating in the network, is given by

$$M(t) = -\frac{1}{N}\sum_{(i,j)} J\,\frac{dA_{ij}}{dH}\ \sin\,(\varphi_j(t)-\varphi_i(t)-A_{ij})\ , \tag{14}$$

in agreement with the equilibrium value of the magnetization

$$M_0 = \langle M(t)\rangle_t = -\frac{1}{N}\,\frac{dF}{dH}\ , \tag{15}$$

where F denotes the free energy. We consider a two-dimensional square lattice with the magnetic field H perpendicular to it, and introduce disorder by choosing the spin locations randomly within circles of radius 0.2 a centered on the lattice sites, where a denotes the lattice constant. Our numerical results are obtained using the standard Monte Carlo method with single spin flip Glauber dynamics for systems of 32×32 spins. We implemented free boundaries and averaged over 32 independent samples.

In the preparation of the system we followed the experimental procedure closely.[7] We start at a high temperature $T_0 = 2J$ and let the system evolve for some time to obtain independence from the initial configuration. The temperature was then lowered typically down to $T_1 = 0.1J$, which is well below the critical temperature $T_c = 1J$ of the pure XY-model. After waiting for a definite time t_w, the magnetic field is either switched from zero to a small value H, or from H to zero, and the evolution of the magnetization with time was followed. The magnitude H of the field jump was chosen to obtain a linear response in the jump of the magnetization; it corresponds to a small number of flux quanta per system.

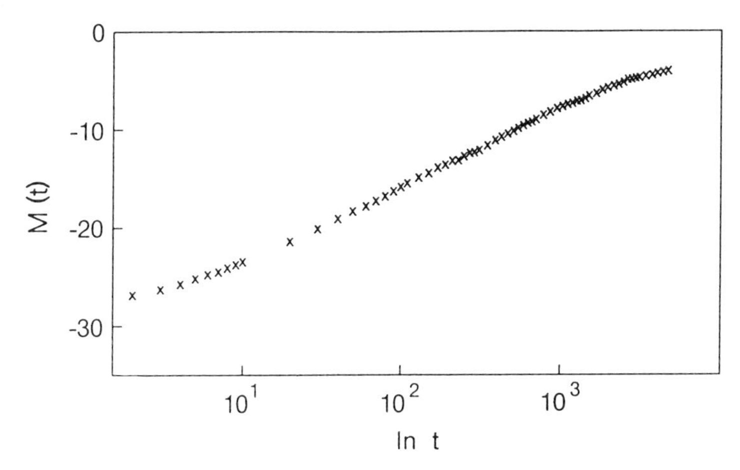

Fig. 4. Decay of the zero-field-cooled magnetization.

The results of a typical ZFC run are depicted in Fig. 4. The relaxation of the magnetization is seen to follow a stretched exponential law in the intermediate regime, which has the form

$$M(t) = M_o \, e^{-\left(\frac{t}{\tau}\right)^{\beta}} . \tag{16}$$

Our fits yield values between 0.3 and 0.4 for β at $T_1 = 0.1J$. By raising the final temperature T_1 (the temperature to which the system is cooled), we find that the jump in the magnetization becomes smaller and the system relaxes more quickly towards its equilibrium value. This behavior is shown in Fig. 5. We also note that the exponent β is almost independent of the waiting time, while the time constant τ changes dramatically.

For longer waiting times the system clearly relaxes more slowly than for short ones, as it is depicted in Fig. 6. This effect disappears smoothly if we increase the temperature T_1 at which the relaxation takes place, beyond the critical temperature $T_c = 1J$. Note that the jump in the magnetization decreases slightly as we reduce the waiting time, indicating a non-equilibrated system. Our observations hold not only in the case of ZFC magnetization but also for the TR magnetization.

In terms of superconducitivity we interpret our results as follows: by switching on the magnetic field, intergranular diamagnetic Josephson currents are induced which screen the interior of the network. However, this current configuration is not stable, and the currents tend to decrease. Consequently, the magnetic field enters the intergranular regions of the specimen until the entire sample has been penetrated and the magnetization has relaxed towards its equilibrium value.

It should be noted that our model is restricted to describing the thermal flow of flux into the sample. In a more realistic description, flux pinning should be taken into account,[19] corresponding to fixed phase differences. We emphasize that the results of our simulation are obtained neglecting the induced magnetic fields. Moreover, well

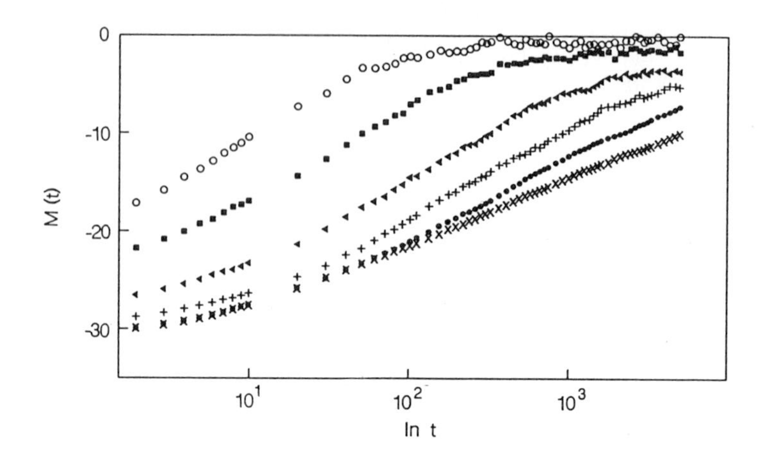

Fig. 5. Temperature dependence of the relaxation of the zero-field-cooled magnetization. The curves correspond to temperature $T = 0.005$ (**x**), $T = 0.05$ (●), $T = 0.3$ (+), $T = 0.5$ (▲), $T = 0.8$ (■) and $T = 1.0$ (○) J.

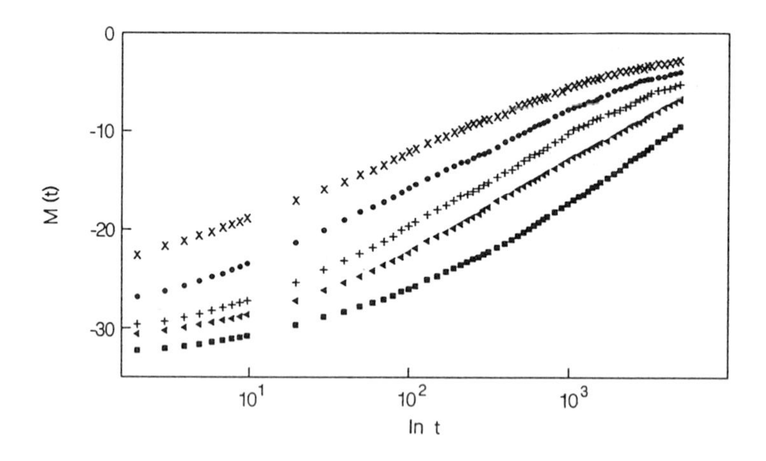

Fig. 6. Waiting time dependence of the relaxation of the zero-field-cooled magnetization. The curves correspond to waiting times $t_W = 10$ (**x**), $t_W = 30$ (●), $t_W = 100$ (+), $t_W = 300$ (▲) and $t_W = 1000$ (■) Monte Carlo steps per spin.

below T_c the system should be treated self-consistently by taking the induced supercurrents into account.

During the waiting time, however, no external magnetic field is present. Thermal activation causes spontaneous changes of the Josephson currents. Due to extremely long relaxation times the system slowly approaches its ground state. Tracing the time development of the spin fluctuations in our simulation implies that domains of fixed current patterns are gradually built up. Changing the magnetic field at this point shifts the ground state to a different current distribution. By approaching the new equilibrium configuration, larger domains of tightly coupled spins have to be broken up if the system has experienced a longer waiting time. This causes a slower relaxation rate.

At present it is not clear what ingredients are needed to obtain aging phenomena. Considering the spin glass case, Koper and Hilhorst[20] derived a waiting time dependence of the response function from a specific time evolution. Alternatively, Sibani and Hoffmann[21] applied a hierarchical model of relaxation in phase space where the low-energy states of the system are organized in a tree structure. Here the probability distribution in phase space changes during the waiting time, resulting in the aging effect.

SUMMARY

The frustrated and disordered XY-model was investigated to model weakly coupled superconducting grains in a magnetic field. Starting from the mean field approximation, we treat the induced field self-consistently to predict the distribution of local fields. We calculate numerically the critical line $T_c(H)$ and demonstrate experimentally that it obeys the $H^{2/3}$ law. Furthermore, studying the relaxation of the magnetization following a field jump, we provide numerical evidence for the aging effect.

ACKNOWLEDGEMENTS

The authors have benefitted from discussions with H. Keller, H. Kinzelbach, L. Lundgren, I. Morgenstern, K.A. Müller and C. Rossel. One of us (R.H.) is grateful to the IBM Research Laboratory, Rüschlikon, for its warm hospitality and for financial support.

REFERENCES

1. K.A. Müller, M. Takshige and J. G. Bednorz, Phys. Rev. Lett. **58,** 1143 (1987)
2. G. Deutscher and K.A. Müller, Phys. Rev. Lett. **59,** 1745 (1987)
3. I. Morgenstern, K.A. Müller and J.G. Bednorz, Z. Phys. B - Condensed Matter **69,** 33 (1987)
4. J.R. Clem, Physica C 153-155, 50 (1988)
5. W.Y. Shih, C. Ebner and D. Stroud, Phys. Rev. B **30,** 134 (1984); C. Ebner and D. Stroud, Phys. Rev. B **31,** 165 (1985)
6. L. Lundgren, P. Svedlindh, P. Nordblad and O. Beckman, Phys. Rev. Lett. **51,** 911 (1983)
7. C. Rossel, Y. Maeno and I. Morgenstern, Phys. Rev. Lett. **62,** 681 (1989)
8. T. Schneider, D. Würtz and R. Hetzel, Z. Phys. B - Condensed Matter **72,** 1 (1988)
9. S. Teitel and C. Jayaprakash, Phys. Rev. B **27** 598 (1983); Phys. Rev. Lett. **51,** 1992 (1983)
10. M. Choi and S. Doniach, Phys. Rev. B **31,** 4516 (1985)
11. H. Keller, B. Pümpin, W. Kündig, W. Odermatt, B.D. Patterson, J.W. Schneider, H. Simmler, K.A. Müller, J.G. Bednorz, K.W. Blazey, C. Rossel, and I. Morgenstern and I.M. Savic, Physica C, **153-155,** 71 (1988)
12. D.R. Hofstadter, Phys. Rev. B **14,** 2239 (1976)
13. R. Hetzel and T. Schneider, Z. Phys. B - Condensed Matter **73,** 303 (1988)
14. R. Hetzel, M. Vanhimbeeck and T. Schneider, Z. Phys. B - Condensed Matter **76,** 259 (1989)

15. J.M. Kosterlitz and D.J. Thouless, J. Phys. C **6,** 1181 (1973); J.M. Kosterlitz, J. Phys. C **7,** 1046 (1974)
16. B. Berge, H.T. Diep, A. Ghazali and P. Lallemand, Phys. Rev. B **34,** 3177 (1986)
17. M.Y. Choi, J.S. Chung and D. Stroud, Phys. Rev. B **35,** 1669 (1987)
18. P. Svedlindh, P. Granberg, P. Norblad, L. Lundgren and H.S. Chen, Phys. Rev. B **35,** 268 (1987)
19. For a current review of flux pinning see, for example, E.H. Brandt, and U. Essmann, Phys. Status Solidi B, **144,** 13 (1987)
20. G.J.M. Koper and H.J. Hilhorst, J. Phys. (Paris) **49,** 429 (1988)
21. P. Sibani and K.H. Hoffmann, Phys. Rev. Lett. **63,** 2853 (1989)

THERMODYNAMIC FLUCTUATIONS AND THEIR DIMENSIONALITY IN CERAMIC SUPERCONDUCTORS OUT OF TRANSPORT PROPERTIES MEASUREMENTS

S. K. Patapis*, M. Ausloos, Ch. Laurent*****

* Physics Department, Solid State Division, University of Athens
104 Solonos Str., Athens 106 80, Greece

** Institut de Physique B5, Universite de Liege, Sart Tilman
B-4000 Liege, Belgium

*** Institut Montefiore B28, Universite de Liege, Sart Tilman
B-4000 Liege Belgium

ABSTRACT

Information concerning the dimensionality of the thermal flustuations of the new high temperature ceramic superconductors, can be derived from the excess conductivity (paraconductivity) near the transition temperature according to Aslamasov-Larkin (AL) theory. Here the above dimensionality is derived from the excess resistivity and thermoelectric power measured in different samples of HTS materials such as YBaCuO and different Bi (PB) SrCaCuO compounds.

INTRODUCTION

It is well known that thermodynamic fluctuations grow in a substance as by changing the temperature the transition temperature Tc is approached. These fluctuations have an influence on the material macroscopic properties such as the transport properties. As a prove the

Dynamics of Magnetic Fluctuations in High-Temperature Superconductors
Edited by G. Reiter *et al.*, Plenum Press, New York, 1991

electrical resistivity of all the new ceramic high temperature superconductors in the vicinity of Tc reveals not a sharp but a rounding rather behavior as the material turns from the normal to the superconducting state. This rounding may possibly come from the inhomegenuity of the polycrystallic samples but the main reason may come from the arised thermodynamic fluctuations. Quite above the transition temperature there is a finite probability of production of short lived "Cooper pairs" which under an electric field have as result the decrease of resistance before reaching the bulk transition temperature.

Precise measurements and detailed analysis of the transport properties (electrical conductivity, thermoelectric power and thermal conductivity) around Tc may give information concernig the dimensionality of the superconducting fluctuations and the critical behavior of these new ceramic high temperature superconducting materials.

Generally in order to probe the dimensionality of the above fluctuations one has to examine in detail the resistivity or equally the conductivity behavior in the transition region and compare this to proper expressions for two or three dimensions (2D or 3D) derived from a theory by Aslamasov and Larkin (AL)[1]. These expressions relate the excess conductivity $\Delta\sigma$ to the reduced temperature ε i.e.

$$\Delta\sigma_3 = \frac{e^2}{32\hbar}\,\frac{1}{\xi(0)}\,\varepsilon^{-1/2} \qquad \text{(for 3D)} \qquad (1)$$

$$\Delta\sigma_2 = \frac{e^2}{16\hbar}\,\frac{1}{d}\,\varepsilon^{-1} \qquad \text{(for 2D)} \qquad (2)$$

By definition we have $\Delta\sigma = \sigma - \sigma_0$ where σ is the electrical conductivity equal to $1/\varrho$ (ϱ, the electrical resistivity) and σ_0 , defined as the normal state conductivity, equals $1/\varrho_0$ where ϱ_0 is a temperature dependant value extrapolated from room temperature behavior of ϱ which at this temperature behaves linear with T. $\varepsilon \cong T - Tc/Tc$ is the reduced temperature deviation from the critical temperatyre value Tc, $\xi(o)$ is the zero temperature coherence length and d is some layer thickness characteristic of the two dimensional system.

Generally $\Delta\sigma$ is analogous to the magnitude of fluctuation effect on conductivity and this effect can be observed here to the new high temperature superconductors because of the value of the normal state resistivity (ϱ_0), ranging mΩ.cm, and the short zero temperature coherence length $\xi(0)$. So conclusions concerning the type of fluctuations (2D or 3D) try to be derived from comparison of data to the above equations (1) and (2) (or better from a log $\Delta\sigma$ - log ε plot) or even more from analogous relations derived from extenstions of the Aslamarov - Larkin theory by Lawrence and Doniach[2] and Tewordt at al[3]. According to the above theory the excess conductivity $\Delta\sigma$ near Tc can be expressed as

$$\Delta\sigma_D = A_D E^{-\lambda} + B_D \qquad (3)$$

where the critical exponent λ is related to the dimensionality D of the order parameter fluctuations by

$$\lambda = 2 - D/2 \qquad (4)$$

Such works based on the above theory have been done by many groups [4] and by our one[5, 6]. Usually from the presentation of the data on a log $\Delta\sigma$ - log ε plot the critical exponent λ and hence the dimensionality D is extracted if a straight line (for D=2 or 3) can fit. A disagreement arises between the different authors on the value of the dimensionality D for the same HTS compound since the above mentioned straight line is plainly the tangent to the data plot and the shape of the graph may allow many slopes in a temperature region (Fig. 1).

In our case we analyse the resistivity data. The excess conductivity $\Delta\sigma = \sigma-\sigma_0$ is not equal to the excess resistivity $\Delta\varrho = \varrho-\varrho_0$. But theoretical work shows that near the transition temperature $\Delta\sigma$ and $\Delta\varrho$ follow a power law behavior

$$\Delta\sigma \simeq \varepsilon^{-\lambda} \qquad (5)$$

with the same singularity[7]. That means

$$\Delta\varrho \simeq \Delta\sigma$$

(but not too close to Tc). The above expressions depends on σ_0 or ϱ_0 (the normal state background) and on the choice of Tc.

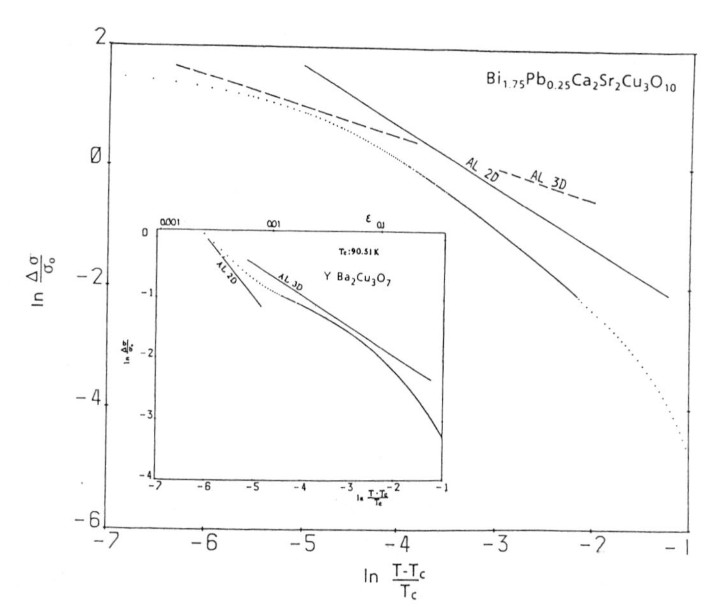

Fig 1 Log-log plot of the excess conductivity $\Delta\sigma$ as a function of ε=T-Tc/Tc. Theoretical slopes of Aslamazov-Larkin theory are given for 2D or 3D dimentionality. For YBaCuO (inset) and one Bi(Pb)CaSrCuO sample (Ref 5 and 6).

As it is observed[5, 6] a power law like

$$\Delta\varrho = A\,\varepsilon^{-\lambda} + \alpha T + \beta \tag{6}$$

can give precise extraction of the value of λ with crossover effects and hence the fluctuation dimensionality. The same power law is valid for the thermoelectric power behavior[8, 9].

In order to eliminate the influence of the choice of the normal state background ϱ_0 the temperature derivative of $\Delta\varrho$ is analysed instead of $\Delta\varrho$ i.e. we look at the relation

$$\frac{d\Delta\varrho}{dT} = \varepsilon^{-(\lambda+1)} \tag{7}$$

To follow the above procedure the data must be quite precise and of high density for a numerical derivative to be safely taken over a small temperature interval. The choice of Tc is also important when ε goes to zero. Tc is left as a free parameter taking its proper value (constrained to $d^2R/dT^2 = 0$) [5, 6, 10,] as such that minimizes the least square deviation in a temperature interval with as many as possible data points[5, 6].

According to the above procedure samples of $Y_1Ba_2Cu_3O_{7-\delta}$ and of different Bi (Pb) SrCaCuO compounds were used in order to measure their transport properties such as electrical resistivity and thermoelectric power in the region of Tc. From the data analysis of these properties the critical exponent λ a cross over and the dimansionanily of the thermodynamic fluctuations are derived.

MEASUREMENTS AND RESULTS

The samples were prepared by the "traditional" solid state reaction used for the new high temperature ceramic superconductors described in detail in Ref. 11 and 12. The polycrystallic samples used for the measurements were parallepiped with approximate dimensious 1x4x10 mm.

Resistivity measurements. For the resistivity measurements a four point technique was used with a dc current density of about 0.2 A/cm^2. Special precautions have been taken for steady state conditions to be achieved. So the temperature of the sample was changing at a slow rate of 3-4 K/h for measurement to measurement while the temperature stability during measuring was better than 5×10^{-3} K/h for several minutes. Even more, special precautions have been taken to eliminate parasitic effects[12, 13].

Precise measurements of the electrical resistance of YBaCuO sample have been taken from room temperature down to the superconducting state without and with the influence of low magnetic field (as high as 4 KG). The ananlysis of the data shows different resistivity regimes. A linear regime starting from higher temperatures is followed, as the transition temperature is approached, by ones succesively governed by exponential, linear and exponential law[10]. Under a weak magnetic field a hysteresis effect is observed in the so called "knee behavior region" of R(T) i.e. between Tc and the temperature T_R where the resistance vanishes[10]. The resistance values for cooling are lower than those for warming up[14]. Such an hysteresis effect was not observed in the Bi - compounds[6].

From a plot of the log of the resistance temperature derivative as a function of ε (Fig. 2) the dimensionanity of the fluctuation is derived. Close to Tc for $\varepsilon < 10^{-2}$ (log ε <-4.6) the slope S is -2 and consequently $\lambda = 1$ (since $S = \lambda + 1$). This in the sence of AL theory and

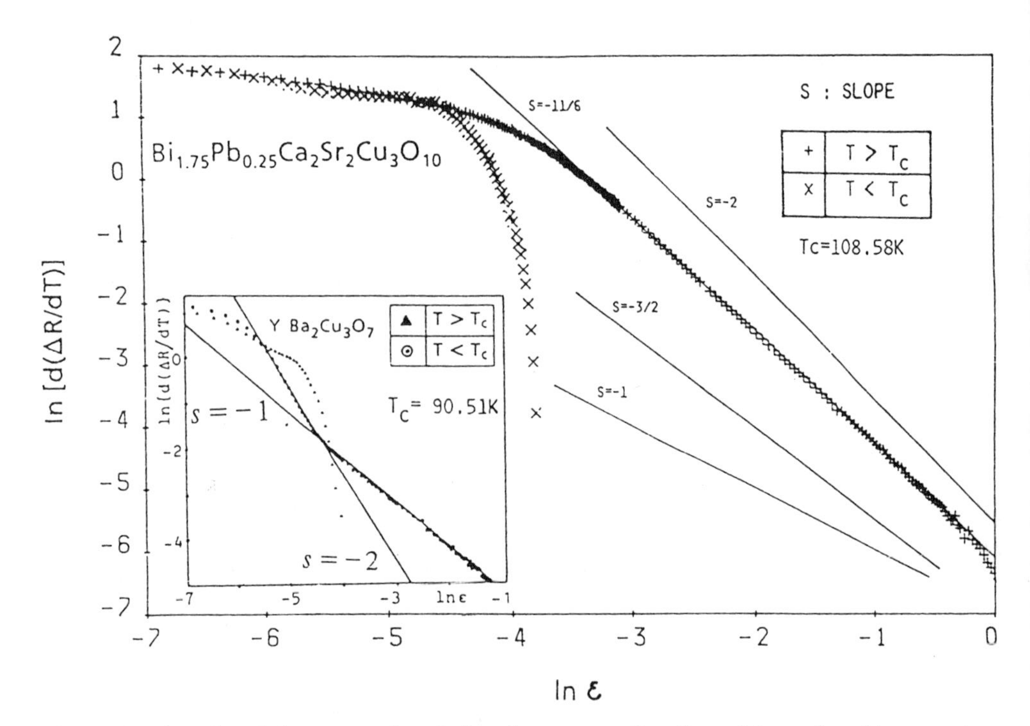

Fig. 2 Log-log plot of the excess electrical resistance as a function of the reduced temperature ε for a BiPbCaSrCuO compound and for YBaCuO (inset).

relation (4) corresponds to two dimensional fluctuations. For higher temperatures up to T=2Tc a logarithmic behaviour can be dinstiguished ($s = -1$ and $\lambda = 0$) which is not predicted from Aslamasov and Larkin[1]. Such a similar behaviour has been observed in the (a,b) plane of a YBaCuO monocrystal[15]. A pair breaking coming from defects in the sample may be the reason of such behaviour[16].

Similar measurements have been made in Bi(Pb) SrCaCuO compounds of ceramic superconductors and a similar analysis has followed. For the temperature interval between ε = 0,015 (T ≃ Tc+1,5K) and ε = 1 (T = 2Tc), a slope value s=11/6 has been found followed by a rounding effect as the temperature approaches closer to Tc (Fig. 2). This same slope has been found for three different Bi - compounds[17]. For 2D or 3D fluctuations we expect correspondingly slopes equal to 2 or 3/2. The slope 11/6 is close but not equal to 2 (Fig. 2). The precision of the data allow such close values to be distinquished. On the other hand data analysis of the thermoelectric power leads to the same value[9]. So this value must be universal[17] and corresponds to a fluctuation dimensionality D = 7/3. Such a value between 2 and 3 is attributed to the fractal nature of the percolation network for the superconducting fluctuations. Through percolation arguments the conductivity mechanism occurs through planar path network[13, 17].

Thermoelectric power measurements. The thermoelectric power of both type superconductors was also measured in different compound samples in absense and under the influence of a small magnetic field. The experimental precautions adopted for resistivity must be strictly valid here since thermoelectric power measurements have much more difficulty[12]. The thermoelectric power of HTS ceramics is quite small of the order of μV/K resulting in

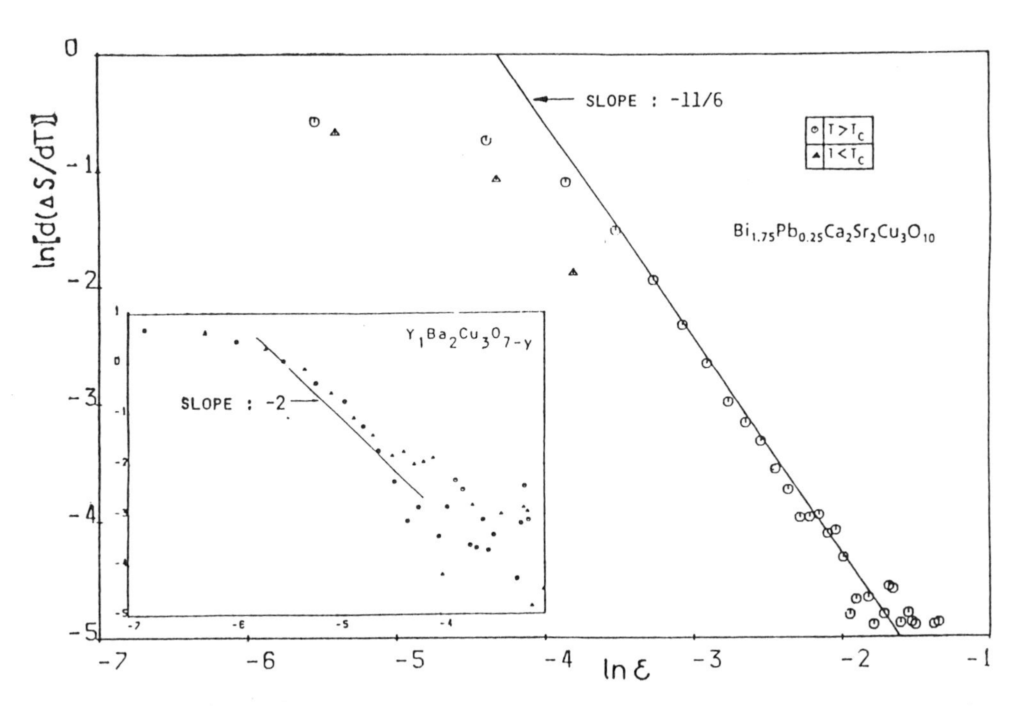

Fig. 3 Log-log plot of the temperature derivative of the "excess" thermoelectric power ΔS for BiPbCaSrCuO (Ref. 8) and dS/dT (Ref. 9) for YBaCuO (inset) as a function of the reduced temperature ε with the best estimated slopes.

small voltage drops. On the other hand spurious voltages is difficult to be controlled and removed[12]. For probing the critical region the thermal gradient must be as small as possible when Tc issmall voltage drops. On the other hand spurious voltages is difficult to be controlled and approached[8, 12]. Data points were taken every 0.3 K and ΔT imposed on the sample was between 0.25 and 2K depending on the run.

On the other hand the intepretation of the data of these measurements is much more difficult. Even the sign of the carriers is not easy to be concluded[18]. For data analysis the normal background behaviour cannot be extracted since generally is not linear as for example, in YBaCuO, the curvature of the thermoelectric power behaviour[8]. Under a magnetic field the thermoelectric behavior of all the samples seems similar to that of the resistivity with the characteristic "knee" under Tc[8, 9]. A proper data analysis similar to that for the resistivity followed in a log-log plot (Fig. 3). From this analysis the same critical exponents were derived[9, 10] leading to the same dimensionality (D) of fluctuations as it is predicted by theory[19].

Aknowledgments: From this place wew would like to thank Prof. C. Politis of Kernfoschungszentrum Karlsruhe, Institut fur Nucleare Festkoperphysik (F.R. Germany), Prof. P. Tarte and Dr. A. Rulmont of the University of Liege, Chemistry Department, Liege (Belgium) and Prof. S. M. Green and Prof. H. L. Luo of the University of California, San Diego, Department of Electrical and Computer Engineering, La Jolla (USA) for the preparation of the samples.

REFERENCES

1. L. G. Aslamazou and A.I. Larkin, Phys. Let 26A:238 (1968)
2. J. Lawrence and S. Doniach in "Proc. 12th Inter. Conference on Low Temperature Physics", E. Kando Ed, Kaigaku, Tokyo p. 361 (1971)
3. L. Teword, D. Fay and Th. Wokkhausen, Sol. St. Comm. 67:301 (1988)
4. P.P. Freitas, C.C. Tsuei and T.S. Plaskett, Phys. Rev. B.36:833 (1987)
   P.P. Freitas and C.C. Tsuei, Phys. Rev.B 37:2074 (1988)
   F. Vidal, J. A. Veira, J. Maza, F. Carcia - Alvarado, E. Moran and M. A. Alario, J. Phys.C, 21:L 599 (1988)
   F. Vidal, J. A. Veira, J. Maza, F. Migueles, E. Moran and M.A. Alario, Sol. St. Comm. 66:421 (1988)
   F. Vidal, J.A. Veira, J. Maza. J. J. Ponte, F. Garcia - Alvarado E. Moran and M. A. Alario, Physica C, 153-155:1371 (1988)
   S. E. Inderhees, M. B. Salamon, N. Goldenfeld, J. P. Rice B.G. Pazol, D. M. Ginsberg,

J. Z. Liu and G.W. Crabtree Phys. Rev. Lett. 60:1178 (1988)

A. Junod, Physica C 153-155:1078 (1988)

H. R. Brand and M.M. Doria, Phys. Rev. B,37:9788 (1988)

B. Oh, K. Char, A. D. Kent, M. Naito, M. R. Beasley, T. H. B. Geballe, R. H. Hammond, A. Kapitulnik and J.M. Graybeal Phys. Rev. B, 37:7861 (1988)

N. Goldenfeld, P. D. Olmsted, T. A. Friedman and D. M. Ginsberg, Sal. St. Comm. 65:465 (1988)

5. M. Ausloos and Ch. Laurent, Phys. Rev. B, 37:611 (1988)
6. M. Ausloos, Ch. Laurent, S. K. Patapis, S. M. Green, H. L. Luo and C. Politis, Mod. Phys. Lett. B, 2:1319 (1988)
7. M. Ausloos, P. Clippe and Ch. Laurent (submitted for publication)
8. Ch. Laurent, S. K. Patapis, M. Laguesse, H. W. Vanderschuren A. Rulmont, P. Tarte and M. Ausloos, Sol. St. Comm. 66:445 (1988)
9. Ch. Laurent S. K. Patapis, S. M. Green, H. L. Luo, C. Politis, K. Durczewski and M. Ausloos, Mod. Phys. Lett. B, 3:241 (1988)
10. Ch. laurent, M. Laguess, S. K. Patapis, H. W. Vanderschueren G. V. Lecompte, A. Rulmont, P. Tarte and M. Ausloos, Z. der Phys. B, 69:435 (1988)
11. S. M. Green, C. Jiang, Vo Mei, H. L. Luo and C. Politis Phys. Rev. B, 38:5016 (1988)
12. Ch. Laurent in Ph. D. Thesis (1988), University of Liege, (Belgium)
13. Ch. Laurent, M. Ausloos, C. Politis and S. K. Patapis in Proceedings of the Nato ASI "Physics and Materials Science of HTS" Bad Windeim, Germany (1989)
14. M. Ausloos, Ch. Laurent, S. K. Patapis, A. Rulmont, P. Tarte Mod. Phys. Lett. B, 3:167 (1987)
15. A. T. Fiory, A. F. Hebard, L. F. Schneemyer, J. V. Waszak in Mat. Res. Soc., Boston (1987)
16. K. Maki and R. S. Thomson, Phys. Rev., 39:2767 (1989)
17. M. Ausloos, Ch. Laurent, S. K. Patapis and C. Politis (submitted for publication)
18. M. Ausloos, K. Durczewski, S. K. Patapis, Ch. Laurent and H. W. Vanderschueren, Sol. St. Com. 65:365 (1988)
19. M. Ausloos in "Magnetic Phase Transitions" M. Ausloos and R. J. Elliott Eds, Springer-Verlag, Berlin, p. 99 (1983)

MODELS OF HIGH TEMPERATURE SUPERCONDUCTORS

V. J. Emery

Department of Physics
Brookhaven National Laboratory
Upton, New York 11973

G. Reiter

Department of Physics
University of Houston
Houston, Texas 77204-5504

ABSTRACT

A central problem in the theory of high-temperature superconductivity is the identification of the simplest model containing the essential physics on an energy scale of a few hundred degrees, where the main phenomena of magnetism, superconductivity and normal charge transport are to be found. A particular issue of importance for this workshop is the nature of the spin fluid probed by neutron scattering, NMR, or μSR. Much of the discussion has addressed the question of whether two spin fluids are required to fit the available data, or one spin fluid will suffice. In our view, neither description is quite accurate – it is better to think in terms of a composite spin fluid.

The widely adopted starting points for discussions of the effects of strong correlations are a single-band Hubbard model[1] or a three-band model involving copper $d_{x^2-y^2}$ and oxygen p_x, p_y states.[2] These are not low-energy Hamiltonians since they involve parameters of the order 1 - 10 eV. They are relevant for various x-ray and UV spectroscopies which clearly show that the three band model is required. However, it is easier to focus on magnetism and superconductivity if high energy states are "*integrated out*" in order to obtain an effective Hamiltonian. It is then important to establish whether there are circumstances of relevance for high-temperature superconductivity in which the one-band and three-band models lead to the same effective Hamiltonian. In examining this question, it has been noticed by a number of people that there is a particularly strong superexchange interaction between holes on neighboring copper and oxygen sites and that it may be a good approximation to assume that they form singlets. Three representations have been used: copper-oxygen bond singlets,[3] copper-centered singlets,[4] and oxygen-centered singlets.[5,6] Once the translational motion of the oxygen hole has been taken into account, all three representations are equivalent and give same results if calculated completely. They differ in that further approximations are usually made in order to simplify the problem.

The physical picture is one of singlets moving through a background of antiferromagnetically correlated spins. The analogous representation of the single-band Hubbard model is known as the t-J model.[1] At this level, some of the strong interactions have been eliminated, but we have not yet arrived at a continuum limit or fixed point Hamiltonian. Zhang and Rice[4] have argued that the copper-centered singlets behave in the same way as missing spins in the t-J model. Their essential point is that the number of degrees of freedom is the same and that therefore the two models may have the same effective Hamiltonian. However, we have shown that the models have different oxygen spin-spin correlation functions[5] and that, in one dimension, the number of degrees of freedom are not the same.[7] Thus we believe that the effective Hamiltonians are different, and more work is needed to determine whether or not their solution leads to the same physical properties for high-temperature superconductors.

In the three-band model considered here, the spins of copper and oxygen holes are strongly correlated and there is no sense in which they may be regarded as independent spin fluids. However, current neutron scattering[8] and NMR data[8] are not easily reconciled on the basis of a single "*immutable*" spin fluid. Proposed explanations[9] of the difference between relaxation rates of copper and oxygen nuclear spins do not rest easily with the observed incommensurate position of antiferromagnetic fluctuations in k-space.[10] Also the behavior of these relaxation rates below T_c is difficult to reconcile with the observation by neutron scattering[8] of spin fluctuations below the superconducting gap. A complete understanding of these observations may well require us to take account of the composite nature of the spin fluid.

This work was supported by the Division of Materials Sciences, Office of Basic Energy Sciences, U.S. Department of Energy under Contract DE-AC02-76CH00016 and by the Texas Center for Superconductivity at the University of Houston under Prime Grant No. MDA 972-8-G-0002 from the U.S. Defense Advanced Research Projects Agency and the State of Texas.

REFERENCES

1. P. W. Anderson, Science 235:1196 (1987).
2. V. J. Emery, Phys. Rev. Lett. 58:2794 (1987).
3. M. Imada, J. Phys. Soc. Jpn 56:3793 (1987).
4. F. C. Zhang and T. M. Rice, Phys. Rep. B37:3759 (1988); ibid B41:2560 (1990); ibid B41:7243 (1990).
5. V. J. Emery and G. Reiter, Phys. Rep. B38:4547 (1988); ibid B38:11938 (1988); ibid B41:7247 (1990).
6. A. Aharony et al., Phys. Rev. Lett. 60:1330 (1988) ;
   E. B. Stechel and D. R. Jennison, Phys. Rep. B38:4632 (1988).
7. V. J. Emery, to be published.
8. See papers in this workshop.
9. See papers by N. Bulut and H. Monien (this workshop).
10. G. Shirane et al., Phys. Rev. Lett. 63:330 (1989).

BCS THEORY EXTENDED TO ANISOTROPIC AND LAYERED HIGH-TEMPERATURE SUPERCONDUCTORS

T. Schneider, M. Frick and M.P. Sörensen

IBM Research Division
Zurich Research Laboratory
8803 Rüschlikon, Switzerland

ABSTRACT

Recent experiments have revealed several key features of the normal and superconducting states of high-temperature superconductors. They impose strong constraints on theoretical models of the phenomenon. Here we explore a tight-binding BCS-type model. The carriers form a narrow and anisotropic band, and are subject to on-site and nearest neighbor intralayer pairing, giving rise to x,y-gap anisotropy. The resulting properties agree remarkably well with experimental results, resolve conflicting interpretations of experiments, offer an easily understandable physical picture, and point to the nature of the pairing mechanism.

1. EXPERIMENTAL CONSTRAINTS

Recent experiments have revealed several key features of the normal and superconducting states of high-temperature superconductors [1,2]. Measurements of the penetration depth [3-6], the coherence length ξ derived from the upper critical fields [7-9], the conductivity σ [10,11], and infrared reflectivity [12-15] clearly point to the presence of strong anisotropy, as expressed by the ratios

$$\frac{\lambda_{ab}}{\lambda_c}, \quad \frac{\xi_\perp}{\xi_\parallel}, \quad \sqrt{\frac{\sigma_\perp}{\sigma_\parallel}}, \quad \sqrt{\frac{m_\parallel}{m_\perp}} \ll 1, \qquad (1)$$

where m is the effective mass, $\parallel$ denotes directions parallel and $\perp$ perpendicular to the layers. Compared to conventional superconductors, the zero-temperature correlation length is very short and the penetration depth much longer. As $\xi_\perp$ turns out to be much smaller than the mean spacing, the interlayer interaction is small and a description in terms of a system of weakly interacting layers appears to be appropriate.

The temperature dependence of the penetration depth [3-6] closely follows BCS-type behavior, consistent with singlet pairing and a nodeless gap. Moreover, the relationship between transition temperature and carrier density points to an unretarded pairing interaction [16]. In contrast to conventional BCS superconductors, there is considerable evidence for an anisotropic gap from tunneling [17-20], infrared reflectivity [13-15] and angular resolved photoemission measurements [21,22]. As far as the normal state is concerned, Hall measurements for H parallel to the c-axis and I parallel to the a,b-plane yield a positive value for the appropriate element of the Hall

Dynamics of Magnetic Fluctuations in High-Temperature Superconductors
Edited by G. Reiter *et al.*, Plenum Press, New York, 1991

tensor, i.e. hole-like, while the elements for H parallel to the a,b-plane are negative [23]. Finally, photoemission experiments [21,22,24-29] provide clear evidence for anisotropic Fermi liquid states, forming narrow bands with low carrier density. These experimental facts place strong constraints on theoretical models.

2. MODEL

Guided by photoemission experiments [21,22,24-29], we assume Fermi liquid states, subjected to an unretarded pairing interaction. To account for narrow bands, the tight-binding description is adopted. The Hamiltonian then reads

$$\mathcal{H} = \sum_{i,j,\sigma} t_{ij}\, c_{i\sigma}^{+}\, c_{j\sigma} - \sum_{i,j} g_{ij}\, c_{i\uparrow}^{+}\, c_{j\downarrow}^{+}\, c_{j\downarrow}\, c_{i\uparrow}\,. \tag{2}$$

t_{ij} are transfer integrals describing the hopping of the carriers between the sites i and j, while g_{ij} is the strength of the pairing interaction between the carriers on these sites. For m layers per unit cell, there will be m bands and their splitting will be controlled by the interlayer transfer integrals. Here, we concentrate on one layer per unit cell, or equivalently, we assume that the splitting is negligibly small, corresponding to an m-fold degenerate band. Thus, the first term describes the band structure of the carriers, while the second term represents the unretarded pairing interaction. For a square lattice (lattice constant a) within the layers, identical layers separated by the lattice constant s, we have one layer per unit cell. Considering then hopping between nearest and next-nearest neighbors within the layers and hopping between adjacent sheets, diagonalization of the first term yields

$$\varepsilon(\mathbf{k}) = A\Big\{-2\big(\cos k_x a + \cos k_y a\big) + 4B\cos k_x a \cos k_y a - 2C\cos k_z s - E_F\Big\}\,, \tag{3}$$

where A, AB and AC correspond to the values of the hopping matrix elements, and E_F denotes the Fermi energy given in terms of the band filling

$$\rho = \frac{1}{N}\sum_{\mathbf{k}}\big(\exp(\beta\varepsilon(\mathbf{k})) + 1\big)^{-1}. \tag{4}$$

N is the number of sites. Guided by the photoemission results [21,22,24-29], the Hall effect measurements [23] and the anisotropy of the coherence length, we fixed the parameters entering Eq. (3) as follows:

$$A = 0.12\,\text{eV}, \quad B = 0.45, \quad C = 0.1. \tag{5}$$

Even though A and AC differ by only one order of magnitude, this choice is shown to lead to a highly anisotropic metallic normal state. In the superconducting state we assume BCS-type pairing. The gap equation is then given by [30]

$$\Delta(\mathbf{k}) = \sum_{\mathbf{q}} V(\mathbf{k}-\mathbf{q})\,\frac{\Delta(\mathbf{q})}{2E(\mathbf{q})}\tanh\frac{\beta E(\mathbf{q})}{2}\,, \tag{6}$$

where

$$E(\mathbf{q}) = \big(\varepsilon^2(\mathbf{q}) + \Delta^2(\mathbf{q})\big)^{1/2} \tag{7}$$

and

$$V(\mathbf{q}) = \sum_{l} g_{l0}\, e^{i\mathbf{q}\cdot\vec{R}_l}\,, \tag{8}$$

where l labels the sites. Here we explore unretarded on-site and intralayer nearest neighbor pairing. Interlayer pairing has been studied previously [2,30]. The pairing interaction is given by

$$V(\mathbf{k}) = g_0 + g_2(\cos k_x a + \cos k_y a) . \tag{9}$$

Assuming singlet pairing, where $\Delta(\mathbf{k}) = \Delta(-\mathbf{k})$, the solutions of the gap equation (6) adopt the anisotropic form

$$\Delta(\mathbf{k}) = \Delta_0 + 2\Delta_2(\cos k_x a + \cos k_y a) . \tag{10}$$

Of particular relevance is the behavior on or close to the Fermi surface $\varepsilon(\mathbf{k}) = 0$, where nodes might appear. Such nodes heavily affect the thermodynamic behavior. In solving Eq. (6), we assume unretarded pairing, as suggested by the empirical relationship between T_c and the carrier density [16]. The k-summation then extends over the full Brillouin zone and the limits of integration with respect to energy correspond to the bottom and the top of the band. All carriers are subject to pairing, and a modest pairing interaction leads to high T_c values.

In summary, the model is of BCS-type, assumes a tight-binding band for the carriers and unretarded pairing, on-site and between nearest neighbors. The tight-binding band of the carriers includes nearest and next-nearest neighbors hopping within the layers and between adjacent sheets. In the weak-coupling limit, T_c is related to the band width and the Fermi energy, yielding high values of T_c for modest coupling strengths. In the following we resort to numerical solutions of the gap equation (6). Due to its nonlinearity, there are several solutions and the one yielding the lowest free energy is followed. It should be kept in mind however, that our present analysis is restricted to a model with one layer per unit cell. An extension to m layers per unit cell will lead to m distinct gap branches. Nevertheless, for nearly degenerate branches, the present model is still applicable. Considerable modifications are expected for the nondegenerate case, which will be treated in a forthcoming paper.

3. MODEL PROPERTIES

In this section, we compare the key properties of the model with current experimental facts. These results should also offer a physical explanation of key features, such as the short coherence length, long penetration depth, and pronounced anisotropy. In doing so, we will now consider normal state properties at zero temperature. The elements of the Hall tensor are given by

$$R_{xyz} = \frac{\sigma_{xyz}}{\sigma_{xx}\sigma_{yy}} = \frac{E_y}{J_x H_z}, \quad R_{yzx} = \frac{\sigma_{yzx}}{\sigma_{yy}\sigma_{zz}} = \frac{E_y}{J_z H_x}, \quad R_{zxy} = \frac{\sigma_{zxy}}{\sigma_{zz}\sigma_{yy}} = \frac{E_z}{J_y H_x} . \tag{11}$$

Assuming an isotropic relaxation time, the transport coefficients read [31]

$$\sigma_{\alpha\alpha} = e^2 \tau n \frac{1}{m_{\alpha\alpha}}, \quad n = \frac{2\rho}{V_0} \tag{12}$$

and

$$\sigma_{xyz} = -2\frac{e^3\tau^2}{\hbar^4 V_0 N} \sum_{\mathbf{k}} \frac{\partial\varepsilon}{\partial k_x} \left(\frac{\partial\varepsilon}{\partial \mathbf{k}} \times \nabla_k\right)_z \frac{\partial\varepsilon}{\partial k_y} \delta(\varepsilon) , \tag{13}$$

where

$$\frac{1}{m_{\alpha\alpha}} = \frac{1}{\rho N \hbar^2} \sum_{\mathbf{k}} \left(\frac{\partial \varepsilon}{\partial k_\alpha} \right)^2 \delta(\varepsilon) \tag{14}$$

is the effective mass, $V_0 = a^2 s$, s denotes the mean spacing of the layers, and the z-direction is perpendicular to the sheets. In this approximation, the R_{xyz} values are independent of the relaxation time. Moreover, it is important to observe that the free electron formula $R = -1/ne$ becomes meaningless, except for a parabolic band, which is not applicable here. In fact, the Fermi surface corresponds to a corrugated column with a cross section depending on the filling, as illustrated in Fig. 1.

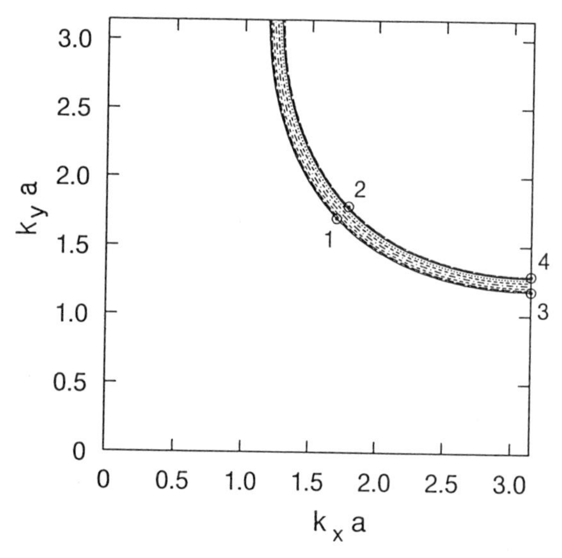

Fig. 1. Cuts parallel to the (k_x,k_y)-plane through the Fermi surface for different k_z-values, $0 \leq k_z \leq \pi$, in steps of $\pi/(10s)$ for $\rho = 0.7$ and the parameters listed in Eq. (5). For $k_z = 0$ the Γ, X and M points are marked. The long-dashed (topmost) curve corresponds to $k_z = 0$ and the numbers 1 to 4 mark particular points on the Fermi surface.

In the London approximation, the zero-temperature value of the penetration depth is also fully determined in terms of normal-state properties by

$$\lambda_{\alpha\alpha}^{-2} = \frac{4\pi n e^2}{c^2} \frac{1}{m_{\alpha\alpha}} . \tag{15}$$

The numerical results listed in Table I agree remarkably well with available experimental data. R_{xyz} is positive, i.e. hole-like, and has the correct order of magnitude as does the carrier density n [23]. The anisotropy and magnitude of the penetration depth can be understood in terms of the low carrier density and the large effective mass. Accordingly, our simple tight-binding parameterization leads to remarkable agreement with the available experimental facts. The narrow anisotropic band implies carriers with both hole and electron-like character and low density, a column-like Fermi surface, and rather large effective masses.

Table I. Normal state properties and zero temperature penetration depth for several band fillings and the parameters listed in Eq. (6); E_F is the Fermi energy, $m_{xx} = m_{yy}$ and m_{zz} are the elements of the effective mass tensor, while R_{xyz} is an element of the Hall tensor in units of 10^{-3} cm^{3}C^{-1}. The other units are: energy in A [Eq. (5)], λ in A and the carrier density in 10^{21} cm^{-3} ($n = 2\rho/V_0$). As characteristic lattice constants we used $a = 3.85$ and $s = 7.76$ Å.

ρ	E_F	$N_N(0)$	$\frac{m_{xx}}{m_e}$	$\frac{m_{zz}}{m_e}$	R_{xyz}	λ_{xx}	λ_{zz}	n
0.7	0.744	0.091	5.8	410	1.18	1149	7007	12.17
0.6	-0.205	0.175	5.0	259	0.88	1159	8239	10.43
0.5	-0.891	0.264	4.7	153	0.70	1229	8703	8.69

Next, we turn to the superconducting state by assuming unretarded on-site and nearest neighbor intralayer pairing. For a treatment of interlayer pairing we refer to Refs. (2) and (30). Keeping the strength of the on-site pairing fixed, we increased the strength of the nearest neighbor interlayer pairing g_2, to explore the effects of the resulting x,y-gap anisotropy.

In Table II, we summarized estimates for the gap parameters and T_c for $\rho = 0.7$ as a function of g_2, the strength of the nearest neighbor intralayer pairing interaction. These estimates, obtained from a numerical solution of Eq. (6), reveal that T_c and gap anisotropy grow with increasing g_2. Of particular interest are the results for $g_2 = 3$ and $g_2 = 4$, yielding transition temperatures of 28 K and 130 K and gap structures varying between $3 \leq \Delta \leq 5$ and $9.4 \leq \Delta \leq 24$ meV, respectively. In fact, these values correspond roughly to experimental estimates.

Table II. Numerical estimates for Δ_0, Δ_2 and T_c obtained from Eq. (6) for $\rho = 0.7$. Energies are in units of $A = 0.12$ eV. Δ_{min} denotes the minimum, Δ_{max} the maximum value of the gap on the Fermi surface.

g_0	g_2	Δ_0	Δ_2	k_BT_c	$\frac{2\Delta_{min}}{k_BT_c}$	$\frac{2\Delta_{max}}{k_BT_c}$	$\frac{\Delta_{min}}{\Delta_{max}}$
1	1	0.0013	-0.0003	0.0009	3.22	3.87	0.83
1	2	0.0044	-0.0027	0.0041	2.82	4.06	0.69
1	3	0.0147	-0.0195	0.0202	2.45	4.24	0.58
1	4	0.0378	-0.1106	0.0932	2.03	4.24	0.48

Next we explore the implications of x,y-anisotropy on superconducting properties. Angular resolved photoemission [21,22], tunneling conductance [17-20] and infrared reflectivity [13-15] measurements yield increasing evidence for a nodeless anisotropic gap, while the temperature dependence of the penetration depth points to standard BCS behavior [6]. To fingerprint this anisotropy, we depict the density of states in the gap region in Fig. 2. This quantity enters various properties of the superconducting state. The standard BCS behavior, a square-root singularity at the gap energies, is removed. For positive energies, the density of states is zero up to Δ_{min}, the minimum value of the gap. The marked discontinuities are Van Hove singularities of the gap, occurring at points 1 to 4 on the Fermi surface (Fig. 1). In any case, there is no longer a unique gap, as its energy varies between Δ_{min} and Δ_{max}. For energies sufficiently far away from the gap, the density of states approaches the normal state behavior, giving rise to the height asymmetry in the gap structure for positive and negative energies. Clearly, for decreasing anisotropy, the modifications will shrink to a small energy interval and standard BCS behavior is recovered.

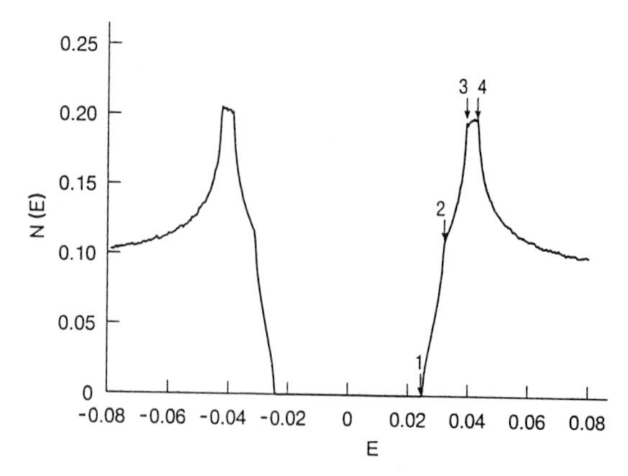

Fig. 2. Gap region of the density of states $N(E)$ for $g_2 = 3$ and $\rho = 0.7$ (see Table II). The arrows mark Van Hove singularities arising from points 1 to 4 on the Fermi surface (Fig. 1). Point 1 corresponds to $\Delta_{min} = 0.025$ while point 4 indicates $\Delta_{max} = 0.043$. Energies are in units of $A = 0.12$ eV.

We are now prepared to explore the effects of x,y-gap anisotropy, mediated by intralayer nearest neighbor pairing, on the temperature dependence of the specific heat, penetration depth, nuclear spin relaxation time, spin susceptibility and the tunneling conductance.

The temperature dependence of the specific heat is depicted in Fig. 3 in terms of $C/\gamma T_c$, where γ is the Sommerfeld constant given by

$$\gamma = \frac{2\pi^2}{3} N_N(0)\, k_B^2 , \tag{16}$$

and $N_N(0) = 0.091$ (Table I). For $g_2 = 1$ the gap anisotropy $\Delta_{min}/\Delta_{max} = 0.83$ is small and the normalized jump $\Delta C/\gamma T_c = 1.45$ corresponds to the standard BCS value. Larger anisotropy, $\Delta_{min}/\Delta_{max} = 0.69$, 0.58 and 0.48, is seen to reduce the jump $\Delta C/\gamma T_c$ to 1.35, 1.25 and 1.02. Recent specific heat measurements on $Tl_2Ba_2CaCu_2O_8$ reveal a very small jump, suggesting even larger anisotropy or close proximity to a node on the Fermi surface [32].

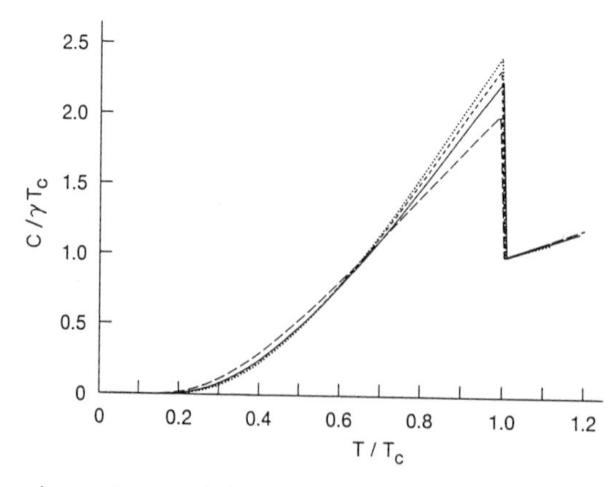

Fig. 3. Temperature dependence of the specific heat for the parameters listed in Table II. Dotted line: $g_2 = 1$, $\Delta_{min}/\Delta_{max} = 0.83$; short-dashed line: $g_2 = 2$, $\Delta_{min}/\Delta_{max} = 0.69$; solid line: $g_1 = 3$, $\Delta_{min}/\Delta_{max} = 0.58$; long-dashed line: $g_2 = 4$, $\Delta_{min}/\Delta_{max} = 0.48$.

To calculate the penetration depth at finite temperature, the use of local electrodynamics is appropriate ($\lambda(0) \gg \xi(0)$). The finite temperature extension of Eq. (7) then reads

$$\lambda_{\alpha\alpha}^{-2} = \frac{4\pi ne^2}{c^2} \sum_{\mathbf{k}} V_\alpha^2(\mathbf{k}) \left[\left(-\frac{\partial f}{\partial \varepsilon} \right) - \left(-\frac{\partial f}{\partial E} \right) \right], \tag{17}$$

where

$$f(x) = (\exp \beta x + 1)^{-1}, \quad V_\alpha(\mathbf{k}) = \frac{1}{\hbar} \frac{\partial \varepsilon}{\partial k_\alpha}, V_o = a^2 s. \tag{18}$$

The temperature dependence of λ_{xx} is shown in Fig. 4 for several values of the coupling constant g_2. It measures the strength of the nearest neighbor intralayer pairing and determines the x,y-anisotropy of the gap. Above $T/T_c > 0.1$, increasing anisotropy leads to a growing reduction of $(\lambda_{\alpha\alpha}(0)/\lambda_{\alpha\alpha}(T))^2$. In particular, the amplitude of the asymptotic behavior $(\lambda_{\alpha\alpha}(0)/\lambda_{\alpha\alpha}(T))^2 = A_{\alpha\alpha}(1 - T/T_c)$ decreases from $A_{\alpha\alpha} = 2$ with increasing anisotropy. It is important to note, however, that larger deviations from the isotropic case require close proximity to nodes in the temperature-dependent gap [2]. Experiments on $BiSr_2CaCu_2O_8$ clearly point to BCS behavior [6], justifying first of all a BCS treatment and setting the lower limit of 0.6 of the anisotropy ratio $\Delta_{min}/\Delta_{max}$. To exclude the presence of anisotropy in this interval, a rather high degree of experimental accuracy would be required.

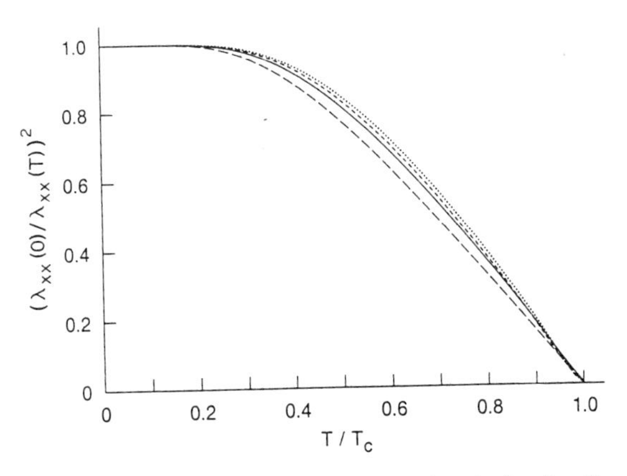

Fig. 4. Temperature dependence of the penetration depth λ_{xx} for the parameters listed in Eq. (5), Table II and $\rho = 0.7$. Dotted line: $g_2 = 1$, $\Delta_{min}/\Delta_{max} = 0.83$; short-dashed line: $g_2 = 2$, $\Delta_{min}/\Delta_{max} = 0.69$; solid line: $g_1 = 3$, $\Delta_{min}/\Delta_{max} = 0.58$; long-dashed line: $g_2 = 4$, $\Delta_{min}/\Delta_{max} = 0.48$.

Next we turn to the spin susceptibility. Neglecting Fermi liquid effects, it is given by [33]

$$\chi = \frac{\beta}{4} \mu_e^2 \sum_{\mathbf{k}} f(E_k)(1 - f(E_k)) . \tag{19}$$

The effect of x,y-gap anisotropy on the temperature dependence is illustrated in Fig. 5. Increasing anisotropy is seen to enhance the spin susceptibility below T_c.

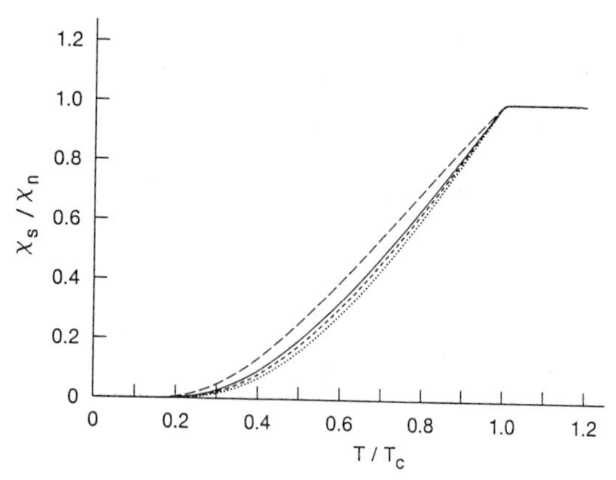

Fig. 5. Temperature dependence of the ratio between the spin susceptibility in the superconducting and normal state for the parameters listed in Eq. (5), Table II and $\rho = 0.7$. Dotted line: $g_2 = 1$, $\Delta_{min}/\Delta_{max} = 0.83$; short-dashed line: $g_2 = 2$, $\Delta_{min}/\Delta_{max} = 0.69$; solid line: $g_1 = 3$, $\Delta_{min}/\Delta_{max} = 0.58$; long-dashed line: $g_2 = 4$, $\Delta_{min}/\Delta_{max} = 0.48$.

Next, we turn to the nuclear spin relaxation rate. The hyperfine contact interaction [34,35] is taken as the coupling between the carriers and nuclei, and is given by

$$\frac{1}{T_1} \simeq \frac{1}{2} \sum_{\mathbf{k},\,\mathbf{k}'} \left(1 + \frac{\varepsilon(k)\varepsilon(k')}{E(k)E(k')} + \frac{\Delta(k)\Delta(k')}{E(k)E(k')}\right) \times f\left(E(k')\right)\left(1 - f\left(E(k)\right)\delta\left(E(k) - E(k')\right)\right). \tag{20}$$

For an isotropic gap, where the density of states exhibits a square-root singularity, the integral is known to diverge. This problem can be resolved by taking into account the lifetime of the quasiparticles and the anisotropy of the gap. Here we neglect the lifetime effect. Figure 6 shows a comparison between the temperature dependences resulting from a small and a large anisotropy. For $T/T_c > 1$, Korringa behavior appears, while the broad peak occurring just below T_c, reminiscent of conventional superconductors, is seen to become weaker with increasing anisotropy.

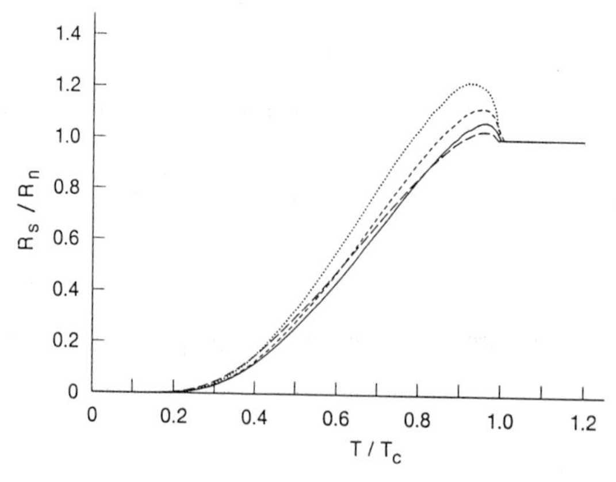

Fig. 6. Temperature dependence of the nuclear spin relaxation rate ratio R_s/R_N for the parameters listed in Eq. (6), Table II and $\rho = 0.7$. R_N refers to the normal state. Dotted line: $g_2 = 1$, $\Delta_{min}/\Delta_{max} = 0.83$; short-dashed line: $g_2 = 2$, $\Delta_{min}/\Delta_{max} = 0.69$; solid line: $g_1 = 3$, $\Delta_{min}/\Delta_{max} = 0.58$; long-dashed line: $g_2 = 4$, $\Delta_{min}/\Delta_{max} = 0.48$.

Recently, the NMR relaxation of O^{17} has been measured in several cuprates [36-38] revealing very little or no enhancement just below T_c and Korringa-type behavior above T_c. This reduction clearly points to the presence of gap anisotropy and a finite lifetime of quasiparticles. Taking $\Delta_{min}/\Delta_{max} \simeq 0.6$, consistent with the infrared, tunneling and penetration depth measurements, an enhancement still remains (Fig. 6) and is further reduced by taking the finite lifetime of the quasiparticles into account. In fact, the enhancement appears close to T_c, which is high, so that lifetime effects play a role. Further work is necessary to disentangle the lifetime and anisotropy-induced reduction of the $1/T_1$ enhancement just below T_c. In any case, we have shown that a reduced or absent enhancement does not contradict extended s-wave pairing.

Gap anisotropy is expected to modify the tunneling conductance of SIN junctions more directly. There are two limiting cases [30]: specular tunneling requires perfect junctions and the transverse momenta are conserved, while the transmission becomes diffuse for corrugated interfaces. Here transverse momenta are no longer conserved. The transfer Hamiltonian approach yields for the zero temperature conductance in an NIS junction [30]

$$\frac{dI}{dV} = \pm 2\pi \sum_{\mathbf{kv}} |T_{\mathbf{kv}}|^2 \left(1 \mp \frac{\varepsilon(\mathbf{k})}{E(\mathbf{k})}\right) \delta(|eV| - E(\mathbf{k})) \times \delta(\xi(\mathbf{v}) - |eV| \pm E(\mathbf{k})), \tag{21}$$

where $\mathbf{k}$ and $\mathbf{v}$ label the states on the superconducting and the normal side, respectively, and $\pm$ = sign (eV) . $\xi(\mathbf{v})$ describes the conduction band on the normal side, which is assumed to have a constant density of states at the Fermi level. $T_{\mathbf{kv}}$ denotes the tunneling matrix element and differs in the two situations mentioned above. Expressions for $T_{\mathbf{kv}}$ have been derived in Refs. [2] and [30]. Here we merely state the results.

For specular transmission, the conductance for a tunneling barrier of height U and thickness d reads in WKB approximation

$$\left(\frac{dI}{dV}\right)_{\text{spec.}} \simeq P\bar{P} \sum_{\mathbf{k}} \left(1 \mp \frac{\varepsilon(\mathbf{k})}{E(\mathbf{k})}\right) \delta(|eV| - E(\mathbf{k})) \left|\frac{\partial \varepsilon(\mathbf{k})}{\partial k_L}\right| \times \exp\left(\frac{\varepsilon(\mathbf{k})}{\varepsilon_0}\right) \exp\left(-\frac{k_T^2}{K^2}\right), \tag{22}$$

where

$$\varepsilon_0 = \frac{1}{d}\sqrt{\frac{\hbar^2 U}{2m}}\,; \quad K^2 = \frac{1}{d\hbar}\sqrt{2mU}\,; \quad \bar{P} = \exp\left(-2d\sqrt{\frac{2mU}{\hbar^2}}\right). \tag{23}$$

m denotes the free electron mass and P a constant prefactor; T and L denote the vector components transverse and longitudinal to the direction of the tunneling current, respectively. In Eq. (22), the barrier height U is assumed to be large. The factor $\exp(-k_T^2/K^2)$ stems from the conservation of transverse momenta and singles out small transverse momenta. Typical barrier parameters are $U = 1$ eV and $d = 10$ Å, which yield $\varepsilon_0 = 0.194$ eV and $K = 0.23$ Å^{-1}.

For diffuse transmission, where transverse momenta are not conserved, the conduction simplifies to [30]

$$\left(\frac{dI}{dV}\right)_{\text{diff.}} \propto \sum_{\mathbf{k}} \left(1 \mp \frac{\varepsilon(\mathbf{k})}{E(\mathbf{k})}\right) \delta(|eV| - E(\mathbf{k})) \left|\frac{\partial \varepsilon(\mathbf{k})}{\partial k_L}\right| \tag{24}$$

To eliminate constant factors, we normalize the conductance in terms of the corresponding values of the NIN junction by setting the gap equal to zero in Eqs. (22) and (24). Thus constant factors representing detailed junction properties cancel out.

Figure 7 shows the normalized conductance for both specular and diffuse tunneling. Neglecting the small contribution of the ε/E-term in expression (24), the conductance for diffuse transmission becomes proportional to the density of states in the superconductor. In fact, the voltage dependence follows rather closely the density of states depicted in Fig. 2 and is nearly independent of the orientation of the tunneling current with respect to the layers. As expected from Eq. (24) and seen in Fig. 7, only specular transmission leads to a pronounced orientation dependence. This dependence has been observed in junctions fabricated on cryogenically cleaved surfaces of epitaxially grown films using the film edge technique [18,19]. Depending on the direction of the tunneling current, i.e. parallel or perpendicular to the layers, the gap estimated from the peak position was found to be anisotropic with the ratio $\Delta_c/\Delta_{ab} \simeq 0.67$. Concerning the quality of these junctions, one expects diffuse tunneling parallel to the layers and proximity to specular transmission for tunneling perpendicular to the layers. In fact, sheets parallel to the layers are known to be cleavage planes. From Fig. 7 it is seen that the peak for specular tunneling perpendicular to the layers, Δ_c, lies close to the minimum value of the gap on the Fermi surface, Δ_{min}. In this case, **k**-vectors around point 2 on the Fermi surface (Fig. 2) are singled out where the gap is close to the minimum value. For specular tunneling parallel to the layers, however, the peak in the conductance stems from the region around points 3 and 4 on the Fermi surface, where the gap reaches its maximum value.

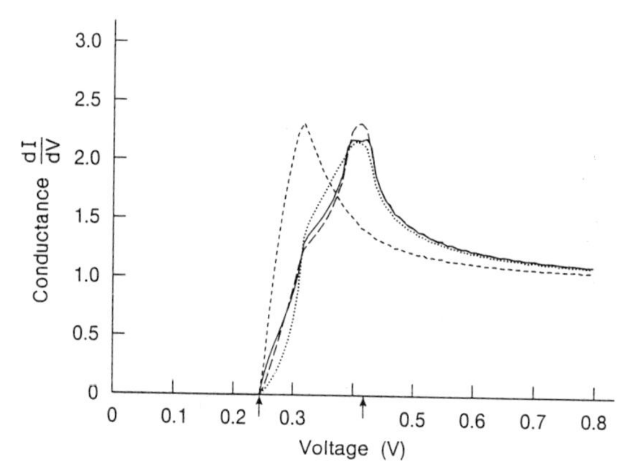

Fig. 7. Normalized tunneling conductance versus voltage for the parameters listed in Eq. (6), Table II , $\rho = 0.7$, $g_2 = 3$ and $\Delta_{min}/\Delta_{max} = 0.58$. The arrows mark the minimum and maximum values of the gap on the Fermi surface. Short-dashed line: specular tunneling in the z-direction (perpendicular to the layers); dotted line: specular tunneling in the x-direction; long-dashed line: diffuse tunneling in the z-direction; solid line: diffuse tunneling in the x-direction.

Thus by invoking x,y-gap anisotropy [Eq. (11)] mediated by nearest neighbor intralayer pairing, a Fermi surface as shown in Fig. 1 and by assuming specular conditions for tunneling perpendicular to the layers, we find remarkable agreement with recent measurements which, estimated from the peak in the conductance for tunneling parallel and perpendicular to the layers, suggests that there is an anisotropic gap. Similarly, the discrepancy between the recent gap estimates obtained from angular resolved photoemission [21,22] can be understood in terms of the variation of the gap on the Fermi surface. In fact, going from points 1 and 2 to 3 and 4 the gap increases from its minimum to its maximum value. Nevertheless, this substantial gap anisotropy does not lead to drastic deviations from standard BCS behavior in the thermodynamic properties, i.e. in the temperature dependence of the penetration depth (Fig. 4).

Finally, we turn to the coherence length in the ground state. It can be obtained from the exponential decay of the correlation function of two quasiparticles, one in the origin, the other at **r**, whose spins are antiparallel. For a spherical Fermi surface and an isotropic gap, it is given by $\xi_0 = \hbar^2 k_F / m\Delta$ [34]. The numerical estimates listed in Table III reflect the strong anisotropy determined by the small interlayer hopping matrix element. In particular, $\xi_\perp$ is found to be smaller than the spacing of the layers and the ratio $\xi_\perp / \xi_\parallel$ is in the range of values estimated from the upper critical fields [7-10]. Because the gap anisotropy does not affect the magnitude of the coherence length, a slight extension of the free electron expression

$$\xi_{\perp,\parallel} = \frac{\hbar}{<\Delta>} \left(\frac{2\tilde{E}_F}{m_{\perp,\parallel}} \right)^{1/2} \tag{25}$$

appears to be appropriate. Thus the anisotropy comes from the effective mass, and the small magnitude is due to the large average gap and mass, and to the small relative Fermi energy. $\xi_{\perp,\parallel}$ are a measure of the size and shape of the Cooper pair. In the present case it is an ellipsoid of volume $V_o = (4\pi/3)\xi_\perp^2 \xi_\parallel$. Thus, within the volume V_o, there are $nV_o = 80$ carriers ($n = 2\rho/V_o$, $\rho = 0.7$, $g_2 = 3$). Even though this number is much lower than in conventional superconductors, where $nV_o = 10^7$, nV_o is still close to the Cooper pairing limit ($nV_o \gg 1$) and very far from the Bose limit ($nV_o \ll 1$) [33].

Table III. Numerical estimates of the zero temperature coherence length for the parameters cited in Eq. (6).

ρ	g_0	g_2	Δ_0	Δ_2	ξ_{xx}/a	ξ_{zz}/s	$\frac{\xi_{zz}}{\xi_{xx}} \frac{a}{s}$
0.7	1	3	0.0147	-0.0195	6.6	0.30	0.10
0.7	1	4	0.0378	-0.1106	5.9	0.31	0.12

4. CONCLUDING REMARKS

Guided by experimental constraints, we explored a BCS-type tight-binding model. In the highly anisotropic metallic normal state the carriers form a narrow and anisotropic band. Normal state properties, including the elements of the Hall tensor, the effective mass, the carrier density and the penetration depth, agree remarkably well with experiment. The carriers having hole and electron-like character are subject to unretarded pairing of on-site and intralayer nearest neighbor origin. Adopting a BCS treatment, the unretarded nature of the pairing interactions leads to a value of T_c which is proportional to the Fermi energy and can be quite large. Fixing the strength of the on-site pairing, we explored the effect of nearest neighbor pairing, which gives rise to x,y-gap anisotropy. For $\Delta_{min}/\Delta_{max} = 0.6$, the superconducting properties reached realistic values, consistent with SIN conductance measurements and the temperature dependence of the penetration depth. It is important to emphasize that nearest-neighbor interlayer pairing which leads to uniaxial gap anisotropy affects the temperature dependence of specific heat, spin susceptibility and NMR relaxation rate quite analogously [2,30]. The effect on the SIN tunneling conductance, however, is fundamentally different. In fact, assuming a uniaxial gap, it was not possible to mimic the experimental facts, namely, a smaller gap for tunneling perpendicular to the layers, taken as the peak conductance, and a larger value for tunneling parallel to the layers. Moreover, we identified the effects of x,y-gap anisotropy on the temperature dependence of the specific heat, penetration depth, spin susceptibility and NMR relaxation rate. For $\Delta_{min}/\Delta_{max} \gtrsim 0.6$, the penetration depth was found to be rather insensitive to gap anisotropy. Thus, the conflict between the penetration depth measurements, which are consistent with isotropic BCS behavior, infrared reflectivity, tunneling and angular resolved photoemission measurements, pointing to an anisotropic gap, has been resolved. A consistent picture requires that $\Delta_{min}/\Delta_{max} \simeq 0.6$. The missing enhancement in $1/T_1$ is then a natural consequence of gap anisotropy and the lifetime of quasiparticles.

The remarkable agreement with the experimental findings points to an unretarded pairing interaction, mediated by on-site and intralayer nearest neighbor coupling. The origin is of electronic nature, yielding a cutoff proportional to the Fermi energy. It should be kept in mind, however, that our analysis was restricted to systems with one layer per unit cell. Accordingly, we assumed single or nearly degenerate gap branches.

ACKNOWLEDGEMENTS

The authors thank A. Aharony, A. Baratoff, S. Ciraci, X. Gedik and K.A. Müller for stimulating discussions and pertinent suggestions.

REFERENCES

1. W.A. Little, Science **242**, 1390 (1989).
2. T. Schneider and M. Frick, in *Strong Correlations and Superconductivity,* H. Fukuyama, S. Maekawa, and A.P. Malozemoff, eds. (Springer-Verlag, Berlin, Heidelberg, New York, 1989), p. 176.
3. Y.J. Uemura, V.J. Emery, A.R. Moodenbaugh, M. Suenaga, D.C. Johnston, A.J. Jacobsen, J.T. Lewandowski, J.H. Brewer, R.F. Kiefel, S.R. Kreitzman, G.M. Luke, T. Riseman, C.E. Stronach, W.J. Kossler, J.R. Kempton, X.H. Yu, D. Opie, and H.E. Schone, Phys. Rev. B **38**, 909 (1988).
4. L. Krusin-Elbaum, R.L. Greene, F. Holtzberg, A.P. Malozemoff, and Y. Yeshurun, Phys. Rev. Lett. **62**, 217 (1989).
5. A.T. Fiory, A.F. Hebard, P.M. Mankiewich, and R.E. Howard, Phys. Rev. Lett. **61**, 1419 (1988).
6. B. Pümpin and H. Keller, preprint.
7. T.T.M. Palstra, B. Batlogg, L.F. Schneemeyer, and R.J. Cava, Phys. Rev. B **39**, 5102 (1988).
8. J.H. Kang, K.E. Gray, R.T. Kampwirth, and D.W. Day, Appl. Phys. Lett. **53**, 2560 (1988).
9. U. Welp, W.K. Kwok, G.W. Crabtree, K.G. Vandervoort, and J.Z. Liu, Phys. Rev. Lett. **62**, 1908 (1989).
10. S. Martin, A.T. Fiory, R.M. Fleming, L.F. Schneemeyer, and J.V. Waszczak, Phys. Rev. Lett. **60**, 2194 (1988).
11. A. Zettl, A. Behorooz, G. Briceno, W.N. Creager, M.F. Crommie, S. Hön, and P. Pinsukanjana, in *Mechanisms of High Temperature Superconductivity*, H. Kamimura and A. Oshiyama, eds. (Springer-Verlag, Berlin, Heidelberg, New York, 1988), p. 249.
12. H. Takagi, H. Matsuyama, H. Katayama-Yoshida, Y. Okabe, S. Hosoya, K. Seki, H. Fujimoto, M. Sato, and H. Inokuchi, Nature **332**, 236 (1988).
13. U. Hofman, J. Keller, K. Renk, J. Schuetzmann and W. Ose, Solid State Commun. **70**, 325 (1989).
14. R.R.T. Collins, Z. Schlesinger, F. Holtzberg, C. Feild, G. Koren, A. Gupta, D.G. Hinks, A.W. Mitchell, Y. Zheng, and B. Dabrowski, in *Strong Correlations and Superconductivity,* H. Fukuyama, S. Maekawa, and A.P. Malozemoff, eds. (Springer-Verlag, Berlin, Heidelberg, New York, 1989), p. 289.
15. G.A. Thomas, J. Orenstein, D.H. Rapkine, M. Capizzi, A.J. Millis, R.N. Bhatt, L.F. Schneemeyer, and J.V. Waszczak, Phys. Rev. Lett. **61**, 1313 (1988).
16. Y.J. Uemura *et al.*, Phys. Rev. Lett. **62,** 2317 (1989).
17. M. Lee, A. Kapitulnik, and M.R. Beasley, in *Mechanisms of High-Temperature Superconductivity*, H. Kamimura and A. Oshiyama, eds. (Springer-Verlag, Berlin, Heidelberg, New York, 1988), p. 220.

18. J.S. Tsai, I. Takeuchi, J. Fujita, T. Yoshitake, S. Miura, S. Tanaka, T. Terashima, Y. Bando, K. Iijima, and K. Yamamoto, in *Mechanisms of High-Temperature Superconductivity*, edited by H. Kamimura and A. Oshiyama (Springer-Verlag, Berlin, Heidelberg, New York, 1988), p. 229.
19. J.S. Tsai, I. Takeuchi, J. Fujita, S. Mura, T. Terashima, Y. Bando, K. Iljiama, and K. Yamamoto, Physica C **157**, 537 (1989).
20. T. Ekino and J. Akimitsu, Phys. Rev. B **40**, 6902 (1989).
21. R. Manzke, T. Buslaps, R. Claessen, and J. Fink, Europhys. Lett. **9**, 477 (1989).
22. C.G. Olson, R. Liu, A.B. Yang, D.W. Lynch, A.J. Arko, R.S. List, B.W. Veal, Y.C. Chang, P.Z. Jiang, and A.P. Paulikas, Science **245**, 731 (1989).
23. Y. Iye, Int. J. Mod. Phys. B **3**, 367 (1989).
24. T. Takahashi, H. Matsuyama, H. Katayama-Yoshida, Y. Okabe, S. Hosoya, K. Seki, H. Fuijmoto, M. Sato and H. Inokuchi, Nature **334**, 691 (1988).
25. J.M. Imer, F. Patthey, B. Darsel, W.D. Schneider, Y. Baer Y. Petroff, and A. Zettl, Phys. Rev. Lett. **62**, 336 (1989).
26. F.J. Himpsel, G.V. Chandrashekar, A.B. McLean, and M.W. Shafer, Phys. Rev. B **38**, 11946 (1988).
27. F. Minami, T. Kimura, and S. Takekawa, Phys. Rev. B **39**, 4788 (1989).
28. T. Takahashi *et al.*, Phys. Rev. B **39**, 6636 (1989).
29. R. Claessen, R. Manzke, H. Carstensen, B. Burandt, T. Buslaps, M. Skibowski, and J. Fink, Phys. Rev. B **39**, 7316 (1989).
30. T. Schneider, H. de Raedt, and M. Frick, Z. Phys. B **76**, 3 (1989).
31. P. B. Allen and W. E. Pickett, Phys. Rev. B **36**, 3926 (1987).
32. F. Seidler, P. Bohm, H. Gens, W. Braunisch, E. Braun, W. Schnelle, Z. Drzaaga, N. Wild, B. Roden, H. Schmidt, and D. Wohlleben, Physica C **157**, 375 (1989).
33. A.J. Legget, Rev. Mod. Phys. **47**, 331 (1975).
34. L.C. Hebel and C.P. Schlichter, Phys. Rev. **113**, 1504 (1959).
35. M. Fibich, Phys. Rev. Lett. **14**, 561 (1965).
36. P.C. Hammel, M. Takigawa, R.H. Heffner, Z. Fisk, and K.C. Ott, Phys. Rev. Lett. **63**, 1992 (1989).
37. Moohee Lee, Y.Q. Song, W.P. Halperin, L.M. Tonge, T.J. Marles, H.O. Marcey, and C.R. Kannewurf, Phys. Rev. B **40**, 817 (1989).
38. E. Oldfield, Ch. Coretsopoulos, S. Yang, L. Reven, H.C. Lee, J. Shore, O. Hee Han, E. Ramli, and D. Hinks, Phys. Rev. B **40**, 6832 (1989).

CORRELATED ELECTRON MOTION, FLUX STATES AND SUPERCONDUCTIVITY

P. Lederer

Physique des Solides, Bât. 510
Université Paris–Sud
91405 Orsay, France

ABSTRACT

When the onsite correlation is strong, electrons can move by usual hopping only on to empty sites but they can exchange position with their neighbors by a correlated motion. The phase in the former process is fixed and it favors Bloch states. When the concentration of empty sites is small then the latter process dominates and we are free to introduce a phase provided it is chosen to be the same for ↑ and ↓–spin electrons. Since for a partly filled band of non–interacting electrons the introduction of a uniform commensurate flux lowers the energy, the correlated motion can lead to a physical mechanism to generate flux states. These states have a collective gauge variable which is the same for ↑ and ↓–spins and superconducting properties are obtained by expanding around the optimum gauge determined by the usual kinetic energy term. If this latter term has singularities at special fillings then these may affect the superconducting properties.

In its present stage the theory predicts orbital currents which result in a distribution of magnetic fields at crystallographically equivalent sites. Such fields are not observed experimentally.

INTRODUCTION

This lecture is based on two recent papers[1,2] which have discussed the stability and properties of flux phases, in a square 2–D lattice as possible novel superconducting phases of strongly correlated electron systems. The name flux phases stems from Affleck and Marston[3], who found that one of the many equivalent forms of the quantum spin liquid state[4] for the undoped two dimensional Mott insulator is a projected determinantal fermion wave function in which the one electron wave functions are solutions of electrons moving in a magnetic field with flux per plaquette of $\Phi = 1/2$ (in units of the universal flux quantum hc/e). Prior to that work, Kalmeyer and Laughlin[5] have shown the equivalence of the Resonating–Valence–Bond and the Fractional Quantum Hall States on a 2–D triangular lattice. Then Laughlin[6] pointed out that the excitations of the 2D Quantum Spin Liquid obey fractional statistics (with fraction $q = 1/2$) ; within this approach, holons can be described as Fermions tied to a flux tube, and the resulting long range force drives a superconducting ground state. The novelty of this mechanism was expressed most clearly in Laughlin's concluding remark[6] : "In summary, it is possible that high–T_c superconductivity can be accounted for by the

Dynamics of Magnetic Fluctuations in High-Temperature Superconductors
Edited by G. Reiter *et al.*, Plenum Press, New York, 1991

following simple idea : "The force mediated by the spins of the Mott insulator is not due to an attractive potential, but rather an attractive vector potential".

Related ideas have also been expressed independently by Wiegmann[7], notably, and others[8]. One central feature is that, in a mean field approach of the gas of particles with fractional statistics (anyons), the ground state is analogous to a Quantum Hall ground state, with exactly filled Landau levels in a spontaneously generated magnetic (mean) field. As the Laughlin state for the fractional quantum Hall effect (FQHE) exhibits ODLRO[9a], it is not too surprising that, when the external field of the FQHE is substituted by a collective self consistent field, the broken symmetry corresponds to superconductivity[9b].

The anyon approach is based on an effective mass approximation for the anyons, and it ignores the discreteness of the lattice.

NON INTERACTING ELECTRONS IN A MAGNETIC FLUX

An indication that the lattice plays an important part in the strongly correlated fermion problem is that the energy of the half–filled band, which is that of the spin 1/2 Heisenberg model on a square lattice is considerably lower in the flux phase than in the "no–flux" phase which is a projected Fermi liquid state[10]. This is contrary to the text book result for non interacting fermions, following which the (spin less) electron system energy is never lower in a magnetic field than in zero field. This point led to the following startling generalization[1] : consider spinless non interacting electrons moving on a square lattice in a nearest neighbor tight binding model ; associate a phase φ_{ij} with each bond $< i,j >$ so that the Hamiltonian is :

$$H_0 = \sum_{<ij>} e^{i\varphi_{ij}} c_i^+ c_j + h.c. \tag{1}$$

then for a density ν, the absolute energy minimum is a uniform flux state i. e. a state in which the sum $\sum_{ij} \varphi_{ij}$ over each plaquette is equal to $2\pi\,\Phi = 2\pi\nu$. This result is supported by accurate numerical calculation[1],[11]. The energy lowering in the uniform flux is of order 20 % at $\nu = 1/2$, in comparison with the no flux state.

The Hamiltonian H_0 has a complex spectrum of eigenvalues. The case of rational $\Phi = p/q$ is easiest to study since the problem may be made periodic. The condition $\Phi = \nu$ corresponds to filling exactly a group of states which correspond to bands split off from the conventional low field lowest Landau level (which has degeneracy Φ). The splitting is the result of the lattice periodic potential superimposed on the periodic orbital motion of electrons. A large gap separates this group of states from higher energy states, except at $\Phi = 1/2 = \nu$. The total energy minimum has a cusp like shape as a function of Φ, analogous to the cusp like minimum of the free electron gas energy around the field values for which a number of Landau levels are exactly filled.

The lowering of electronic energy in the presence of flux in a periodic lattice can be considered as a generalization of the 1–D Peierls instability to the 2–D square lattice. The flux adjusts to the density so as to open up the largest gap at the Fermi level. Note that the Hall number of the state with $\Phi = \nu$ is 1, in units of h/e^2. The result described above suggests that it is important, when studying correlated electron systems and the possible occurrence of flux phases as superconducting ground states, to take into account explicitly the tight binding periodic potential of the lattice since it lifts the degeneracy associated with the Landau level structure of the free electrons. This is discussed in the next section, along with some problems which this unveils in the flux phase approach to superconductivity.

FLUX PHASES IN THE t–J MODEL

The t–J model has been actively discussed in this conference ; there is growing numerical and empirical evidence that it is a correct simplified Hamiltonian to describe the strongly correlated CuO_2 planes of the high T superconducting ceramics[12,13,14]

$$H = P\left\{ \sum_{<ij>,\sigma} t\, c^{+}_{i\sigma} c_{j\sigma} + h.c. + J \sum_{<ij>} \vec{S}_i \vec{S}_j \right\} P$$
$$= H_t + H_J \tag{2}$$

the projector $P = \prod_i (1 - n_{i\uparrow} n_{i\downarrow})$ is on the subspace with no doubly occupied site. Note that the Hamiltonian is invariant with respect to a gauge transformation $c^{+}_{i\sigma} \to e^{i\theta_i} c^{+}_{i\sigma}$ together with $t_{ij} \to t\, e^{-i(\theta_i - \theta_j)}$. Total number conservation is expressed by the Hamiltonian invariance with respect to a global gauge transformation $c^{+}_{i\sigma} \to e^{i\theta} c^{+}_{i\sigma}$.

According to the remarks in the previous section, it is natural to consider as a class of low energy variationnal states the following wave functions :

$$\left| \psi(\Phi) \right> = P \left| \psi_0(\Phi) \right\rangle = P \prod_{\ell}^{occ} \tilde{c}^{+}_{\ell\uparrow} \prod_{\ell'}^{occ} \tilde{c}^{+}_{\ell'\downarrow} \left| vac \right\rangle \tag{3}$$

where $[H_0, \tilde{c}^{+}_{\ell}] = \epsilon_\ell \tilde{c}^{+}_{\ell}$
The eigenvalues ϵ_ℓ depend only on Φ, but the eigenstates are gauge dependent, since they depend on the phases {fig. }.
It is easily shown that the choice $\{\ell\} \equiv \{\ell'\}$ is the low energy one, and makes (3) a singlet wave function. The class of wave functions in eq. (3) was proposed, in a different form, by Anderson Shastry and Hristopoulos[15,16] as a generalization of the Affleck–Marston 1/2–flux state.
A simple and reasonably accurate way to treat the projection operator P is, following Zhang et al. [10], to introduce Gutzwiller renormalization factors :

$$< H > = \langle \psi \mid H_t + H_J \mid \psi \rangle \simeq g_t \langle H_t \rangle_0 + g_J \langle H_J \rangle_0 \tag{4}$$

where $g_t = 4\delta/(1 - 2\delta)$ and $g_J = 4/(1 + 2\delta)^2$ in terms of the hole concentration. In the calculation of the exchange term we can view it as arising from correlated hopping processes i. e., in a singlet wave function with no broken spin rotational invariance

$$\langle \vec{S}_i . \vec{S}_j \rangle = \frac{3}{2} \langle S_{i+} . S_{j-} \rangle \simeq \frac{3}{2} g_J < S_{i+} S_{j-} >_0$$

and

$$\begin{aligned} \langle S_{i+} . S_{j-} \rangle_0 &= \langle c^{+}_{i\uparrow} c_{i\downarrow} c^{+}_{j\downarrow} c_{j\uparrow} \rangle_0 \\ &= - \langle c^{+}_{i\uparrow} c_{j\uparrow} c^{+}_{j\downarrow} c_{i\downarrow} \rangle_0 \\ &= - \langle e^{i\varphi_{ij}} c^{+}_{i\uparrow} c_{j\uparrow} \rangle_0 \langle e^{-i\varphi_{ij}} c^{+}_{j\downarrow} c_{i\downarrow} \rangle_0 \end{aligned} \tag{5}$$

The next step is to realize that if the phases φ_{ij} in the last equation are precisely those in eq. (3), then the symmetry of the lattice, together with gauge invariance, ensure that $\langle e^{i\varphi_{ij}} c^{+}_{i\sigma} c_{j\sigma} \rangle$ is a constant independent of i, j and σ (for i and j nearest neighbours), so that

$$\langle e^{i\varphi_{ij}} c^{+}_{i\sigma} c_{j\sigma} \rangle_0 = \frac{1}{4N} \left\langle \sum_{\ell m} e^{i\varphi \ell m} c^{+}_{\ell} c_m \right\rangle_0$$

$$= \frac{1}{4N} E_T(\Phi) \qquad (6)$$

where $E_T(\Phi)$ is the total electronic energy associated with eq. (1).

It is clear that the exchange energy in (5) is minimized if we choose the same phase factors for ↑ and ↓ spins so that ↑ and ↓ spins see effective one electron Hamiltonian of the form (1) but the phases for the ↑ and ↓ are linked by the correlated motion[2].

The kinetic term is evaluated using the same trick :

$$\langle c^{+}_{i\sigma} c_{j\sigma} \rangle_0 = \langle e^{i\varphi_{ij}} c^{+}_{i\sigma} c_{j\sigma} \rangle_0 \, e^{-i\varphi_{ij}}.$$

so that

$$\sum_{ij} \langle c^{+}_{i\sigma} c_{j\sigma} \rangle_0 = 2E_T(\Phi) \, \overline{\cos \; \varphi_{ij}} \qquad (7)$$

where the overline means average over all bonds. Since, for any given flux Φ, $\overline{\cos \; \varphi_{ij}}$ depends on the choice of gauge, the lowest kinetic energy is $2E_T(\Phi)\, K(\Phi)$ where $K(\Phi)$ results from optimizing the gauge choice. It is clear from eq. (7) that the introduction of phase factors corresponding to a finite flux is not favoured by the kinetic term H_t. This term is diagonalized by Bloch states and it prefers a state without flux. However the number of hops involving H_t varies as the number of holes or empty sites i. e. $\propto \delta$. Therefore the correlated hops should dominate over single particle hops roughly if $J \gtrsim \delta t$. Then the energy is[2] :

$$\langle\, H_t + H_J \,\rangle = 2tg_t \, E_T(\Phi) \, K(\Phi) - \frac{3J}{16} g_J \left[E_T(\Phi) \right]^2 \qquad (8)$$

The occurrence of $K(\Phi)$ in eq. (8) is the single most important difference with the effective mass approximation underlying the anyon approach, as it points out a competition between the exchange term, which always favours a flux phase through its cusp like minimum at $\Phi = \nu$, and the kinetic term which for a given J/t ratio favours $\Phi = 0$ at large δ, $\Phi = 1/2$ at small δ. The function $K(\Phi)$ has been studied in the past in connection with the ground state energy of coupled arrays of Josephson junctions in the presence of an external field[17,18]. The Hamiltonian of the latter system is precisely :

$$H_{JJ} = -J \sum_{ij} \cos(\theta_i - \theta_j - A_{ij}) \qquad (9)$$

where θ_i denotes the phase of the superconducting order parameter (in our case a choice of gauge) and A_{ij} is proportional to the line integral of the vector potential $\vec{A}$ between the ith and the jth site (as in our case).

For the case $1/3 \lesssim \Phi \lesssim 1/2$, Halsey[18] has argued that for any rational value the ground state is a stripe phase he called the staircase state, such that the phase difference on a bond is a constant along a path which is a regular infinite staircase. The unit cell for a flux p/q is a square supercell of $q \times q$ elementary plaquettes. Following Halsey, $K(\Phi)$ is never smaller than $2/\pi$, which might be its actual value for

irrational Φ, and is likely an irregular function, with cusps at all rational values of Φ. Thus we can establish that fractional values of Φ correspond to the lowest state of eq. (8) provided the ratio $J/\delta t$ is sufficiently large.
For instance, ref. (2) established that $\Phi = \nu = 3/8$ is the most stable flux for $J/t > 3$ and $\Phi = \nu = 7/16$ is stable for $J/t > 1.8$. Depending on the behaviour of $K(\Phi)$ around a rational value, two different things may happen. Either the cusp in $E_T(\Phi)$ ensures the stability of a Commensurate Flux Phase (CFP) with $\Phi = \nu$, or the singularity of $K(\Phi)$ ensures the stability of a flux phase with $\Phi = p/q$ for a continuous range $\delta\nu$ of values of $\nu = \nu_r + \delta\nu$ around each rational $\nu_r = p/q$. In the first case we have a state with a gap at the Fermi level. In the second case we have a finite density of states at the Fermi level. Our poor knowledge of $K(\Phi)$ does not allow at this stage to decide which is the actual situation.

PROPERTIES OF FLUX PHASES

First consider the CFP, defined by the condition $\Phi = \nu$ in an interval of ν values. The stability of such a state is driven by the cusp behaviour of $E_T(\Phi)$[2]. Consider applying a small external magnetic field. The total change ΔE_0 of the ground state energy is :

$$\Delta E_0 = 2tg_t\, E_T(\Phi)\, [K(\Phi - \Phi_{ext}) - K(\Phi)] \tag{10}$$

Provided $K(\Phi - \Phi_{ext}) - K(\Phi)$ is small enough, the cusp in $E_T(\Phi)$ at $\Phi = \nu$ locks the flux inside the sample at its zero field value. So the wave function is rigid in a magnetic field, and it was argued in ref. 2 that this is a Meissner state, and the CFP is a superconducting state. This reasoning seems to rely on the structure of the CFP ground state, and it would seem that the flux state at fixed rational flux would not be superconducting. However the following argument suggests that the flux state is superconducting, CFP or not CFP : the ground state of H_{JJ} (eq. (9)) exhibits two broken symmetries[17,19]. The first of these symmetries is the global gauge symmetry of phase rotation. The phases in any ground state given by a set of gauge angles $\{\theta_i\}$ can be rotated into another ground state $\{\theta_i + \varphi\}$. This broken gauge symmetry is obviously a broken global electromagnetic gauge symmetry, to which a Goldstone mode is associated by considering a slowly varying $\varphi(r)$. This broken symmetry is the signature of superconductivity. An immediate objection is that this broken global gauge symmetry drives the fluctuation of the particle number N, while the state in eq. (3) has a fixed number of particles[20]. However since the broken symmetry of the ground state of H_{JJ} (eq. (9)) is obtained only in the thermodynamic limit, this means that in the thermodynamic limit particle number fluctuation is allowed, and only the density ν is fixed.

The notion that CFP states are superconducting is confirmed also by the work on the Quantum Hall States[9a], following which the latter have ODLRO, the order parameter being the non zero expectation value of a composite particle operator describing a fermion tied to a flux tube. The Quantum Hall State is not a superconducting state because it is incompressible, having a fixed external magnetic field. When the magnetic field is a statistical field, the fluid is no longer incompressible (since $\Phi = \nu$) and the ODLRO of the Quantum Hall State becomes that of superconductivity, as suggested in the context of the anyon approach in ref. (9b).

An obvious broken symmetry in the Flux Phases, by construction, is time reversal T, since our trial wave function is solution of eq. (7). As a result the expectation value $E_{123} \equiv \langle \vec{\sigma}_1 \cdot (\vec{\sigma}_2 \wedge \vec{\sigma}_3) \rangle$, where 123 label lattice sites, is non zero, and easily deduced from the set $\{\theta_i\}$ and $\{\varphi_{ij}\}$[21].

Finally translation invariance is also broken. The unit cell of the problem with flux p/q is a q x q supercell, as mentioned above. As a result, in the presence of holes, each supercell exhibits a certain pattern of orbital currents[22] easily computed from $\{\theta_i\}$ and $\{\varphi_{ij}\}$: the current on a bond J_{ij} is

$$J_{ij} = i \frac{eat}{\hbar} \sum_{\sigma} [< c^{+}_{i\sigma} c_{j\sigma} > - < c^{+}_{j\sigma} c_{i\sigma} >]$$

The current pattern produces magnetic fields which are a specific prediction of the Flux Phase picture, since these occur in the superconducting ground state[23]. In contrast with the anyon picture[23], the magnetic field is not uniform in the tight binding lattice, and for any given supercell, there exist various inequivalent sites (although crystallographically equivalent) with different local magnetic fields. In the superconducting ground state, the Meissner effect results in a zero average field in the bulk of a sample, so that the model predicts a distribution of static magnetic fields centered around zero at crystallographically equivalent Cu (or O) sites. The typical order of magnitude of the width of the distribution is of order ($450\ \delta$) gauss where δ is the hole concentration. The μ SR experiments reported so far (see for instance the report by Lukes in these Proceedings, or ref.[24]) do not offer evidence in favour of such a field distribution in $YBa_2Cu_3O_7$. It is not clear that in the latter compound the muon stopping sites have the right symmetry, (for instance chain sites might have zero field anyway) ; but this argument does not apply in $La_{1.85}Sr_{0.85}CuO_4$ where no static electronic moments seem to be observed either[24]. This might mean that Flux states do not describe the superconducting cuprates, or that the mean field is too crude an approximation as far as local magnetic fields are concerned.

It is likely that various perturbations, such as lattice imperfections, impurities, interlayer couplings, etc..., smooth out the singularities in $K(\Phi)$ associated with rationals with large denominators. Such perturbations do not suppress the cusps in the total energy as long as the characteristic energy scales are small compared with the Hofstadter gap between the two lowest Landau–Hofstadter groups of states. Therefore I expect that the gapless superconducting state would in general be suppressed in favour of the CFP state.

DISCUSSION

Within the effective mean field approximation I have discussed, Flux Phases (with non zero flux) are stable above a critical value of J/t which depends on ν. This result is true in the restricted class of variational states we have chosen. It is obviously interesting to investigate the stability of the Flux Phases in a broader parameter space.

In agreement with the anyon approach, it appears that Flux Phases of the t–J model are superconducting phases. They display the equivalent of full Landau levels for the anyon picture, i. e. full Hofstadter band groups with Hall number 1. In contradiction with the anyon picture, the lattice structure brings about a frustration of the kinetic energy term, evidenced by the sharp reduction factor $K(\Phi)$ and its irregular behavior as a function of flux. As a result, Flux Phases with uniform field are stable only for $J/t \gtrsim 1$, a limit a priori contrary to what is commonly accepted for the actual superconducting oxides[25]. It is straightforward to see that a staggered flux phase, with the flux alternatively positive or negative on elementary plaquettes is more stable than a CFP at $t/J \gtrsim 1$, since the kinetic energy is much less reduced in the staggered flux state near half filling being proportional to $\cos \frac{\pi\varphi}{2}$, while the exchange term is, near half filling, only slightly reduced compared to its value in the CFP. This idea has been checked numerically in ref.[26]. The staggered flux phase, in turn is unstable with respect to d–wave superconductivity[27].

The results discussed in this paper can be interpreted in various ways : on one hand they confirm the view that strongly correlated electrons can have a superconducting ground state which has strong analogies with Quantum Hall States in a lattice. On the other hand they point out difficulties in the anyon approach (in particular the effective mass approximation) inasmuch as the kinetic term is strongly reduced by the lattice effects, so that flux phases with uniform statistical field need large J/t ratios. Further work is needed to ascertain whether the mean field approach discussed so far can be improved to deal with the kinetic energy problem, and whether Flux Phases have something to do with the superconducting oxides in spite of the negative evidence regarding the internal magnetic fields.

ACKNOWLEDGEMENTS

This work is supported in part by ESPRIT contract P 3041–MESH.

REFERENCES

1. Y. Hasegawa, P. Lederer, T. M. Rice and P. B. Wiegmann, Phys. Rev. Lett., 63, 907, (1989)
2. P. Lederer, D. Poilblanc and T. M. Rice, Phys. Rev. Lett., 63, 1519, (1989)
3. I. Affleck and B. J. Marston, Phys. Rev. B 37, 3774, (1988)
4. I. Affleck, Z. Zou, T. Hsu and P. W. Anderson, Phys. Rev. B 38, 745, (1988)
5. V. Kalmeyer and R. B. Laughlin, Phys. Rev. Lett. 59, 2995, (1987)
6. R. B. Laughlin, Phys. Rev. Lett. 60, 2677, (1988)
7. P. B. Wiegmann, Phys. Rev. Lett. 60, 821, (1988) ; Physica Scripta T 27, 160 C
8. X. G. Wen and A. Zee, ITP preprint (1989)
9. a) N. Read, Phys. Rev. Lett. 62, 86, (1989) ; see also S. M. Girvin and A. H. Mac Donald, Phys. Rev. Lett. 58, 1252, (1987)
   b) Dung–Hai Lee and Matthew P. A. Fisher, Phys. Rev. Lett. 63, 903, (1989)
10. F. C. Zhang, C. Gros, T. M. Rice and H. Shiba, Supercond. Sci. Technol. 1, 36, (1988)
11. G. Montambaux, Phys. Rev. Lett. (C), 63, 1657, (1989)
12. P. W. Anderson, Science 235, 1196, (1987)
13. F. C. Zhang and T. M. Rice, Phys. Rev. B 37, 3759, (1988)
14. Various authors in those Proceedings discuss the phase separation which occurs in the t–J model. This phenomenon was discussed some time ago in connection with the physics of bcc ^{3}He in the mK range. The point of view in this paper is that phase separation, which is possible for uncharged fermion crystals is inhibited by Coulomb forces in doped Mott insulators. See G. Montambaux, M. Héritier and P. Lederer, J. Low Temp. Phys., 47, 39, (1982)
15. P. W. Anderson, B. S. Shastry and D. Hristopoulos, Princeton preprint (1989)
16. D. Poilblanc, ETH, preprint (1989)
17. S. Teitel and C. Jayaprakash
   – Phys. Rev. Lett. 51, 1999, (1983)
   – Phys. Rev. B 27, 598, (1983)
18. T. C. Halsey, Phys. Rev. B 31, 5728, (1985)
19. J. Villain, J. Phys. C : Solid State Phys. 10, 1717, (1977)
20. I am indebted to N. Papanicolaou for a helpful remark on that point
21. X. G. Wen, F. Wylczek, A. Zee, Phys. Rev. B 39, 11413, (1989)
22. T. M. Rice, P. Lederer, D. Poilblanc, ETH Preprint, (1989)
23. Y. H. Chen, F. Wilczek, E. Witten, B. I. Halperin, IAZSSNS–HEP–89/27, preprint
   B. I. Halperin, J. March–Russel and F. Wilczek, HUTP–89/A010, preprint
24. R. F. Kiefl et al., Phys. Rev. Lett. 63, 2136, (1989)
25. See also the variational Monte–Carlo results by Shoudan Liang and Nandini Trivedi, preprint
26. Y. Hasegawa and D. Poilblanc, preprint
27. F. C. Zhang, preprint

Orbital Dynamics and Spin Fluctuations in Cuprates

J. ZAANEN*, A. M. OLEŚ† AND L. F. FEINER‡

*Max-Planck-Institut für Festkörperforschung, Heisenbergstr. 1, D-7000 Stuttgart 80, F.R.G.; †Institute of Physics, Jagellonian University, 30-059 Kraków, Poland; ‡Philips Research Laboratories, P.O. Box 80000, 5600 JA Eindhoven, The Netherlands

Recently it has become clear that the orbital momentum is not completely quenched in the High Tc superconductors. In this paper we focus on the role of orbital degrees of freedom in Mott-Hubbard insulators. We first rederive the molecular field (Kugel-Khomskii) picture for d^9 systems and we present new results on the excitation spectra. We show that this classical picture is incomplete: *low lying orbital excitations give rise to a strong enhancement of the quantum spin fluctuations.* Further, we comment on Weber's suggestion that observed phonon-anomalies might be due to Jahn-Teller physics.

I. Introduction

The theoretical modelling of high Tc superconductors usually involves one or the other spin-degenerate model, like the $t - J$ model or the three band model. The motivation to neglect (3d) orbital degeneracy is the expected largeness of crystal field splittings in these relatively ionic materials, leading to quenching of the orbital angular momentum at half filling. Away from half filling, things are a bit less obvious. It could be imagined that holes would tend to form a high spin ($\sim d_{x^2-y^2\uparrow}d_{3z^2-1\uparrow}$) 'Cu(III)' state. However, at least the theoretical evidence seems to point at a low spin state ('Zhang-Rice singlet'[1]), basically because of the strong covalency in these materials[2].

However, recently experimental evidence was presented, showing that orbital momentum is not completely quenched. This is based on polarized XAS[3] and EELS[4] measurements, showing Cu d-weight in the c-direction, pointing at the presence of (most probably) d_{3z^2-1} holes in the ground state. Although this admixing is small at half filling ($\simeq$ 3%), it becomes appreciable in the superconductors ($\simeq$ 20%, even 40% has been claimed[3]).

In correlated systems, orbital degrees of freedom behave manifestly different from the familiar ('$sp)^3$', '$d_{x^2-y^2}$') single particle picture. In Mott-Hubbard insulators one can, at energies below the correlation gap, separate the orbital momentum carried by the electron from its charge. The resulting objects behave very much like spins, a concept which has flourished in the context of the Jahn-Teller problem[5]. The next step is to consider how these orbital pseudo-spins and the normal spins interact. As we will discuss in Section II, spins and orbitals form one dynamical entity, described by rather spectacular Hamiltonians (see Eq. (2)), which we name 'Kugel-Khomskii (KK) models' after their inventors[6,7].

In fact, only little is known about the physics contained in KK-models. First, in Section III we will rederive[7] the classical (molecular field) phase diagram. On this level of approximation, the dependency of spin and orbital degrees of freedom already shows up; the spin order depends on the orbital order and vice versa. In the same section we present new results for the classical excitation spectra. These are remarkably rich; for instance, up to four different magnon branches can occur.

In section IV we will present our main result: low lying orbital excitations enhance the quantum spin fluctuations (QSF). We will extract the basic physics from a simple example and we will use Schwinger boson theory to study the lattice problem. These findings are further discussed in Section V, where we also comment on the role of holes (we do not pay explicit attention to superconductivity[8,9]) and the role of phonons, in connection with recently observed phonon anomalies[9,10].

II. The Kugel-Khomskii model

In an atom only J or $L.S$ makes sense and not L or S separately. In the presence of electron-electron interactions, the same is true in solids. It is therefore not possible to treat the rotations in spin and orbital space as independent, and this is the reason that KK-models are rather complicated. Group theoretically, this implies larger Lie algebra's than $\otimes_i su(2)_i$. For instance, in the d^9 problem there are four fermion families ($x\uparrow, x\downarrow, z\uparrow, z\downarrow; x, z \propto x^2-y^2, 3z^2-1$) and the low energy sector in the insulator is described by the 16 bilinears $A_{m\sigma}^{m'\sigma'} = d_{m\sigma}^{\dagger} d_{m'\sigma'}$, ($m \sim x, z; \sigma \sim \uparrow, \downarrow$), having the commutation relations

$$[A_{m_1\sigma_1}^{m_2\sigma_2}, A_{m_3\sigma_3}^{m_4\sigma_4}] = \delta_{m_2\sigma_2}^{m_3\sigma_3} A_{m_1\sigma_1}^{m_4\sigma_4} - \delta_{m_1\sigma_1}^{m_4\sigma_4} A_{m_3\sigma_3}^{m_2\sigma_2}, \qquad (1)$$

defining the Lie algebra su(4), i.e. one is dealing with $\otimes_i su(4)_i$ instead.

The same kinetic exchange mechanism is present as in the case of the spin-only problem and one ends up with a generalized Heisenberg Hamiltonian in which the orbital-, spin- and 'mixed' degrees of freedom on different sites are coupled by superexchange interactions. These KK Hamiltonians are derived using strong coupling perturbation theory[11], starting from the degenerate, single band (to avoid tedious algebra) Hubbard Hamiltonian. The procedure is illustrated in Fig. 1 for the d^9 case. One aims at a description of the low energy sector containing d^9 states with internal spin (arrows) and orbital (x,z) degrees of freedom (split by the crystal field energy ε_z). These internal degrees of freedom are coupled by virtual fluctuations to intermediate states : $d_i^9 d_j^9 \rightleftharpoons d_i^8 d_j^{10}$. In case that different orbitals on n.n. sites mix (as for $d_{x^2-y^2}$ and d_{3z^2-1} states in a square lattice, $t_{xz} = \pm t_{xx}/\sqrt{3}, t_{zz} = t_{xx}/3$), one finds terms containing orbital flips. For instance, the process indicated in the figure would give rise to a term in the Hamiltonian of the form $\pm t_{xx}^2/(3\sqrt{3}U)(d_{iz\downarrow}^{\dagger} d_{ix\uparrow})(d_{jx\uparrow}^{\dagger} d_{jx\downarrow})$: on site j the spin in the x orbital is flipped, and on site i the spin is flipped accompanied by an orbital flip. Finally, it turns out that the Hundts rule splitting (J_H) in the intermediate d^8 state can play an important role. We derived the strong coupling Hamiltonian using a model multiplet splitting as indicated in the figure, with $T = |x \uparrow z \uparrow>$ etc. at $U - J_H$, $S_1 = (|x \uparrow z \downarrow> - |x \downarrow z \uparrow>)/\sqrt{2}$ at U and $S_2 = |x \uparrow x \downarrow>, |z \uparrow z \downarrow>$ at $U + 2J_H$, i.e. somewhat more general than Kugel and Khomskii who neglected the $S_1 - S_2$ splitting[6,7].

In this way we derive the effective Hamiltonian

$$
\begin{aligned}
H = \frac{1}{9} \sum_{<ij>} (J^0_{ij} [4 (3\vec{S}_{ix} + \vec{S}_{iz} + \sqrt{3}(-1)^{\hat{y}\hat{b}_{ij}} (\vec{T}_{ir} + \vec{T}_{il})) \\
\times (3\vec{S}_{jx} + \vec{S}_{jz} + \sqrt{3}(-1)^{\hat{y}\hat{b}_{ij}} (\vec{T}_{jr} + \vec{T}_{jl})) \\
+ 4 (n_{-i} + \sqrt{3}(-1)^{\hat{y}\hat{b}_{ij}} (T_{ir} + T_{il})) (n_{-j} + \sqrt{3}(-1)^{\hat{y}\hat{b}_{ij}} (T_{jr} + T_{jl}))] \\
+ J^1_{ij} [-8 (\vec{S}_{ix} + \vec{S}_{iz}) (\vec{S}_{jx} + \vec{S}_{jz}) \\
+ 2 (\vec{S}_{ix} - \vec{S}_{iz} + \sqrt{3}(-1)^{\hat{y}\hat{b}_{ij}} (\vec{T}_{ir} + \vec{T}_{il})) (\vec{S}_{jx} - \vec{S}_{jz} + \sqrt{3}(-1)^{\hat{y}\hat{b}_{ij}} (\vec{T}_{jr} + \vec{T}_{jl})) \\
+ 6 (n_{-i} + \sqrt{3}(-1)^{\hat{y}\hat{b}_{ij}} (T_{ir} + T_{il})) (n_{-j} + \sqrt{3}(-1)^{\hat{y}\hat{b}_{ij}} (T_{jr} + T_{jl}))] \\
+ J^2_{ij} [-8 (3\vec{S}_{ix} + \vec{S}_{iz} + \sqrt{3}(-1)^{\hat{y}\hat{b}_{ij}} (\vec{T}_{ir} + \vec{T}_{il})) (3\vec{S}_{jx} + \vec{S}_{jz} + \sqrt{3}(-1)^{\hat{y}\hat{b}_{ij}} (\vec{T}_{jr} + \vec{T}_{jl})) \\
+ 8 (n_{-i} + \sqrt{3}(-1)^{\hat{y}\hat{b}_{ij}} (T_{ir} + T_{il})) (n_{-j} + \sqrt{3}(-1)^{\hat{y}\hat{b}_{ij}} (T_{jr} + T_{jl}))] \\
+ \sum_i [(-\varepsilon_z - \sum_{j(i)} \frac{4}{9} (J^0_{ij} - 2J^2_{ij})) n_{-i} + \sum_{j(i)} \frac{4}{9} \sqrt{3}(-1)^{\hat{y}\hat{b}_{ij}} (-J^0_{ij} + 2J^2_{ij}) (T_{ir} + T_{il})], \quad (2)
\end{aligned}
$$

plus some constant terms. In Eq. (2) we introduce the following notation for the su(4) generators; Spin dependent operators ($\vec{X} = (X^x, X^y, X^z)$ with $X^{\pm} = X^x \pm iX^y$):

$$
S^+_{ix} = d^\dagger_{ix\uparrow} d_{ix\downarrow}, \; S^+_{iz} = d^\dagger_{iz\uparrow} d_{iz\downarrow}, \; T^+_{ir} = d^\dagger_{ix\uparrow} d_{iz\downarrow}, \; T^+_{il} = d^\dagger_{iz\uparrow} d_{ix\downarrow}. \qquad (3a)
$$

The $\vec{S}$ operators represent pure (x- or z-) spin-flips and the $\vec{T}$ operators spin-flips, accompanied with a (raising (r) or lowering (l)) 'orbital-flip'. Spin independent operators:

$$
n_{-i} = (n_{ix\uparrow} + n_{ix\downarrow} - n_{iz\uparrow} - n_{iz\downarrow})/2,
$$

$$
T_{ir} = (d^\dagger_{ix\uparrow} d_{iz\uparrow} + d^\dagger_{ix\downarrow} d_{iz\downarrow})/2, \; T_{il} = (d^\dagger_{iz\uparrow} d_{ix\uparrow} + d^\dagger_{iz\downarrow} d_{ix\downarrow})/2. \qquad (3b)
$$

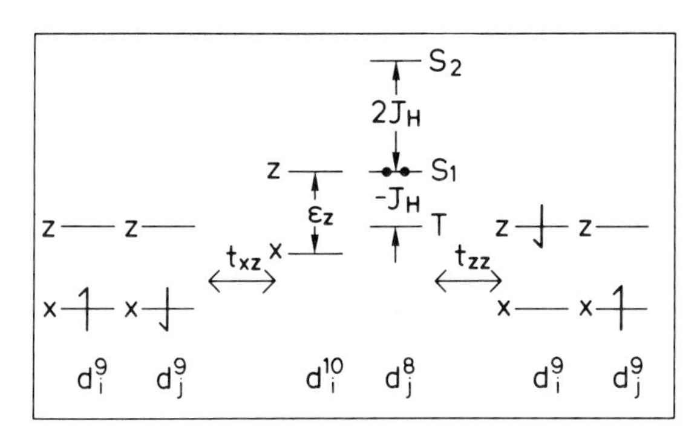

Fig. 1. Fluctuations leading to coupling of spin- and orbital degrees of freedom in the strong coupling picture.

We introduced the coupling constants $J^0 = |t(xx)|^2/U$ ($J = 2J^0$), $J^1 = |t(xx)|^2/(U-J_H) - J^0$ and $2J^2 = J^0 - |t(xx)|^2/(U+2J_H)$, i.e. J^0 would be the only coupling constant in the absence of the Hundt's rule splitting of the d^8 intermediate state, and J^1 and J^2 arise from the singlet-triplet splitting and the S1-S2 splitting (Fig. 1), respectively. By using the '$\sigma.\tau$' representation[6,7] and setting $J^2 = 0$ and $J^1 = J_H/U$, it can be shown that Eq. (2) is equivalent to Eq. (18) in ref.[6]. Compared to the '$\sigma.\tau$' representation, ours is more convenient for the computation of excitation- and fluctuation properties.

III. The Classical Picture

The simplest treatment of the Kugel-Khomskii Hamiltonian is by taking a molecular field approximation (MFA) for the ground state, followed by a classical (and linear) treatment of the excitations. In MFA the orbital and spin degrees of freedom in Eq. (2) decouple. This allows us to introduce the single-particle like orbitals

$$a^{\dagger}_{i\sigma} = \cos(\theta_i)d^{\dagger}_{ix\sigma} + \sin(\theta_i)d^{\dagger}_{iz\sigma}. \tag{4}$$

Inserting these in Eq.(2) yields an effective Hamiltonian of the form $\sim \sum_{<ij>}[f(\theta_i,\theta_j) \times \vec{S}_i\vec{S}_j + g(\theta_i,\theta_j)]$ where $\vec{S}$ is a pure spin degree of freedom ($S^+ = a^{\dagger}_{\uparrow}a_{\downarrow}$). Assuming a two sublattice (A,B) solution for the square lattice, minimizing with respect to the orbital mixing angles on the two sublattices (θ_A, θ_B) results in the self consistency equations (neglecting J_H)

$$\tan(2\theta_{A,B}) = \frac{3\sin(2\theta_{B,A})(4\vec{S}_A\vec{S}_B + 1)}{2(4\vec{S}_A\vec{S}_B - 1) + \cos(2\theta_{B,A})(4\vec{S}_A\vec{S}_B + 1) - 9\varepsilon_z/8J^0}. \tag{5}$$

Assuming magnetic Néel order, $< \vec{S}_A\vec{S}_B > = -\frac{1}{4}$, it follows directly from Eq. (5) that either $\theta_{A,B} = 0$, corresponding with the pure x^2-y^2 AFM with energy $E(AFM,x) = -8J^0/3 - \varepsilon_z/2$, or $\theta_{A,B} = \pi/2$, the pure $3z^2-1$ AFM at $E(AFM,z) = 8J^0/9 + \varepsilon_z/2$ Apparently, in MFA *the antiferromagnetism of the spin system leads to a complete suppression of orbital ordering.* This can be understood qualitatively. First, if only

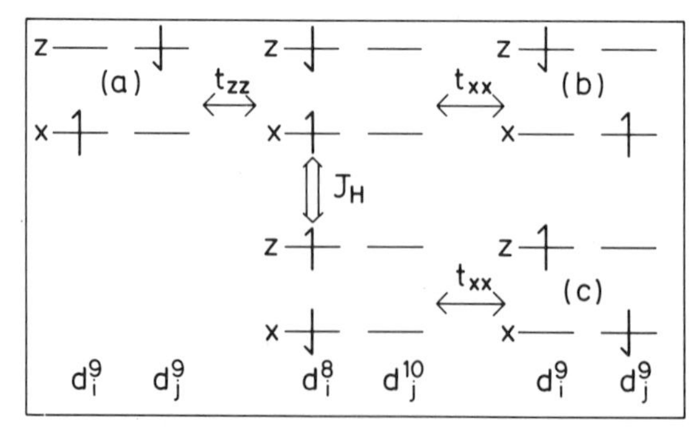

Fig. 2. In the absence of the Hundt's rule coupling J_H an orbital excitation can only propagate if it is accompanied by a spin flip (a → b). For non-zero J_H it can propagate without spin flip (a → c).

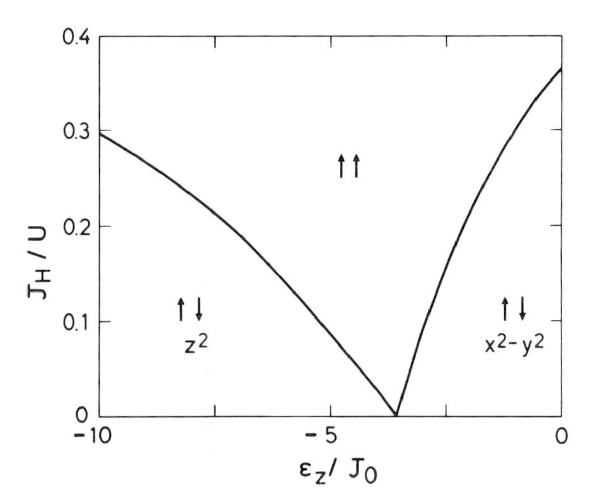

Fig. 3. Ground state molecular field phase diagram of the Kugel-Khomskii model, as a function of ε_z/J^0 and J_H/U, indicating the regions where the pure $x^2 - y2$- and $3z^2 - 1$ Néel phases and the ferromagnetic quadrupolar phases are stable.

on-site amplitudes can develop, the mean field Hamiltonian is of the form $< S_B > S_A + const. < T_B > T_A$ (S and T being diagonal and off-diagonal in orbital space, respectively), because the cross terms ($< S_B > T_A$, etc.) are cancelled by the different signs in the x and y directions. Secondly, orbital order could be driven by the $T.T$ term; however, this term does not contribute for anti-parallel spins (if $J_H = 0$), as can be inferred from the example sketched in Fig. 2.

For ferromagnetic spin order, orbital order can exist and we infer from Eq. (5) $cos(2\theta) = 9\varepsilon_z/(64J^0)$, with anti-ferro orbital order (AFO, $\theta = \theta_A = -\theta_B$) and energy $E(FM, AFO) = -4J^0/9 - 9\varepsilon_z^2/(256J^0)$. By comparing the energies, we find that this FM-AFO state is stable precisely at the cross-over point (at $\varepsilon_z/J^0 = -32/9$) from the x-AFM to the z-AFM (together with two more phases !). Not surprisingly, switching on J_H stabilizes the ferromagnetic state more than any of the AFM states (Hundt's rule !). In Fig. 3 we show the resulting phase diagram as a function of ε_z/J^0 and J_H/U.

How to picture this FM-AFO state? According to Eq. (4), hybrids are formed from the $x^2 - y^2$ and $3z^2 - 1$ orbitals with alternating phasing on the two sublattices

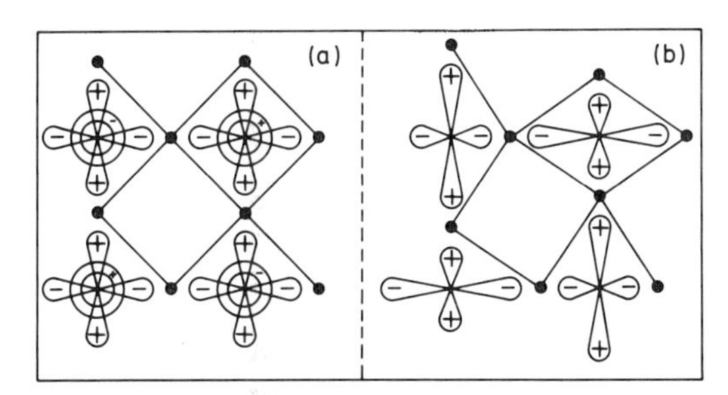

Fig. 4. Sketch of the ferromagnetic orbital ordered phase. (a) The electronic instability and (b) the distortion in the oxygen lattice (dots), driven by the electronic instability.

(Fig. 4). This will give rise to a structural instability in the oxygen lattice; the Cu-O bonds will strengthen in the direction where the lobes of the $x^2 - y^2$ orbital have the same sign as the in plane component of the $3z^2 - 1$ orbital and in the other direction the bond will be weakened. The net result is a quadrupolar distortion as indicated in Fig. 4. In fact, using these arguments, Kugel and Khomskii predicted the existence of such a structural distortion in the FM quasi-2D compound K_2CuF_4[12]. This prediction has been later confirmed by experiment[13]. We note that also in the present case we find continuity between weak coupling and strong coupling. It can be shown that for any finite $3z^2 - 1$ occupancy, the Fermi surface of the FM CuO perovskite plane (or non-magnetic NiO plane[14]) is subject to a perfect nesting condition under a quadrupolar distortion[15]. This nesting is very unusual; it does not relate to any high symmetry direction in the Brillioun zone.

Following the spirit of classical linear spin wave theory we calculated the excitation spectra in the different classical phases. Starting point are the equations of motion for the operators defined in Eq. (3)

$$\dot{X}^{\dagger}_{mi} = -\frac{i}{\hbar}[X^{\dagger}_{mi}, H]. \tag{6}$$

Averaging these over the classical ground state and transforming the resulting linear system to k-space ($X^{\dagger}_{mi} = \frac{1}{\sqrt{N}}\sum_{\vec{R}_i} u^*_{m\vec{k}} \exp i(\vec{k}\vec{R}_i - \frac{E_{\vec{k}}}{\hbar}t))$ leads to the eigenvalue problem

$$\sum_{m'}(A_{\vec{k}}(m,m') - \delta^m_{m'}E_{\vec{k}})u^*_{m'\vec{k}} = 0, \tag{7}$$

from which the 'magnon' dispersion ($E_{\vec{k}}$) can be determined.

In general, this is a rather tedious problem. To determine the transverse (with respect to the spin) excitations the size of the matrix Eq. (7) is 8 × 8, i.e. the degrees of freedom per sublattice are $S^+_{xi}, S^+_{zi}, T^+_{ir}, T^+_{il}$. This problem is simpler for the x-AFM state, because the S^+_{zi} and T^+_{ir} do not contribute. Ignoring J_H we find the dynamical matrix

$$\begin{pmatrix} \lambda - E_{\vec{k}} & 0 & Q_{+\vec{k}} & Q_{-\vec{k}}/\sqrt{3} \\ 0 & \lambda + \varepsilon_z - E_{\vec{k}} & Q_{-\vec{k}}/\sqrt{3} & Q_{+\vec{k}}/3 \\ -Q_{+\vec{k}} & -Q_{-\vec{k}}/\sqrt{3} & -\lambda - E_{\vec{k}} & 0 \\ -Q_{-\vec{k}}/\sqrt{3} & -Q_{+\vec{k}}/3 & 0 & -\lambda - \varepsilon_z - E_{\vec{k}} \end{pmatrix} \begin{pmatrix} u_{\vec{k}}(x,A) \\ u_{\vec{k}}(l,A) \\ u_{\vec{k}}(x,B) \\ u_{\vec{k}}(l,B) \end{pmatrix} = 0, \tag{8}$$

with

$$Q_{\pm\vec{k}} = J\gamma_{\pm\vec{k}}, \ \gamma_{\pm\vec{k}} = \cos(k_x) \pm \cos(k_y),$$
$$\lambda = 4J. \tag{9}$$

We find for the dispersions

$$\begin{aligned} E_{\pm\vec{k}} &= \frac{1}{\sqrt{2}}[2\lambda^2 + 2\varepsilon_z\lambda + \varepsilon_z^2 - 10Q^2_{+\vec{k}} - 6Q^2_{-\vec{k}} \\ &\pm (192Q^2_{+\vec{k}}Q^2_{-\vec{k}} + (16Q^2_{+\vec{k}} - 12Q^2_{-\vec{k}})\varepsilon_z^2 + 64Q^4_{+\vec{k}} + 32Q^2_{+\vec{k}}\varepsilon_z\lambda \\ &+\varepsilon_z^4 + 4\varepsilon_z^3\lambda + 4\varepsilon_z^2\lambda^2)^{\frac{1}{2}}]^{\frac{1}{2}}. \end{aligned} \tag{10}$$

Inclusion of J_H does not change these dispersions significantly (the main effect is that J gets effectively smaller) and in Fig. 5 we show the magnon spectrum for $J_H/U = 0.2$ and $\varepsilon_z = 0$. As is evident from Eq.'s (8-10), an acoustic- and optical

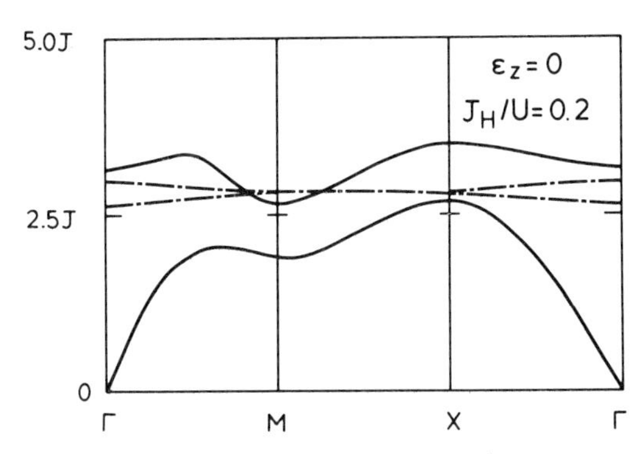

Fig. 5. Dispersions of the (with respect to the spin) transverse- (full lines) and longitudinal (dashed lines) excitations in the pure $x^2 - y^2$ Néel state, for the parameters as indicated.

magnon branch are found. At Γ, the acoustical magnon is of the usual (S_x^+) sort, but the optical branch is new: it corresponds to a *spin-flip accompanied by an orbital flip* (T_l^+). Along the $\Gamma - X$ direction these two modes do not mix for symmetry reasons. However, for all other directions they do interact (e.g. compare $\Gamma - X$ and $\Gamma - M$) and this coupling is rather strong. Approaching the Γ point, this coupling decreases and one can show from Eq. (10) that the corrections show up in $O(k^2)$, i.e. the spin wave velocity is not affected and takes the Heisenberg value.

The same calculation can be done for (with respect to the spin) longitudinal modes. The final result is simple; on each sublattice a single physical mode is found, corresponding with the simple orbital excitation $d_{z\sigma}^{\dagger} d_{x\sigma}$. This mode is in the absence of Hundt's rule coupling dispersionless. This could have been anticipated from Fig. 2; an orbital excitation cannot propagate in a pure Néel spin background. However, the Hundt's rule coupling can mediate the propagation of this excitation and we find a width $\sim 8J^1/3$. We note that QSF will have a strong influence on this width. In the quantum Heisenberg ground state, neighbouring spins have a probability of ~ 0.28 to be parallel[16], and from this number we estimate a lower bound for the exciton dispersion $\simeq J^0 \simeq 0.06eV$.

The most interesting (transverse) excitation spectrum is obtained in the FM-AFO (K_2CuF_4) phase. It is now possible to flip both the $x^2 - y^2$- as well as the $3z^2 - 1$ spin and together with the spin flip one can flip the orbital in both directions. Therefore, there are in total four (strongly mixed) transverse branches and in Fig. 6 we show

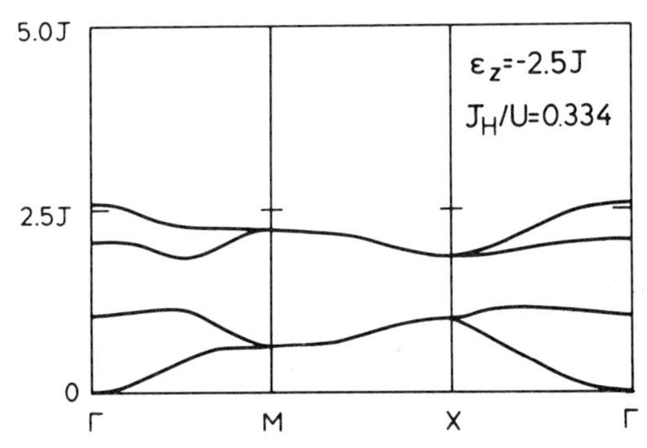

Fig. 6. Transverse excitations (magnons) in the orbital-ordered, ferromagnetic phase.

Table I. The lowering of the two site energy (Δ) due to orbital fluctuations for the (spin) singlet (S), triplet (T) and Néel (N) states in units of the $\varepsilon_z \to \infty$ singlet-triplet splitting ($J_H = 0$). $< n_z >$'s are the the z-occupancies in the singlet and triplet ground states.

ε_z/J^0	Δ_S	Δ_N	Δ_T	$< n_z >_S$	$< n_z >_T$
0	0.611	0.222	0.278	0.25	0.50
2	0.420	0.096	0.067	0.14	0.06
4	0.308	0.059	0.034	0.09	0.02
8	0.194	0.033	0.017	0.04	0.01
16	0.109	0.017	0.008	0.01	0

a representative example. The single acoustical branch has $E_k \sim k^2$ for small k, reflecting the FM character of the ground state. However, the overall spectrum does not look like that of a FM. For instance, the down bending of the lowest lying optical mode and the tendency towards degeneracy of the two high lying optical branches, reflect clearly the symmetry breaking in orbital space. It would be interesting to study K_2CuF_4 experimentally by neutron scattering.

IV. The Role of Quantum Fluctuations

In atomic multiplet theory, it is a common experience that orbital configuration mixing becomes more important if the spin gets smaller. The same happens in the present case, as can be checked easily by considering the exact solution of the two site problem. We considered the (spin) singlet- and triplet sectors, and for comparison also the 'Néel' ($\uparrow\downarrow$) spin configuration. In all cases, the ground state is a linear combination of three orbital configurations $\sim |x_i x_j>, (|x_i z_j> + |z_i x_j>)/\sqrt{2}, |z_i z_j>$. In table I we compare the lowering of the ground state energy, with respect to the energy of the pure $|x_i x_j>$ state, due to the mixing of the orbital configurations, for the different spin configurations. We see that indeed the orbital configuration mixing effects are much stronger (compare with $< n_z >$!) in the singlet- than in the triplet sector. The comparison with the 'Néel' result makes clear that the major contribution in the singlet sector comes from orbital flips *accompanied by spin flips*, which are completely neglected in the classical theory of the last section. These findings suggests that *low lying orbital excitations will have a profound influence on the quantum zero point motion of the spins.*

Recently, Arovas and Auerbach showed that mean-field theories based on the Schwinger boson representation of the Heisenberg (H) model are remarkably successful in catching the quantum effects for small spin[17]. In this picture, the long range AFM order at zero temperature in 2D is due to Bose condensation of the Schwinger bosons[18]. At finite temperatures a gap is present in the boson excitation spectrum and the spins have a finite coherence length (ξ). The temperature dependence of ξ is strongly renormalized by quantum effects and is in good agreement with non-linear σ model results[19] and experiment[20]. It is therefore expected that the Schwinger boson mean field theory also can account for the *additional* quantum fluctuations induced by the orbital degrees of freedom. It is straightforward to generalize this technique to our case; the idea is to represent the commutation relations of $su(4)$ (Eq. (1)) by bosons which is easily done by $(d^\dagger_{m\sigma}, d_{m,\sigma}) \to (b^\dagger_{m\sigma}, b_{m\sigma})$, where the b's obey boson commutation relations. In terms of these bosons, the KK-model Eq. (2) takes the form (neglecting J_H)

$$H = - \sum_{<ij>} 2J^0_{ij}[\Delta^*_{ij}(xx) + \frac{\Delta^*_{ij}(zz)}{3} + \frac{(-1)^{\hat{b}_{ij}\hat{y}}}{\sqrt(3)}(\Delta^*_{ij}(zx) + \Delta^*_{ij}(xz))]$$

$$\times [\Delta_{ji}(xx) + \frac{\Delta_{ji}(zz)}{3} + \frac{(-1)^{\hat{b}_{ij}\hat{y}}}{\sqrt(3)}(\Delta_{ji}(xz) + \Delta_{ji}(zx))]$$

$$+ \sum_{<ij>} 8J^0_{ij}[\frac{n_{-i}}{3} + \frac{(-1)^{\hat{b}_{ij}\hat{y}}}{\sqrt(3)}(T_{ir} + T_{il})][\frac{n_{-j}}{3} + \frac{(-1)^{\hat{b}_{ij}\hat{y}}}{\sqrt(3)}(T_{jr} + T_{jl})]$$

$$- \sum_i \varepsilon_z n_{-i} + \sum_i \lambda_i (\sum_n b^\dagger_{in} b_{in} - 2S) + \sum_{<ij>} \frac{4J^0_{ij}}{9}. \tag{11}$$

In Eq. (11), the spin-independent operators n_- and T have the same meaning as in the fermion representation. We defined the bond fields

$$\Delta^*_{ij}(\mu\nu) = b^\dagger_{i\mu\uparrow} b^\dagger_{j\nu\downarrow} - b^\dagger_{i\mu\downarrow} b^\dagger_{j\nu\uparrow}, \tag{12}$$

measuring the strength of a singlet on bond ij. Finally, we included the Lagrange multipliers λ_i to enforce the constraint that the total number of bosons on each site be 2S. We note that in contrast to the H model, $S \neq \frac{1}{2}$ is not physical. As we will show elsewhere, Eq. (2) is only of physical significance for d^9 (and, hence, $S = \frac{1}{2}$), while for other occupancies very different 'Kugel-Khomskii' Hamiltonians are found[21]. Nevertheless, it is usefull to keep S as a free parameter to study the crossover to the classical limit.

We construct the following mean field approximation to Eq. (11); first, we treat $\lambda = \overline{\lambda_i}$ as a constant which is independent of position, as usually. Secondly, we decouple the spin dependent part in the same (Hartree) spirit as done for the H model by introducing the order parameters

$$Q_{x,y} = \overline{\Delta_{ji}(xx) + \Delta_{ji}(zz)/3 \pm (\Delta_{ji}(xz) + \Delta_{ji}(zx))/\sqrt{3}}. \tag{13}$$

In contrast to the (on-site) molecular field decoupling (where the '$S.T$' terms dropped out by construction), these can be different for the x and y direction. For the moment we ignore this possibility and take $Q = Q_x = Q_y$. Finally, we could also consider order parameters arising from the spin-independent part of Eq. (11). However, for not too negative ε_z, these Ising-like degrees of freedom are of secondary importance and we neglect them.

The resulting mean field Hamiltonian, after transformation to k-space, yields a Bogoliubov-de Gennes equation of the form Eq. (8) with $Q_{\pm\vec{k}} = Q\gamma_{\pm,\vec{k}}$ and λ being the average Lagrange multiplier. We end up with the mean field free energy

$$F(\lambda, Q) = 2Q^2 - \frac{1}{2}(2S+2)\lambda - \frac{1}{2}\varepsilon_z + \frac{1}{\beta}\sum_\pm \int \frac{d^2k}{(2\pi)^2} \ln(2\sinh(\frac{\beta}{2}E_{\pm\vec{k}})), \tag{14}$$

with $\beta = 1/k_BT$ and $E_{\pm\vec{k}}$ defined by Eq. (9) and the integral taken over the full Brillioun zone. For simplicity we restrict ourselves to the 'symmetric' case $\varepsilon_z = 0$ (Fig. (5)). In this case, the dispersion relations Eq. (10) take the simple form

$$E_{\pm\vec{k}} = \sqrt{\lambda^2 - 16Q^2\Gamma^2_{\pm\vec{k}}}, \tag{15a}$$

with

Table II. The quantum renormalization factor for the free-energy in the Heisenberg ($\delta_F(H)$) and the Kugel-Khomskii ($\delta_F(K)$) systems and the same for the spin-wave velocity ($Z_c(H), Z_c(K)$) as a function of the spin magnitude (S).

S	$\delta_F(H)$	$\delta_F(K)$	$Z_c(H)$	$Z_c(K)$	$\delta_F(K)/\delta_F(H)$
1/2	0.341	0.601	1.158	1.265	1.76
1	0.164	0.283	1.079	1.132	1.72
2	0.080	0.137	1.039	1.066	1.70
3	0.053	0.090	1.026	1.044	1.69
4	0.040	0.067	1.020	1.033	1.69
5	0.032	0.054	1.016	1.026	1.69

$$\Gamma^2_{\pm\vec{k}} = \frac{1}{9}[2\cos^2(k_x) + \cos(k_x)\cos(k_y) + 2cos^2(k_y)$$

$$\pm 2\sqrt{\cos^4(k_x) + \cos(k_x)\cos(k_y)(cos^2(k_x) + cos^2(k_y)) + \cos^4(k_y)}], \qquad (15b)$$

We redefine $\lambda = 2\Lambda$ and $Q = \frac{1}{2}\Lambda\eta$ and minimizing Eq. (14) to Λ and η yields the saddle point equations

$$2S + 2 = \sum_{\pm}\int \frac{d^2k}{(2\pi)^2}\coth[\beta\Lambda(1-\eta^2\Gamma^2_{\pm\vec{k}})^{\frac{1}{2}}](1-\eta^2\Gamma^2_{\pm\vec{k}})^{-\frac{1}{2}}, \qquad (16a)$$

$$\Lambda = \sum_{\pm}\int \frac{d^2k}{(2\pi)^2}\coth[\beta\Lambda(1-\eta^2\Gamma^2_{\pm\vec{k}})^{\frac{1}{2}}](1-\eta^2\Gamma^2_{\pm\vec{k}})^{-\frac{1}{2}}\Gamma^2_{\pm\vec{k}}, \qquad (16b)$$

and the free energy

$$F(\Lambda,\eta) = \frac{1}{2}\Lambda^2\eta^2 - (2S+2)\Lambda + \frac{1}{\beta}\sum_{\pm}\int \frac{d^2k}{(2\pi)^2}\ln(2\sinh[\frac{\beta}{2}(1-\eta^2\Gamma^2_{\pm\vec{k}})^{\frac{1}{2}}]). \quad (17)$$

We note that Eq.'s (16-17) are similar to the ones for the Heisenberg model[17]. The latter are obtained by setting $2S+2 \to 2S+1$ in Eq.'s (17a),(18), omitting the optical branch and taking for the acoustical branch $\Gamma_{+\vec{k}} \to \frac{1}{2}\gamma_{+\vec{k}}$.

From the discussion of the excitation spectra in the previous section it might be clear that if only Λ and η are non-zero the problem is similar to the H model (for instance, the minimum of $E_{+\vec{k}}$ is at $\vec{k} = 0$), although the renormalizations may be quantitatively different. We solved Eq.'s (16,17) numerically and in table II we compare some results with the outcomes of the Schwinger boson mean field theory for the H model. We consider the mean field free energy and the spin wave velocity at zero temperature. We define the free energy and spin wave velocity as

$$F_x = -4JS^2(1+\delta_F(x)), \qquad (18.a)$$

$$\hbar c_x = \sqrt{8}JSaZ_c(x), \qquad (18.b)$$

with $x = H$ for the H- and K for the KK-model. δ measures the lowering of the free energy by quantum fluctuations, relative to the classical value $(-4JS^2)$ and $Z_c(X)$ contains the quantum enhancement of the spin-wave velocity (classically $Z_c = 1$). From the table we see that *the quantum corrections to the free energy and the spin wave velocity are nearly twice as large for the Kugel-Khomskii model (for $\varepsilon_z = 0$), as for the Heisenberg model.*

V. Discussion and Outlook

In this paper we presented our finding that low lying orbital excitations give rise to a strong increase of the quantum-spin fluctuations. In the last section we investigated this in the context of the (near) Néel state, and the question comes to the mind if these orbital fluctuations can stabilize a different ground state. We already mentioned the possibility that the order parameter Q is different in the x- and y-directions, corresponding with (quantum) Ferro type orbital order. For $\varepsilon_z = 0$ this is not the case, but we cannot exclude that it happens for lower values of this parameter. One can also think in terms of more exotic ground states. One possibility is the Affleck-Marston flux-phase[22], which has been shown to be unstable with respect to the uniform (Néel) configuration in the framework of the Schwinger boson theory for the Heisenberg model[17]. We have calculated the relative stability of the flux phase with respect to the uniform state and, although (for $\varepsilon_z = 0$) the latter has a lower energy, *the orbital excitations push down the flux phase relative to the uniform one.* We find that the former lies $\sim$ 35% higher in energy for the Heisenberg model, but only $\sim$ 12% for the KK-model. Clearly, this problem deserves further investigation.

What is the relevance of our findings for high Tc superconductivity? Clearly, at half filling the $3z^2 - 1$ occupancy is too low to change the Heisenberg picture significantly and the issue is what happens in the superconductors where much more $3z^2 - 1$ character is present. Can we write down a $J - t$ model where the spin part is of the KK-type, or do we create $3z^2 - 1$ holes by doping? At least from the classical point of view, also in the first scenario an increase of the $3z^2 - 1$ character under doping will occur. The hole will give rise to missing bonds, and this leads to a quadrupolar (orbital) polarization cloud surrounding the hole[21]. This effect seems to be confirmed by a recent cluster-diagonalization study[23].

Finally, *if orbital degrees of freedom are important one cannot neglect the phonons*.[24] In this respect, we want to comment on Weber's suggestion[8] that the phonon anomaly detected by Pintschovius and coworkers[10] in La_2CuO_4 is of Jahn-Teller origin. Experimentally, a high lying phonon mode seems to split in two branches, suggesting a crossing and mode-coupling of this phonon with another excitation of unknown character. We derive for the Jahn-Teller coupling in the tetragonal unit cell[21]

$$H_{J-T} \sim \frac{1}{\sqrt{N}} \sum_{\vec{q}} \sum_{s} \sum_{i} 2[\xi_s^b(\vec{q}))(T_{ir} + T_{il}) + \xi_s^a(\vec{q}) n_{-i}] \exp(i\vec{q}\vec{R}_i)(a_{\vec{q}}^{(s)} + a_{-\vec{q}}^{(s)\dagger}), \quad (20)$$

where the electronic operators are defined in Eq. (3b) and the ξ's measure the coupling strength between the electronic excitations and the phonon branch s. Further, $\xi_s^a(\vec{q})$ is proportional to the local *breathing* character- and $\xi_s^b(\vec{q})$ to the local *quadrupolar* character in the phonon s. Including Eq. (20) in the equations of motion Eq. (7) for the longitudinal excitations and linearizing afterwards shows that the breathing mode drops out but *there is a harmonic coupling left between the orbital excitation* ($d_{z\sigma}^\dagger d_{x\sigma}$) *and the quadrupolar local phonon*[21]. Further, from symmetry it follows that ξ^b vanishes at the zone center and is at maximum at the zone boundary, consistent with experiment. Thus, we predict that if the phonon anomaly is of Jahn-Teller origin (and due to harmonic mode coupling), the phonon which is involved is of quadrupolar character and not of the breathing sort.

We are left with some quantitative problems: first, we derived in section III a lower bound for the width of the longitudinal orbital excitation of $\sim$ 0.05 eV, of the order of the total phonon bandwidth! Further, from optical spectroscopy it is known that the bare $\xi's$ are also quite large ($\sim$ 0.1eV). Finally, in the light of the strong tetragonal distortion it seems unlikely that the bare orbital excitation is at such a low energy. There is a way out: the orbital excitation will dress itself with breathing mode phonons, giving rise to a Franck-Condon (breathing) phonon progression. The zero-phonon mode will correspond with a small polaron, where the distorted octahedron is de-elongated, and one might assert that this is the unidentified object. Both the bandwidth, as well as the coupling with the quadrupolar phonon will be strongly reduced, solving the three problems we mentioned. In other words, according to this picture, investing ~ 0.1 eV per unit cell could stabilize the K_2CuF_4 type distortion (Fig. 4) in La_2CuO_4!

Still, this picture is not completely satisfying. How to explain that the phenomenon is rather universal ($YBa_2Cu_3O_7$, La_2NiO_4)[10]? Further, what to do with recently reported splittings of lower lying phonons[25]?

We acknowledge helpful discussions with O. K. Andersen, A. Bianconi, J. Fink, D. Foerster, P. Horsch, G. A. Sawatzky and W. Weber. We are especially gratefull to T. Dombre who introduced us to Schwinger boson theory. One of us (JZ) would like to thank the Institute for Scientific Interchange (ISI), Torino, Italy, where part of this work was done, for their hospitality and support.

REFERENCES

1. F. C. Zhang and T. M. Rice, Phys. Rev. B 37, 3759 (1988).
2. H. Eskes and G. A. Sawatzky, Phys. Rev. Lett. 61, 1415 (1989).
3. A. Bianconi et al, Physica C, in press.
4. N. Nücker et al, Phys. Rev. B 39, 6619 (1989).
5. G. A. Gehring and K. A. Gehring, Rep. Prog. Phys. 38, 1 (1975).
6. K. I. Kugel and D. I. Khomskii, Sov. Phys. Usp. 25, 231 (1982).
7. K. I. Kugel and D. I. Khomskii, Sov. Phys.-JETP 37, 725 (1973).
8. W. Weber, Physica C, in press.
9. D. L. Cox et al, Phys. Rev. Lett. 62, 2188 (1989).
10. H. Rietschel, L. Pintschovius and W. Reichardt, Physica C, in press; W. Reichardt et al, ibid; L. Pintschovius et al, Europhys. Lett. 5, 247 (1988).
11. K. A. Chao, J. Spalek and A. M. Oleś, Phys. Stat. Sol. B 84, 747 (1977).
12. D. I. Khomskii and K. I. Kugel, Solid State Commun. 13, 763 (1973).
13. Y. Ito and J. Akimutsu, J. Phys. Soc. Jpn. 33, 1333 (1976); L. D. Khoi and P. Veillet, Phys. Rev. B 11, 4128 (1975); W. Kleemann and Y. Farge, J. Phys. (Paris) 36, 1293 (1975).
14. J. Zaanen, Proc. Int. Symp. High Temperature Superconductivity, Jaipur, India, ed. K. B. Garg, pp. 31 (1989).
15. O. K. Andersen, unpublished.
16. P. Horsch and W. von der Linden, Z. Phys. B 72, 181 (1988).
17. D. P. Arovas and A. Auerbach, Phys. Rev. B 38, 316 (1988); ibid., Phys. Rev. Lett. 61, 617 (1988).
18. D. Yoshioka, J. Phys. Soc. Jpn. 58, 5978 (1988).
19. S. Chakravarty, D. R. Nelson and B. I. Halperin, Phys. Rev. Lett. 60, 1057 (1988).
20. G. Shirane et al, Phys. Rev. Lett. 59, 1613 (1987).
21. J. Zaanen, L. F. Feiner and A. M. Oleś, unpublished.
22. I. Affleck and J. B. Marston, Phys. Rev. B 37, 3774 (1988).
23. H. Eskes, G. A. Sawatzky and L. F. Feiner, Physica C 160, 424 (1989).
24. K. I. Kugel and D. I. Khomskii, Sov. Phys.-JETP 52, 501 (1980).
25. W. Reichardt et al, unpublished.

MAGNETIC FRUSTRATION MODEL AND SUPERCONDUCTIVITY ON DOPED LAMELLAR CuO_2 SYSTEMS

Amnon Aharony

Raymond and Beverly Sackler Faculty of Exact Sciences
School of Physics and Astronomy
Tel Aviv University, Tel Aviv 69978, Israel

EXTENDED ABSTRACT

In most of the doped lamellar CuO_2 based high temperature superconductors, the behavior varies strongly as function of the dopant concentration, which is directly related to the concentration of the electronic holes. In the present paper we emphasize the interplay among the various phases which appear in the temperature - concentration phase diagram.

Without doping, these systems are insulating antiferromagnets. The antiferromagnetic exchange interactions between the localized spin-1/2 copper ions are strong in the CuO_2 planes, and weak among the planes. The quantum spin fluctuations can be renormalized, and the behavior in each plane is excellently described by an effective classical two-dimensional Heisenberg antiferromagnetic model.[1] Recently,[2] we have shown that the weak coupling between the planes is very well described by a mean field theory. In La_2CuO_4, the orthorhombic rotation of the CuO_6 octahedra generates an additional antisymmetric Dzyaloshinskii-Moriya interaction between the Cu spins in the planes. This results in a bilinear coupling between the staggered magnetization perpendicular to the plane, yielding a canting of the magnetic moments out of the planes. At zero magnetic field, the ferromagnetic moments of neighboring planes order antiparallel to each other. However, external magnetic fields cause a variety of spin flip transitions.[2,3] The bilinear coupling between the staggered and the uniform magnetizations implies an indirect coupling between the external uniform field and the staggered magnetization, explaining the unusual sharp peak in the uniform susceptibility at the Neél point.[2] Our mean field theory for weakly coupled planar Heisenberg models fits the susceptibility data, with practically no adjsutable parameters.[2] Similar theories should work for all other properties of these weakly coupled planar systems, e.g. in the superconducting phase.

Doping introduces quenched randomness, as well as deviations from stoichiometry, usually adding electronic holes, which reside mainly on the oxygen ions in the planes. At low concentration, these holes are localized, exhibiting variable range hopping conductivity.[4] The holes always remain localized in the planes, with a small perpendicular localization length. Within the plane, the localization length at low doping is of order 10Å (about 2.5 lattice constants). In La_2CuO_4, the hole conductivity is very sensitive to the external magnetic field, changing by about a factor 2 at the spin flip transition.[2,3] This shows a strong coupling between the hole states and the magnetic ordering.

In order to discuss the coupling of the hole to the underlying magnetic ordering of the copper spins, we considered[5] a single hole, localized at an oxygen site. There is a strong superexchange coupling of the spin of this hole to those of the two neighboring copper ions,

Dynamics of Magnetic Fluctuations in High-Temperature Superconductors
Edited by G. Reiter *et al.*, Plenum Press, New York, 1991

resulting in an effective ferromagnetic exchange interaction between these copper spins. The exchange interaction between the hole spin and the copper spins will be ferromagnetic (antiferromagnetic) if the hole sits mainly on a p_π (p_σ) oxygen state, with the wave function mainly perpendicular (parallel) to the bond. The resulting three-spin state will have total spin 3/2 (or 1/2). It is not yet completely clear which state wins. However, both situations create ferromagnetic Cu-Cu interactions, and frustrate the antiferromagnetic copper state. We believe that this frustration is responsible to the fast decrease of the Neél temperature with doping. We also predicted[5] that at higher doping, the antiferromagnetic phase should be replaced by a spin glass phase, as indeed confirmed by many experiments. Alternative pictures, which couple the hole into a singlet state with one copper ion, do not contain the frustration necessary for the spin glass.

The frustration model implies that the hole spin prefers to be perpendicular to the copper spins, and may help in understanding the sensitivity of the hole mobility to the spin ordering. In addition, the disturbance to the copper antiferromagnetic ordering is smaller if two holes sit on neighboring or next nearest neighboring (nnn) oxygens. Since the former seems to be excluded by Coulomb repulsion, we considered the attractive pairing potential between two holes on nnn oxygens.[6] This potential decays with increasing doping, as the antiferromagnetic correlation length decreases. Assuming that the holes move freely in the p-band, we used a BCS theory with the cutoff set at the Fermi energy (proportional to the number of mobile holes).[6] Adjusting only the temperature scale, we were able to reproduce the concentration dependence of the superconducting transition temperature of $La_{2-x}Sr_xCuO_4$, including the peak at around $x \simeq 0.15$ and the decay for larger x.[7] More recently,[8] we also considered the tight binding band structure for the motion of the Cu-O-Cu "polaron", and the results are qualitatively similar.

ACKNOWLEDGEMENTS

Work supported by grants from the U.S.-Israel Binational Science Foundation and from IBM.

REFERENCES

1. S. Chakrvarty et al, Phys. Rev. Lett. 60, 1057 (1988).
2. T. Thio et al, Phys. Rev. B38, 905 (1988).
3. T. Thio et al, to be published.
4. M. A. Kastner et al, Phys. Rcv. B37, 111 (1988);
   N. W. Preyer et al,, Phys. Rev. B39, 11563 (1989).
5. A. Aharony et al, Phys. Rev. Lett. 60, 1330 (1988).
6. R. J. Birgeneau et al, Z. Phys. B71, 57 (1988).
7. A. Aharony et al, IBM J. Res. Develop. 33, 287 (1989).
8. L. Klein and A. Aharony, unpublished.

STRONG COUPLING REGIME IN THE HUBBARD MODEL AT LOW DENSITIES

Alberto Parola[1], Sandro Sorella[1],
Michele Parrinello[2] and Erio Tosatti[1]

[1]SISSA, Strada Costiera 11, Trieste Italy
[2]IBM Research Division, Zurich Research Laboratory
8803 Rüschlikon, Switzerland

ABSTRACT

The exact solution of the problem of 2 electrons in the Hubbard model is shown to be characterized by strong correlations both in one and two spatial dimensions even for arbitrary values of the coupling constant U both in one and in two dimensions. This result supports the conjecture that the low density regime of the Hubbard model is described by the strong coupling limit for every $U > 0$.

INTRODUCTION

The physical properties of two dimensional interacting electron systems are of great current interest because of their possible relevance for understanding the mechanisms of high temperature superconductivity[1]. An open question in the physics of low–dimensional fermionic models is whether their properties are described by standard Fermi Liquid theory or whether instead they show unconventional behavior. In the latter case, the system is said to be dominated by the strong coupling regime, in that its physical properties at every coupling U are qualitatively similar to the $U = \infty$ limit rather than to the $U = 0$ free electron case. In 1D, renormalization group studies[2], together with the exact solution of selected models[3], have shown that the $U = 0$ case is indeed a singular point in the phase diagram of the system. In 2D, no exact (or even just plausible) results are available and this question has no answer so far. One of the simplest systems which have been studied in this context is the Hubbard model (HM), defined by the hamiltonian:

$$H = -\sum_{<i,j>,\sigma} c^{\dagger}_{j,\sigma} c_{i,\sigma} + U \sum_{i} n_{i,\uparrow} n_{i,\downarrow} \tag{1}$$

Dynamics of Magnetic Fluctuations in High-Temperature Superconductors
Edited by G. Reiter *et al.*, Plenum Press, New York, 1991

where periodic boundary conditions are understood and the sum in the kinetic term is restricted to nearest neighbors.

In this note we show that the basic features of the known 1D behavior of the HM could have been anticipated by a careful analysis of the exactly soluble problem of two electrons in the otherwise empty lattice[4,5]. A similar study is then carried out also in 2D. The results suggest the conjecture that, even in this case, the low density HM is always in the strong coupling regime for every non–zero value of the Coulomb repulsion U.

ONE DIMENSIONAL SYSTEM

For simplicity we limit our analysis to the singlet and zero momentum subspace, which contains the actual ground state of the system. The most general expression of the eigenvector of H is given by:

$$|\psi>=\sum_{r_1,r_2}\psi(r_1-r_2)c^{\dagger\uparrow}_{r_1}c^{\dagger\downarrow}_{r_2}|0> \tag{2}$$

where $\psi(x)=\psi(-x)$ and $\psi(x)=\psi(x+L)$ due to the periodic boundary conditions. The 1D lattice is defined by the coordinates r of the sites ranging from $-L+1$ to L. The eigenvalue equation for the wavefunction (2) can be easily obtained by direct substitution into (1):

$$-2[\psi(x+1)+\psi(x-1)]+U\psi(0)\delta_{x,0}=E\psi(x) \tag{3}$$

If $\psi(0)=0$ no singlet solution can be found. Therefore we can always assume $\psi(0)=1$. In this case, the solution of Eq. (3) is immediately obtained by Fourier transform:

$$\psi(x)=\frac{U}{2L}\sum_{n=-L+1}^{L}\frac{\exp(i\pi nx/L)}{E+4\cos(\pi n/L)} \tag{4}$$

and the consistency condition $\psi(0)=1$ becomes the eigenvalue equation for E. The finite sum in Eq. (4) can be carried out explicitly by contour integration giving a more compact form of the eigenvalue equation:

$$\tan(\theta L)\sin\theta=\frac{U}{4} \tag{5}$$

where $\cos\theta=-E/4$. Eq (5) has L solutions for *real* θ, the mth one ($m=0...L-1$) belonging to the interval $\frac{\pi}{2L}2m<\theta_m<\frac{\pi}{2L}(2m+1)$. It is interesting to notice that the equation for the ground state eigenvalue $\theta_0=\frac{\pi}{2L}\alpha$ ($0<\alpha<1$) simplifies in the large L limit:

$$\tan\left(\frac{\pi}{2}\alpha\right)=\frac{UL}{2\pi\alpha} \tag{6}$$

In this equation, the coupling constant U and the system size L only appears through their product UL. Therefore, whatever is the strength of the interaction $U > 0$, the right hand side of Eq. (6) tends to ∞ in the large L limit yielding $\alpha = 1$ and then a ground state energy $E = -4(1 - \frac{\pi^2}{8L^2})$.

The explicit form of the wavefunction $\psi(x)$ corresponding to a given eigenvalue θ_m can be similarly obtained from Eq. (4) and, when properly normalized, is

$$\psi_m(x) = \frac{|\cos(\theta_m(L-x))|}{\sqrt{L + \frac{1}{2}\sin(2\theta_m L)/\tan\theta_m}} \tag{7}$$

Again it is instructive to look at the large L limit of the ground state wavefunction in Eq. (7)

$$\psi_0 = \frac{1}{\sqrt{L}}|\sin(\frac{\pi}{2L}x)| \tag{8}$$

which does not depend on the value of the coupling U and vanishes at the origin with a linear increase at small distances $x \ll L$. Eq. (8) shows that two particles in the 1D HM tend to stay as far as possible, independent of the strength of the Coulomb repulsion $U > 0$ if the lattice is sufficiently large. This result has to be contrasted with the $U = 0$ case where the wavefunction is just a constant. Another interesting property which can be easily obtained from this solution, is the momentum distribution $n(k)$ whose limiting behavior for large lattice size is again universal

$$n(k) = \frac{8}{\pi^2}\frac{1}{(4n^2-1)^2} \tag{9}$$

where $k \equiv \frac{2\pi}{2L}n$. Also this quantity tends to a non–trivial limit for large L which is independent of $U > 0$ and different from the free electron case $n(k) = \delta_{k,0}$. In particular, it shows a 20% depletion in the occupation number at $k = 0$ and a corresponding spread of the wavefunction over the whole k–space.

Although the solution of the two electron problem in a large system does not necessarily give information about the low density behavior of the model, in the 1D case we can compare our findings to the exact solution of the model obtained by Lieb and Wu at finite density[3]. In particular the first terms of the low density expansion of the ground state energy turn out to be *universal*, that is different from the free electron limit and independent of U :

$$\left(\frac{E}{2L}\right)_{U\neq 0} \rightarrow -2\rho + \frac{\pi^2}{3}\rho^3 + \cdots \tag{10}$$

Moreover, the change in E due to the interaction is given by

$$\left(\frac{E}{2L}\right)_{U\neq 0} - \left(\frac{E}{2L}\right)_{U=0} \rightarrow \frac{\pi^2}{4}\rho^3 \tag{11}$$

exactly as in our two electrons calculation where $\rho = \frac{2}{2L}$.

For completeness, let us add that Eq. (5) does not exhaust the singlet spectrum of the system. In fact, however small is U, there is an additional state at the top of the energy spectrum characterized by an imaginary value of θ. This corresponds to a bound state with an exponentially localized wavefunction. The existence of this bound state (which corresponds to the ground state for $U < 0$) is clearly at the origin of the singular behavior of the above results for $U > 0$.

TWO DIMENSIONAL SYSTEM

The same calculation can be performed in 2D where there is no exact solution at finite density to compare with. The main equations are basically unchanged but the elegant analysis leading to Eq. (5) cannot be extended to the 2D case. Therefore we have to deal with the rather involved eigenvalue equation:

$$\sum_{n=-L+1}^{L} \sum_{m=-L+1}^{L} \frac{1}{z + \cos(\frac{\pi}{L}n) + \cos(\frac{\pi}{L}m)} = \frac{16L^2}{U} \tag{12}$$

where $z = E/4$. In the dilute limit $L \to \infty$ the lowest eigenvalue z will be close to the free value $z = -2$. So we let

$$z = -2 + \frac{1}{2}\left(\frac{\pi\alpha}{L}\right)^2 \tag{13}$$

Splitting the sum in Eq. (12) into a "low momentum" and a "high momentum" part, we find that the eigenvalue equation reduces to the form

$$\frac{1}{2\pi^2} \sum_{n,m} \frac{1}{\alpha^2 - (n^2 + m^2)} = const > 0 \tag{14}$$

where the right hand side has a finite limit as $L \to \infty$ which depends on the value of U. The sum is restricted to $|n|, |m| < Lq_0/\pi$ where q_0 is a small but otherwise arbitrary cut–off wavevector. The crucial feature of Eq. (14) is that the sum involved *does not* have a finite limit for large L but instead goes like $-2\pi \ln(L)$. On the other hand the only *positive* term which can compensate this large and negative contribution comes from the zero momentum component $1/\alpha^2$. The balance between these two terms gives the large L behavior of the eigenvalue:

$$\alpha = \frac{1}{\sqrt{2\pi \ln L}} \tag{15}$$

which again does not depend on U to leading order in L. From this calculation we conclude that the ground state energy of the two electrons system has a logarithmic dependence on the size of the lattice

$$E = -8\left(1 - \frac{1}{8L^2 \ln L}\right) \tag{16}$$

in the dilute limit. This might have implications on the virial expansion of the ground state energy of the 2D HM. In fact, in analogy with the 1D case, it is tempting to conclude that the first term in the low density expansion of E might be given by

$$\left(\frac{E}{V}\right)_{U\neq 0} - \left(\frac{E}{V}\right)_{U=0} \rightarrow \frac{\pi}{|\ln \rho|}\rho^2 \tag{17}$$

Although it should be stressed that the above result for two electrons constitute no proof for the many–electron case, note that this expression agrees with the bounds given by Rudin and Mattis[6] in their analysis of the HM at low density.

Turning to the wavefunction, a similar analysis shows that it grows with distance as

$$\psi_0(r) \sim \ln r \tag{18}$$

for $1 \ll r \ll L$. The effect of repulsion between the two electrons is therefore not as strong as in the 1D case. Yet it still is independent of the coupling U, suggesting that the 2D HM could also be dominated by the strong coupling regime at low density. A similar conclusion can be reached by looking at the two–electron momentum distribution

$$n(k) \propto \left(\frac{1}{1-2\pi(n^2+m^2)\ln L}\right)^2 \tag{19}$$

showing (marginal) logarithmic corrections to the free electron result.

Finally, we notice that also in 2D these anomalies of the ground state energy and wavefunction are somehow in correspondence with the presence of a bound state at the top of the spectrum for arbitrarily small values of U. This feature is *absent* in more than 2D, and also are the universal properties of the ground state which have been discussed so far in one and two dimensions. This fact suggests that for electronic systems, D=2 is the marginal dimension separating the strong coupling regime from the standard Fermi Liquid behavior of three dimensional physics, in agreement with a conjecture put forward by P.W. Anderson[1].

REFERENCES

1. See for instance: Anderson P.W., *Frontiers and Borderlines in Many Particles Physics* E. Fermi School, Varenna, Norh–Holland (1988).
2. J. Solyom, Adv. in Phys. **28**, 201 (1979).
3. E. H. Lieb and F.Y. Wu, Phys. Rev. Lett. **20**, 1445 (1968); Lieb E.H. and Mattis D.C. *Mathematical Physics in One Dimension*, Academic Press (New York), 1966.
4. C. Mei, L. Chen, Z. Phys. B **72**, 429 (1988); L. Chen, C. Mei, Phys. Rev. B **39**, 9006 (1989).
5. D.C. Mattis, *Rev. Mod. Phys.* **58**, 361 (1986).
6. S. Rudin and D.C. Mattis, *Phys. Lett*, **110A**, 273 (1985).

HOW GOOD IS THE STRONG COUPLING EXPANSION OF THE TWO DIMENSIONAL HUBBARD MODEL?

B. Friedman, X. Y. Chen and W. P. Su

Department of Physics and Texas Center for Superconductivity
University of Houston
Houston, Texas 77204-5504

ABSTRACT

Validity of the strong coupling expansion of the two dimensional Hubbard model is questioned. We present evidence of disagreement between the strong coupling expansion and the exact solution of the Hubbard model on small lattices.

With the discovery of high temperature superconductivity [1,2] the study of strongly correlated electron systems has gained a new impetus. [3] Recall the Hubbard is the "simplest" such model with Hamiltonian given by

$$H = -t \sum_{<i,j>\sigma} (c^{+}_{i,\sigma} c_{j,\sigma} + h.c.) + U \sum_{i} n_{i\uparrow} n_{i\downarrow} \tag{1}$$

Unfortunately the Hubbard model is a very hard problem! It is then tempting to try to simplify the Hubbard model through various transformations, restrictions etc. Since in the copper oxide superconductors it seems that the hopping integral t is much smaller than the onsite repulsion U it is natural to consider a strong coupling approximation in t/U.[4]

The result of such an expansion in t/U to second order yields the strong-coupling Hamiltonian

$$H_s = H_1 + H_2 + H_3 \tag{2}$$

$$H_1 = -t \sum_{<i,j>\sigma} (c^{+}_{i,\sigma} c_{j,\sigma} + h.c.)$$

$$H_2 = -\frac{2t^2}{U} \sum_{<i,j>\sigma} (c^{+}_{j,\sigma} c_{i,\sigma} n_{i,-\sigma} c^{+}_{i,\sigma} c_{j,\sigma} + c^{+}_{j,-\sigma} c_{i,-\sigma} c^{+}_{i,\sigma} c_{j,\sigma})$$

$$H_3 = -\frac{t^2}{U} \sum_{<i,j,k>\sigma} (c^{+}_{k,\sigma} c_{j,\sigma} n_{j,-\sigma} c^{+}_{j,\sigma} c_{i,\sigma} + c^{+}_{k,-\sigma} c_{j,-\sigma} c^{+}_{j,\sigma} c_{i,\sigma} + (i \longleftrightarrow k))$$

with a state space consisting of singly occupied states only. For either of these two Hamiltonians, H or H_s there is (to date) no systematic analytic means of solution for dimensions greater than or equal to two. Consequently, it seems reasonable to attempt various numerical approaches. One numerical approach, the approach we

Dynamics of Magnetic Fluctuations in High-Temperature Superconductors
Edited by G. Reiter *et al.*, Plenum Press, New York, 1991

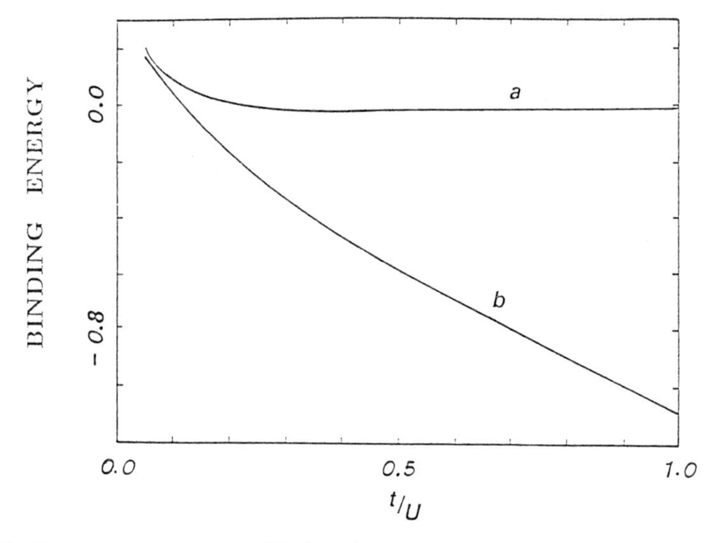

Fig. 1. Binding energy versus t/U for the four site square. Curve a is calculated using the Hubbard model; curve b is calculated using the strong coupling Hamiltonian.

shall adopt here, is the direct diagonalization of small system. In such a direct diagonalization one sees the advantage of the strong coupling Hamiltonian in that even though the Hamiltonian is more complicated the state space is smaller facilitating direct numerical diagonalization. However, there is no a priori way to assess at what values of t/U that the strong coupling expansion is an accurate representation of the physics of the Hubbard model.

In this note, we wish to raise questions about the range of validity of the above strong coupling expansion to the two dimensional Hubbard model. (For a more complete version of this work see [5].) Note that "range of validity" is definitely not the same as the "radius of convergence". It is well known that low order truncations of a divergent series may well represent the physics much better than a low order truncation of a convergent series. In the following pages we shall compare results obtained for H and H_s for four and eight site clusters all in two dimensions. Our study is motivated by work ultimately published in references [6] and [7].

We first consider the square lattice with four sites. Fig.1 is a plot of binding energy versus t/U. We define the biding energy as $(\varepsilon(2) - \varepsilon(1)) - (\varepsilon(1) - \varepsilon(0)))/2$ where $\varepsilon(i)$ is the energy of the ground state with i holes. (No holes corresponds

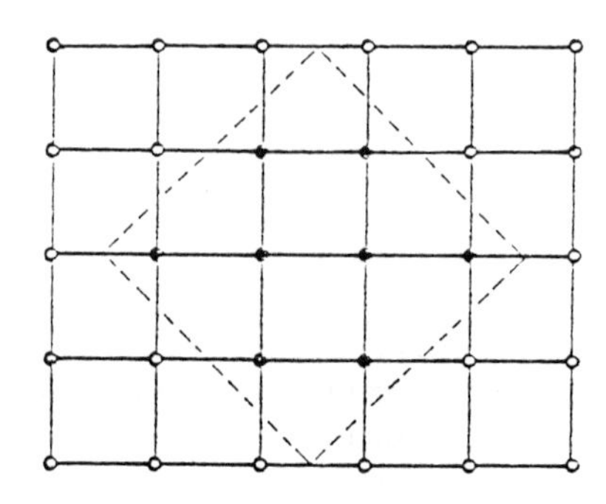

Fig. 2. The eight site tilted square.

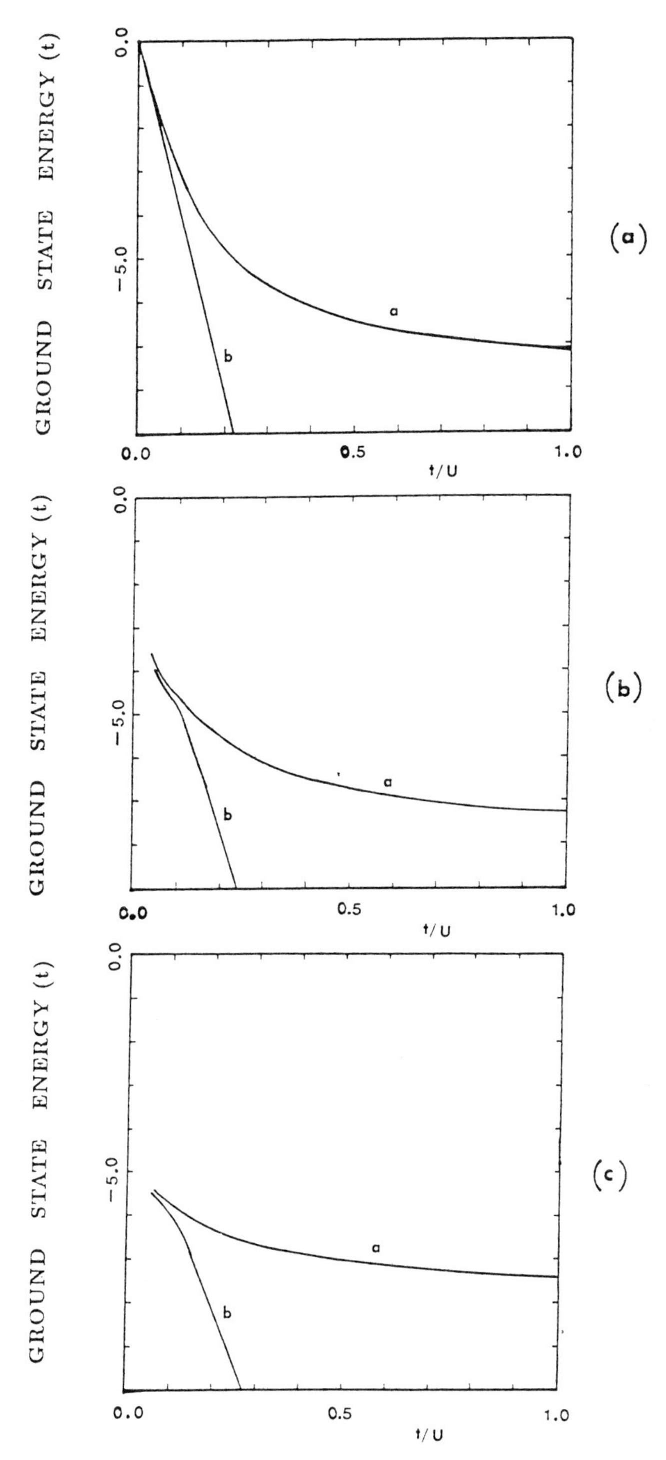

Fig. 3. Ground state energy versus t/U for the eight site tilted square. Fig. 3(a) is the half-filled case. Fig. 3(b) is for one hole. Fig. 3(c) is for two holes. Within each figure the curve marked a is calculated using the Hubbard model; the curve b is calculated using the strong coupling Hamiltonian.

to one electron per site). We work in units where $t = 1$. The Hubbard model has a relatively shallow minimum in the binding energy while H_s has a large binding energy that increases rapidly with t/U. For $t/U < 0.1$ there is qualitative agreement between H and H_s for the binding energy. However, in this region there is no binding.

We now turn to the eight site model. We consider the eight site tilted square (see Fig. 2). In Fig. 3 we have plotted the total energy versus t/U for H and H_s for zero, one and two holes. From these figures one concludes that as far as the total energy is concerned H_s is probably a quantitatively accurate approximation for H for $U > 20$.

So far our results have concerned ground state energies only. How does the strong coupling expansion fare for properties of the ground state wave function? Fig. 4 is a plot of the staggered magnetization distribution [8] for the eight site tilted square with one hole. From Fig. 4 we observe no evidence in the ground state of antiferromagnetic ordering for H. That is, the staggered magnetization distribution is a trapezoidal shaped curve peaked at zero. (upper solid curve). On the other hand, for H_s there is the generally flat distribution characteristic of antiferromagnetic ordering. (dashed dotted curve). We also find that the first excited state (at $t/U = 0.15$) of the Hubbard model has a flat distribution, (lower solid curve). i.e. an excited state of the Hubbard model close in energy to the ground state is antiferromagnetic while the ground state is not.

It thus appears that the ground state wave function obtained from H_s and the

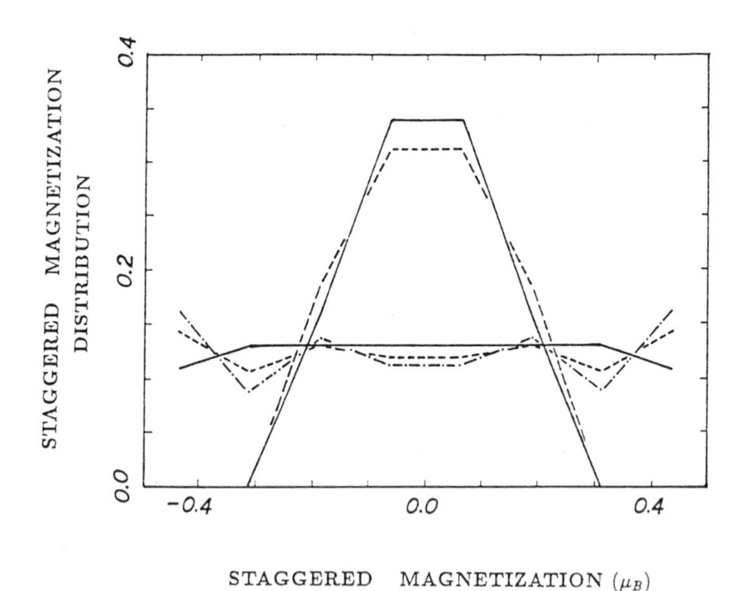

Fig. 4. Staggered magnetization distribution versus staggered magnetization for the tilted square. The upper solid curve is for the Hubbard model ground state. The upper dashed curve is for the Hubbard model ground state without double occupied states. The lower solid curve is for a low lying excited state of the Hubbard model. The lower dashed curve is for the same low lying state without double occupied states. The dashed dotted curve is for the ground state of the strong coupling Hamiltonian.

ground state wave function obtained from H are qualitatively quite different for the eight site tilted square. One can rationalise this difference by noting that the ground state wave function for H_s contains no doubly occupied states while systematic perturbation theory to even first order for the ground state would contain some doubly occupied states. We expect intuitively that the doubly occupied states would tend to suppress staggered magnetization. The upper dashed curve in Fig. 4 is the ground state magnetization distribution without including doubly occupied states in the wave function. We see that the resulting magnetization distribution is closer to that given by the ground state for H_s, however it is still quite different. It appears that the difference in the ground state wave function is more subtle than the absence of the doubly occupied states.

To better characterize the difference between the ground state in the Hubbard model and H_s we have calculated the total spin S for the $t/U = 0.15$ and 1 hole. For the ground state of the Hubbard model we find $S = 3/2$; for the first excited state we find $S = 1/2$. On the other hand, for H_s we find the ground state has $S = 1/2$. It appears that the strong coupling Hamiltonian has inverted the order of a low lying excited state and the ground state. Our result for the Hubbard model (on the tilted square) that the first excited state has a tendency toward order while the ground state shows no such tendency, reminds us of the fractional quantized Hall effect. (That such an effect should occur for high T_c materials was first suggested by Anderson.[3,9])

We see for small lattices and moderately large values of U the strong coupling Hubbard model is not a good approximation to the Hubbard model. Recently a very nice work[10] has been published comparing H and H_s for a ten site cluster. It appears for 10 sites the difference between H and H_s is not so dramatic as in the 4 and 8 site cases. However it seems 4 and 8 site clusters differ quite a bit from 10 site clusters (comparing results for H only). It is difficult to tell what is the true state of affairs for a large system. There are indications[11], however, that the 16 site cluster more closely resembles 4 and 8 sites than the 10 site cluster. It would be interesting to see to what extent the ground state and low lying states of the 16 site cluster are antiferromagnetically ordered when a small number of holes are present.

This work was supported by the Texas Advanced Research Program under grant No. 1053 and by the Texas Center for Superconductivity at the University of Houston under prime Grant No. MDA 972-88-0002 to the University of Houston from Defense Advanced Projects Agency and the State of Texas.

REFERENCES

1. J.G. Bednorz and K.A. Muller, Z. Phys. B**64**, 188(1986).
2. C.W. Chu, P.H. Hor, R.L. Meng, L. Gao, Z.J. Huang and Y.Q. Wang, Phys. Rev. Lett. **58**, 405(1987).
3. P.W. Anderson, Science **235**, 1196(1987).
4. K. Huang and E. Manousakis, Phys. Rev. B**36**, 8302(1987).
5. B. Friedman, X.Y. Chen and W.P. Su, Phys. Rev. B**40**, 4431(1989).
6. E. Kaxiras and E. Manousakis, Phys. B**40**, 2596(1989).
7. S. Tang and J.E. Hirsch, Phys. Rev. B**40**, 2594(1989).
8. E. Kaxiras and E. Manousakis, Phys. B**37**, 659(1988).
9. R. Laughlin, Science **242**, 525(1988).
10. M. Ogata and H. Shiba, J. Phys. Soc. Jpn. **58**, 2836(1989).
11. E. Dagotto, R. Joyrt, A. Moreo, S. Bacci and E. Gagliano preprint.

THE HUBBARD MODEL FOR $n \neq 1.0$: NEW PRELIMINARY RESULTS

A.N. Andriotis, Qiming Li*, C.M. Soukoulis* and E.N. Economou

Foundation for Research and Technology - Hellas (FO.R.T.H.)
Institute of Electronic Structure and Laser
P.O. Box 1527, Heraklion 711 10, Crete, Greece

INTRODUCTION

We consider the one band Hubbard Hamiltonian

$$H = \sum_{i\sigma} \varepsilon_0 n_{i\sigma} + \sum_{\substack{i,j,\sigma \\ i \neq j}} V_{ij} \alpha_{i\sigma}^{+} \alpha_{j\sigma} + U \sum_{i} n_{i\sigma} n_{i-\sigma} \tag{1}$$

where the sites {i} form a periodic lattice; σ is taken +1 for spin up and -1 for spin down; ε_0 is a constant which can be taken as zero; V_{ij} is the hopping integral and in the present work is taken to be a constant V for i,j being nearest neighbours and zero otherwise; U is the on-site Coulomb repulsion and finally $n_{i\sigma} = \alpha^{+}_{i\sigma} \alpha_{i\sigma}$ with $\alpha^{+}_{i\sigma}$, $\alpha_{i\sigma}$ being the creation and annihilation operators respectively.

The physical parameters of the model are :

(i) the ratio U/V
(ii) the average number, n, of electrons per lattice site
(iii) the type of lattice

Due to particle-hole symmetry one obtains identical results for n and 2-n. Thus we can restrict ourselves in the range $0 \leq n \leq 1$.

In our earlier investigations[1] we found solutions of the Hamiltonian of eqn.(1) in the random field Hartree Fock approximation. The approximation we have employed reduces the original Hamiltonian to one equivalent to a random binary alloy incorporating short range order. We applied then a sophisticated averaging scheme which is based on a conditional Coherent Potential Approximation[2] (CPA), to be outlined below.

The results of our earlier studies were in very good agreement with exact results in 1-D and with results obtained by other investigators who had employed different approaches.[3] Thus it became clear that the basic features of our approximation are sound and do not reflect an artifact of a particular calculational method. Recently, it has been suggested that some of the features of the Hubbard Hamiltonian may be related to the new high T_c superconductors.[4-5] This created an intense interest in reexamining and extending our knowledge on the Hubbard model. In the present work we report preliminary results for $n \neq 1$ based on our random field approximation.

* Ames Laboratory and Department of Physics, Iowa State University, Ames, Iowa 50011.

The random field approximation of Hamiltonian (1) is associated with ground states (for the various values of n and U/V), which exhibit non-integral moments in ordered phases as well as disordered phase which could be called "spin-glass" or "spin-liquid" states. Such states can be identified with the spin-fluid state of Anderson which has attracted a great recent interest in relation with high T_c superconductors.

In the present work, we reexamine and extent our earlier work,[1] in particular the disordered spin-glass state. Furthermore we discuss possible ways which allow us to relax some of the restrictions and assumptions or our earlier investigations. In section 2 we briefly review our calculational method; in section 3 we present and discuss our results.

SOLUTION OF THE ONE BAND HUBBARD HAMILTONIAN WITHIN THE RANDOM FIELD APPROXIMATION AND THE CONDITIONAL CPA

Our approximation is based on two steps:

(i) We follow Hubbard's original suggestion and replace the cumbersome many body U-term of eqn.(1) by a random one body term, i.e.

$$U\, n_{i\sigma}\, n_{i-\sigma} \approx \varepsilon_{i\sigma}\, n_{i\sigma} \tag{2}$$

where $\varepsilon_{i\sigma}$ are correlated random variables, the distribution of which is determined self-consistently.

(ii) Assuming that the repulsive on-site interaction (U-term in (1)) create local moments of fixed size μ, which can be oriented either up or down, we restrict the $\{\varepsilon_{i\sigma}\}$ distribution to a binary one. As a result, the solutions to eqn.(1) under the approximation of eqn.(2) can be obtained with techniques already developed for random binary alloys. It should be noted, however, that within the present solution the size of the local moment μ has to be determined self-consistently.[6]

Under the present assumptions, (their validity and/or their relaxation is discussed in section 3), we can take the following binary distribution for $\{\varepsilon_{i\sigma}\}$:

$$p(\varepsilon_{i\sigma},\varepsilon_{i-\sigma}) = \chi_A\ \delta(\varepsilon_{i\sigma} - \varepsilon_{A\sigma})\delta(\varepsilon_{i-\sigma} - \varepsilon_{A-\sigma}) + \chi_B\ \delta(\varepsilon_{i\sigma} - \varepsilon_{B\sigma})\delta(\varepsilon_{i-\sigma} - \varepsilon_{B-\sigma}) \tag{3}$$

where (in the absence of magnetic field) $\chi_A = \chi_B = 1/2$ and

$$\varepsilon_{A\sigma} = \varepsilon_{B-\sigma} = \frac{1}{2}\ U(n-\mu) \tag{4}$$

$$\varepsilon_{A-\sigma} = \varepsilon_{B\sigma} = \frac{1}{2}\ U(n+\mu) \tag{5}$$

The local moment μ is determined by the self-consistency condition[6]

$$\mu = \int_{-\infty}^{E_F} [\varrho_{\sigma}^{A}(E) - \varrho_{-\sigma}^{A}(E)]\, dE \tag{6}$$

For the evaluation of μ it is obvious that we must have the electronic density of states (DOS) $\varrho^A{}_\sigma$ and $\varrho^A{}_{-\sigma}$ which represent the average DOS for the σ and $-\sigma$ spin state respectively, at a site where $\varepsilon_{i\sigma} = \varepsilon_{A\sigma}$ and $\varepsilon_{i-\sigma} = \varepsilon_{A-\sigma}$. The Fermi level E_F which appears in eqn.(6), is determined by the input parameter n which specifies the average number of electrons per lattice site, i.e.

$$n = \int_{-\infty}^{E_F} (\varrho_\sigma^A + \varrho_{-\sigma}^A) dE \tag{7}$$

The electron DOS's $\varrho^A{}_\sigma = \varrho^B{}_{-\sigma}$ and $\varrho^A{}_{-\sigma} = \varrho^B{}_\sigma$ can be obtained from the following equations:

$$\varrho_\sigma^\alpha (E) = -\frac{1}{\pi} \lim_{s\to 0} \text{Im} <G_{j\sigma}(E+is)>_{j=a} \quad , \quad \alpha = A \text{ or } B \tag{8}$$

$$G_{j\sigma}(E+is) = <j\sigma \left| \frac{1}{E+is-H_\sigma} \right| j\sigma> \tag{9}$$

where H_σ results from eqn.(1) with the approximation of eqn.(2) and without the spin summations.

In the averaging process implied by eqn.(8), for which we employ the conditional CPA[2], off-site correlations are incorporated in H_σ through a parameter $P_{A/B}$ which gives the probability a lattice site being A under the condition that a given nearest neighbour site is B. Thus for $P_{A/B}=1$ we have long range order of antiferromagnetic order. All other intermediate cases for $P_{A/B}$ are allowed as potentialities. The conditional CPA calculates conditionally averaged Green's functions corresponding to one or more site energies ε_i kept fixed. The other site energies ε_j ($j \neq i$) over which the average is performed are replaced by two self-energies Σ_1 and Σ_2 periodically arranged. Σ_1 and Σ_2 are determined by the CPA condition which ensures that this replacement creates no scattering on the average in the vicinity of the site i.

Finally, the ground state energy per lattice site is obtained from the expression

$$E = \frac{U}{4} (n^2+\mu^2) + \int_{-\infty}^{E_F} [\varrho_\sigma^A(E') + \varrho_{-\sigma}^A(E')] E' \, dE' \tag{10}$$

where n and μ are the results of eqns.(7) and (6) respectively.

It should be noted that the energy E depends on $P_{A/B}$. For given U/V and n the energetically more favorable state can be obtained by minimizing E with respect to $P_{A/B}$ (and μ; the minimization with respect to μ is equivalent to the self-consistency eqn.(6)).

The numerics of this model become much easier if a Bethe lattice approach is employed to describe the real lattice. For the Bethe lattice analytic expressions exist for the matrix elements of the Green's function in the site representation[7]. When these expressions are used, the CPA condition takes the form of two polynomial equations (with respect to the self-energies) which are solved using Newton-Ramson's iterative scheme.[2,8]

The results which are presented in the present work refer to a Bethe lattice with connectivity K=5, i.e. with Z=K+1=6 nearest neighbours. The half-bandwidth $B=2V\sqrt{K}$ and is taken equal to 2.

RESULTS AND DISCUSSION

Following the method we have described in the previous section we calculate the ground state energy, $E=E(n, U/V, P_{A/B})$ of the system according to eqn.(10) for the given input parameters n, U/V and $P_{A/B}$. The ground state for a given n and U/V is the one for which $E(n, U/V, P_{A/B})$ is minimum as a function of $P_{A/B}$. The ground state energy for a given n and U/V is denoted by E_o and the corresponding $P_{A/B}$ by $P^*{}_{A/B}$.

The results we have obtained for the ground state of our system in terms of U/V and n can be summarized in a phase diagram shown in Fig.1. The dotted lines which coincide with the solid lines ab and part of ad in Fig.1 separate the (U/V-n)-plane into three regions: the Paramagnetic or Pauli (P), the Ferromagnetic (F) and the Antiferromagnetic (AF) region. In each region the indicated state is the energetically lower than the other two. The solid lines separate the three regions from a short range or a spin-glass (SG) phase. The SG phase is characterized by $0<P^*_{A/B}<0.85$ and non-zero magnetic moments. The shaded area corresponds to $P^*_{A/B}=0.5$. On the other hand, the regions between the shaded area and the solid lines exhibit short range order characterized by $P^*_{A/B}\approx 0.85$ to the right and by $P_{A/B}\approx 0.2$ to the left of the shaded area. This allows us to consider the region to the right of the shaded area as an AF phase with no perfect order. In this short range region the electron DOS of the system develops "pseudogaps" and the system becomes stable by occupying the lowest bands. The Fermi level of the system is located at a "pseudogap" for such states.

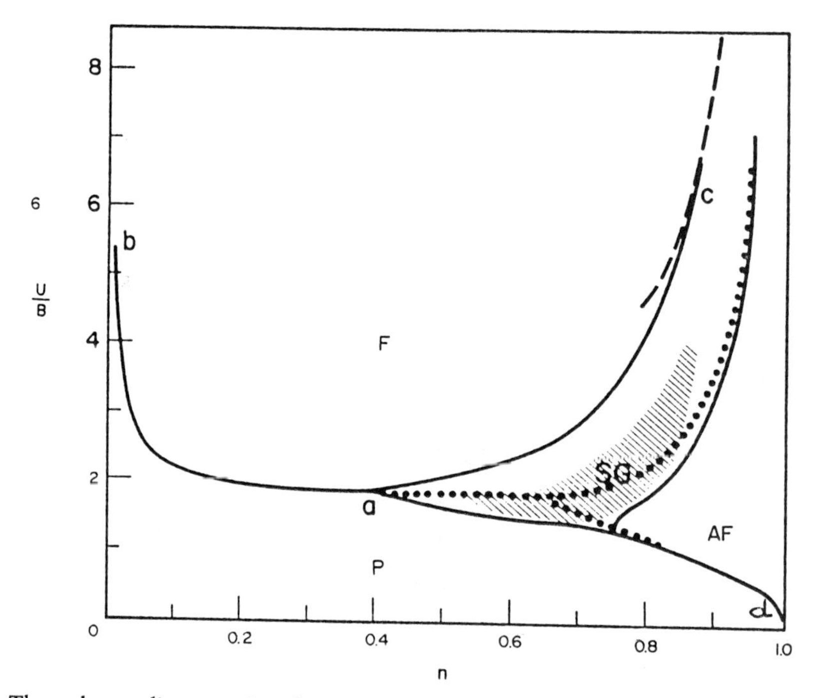

Fig.1. The phase diagram in the U-n plane of the Hubbard model for a Bethe lattice of connectivity K=5 and half bandwidth B=2. In addition to the usual Paramagnetic (P), Ferromagnetic (F) and Antiferromagnetic (AF) phases, a short range order phase is found which exhibits regions where a spin-glass (SG) or spin-liquid phase (shaded region) appears as a possible ground state of the system (see text).

It should be emphasized that the boundaries of the short range and SG regions are interfaces where we observe a gradual transition from one type of state to the other.

In order to examine more systematically the results of our calculations we have focused to a system with a specific value of U/B and a number of electrons per site which can be varied. This means that we have concentrated on a horizontal line (at the specific U/B) on the phase diagram; namely we examined the case for U/B=2.5. Our results are summarized in Figures 2-5.

In Fig.2 we present the calculated values $P^*_{A/B}$ as a function of the electrons per lattice site for U/B=2.5. According to Fig.2 the off site correlations exhibit a more or less abrupt change at $n \approx 0.65$ from a Ferromagnetic to a short range state. The latter state is found to have $\mu \neq 0$ and $P^*_{A/B} \in [0.4\text{-}0.6]$. As n increases we observe a gradual transition to the antiferromagnetic state.

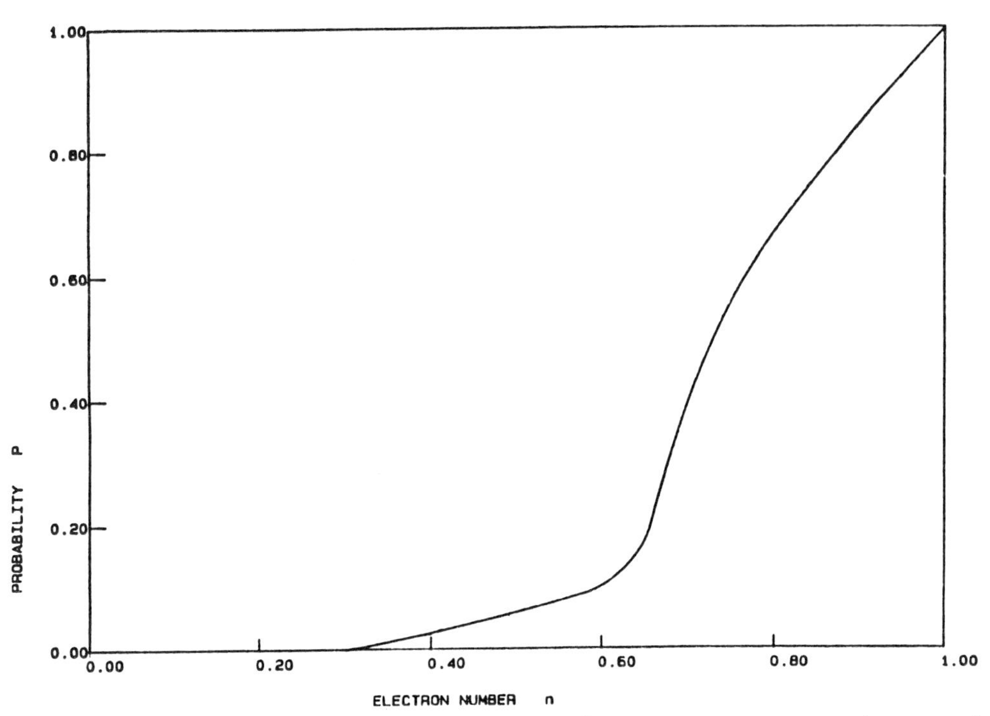

Fig.2. Variation of the short range parameter $P^*_{A/B}$ (which minimizes the ground state energy for given n and U/B) as a function on the electron number n for a Bethe lattice of connectivity K=5 and U/B=2.5.

In Fig.3 we show the coupling constant J of an equivalent Ising model as a function of the electrons per lattice site for U/B=2.5. The J constant has been calculated at T=0 according to the formula[9]

$$J = \frac{1}{Z} \left(\frac{\partial E}{\partial P^*_{A/B}}\right)_{T,U} \tag{11}$$

The dashed parts of the graph, where J becomes negative, denote regions where our preliminary results are calculated with large uncertainty. Nevertheless, the main feature of this graph is the appearance of significant J values for $n \geq 0.85$ where (according to Fig.2) the ground state exhibits short range correlations of near AF order.

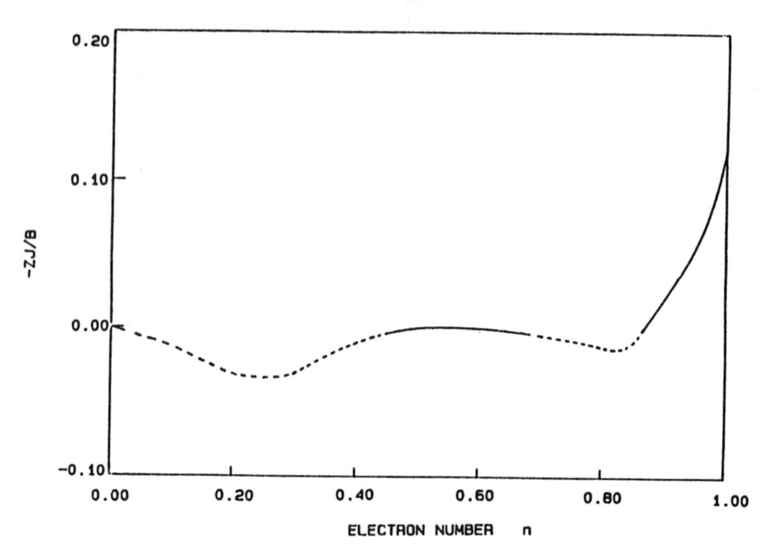

Fig.3. Variation of the coupling constant J of an equivalent Ising model as a function of the electron number n for a Bethe lattice of connectivity K=5, U/B=2.5 and corresponding $P^*_{A/B}$ shown in Fig.2. The dotted parts of the graph indicate numerical uncertainties currently under investigation.

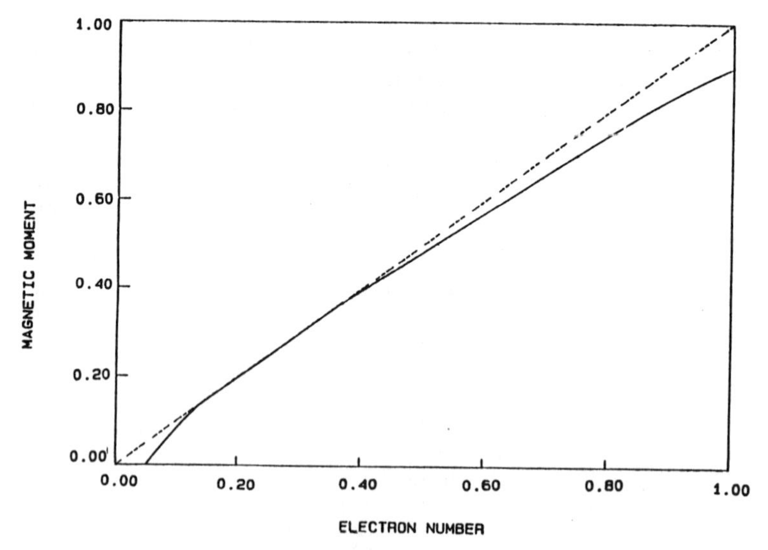

Fig.4. Variation of the magnetic moment μ per lattice site as a function of the electron number n for a Bethe lattice of connectivity K=5, U/B=2.5 and corresponding $P^*_{A/B}$ shown in Fig.2.

The behaviour of the local moment μ (without quantum fluctuations) as a function of n for U/B=2.5 is shown in Fig.4. Here we observe that $\mu(n)$ follows the μ=n line (dotted line indicating no double occupancy) rather closely, departing from it for small n (where the system becomes paramagnetic) and to a small but increasing extent as n exceeds 0.5 and approaches unity.

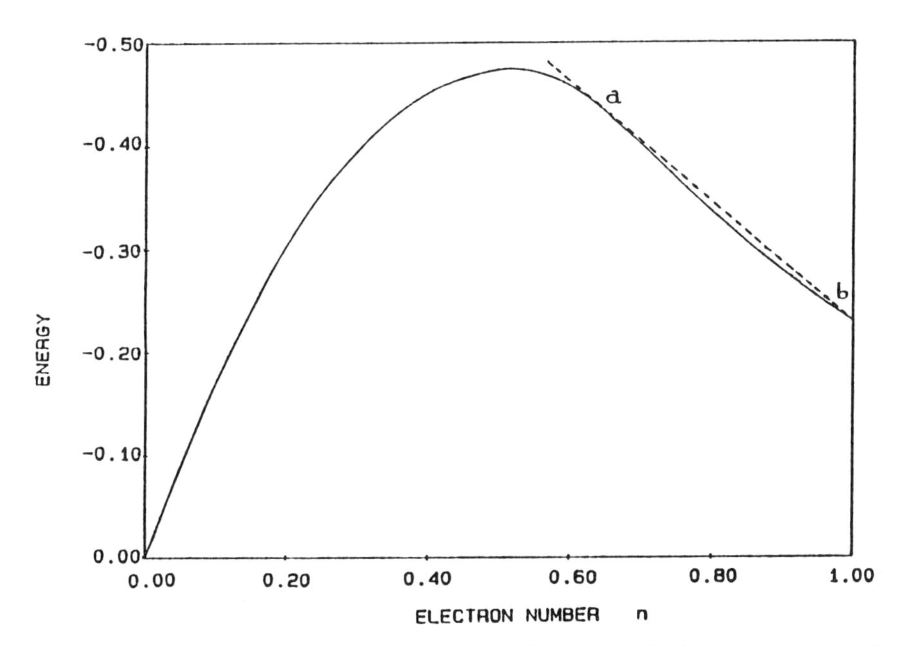

Fig.5. The ground state energy E_0 as a function of the electron number n corresponding to $P^*_{A/B}$ and μ shown in figures 2 and 4 respectively.

Finally, in Fig.5 we present the results for the ground state energy E_0 as a function of n for U/B=2.5. One of the major characteristics of this graph is that for n between roughly 0.60 and 1.00 the ground state of the system seems to become unstable towards phase separation. Indeed, the dashed line ab describing a two phase system (phase a with $P^*_{A/B}\approx 0.2$ and phase b with $P^*_{A/B}\approx 1.00$) seems to correspond to lower energy that the single phase system. Our preliminary results although indicative of phase separation cannot be considered as conclusive due to numerical uncertainties. We point out, however, that similar results for phase separation have been obtained by Marder et al (see elsewhere in the present volume) for the t-J model. Preliminary results for larger U/B indicate that this effect of phase separation becomes more pronounced. We have also found similar trends in the 1-D case. The results of these calculations which are still in progress will be discussed elsewhere.

REFERENCES

1. A. N. Andriotis, P.N. Poulopoulos and E.N. Economou, Solid State Communications 39, 1175 (1981).
2. C. T. White and E.N. Economou, Phys. Rev. B15, 3742 (1977).
3. S. H. Liu, Phys. Rev. B17, 3629 (1978).
4. P. W. Anderson, G. Baskaran, Z. Su and T. Hsu, Phys. Rev. Letters 58, 2970 (1987).
5. V. J. Emery, Phys. Rev. Letters 58, 2974 (1987).
6. M. Cyrot, Phil. Mag. 25, 1031 (1972).
7. E. N. Economou, "Green's Functions in Quantum Physics", Springer-Verlag 1983.
8. E. N. Andriotis and J.E. Lowther, J. Phys. F: Metal Phys. 16, 1189 (1986).
9. E. N. Economou, C.T. White and R.R. DeMarco, Phys. Rev. B18, 3946 (1978); C. T. White and E.N. Economou, Phys. Rev. B18, 3959 (1978).

EXACT MICROSCOPIC CALCULATION OF SPIN WAVE FREQUENCIES AND LINEWIDTHS IN THE TWO-DIMENSIONAL HEISENBERG ANTIFERROMAGNET AT LOW TEMPERATURE

Thomas Becher and George Reiter

Physics Department and Texas Center for Superconductivity
University of Houston
Houston, TX 77204-5504

INTRODUCTION

We present here exact results for the temperature dependent spin wave frequencies and damping in the 2-D Heisenberg antiferromagnet. Results for the 2-D ferromagnet have been presented previously.[1] We use methods that were first developed in 1-D classical systems,[2] where the dynamics at all wavelengths were obtained, and that have been used to discuss the 2-D antiferromagnet at T=0.[3] We find that in 2-D, the longest wavelength behavior is not readily determined by these methods, and so we will restrict ourselves to wavelengths such that $q\xi >> 1$, where ξ is the coherence length.

The Hamiltonian is

$$H = J \sum_{i,\delta} \vec{S}_i \cdot \vec{S}_{i+\delta} \tag{1}$$

We wish to calculate

$$R(q,\omega) = \mathrm{Re} \int_0^{\infty} e^{i\omega t} <\vec{S}q(t)|\vec{S}\text{-}q>dt \tag{2}$$

It may be shown that

$$R(q,\omega) = \mathrm{Re}\, i\, \chi_q \frac{1}{\omega - \dfrac{\omega_q^2}{\omega + \gamma_q(\omega)}} \tag{3}$$

where $\chi_q = <\vec{S}_q|\vec{S}_q>$, $\omega_q^2 = <\dot{\vec{S}}_q|\dot{\vec{S}}_q>/\chi_{q_1}$, $\dot{\vec{S}}_q = \partial\vec{S}_q/\partial t$, $<A|B> \equiv <\int_0^{\beta} e^{\tau H}A^{+}e^{-\tau H}B d\tau>$

and $\beta = (1/KT)$. The function $\gamma_q(\omega)$ is physically[2] the decay rate for a spin current of

Dynamics of Magnetic Fluctuations in High-Temperature Superconductors
Edited by G. Reiter *et al.*, Plenum Press, New York, 1991

wavevector q and frequency ω. In terms of the real and imaginary parts of $\gamma_q(\omega) = \gamma_q'(\omega) + i\gamma_q''(\omega)$

$$R(q,\omega) = \frac{\chi_q \omega_q^2 \gamma_q''(\omega)}{[\omega^2 - \omega_q^2 + \omega\gamma_q'(\omega)] + [\omega\gamma_q''(\omega)]^2} \tag{4}$$

The spin wave frequency $\omega_q(T)$ will be defined as the zero of $[\omega^2 - \omega_q^2 + \omega\gamma_q'(\omega)]$, and the full width at half maximum is then $\gamma_q''(\omega_T)$. These definitions make sense when the variation with frequency of $\gamma_q(\omega)$ over the width of the line is negligible. One can show that this will be the case if the temperature is sufficiently low. The response function (4) is then that of a damped harmonic oscillator with zero initial velocity and the lineshape can be described as a product of Lorentzians, a fact that has already been noticed by Wysin and Bishop in their simulations. If the system were classical $\gamma_q(\omega)$ would vanish at zero temperature, where the coherence length is infinite and the excitation would be perfectly well defined at all wavelengths. For the quantum system, it will have a non-vanishing value, proportional to 1/S at T=0. We will obtain $\gamma_q(\omega)$ as a perturbation expansion in the number of spin wave excitations, which corresponds to an expansion in 1/S at T=0 and KT/JS^2 in the classical limit.

It is not altogether obvious that a spin wave expansion can be carried out at finite temperature where there is no long range order. That it is possible to do so was conjectured independently by G. Reiter[1] and S. Elitzur,[5] and proven by F. David.[6] It is only necessary to calculate rotationally invariant correlation functions, and the spin wave perturbation theory will be finite term by term.

It is then straightforward to calculate the equilibrium averages that appear in Eq. (3) We find

$$<\dot{S}_q|\dot{S}_{-q}> = 2(J_0 - J_q)<\vec{S}_i \cdot \vec{S}_{i+1}> = \begin{cases} 2(J_0 - J_q)(1 + .079/S)^2 & T=0 \\ 2(J_0 - J_q)(1 - KT/4JS)^2 & \text{Classical} \end{cases} \tag{5}$$

and

$<S_q|S_q> =$

$$\frac{2}{C_1(J_0+J_q)}\left(1-\frac{1}{2NS}\sum_{q_1}\frac{J_0-\sqrt{J_0^2-J_{q_1}^2}}{\sqrt{J_0^2-J_{q_1}^2}}\right)-\frac{1}{2NS}\sum_{q_1}\frac{J_{q_1}J_{q-q_1}-J_0^2+\sqrt{J_0^2-J_{q_1}^2}\sqrt{J_0^2-J_{q-q_1}^2}}{\sqrt{J_0^2-J_{q_1}^2}\sqrt{J_0^2-J_{q-q_1}^2}\left(\sqrt{J_0^2-J_{q_1}^2}+\sqrt{J_0^2-J_{q-q_1}^2}\right)} \quad T=0$$

$$\frac{2}{\left(1-\frac{KT}{4JS^2}\right)(J_0+J_q)}\left(1-\frac{KT}{NS^2}\sum_{q_1}\frac{J_0}{J_0^2-J_{q_1}^2}\right)+\frac{KT}{NS^2}\sum_{q_1}\frac{\left(J_0^2-J_{q-q_1}J_{q_1}\right)}{\left(J_0^2-J_{q_1}^2\right)\left(J_0^2-J_{q-q_1}^2\right)} \quad \text{Classical} \tag{6}$$

where $C_1 = [1+.079/S]$ and $J_q = 2J[\cos q_x + \cos q_y]$. The first term in the above expression is the transverse susceptibility, the second the longitudinal susceptibility with longitudinal and transverse defined with respect to a particular (arbitrary) choice of the spin direction about which the spin wave theory is done. Although the terms are separately finite at T=0, they are not at finite temperature and we see the necessity for calculating rotationally invariant correlation functions. The prefactor C_1 or $(1\text{-}KT/4JS^2)$ is the lowest order Hartree-Fock contribution to the magnon self energy in the T=0 or classical limits, respectively. More generally, the magnon energy, to lowest order in the spin wave expansion is

$$\varepsilon_q = \varepsilon_q^o \left(1 + \frac{1}{2S}\left(1 - \frac{1}{2N}\sum_{q_i} (2n_{q_1} + 1)\sqrt{4 - (\cos q_x + \cos q_y)^2}\right.\right. \tag{7}$$

where $\varepsilon_q^o = 2JS\sqrt{4 - (\cos q_x + \cos q_y)^2}$.

We will make a distinction between the magnon propagator, which will be defined as $<a_q(t)|a_q>$, where a_q is an annihilation operator for a magnon, and the spin wave propagator, Eq. (4). They are the same to the lowest order in the spin wave expansion. The leading term in the expansion of $\gamma_q(\omega)$ has been derived in Ref. (3) and is

$$\gamma_q(\omega) = - \frac{S^4}{4<\vec{S}_q|\vec{S}_q>}\left\{\sum_{q_1 q_2} \left(\Gamma^+_{q_1 q_2}\right)^2 \frac{\omega}{\omega^2 - (\varepsilon_{q_1} + \varepsilon_{q_2})^2}\frac{\left[n_{q_1} + n_{q_2} + 1\right]}{\varepsilon_{q_1} + \varepsilon_{q_2}}\delta(\vec{q}\text{-}\vec{q}_1\text{-}\vec{q}_2)\right.$$

$$\left. + \sum_{q_1 q_2} \left(\Gamma^-_{q_1 q_2}\right)^2 \frac{\omega}{\omega^2 - (\varepsilon_{q_1} + \varepsilon_{q_2})^2}\frac{n_{q_2} - n_{q_1}}{\varepsilon_{q_1} + \varepsilon_{q_2}}\right\}\delta(\vec{q}\text{-}\vec{q}_1\text{-}\vec{q}_2) \tag{8}$$

where

$$\Gamma^{\pm}_{q_1 q_2} = (J_0\text{-}J_q)^{1/2}(J_0\text{+}J_{q_2})^{1/2}[2J_0\text{-}J_{q_2}\text{+}J_{q_1}](J_{q_2}\text{-}J_{q_1})$$

$$\pm (J_0\text{-}J_q)^{1/2}(J_0\text{-}J_{q_2})^{1/2}[2J_0\text{+}J_{q_2}\text{-}J_{q_1}](J_{q_2}\text{-}J_{q_1})$$

$$- (J_0^2\text{-}J_q^2)\,[(J_0\text{-}J_{q_1})^{1/2}(J_0\text{+}J_{q_2})^{1/2} \pm (J_0\text{+}J_{q_1})^{1/2}(J_0\text{-}J_{q_2})^{1/2}] \big/ (J_0^2\text{-}J_{q_1}^2)^{1/4}(J_0^2\text{-}J_{q_2}^2)^{1/4} \tag{9}$$

and n_q is the Bose occupation factor. This term arises from the longitudinal part of the spin wave expansion for the rotationally invariant correlation function that defines $\gamma_q(\omega)$. The transverse part is of higher order in 1/S or KT/JS^2.

The expression is correct in either one or two dimensions. One need only use the appropriate expression for J_q in $\Gamma^{\pm}$ and Eq. (5,6,7). The one dimensional result leads to a damping and frequency shift proportional to KT, which is also ξ^{-1} in this case. To the extent that higher order terms in the expansion of γ are not singular at the spin wave frequency, this suffices to obtain the damping and frequency shift to leading order exactly. The lowest order term is insufficient in two dimensions, where, as we shall see, it is necessary to go to next order to obtain the damping.

We will restrict ourselves subsequently to wavelengths near $\vec{q} = (\pi,\pi)$. With $\vec{k}=\vec{q}\text{-}\vec{\pi}$, we find for the imaginary part of $\gamma_q(\omega)$

$$\gamma''_q(\omega) = \frac{1}{16JS^2} \frac{(\omega^2 - c^2k^2)^{3/2}}{\omega} \theta(\omega^2 - c^2k^2) \qquad T=0$$

$$\gamma''_q(\omega) = \frac{KT}{\sqrt{2}JS^2\omega} \Big((\omega^2 - (ck)^2)\,\theta(\omega^2 - c_T^2k^2) \qquad (10)$$

$$+ (c_T^2k^2 - \omega^2)\theta(c_T^2k^2 - \omega^2) 2\tan^{-1}\big(|\omega|/(c_T^2k^2 - \omega^2)^{1/2}\big)\Big) \qquad \text{Classical}$$

Here c is the magnon velocity determined from Eq. (7), at T=0, $2\sqrt{2}JS(1+.079/S)$ and c_T is the classical value obtained from (7), $2\sqrt{2}JS(1\text{-}KT/4JS^2)$.

Note that $\gamma''_q(\omega)$ vanishes at the magnon frequency, ck or c_Tk, in both cases, so that this leading term does not appear to contribute any damping. This is in distinction to the case in one dimension, where the lowest order term does not vanish at the magnon frequency. In fact, the situation is not so simple. $\gamma''_q(\omega)$ should be evaluated at the actual spin wave resonance frequency, $\omega_q(T)$ as defined after Eq. (4), and it is not obvious to us, that this is the same as the magnon frequency, as determined to lowest order by Eq. (7). $\omega_q(T)$ is determined by using (5), (6) and (8)

$$\omega_q^2 = (\omega_q^0)^2 \Bigg[\left(1 + \frac{.079}{S}\right)^3 + \frac{1}{2NS} \sum_{q_1} \frac{J_0 - \sqrt{J_0^2 - J_{q_1}^2}}{\sqrt{J_0^2 - J_{q_1}^2}}$$

$$+ \frac{S^2}{8(J_0 - J_q)} \sum_{q_1q_2} \left(\Gamma^+_{q_1q_2}\right)^2 \frac{\delta(q\text{-}q_1\text{-}q_2)}{\left[\varepsilon_q^2 - (\varepsilon_{q_1} + \varepsilon_{q_2})^2\right]\left[\varepsilon_{q_1} + \varepsilon_{q_2}\right]}\Bigg] \qquad T=0$$

(11)

$$= (\omega_q^0)^2 \Bigg[\left(1 - \frac{KT}{4JS^2}\right)^3 + \frac{KT}{NS^2} \sum_{q_1} \frac{J_0}{J_0^2 - J_{q_1}^2} - \frac{\frac{J_0 + J_q}{2}\left(J_0^2 - J_{q_1}J_{q-q_1}\right)}{\left(J_0^2 - J_{q_1}\right)\left(J_0^2 - J_{q-q_1}\right)}$$

$$+ \frac{KTS^2}{8(J_0 - J_q)N} \sum_{q_1q_2} \frac{\delta q\text{-}q_1\text{-}q_2}{\varepsilon_{q_1}\varepsilon_{q_2}} \left(\frac{\left(\Gamma^+_{q_1q_2}\right)^2}{\varepsilon_q^2 - (\varepsilon_{q_1} + \varepsilon_{q_2})^2} + \frac{\left(\Gamma^-_{q_1q_2}\right)^2}{\varepsilon_q^2 - (\varepsilon_{q_1} - \varepsilon_{q_2})^2}\right)\Bigg] \qquad \text{Classical}$$

We have omitted the term coming from the longitudinal susceptibility in Eq. (11) for T=0, as it is of order (1/k) and hence negligible for the case we will consider.

The limit $k \to 0$ of $\left(\Gamma^+_{q_1q_2}\right)^2$ is $64(J_0^2 - J_{q_1}^2)J_{q_1}^2$ with $q_2 = \vec{\pi}\text{-}\vec{k}\text{-}q_1$, while $\left(\Gamma^-_{q_1q_2}\right)^2$ vanishes as k^4. The contribution from the last term in the classical expression in Eq. (10) is therefore zero. Thus only the terms corresponding to the decay into a pair of magnons contribute to the frequency shift. We find for the T=0 case, that the factor multiplying $(\omega_q^0)^2$ in (11) is

$$\left(1+\frac{.079}{S}\right)^3+\frac{1}{2NS}\sum_q\left[\frac{\sqrt{J_0^2-J_q^2}}{J_0}-1\right] \tag{12}$$

The integral is (-.079/S). Thus, to first order in 1/S, $\omega_q(T=0)=\varepsilon_q^0(1+.079/S)$. The frequency shift defined by (4) for the spin wave and by (7) for the magnon are identical. As a consequence, there is no contribution to the damping from (10), as $k\rightarrow 0$ and T=0. This was reported incorrectly in Ref. (3). The cancellation that occurs above was not noted, and the shift was evaluated numerically with a faulty program. The same sort of cancellation occurs in the classical limit. The prefactor becomes

$$\left(1-\frac{KT}{4JS^2}\right)^3+\frac{KT}{J_0NS^2}\sum_q(1)=1-2\left(\frac{KT}{4JS^2}\right)+0\left(\frac{KT}{4JS^2}\right)^2 \tag{13}$$

and hence the renormalization frequency is, for $\vec{q}\simeq\vec{\pi}$

$$\omega_q(T)=\varepsilon_q^0(1-KT/4JS^2) \tag{14}$$

again agreeing with the result of Eq. (7) in the appropriate limit. There is, therefore, no contribution from (10) to the linewidth. For general values of T, it can be shown that Eq. (7) for the spin wave frequency holds, near k=0, for all temperatures from T=0 to temperatures sufficiently high that the classical limit is appropriate.

That the lowest order expression for the damping, arising from the contribution of the longitudinal spin fluctuations to $\gamma_q(\omega)$, vanishes, is distinctly different from the result in one dimension. There, $\gamma_q''(\omega_q)$ is not generally zero, and the frequency shift is irrelevant in determining the damping. The damping of the spin wave is then proportional to T.[7] The magnon frequency shift turns out to agree with that obtained from the spin wave dispersion relation, but the magnon linewidth must be of order T^2, that is, much smaller than the calculated and observed value for the spin waves.

We know of no requirement that the pole in the magnon propagator and the pole in the spin wave propagator, Eq. (4), have the same location for the real or the imaginary parts. The frequencies that go into the perturbation expansion of (4) are those that appear in the magnon propagator, but no self consistency is required with the values $\omega_q(T)$, and certainly in one dimension, no self consistency is possible for the imaginary parts. It seems possible, therefore, that there is a finite damping for the spin waves at T=0, arising from the longitudinal fluctuations and a shift of the spin wave resonance frequency above the magnon frequency due to the higher order anharmonicity. This shift would have to be or order $(1/S)^2$, however, or in the classical limit $(KT/JS^2)^2$, and so would produce a damping of order $(1/S)^3$ or $(KT/JS^2)^3$ when inserted in (10). To calculate the leading term in the expression for the damping, it suffices then, to calculate $\gamma_q''(\omega)$ to second order.

The exact expression for $\gamma_q(\omega)$ involves the correlation function[2]

$$<\delta\vec{S}_{q_1}\cdot\vec{S}_{q_2}\vec{S}_{q_3}(t)Q|Q\vec{S}_{q_3}'\delta\vec{S}_{q_2}'\cdot\vec{S}_{q_1}''> \tag{15}$$

where the projection operator Q projects out the part of the operator that is proportional to $\vec{S}_q$, and $\delta\vec{S}_{q_1}\cdot\vec{S}_{q_2}=[\vec{S}_{q_1}\cdot\vec{S}_{q_2}-<\vec{S}_{q_1}\cdot\vec{S}_{q_2}>]$. To obtain (8), the operator $\vec{S}_{q_3}(t)\cdot S_{q_3}'$ was replaced by $S^2 S_{q_3-\pi}S_{q_3'-\pi}$, and the remaining operators calculated to lowest order in spin wave theory. This is the basis for the statement that (8) arises from the contribution of longitudinal fluctuations to $\gamma_q(\omega)$. In next order, this term is replaced by the spin wave expansion of the transverse components, and the other operators are expanded as before. There are other possibilities that contribute to the same order in the number of spin waves, but these are all corrections to (8), give contributions that will vanish at the spin wave frequency, and can be neglected.

The effect of the projection operator is to eliminate all equal time contractions so that the remaining averages all involve three spin wave operators. The terms that correspond to the creation or annihilation of three spin waves give a contribution that vanishes at the spin wave frequency because of the convexity of the frequency spectrum, which implies that $\omega_q \leq \omega_{q_1}+\omega_{q_2}+\omega_{q_3}$ when $\vec{q}=\vec{q}_1+\vec{q}_2+\vec{q}_3$. These are the only terms that survive at T=0, so there is no damping at T=0, to this order.

The classical expression, taking the long wavelength limit for all the magnon dispersion relation is

$$\gamma_q(\omega)=\frac{\pi}{2}J\left(\frac{KT}{JS^2}\right)^2\int dq_1dq_2dq_3\delta(q-q_1-q_2-q_3)\left[\delta\left(\frac{\omega}{c}+(q_1+q_2-q_3)\right)+\delta\left(\frac{\omega}{c}-(q_1+q_2-q_3)\right)\right]$$

$$\frac{\left[qq_1-\vec{q}\cdot\vec{q}_1\right]^2+\left[qq_2+q\cdot q_2\right]^2+\left[qq_3+q\cdot\vec{q}_3\right]^2}{q_1^2q_2^2q_3^2} \tag{16}$$

This result may be shown to be exactly that of Tyc and Halperin,[8] at $\omega=\varepsilon_q$. They derived their result by assuming that the expressions for the damping derived for the ordered state in three dimensions, could be applied to the two dimensional problem if $q\xi>>1$. The result above provides a proof (if the details were spelled out) of this assumption. We refer the reader to their paper for the evaluation of the integrals. The result is that the damping can be expressed as, near $\vec{q}=\vec{\pi}-\vec{k}$

$$\gamma_q''(\omega_q)=ck\left(\frac{KT}{JS^2}\right)^2(a_1+a_2\ln q) \tag{17}$$

with the parameters a_1 and a_2 depend upon whether or not the magnitude of the damping of the intermediate magnons at the zone boundaries exceeds the incoming spin wave frequency. Self consistency effects are included in Tyc and Halperin's work that are not contained in Eq. (16), but may easily be added.

The coefficients a_1 and a_2 are also dependent upon having made the long wavelength approximation to obtain (16), which is not essential to do. Exact values will be presented elsewhere.

We stated erroneously previously,[3] based upon an approximation to the exact second order calculation sketched above, that the damping did not vanish as $k \rightarrow 0$.

Equation (17) is, we claim, an exact result as $T \rightarrow 0$, and the values of a_1 and a_2 can be obtained by doing numerically the integrals involved. It differs qualitatively from the results due to Grempel[9] in the present volume, and from previous derivations by Arovas and Auerbach,[10] based upon uncontrolled approximations.

Conclusion

The damping and frequency shifts in the 2D Heisenberg antiferromagnet can be obtained by a systematic spin wave perturbation theory, free of divergences. The lowest order frequency shift is linear in temperature in the classical regime, and proportional to 1/S at T=0. It is given by Eqs. (12) and (14), in these limits, and more generally, by Eq. (7), at least for $\vec{q} \simeq (\pi,\pi)$. The lowest order damping is proportional to $(KT/JS^2)^2$ in the classical limit, and vanishes at least to second order in $(1/S^2)$ at T=0. It is given by Eq. (17) for the classical case. These results are distinctly different from the one dimensional case, where the damping function $\gamma_q(\omega_q)$ is proportional to (KT/JS^2) classically, and leads to a power law singularity at T=0 for finite spin values.[11]

Acknowledgement

This work was supported by the Texas Center for Superconductivity, under Prime Grant No. MDS972-8-G-0002 from the U. S. Defense Advanced Research Projects Agency and the State of Texas, and the Division of Materials Sciences, U.S. Department of Energy under Contract No. DE-AC02-76CH00016.

References

1. G. Reiter, Phys. Rev. B. 21, 5356 (1980).
2. G. Reiter, A. Sjolander, Phys. Rev. Letts. 39, 1047 (1977); G. Reiter and A. Sjolander, J. Phys. C 13, 3027 (1980).
3. T. Becher and G. Reiter, Phys. Rev. Letts. 63, 1004 (1989).
4. G. Wysin and A. R. Bishop (preprint).
5. S. Elitzur, Nucl. Phys. B 212, 501 (1983).
6. F. David, Commun. Math. Phys. 81, 149 (1981).
7. The exception to this occurs in the ferromagnet at $q = \pi$. There $\gamma''_q(\omega_q)$ does not vanish. The damping is non zero because the frequency is shifted, and is actually proportional to $T^{3/2}$.
8. S. Tyc and B. I. Halperin (preprint).
9. D. Grempel (this volume).
10. A. Auerbach and D. P. Arovas, Phys. Rev. Lett. 61, 617 (1988).
11. K. Stuart, Thesis, Texas A&M University.

MAGNETIC EXCITATIONS IN THE DISORDERED PHASE OF THE 2-D HEISENBERG ANTIFERROMAGNET

D.R. Grempel

Centre d'Etudes Nucléaires, DRF/SPh-MDN, F-38041 Grenoble
France

I-INTRODUCTION

The discovery of the new high temperature superconductors has been at the origin of renewed interest in the physics of the 2-D quantum antiferromagnet. At low temperatures La_2CuO_4 and $YBa_2Cu_3O_6$ are anisotropic insulating antiferromagnets with Néel temperatures of 245 and 415 °K, respectively[1,2]. The in-plane exchange constant is very large, $J_{\parallel} \sim 2000°K$, and the anisotropy ratio $J_{\perp}/J_{\parallel} \sim 10^{-5}$ [2]. Due to this large anisotropy these compounds are very accurately bidimensional.

The magnetic excitations of $La_{2-x}Ba_xCuO_4$ [3] and $YBa_2Cu_3O_{6+x}$ [4] have been studied in great detail by inelastic neutron scattering. The important experimental observation is that the dispersion and amplitude of the magnetic excitations are very close to those of 2-D spin waves even above the 3-D Néel temperature. These excitations are very sharp in the undoped compounds and very sensitive to the presence of disorder that leads to softening and damping. The purpose of this paper is to describe the nature of these excitations and to account for some of the experimental observations.

The existence of well defined propagating excitations in the disordered phase at low temperatures is characteristic of low dimensional systems. It arises because, at a temperature $T \ll J$, spin wave states of wavevector $q \gtrsim q_c = kT/hc$ are essentially unpopulated and, as a result, an excitation with $q \gtrsim q_c$ can propagate without encountering others of comparable energy to interact with. Since spin wave interactions are effective mostly among those of nearby energy, we expect that short wavelength spin waves be underdamped. Notice that in the three dimensional case the energy of a spin wave at the top of the spectrum is of the order of Tc. Thus one may say that by the time we reach a temperature $T \gg Tc$ all spin waves have already been excited and the picture just described is no longer applicable.

The above arguments do not mean that the finite temperature excitations are entirely equivalent to those at T = 0. After all, they propagate in a magnetically distorted background and, as a consequence, they must interact with fluctuations whose typical wavelength is the scale of the disorder, the correlation length ξ. However, since at low temperatures the latter is very large, the relevant modes are much slower than the spin waves that we study and this separation of time scales makes it possible to

Dynamics of Magnetic Fluctuations in High-Temperature Superconductors
Edited by G. Reiter *et al.*, Plenum Press, New York, 1991

carry the analysis quite far. One may imagine proceeding in three stages : first, one works out the parameters of the effective Hamiltonian that controls the dynamics of the slow modes by integrating out the spin waves as in usual renormalisation group calculations; then, one computes the response of the faster modes to an external probe in the presence of a generic slow fluctuation; and, finally, one performs an appropriate statistical average over all the possible configurations of the slow fluctuations. Although this seems quite general it is usually impossible to carry out the second step. Here, the clear separation of time scales refered to allows us to use a form of the adiabatic approximation to go over the difficulties.

The dynamics of the quantum 2-D Heisenberg antiferromagnet in the paramagnetic phase has been recently studied by Tyc, Halperin, and Chakravarty [5] by numerical simulation supplemented by considerations of dynamic scaling. Their paper deals with the case $\omega \ll kT$, where the mapping of the 2-D Heisenberg antiferromagnet on to the classical lattice rotor model is valid. In their case spin wave interactions are the main reason for damping. Dealing with the opposite limit, this paper is, in a sense, complementary to theirs.

Ideas closely related the ones presented here have been developed in the past to discuss the magnetic excitations of a 2-D ferromagnet [6] and the paramagnetic fluctuations in 3-D iron and nickel [7]. The following calculation is in the same spirit, although it differs in methodology and detail.

II- THE MODEL

We follow Chakravarty, Halperin, and Nelson [8] (CHN) who have shown how to map the antiferromagnetic quantum Heisenberg model on to the non-linear σ model in two space, one time dimensions. The partition function of the latter model is given by

$$Z = \int \mathcal{D}\,\mathbf{n}(\mathbf{x},\tau)\, \exp\left[-\frac{\rho_o}{2}\int_0^\beta d\tau \int d^2\mathbf{x} \left(\frac{\partial \mathbf{n}}{\partial x_i}\frac{\partial \mathbf{n}}{\partial x_i} + \frac{1}{c^2}\frac{\partial \mathbf{n}}{\partial \tau}\frac{\partial \mathbf{n}}{\partial \tau} \right)\right] \qquad (1)$$

where the unit vector $\mathbf{n}(\mathbf{x},\tau)$ is the local order parameter, the space and time varying staggered magnetisation. ρ_o is the bare stiffness constant, and c the bare spin wave velocity. CHN have shown how to determine these parameters from the observed spin wave velocity and uniform field susceptibility. We have chosen units in which h = 1.

Making use of the periodicity of $\mathbf{n}(\mathbf{x},\tau)$ in imaginary time we write

$$\mathbf{n}(x,\tau) = \frac{1}{\beta}\int \frac{d^2q}{(2\pi)^2} \sum_{\omega_n} \mathbf{n}(\mathbf{q},\omega_n)\, \exp(i\mathbf{q}.\mathbf{x} - \omega_n\tau) \qquad (2)$$

with $\omega_n = 2\pi\, n\, kT$. The dynamic magnetic structure factor is given by the analytic continuation in frequency of the order parameter correlation function :

$$S(q,\omega) = \frac{2}{1-\exp(-\beta\omega)}\, \Im m[\langle \mathbf{n}(q,\omega_n).\mathbf{n}(-q,-\omega_n)\rangle]_{\, i\omega_n \to \omega+i\eta} \qquad (3)$$

$$\equiv \frac{2}{1-\exp(-\beta\omega)} \; \chi''(\omega)$$

where the angular brackets stand for the quantum statistical average and the second equality in (3) defines the absortive part of the susceptibility.

A convenient way of performing the separation of time scales refered to in the introduction is to parametrise the configurations of the local order parameter in terms of the local rotation that makes $\mathbf{n}(\mathbf{x},\tau)$ point into the direction of the z-axis:

$$\mathbf{n}(\mathbf{x},\tau) = \mathbb{R}(\mathbf{x},\tau)\begin{pmatrix}0\\0\\1\end{pmatrix} \tag{4}$$

Following Polyakov[9] we write $\mathbb{R}$ as the product of two matrices containing the slow and fast components of the rotation respectively, i.e. :

$$\mathbf{n}(\mathbf{x},\tau) = \mathbb{R}_s(\mathbf{x},\tau)\mathbb{R}_f(\mathbf{x},\tau)\begin{pmatrix}0\\0\\1\end{pmatrix} \equiv \mathbb{R}_s(\mathbf{x},\tau)\mathbf{s}(\mathbf{x},\tau) \tag{5}$$

The Fourier components of the angles that parametrise $\mathbb{R}_s$ ($\mathbb{R}_f$) only include wavevectors smaller (larger) than a cutoff κ_o of the order of the inverse correlation length. The vector field $\mathbf{s}(\mathbf{x},\tau)$ is then, by construction, that part of $\mathbf{n}$ that couples to external perturbations whose wavelength is shorter than κ_o^{-1} . By substituting (5) into (1) we obtain the action

$$\mathcal{A}[\mathbf{n}(x)] = \frac{\rho_o}{2}\int d^2\mathbf{x}\int_0^\beta d\tau \;\left|\left(\partial_\mu + \mathbb{R}^T(x)\partial_\mu\mathbb{R}(x)\right)\mathbf{s}\right|^2 \tag{6}$$

We have used a compact notation where the index μ runs from 0 to 2 and $x = (\mathbf{x}, c\tau)$. It may be shown that

$$\mathbb{R}_s^T(x)\partial_\mu\mathbb{R}_s(x)\mathbf{s} = \mathbf{A}_\mu(x) \wedge \mathbf{s}(x) \tag{7}$$

where the A_μ are vectors in spin space with components ($A_\mu^\pm = A_\mu^x \pm i\, A_\mu^y$)

$$A_\mu^o(x) = \cos\theta\; \partial_\mu\varphi - \partial_\mu\psi \tag{8}$$

$$A_\mu^\pm(x) = -\frac{1}{\sqrt{2}}\,(\sin\theta\; \partial_\mu\varphi \mp i\; \partial_\mu\theta)\; \exp(\pm i\; \psi) \tag{9}$$

θ, φ, and ψ are the Euler angles that parametrise the slow rotation $\mathbb{R}_s$. With the use of (7), (8), and (9), (6) becomes:

$$\mathcal{A}[\mathbf{n}(x)] = \mathcal{A}_0[\mathbf{A}_\mu] + \mathcal{A}_1[\mathbf{s};\mathbf{A}_\mu] \tag{10}$$

$$\mathcal{A}_0[\mathbf{A}_\mu] = \rho_0 \int d^2\mathbf{x} \int_0^\beta d\tau \; A_\mu^+(x)\; A_\mu^-(x) \tag{11}$$

$$\begin{aligned}\mathcal{A}_1[\mathbf{s};\mathbf{A}_\mu] = & \frac{\rho_0}{2} \int d^2\mathbf{x} \int_0^\beta d\tau \; \left|\left(\partial_\mu - i\, A_\mu^0\right)s^-(x)\right|^2 \\ & - \frac{\rho_0}{4} \int d^2\mathbf{x} \int_0^\beta d\tau \; \left[A_\mu^+\, s^-(x) + c.c\right]^2 \\ & - i\sqrt{2}\,\frac{\rho_0}{2} \int d^2x \int_0^\beta d\tau \; \left[\left(\partial_\mu - i\, A_\mu^0\right)s^- s^0 A_\mu^+ - c.c\right]\end{aligned} \tag{12}$$

where we have used the decomposition $\mathbf{s} = s^0\, \mathbf{e}_0 + 1/\sqrt{2}\left(s^+\, \mathbf{e}_- + s^-\, \mathbf{e}_+\right)$, $\mathbf{e}_\pm = 1/\sqrt{2}(\mathbf{e}_x \pm i\, \mathbf{e}_y)$

It may be easily shown that (11) is equivalent to

$$\mathcal{A}_0[\mathbf{A}_\mu] = \frac{\rho_0}{2} \int d^2\mathbf{x} \int_0^\beta d\tau \; |\partial_\mu \mathbf{n}_B(x)|^2 \tag{13}$$

where $\mathbf{n}_B$, the background order parameter, is defined by

$$\mathbf{n}_B(x) = \mathbb{R}_S(x) \begin{pmatrix} 0 \\ 0 \\ 1 \end{pmatrix} = \begin{pmatrix} \sin\theta(x)\cos\varphi(x) \\ \sin\theta(x)\sin\varphi(x) \\ \cos\theta(x) \end{pmatrix} \tag{14}$$

The action (12) describes the dynamics of the fast modes in the presence of the slow disturbance $\mathbf{n}_B$. The dynamics of the latter is described by a renormalised Hamiltonian obtained from (1) by integrating out the varibles $\mathbf{s}$. It can be shown that, to one loop order, the parameters of the effective Hamiltonian obey the renormalisation group equations derived by CHN. Notice that both (13) and the effective Hamiltonian are independent of ψ, the third Euler angle, as they should.

In terms of the new variables the order parameter correlation function may be written as

$$\langle \mathbf{n}(x).\mathbf{n}(0) \rangle = \frac{1}{Z[\mathbf{A}]} \int \mathcal{D}\mathbf{A}_\mu \; \exp[-\mathcal{A}(\mathbf{A}_\mu)] \; \langle s^\alpha(x) s^\beta(0) \rangle_{\mathbf{A}} \left(\mathbb{R}_S^T(0)\mathbb{R}_S(x)\right)_{\beta\alpha} \tag{15}$$

The spin correlation function in the integrand is evaluated using the action (12).

It may be shown that for distances $|\mathbf{x}| \gg \kappa_0^{-1}$ the effect of the fast modes reduces a renormalisation of the amplitude of the slow fluctuations:

$$\langle \mathbf{n}(x).\mathbf{n}(0) \rangle = \left[1 + \frac{kT}{2\pi\, \rho_0} \ln(\kappa_0)\right]^2 \langle \mathbf{n}_B(x)\mathbf{n}_B(0) \rangle \tag{16}$$

If we want to probe shorter distances, however, it is necessary to evaluate the correlation function in the local frame.

III- SPIN WAVES IN THE LOCAL FRAME

The transverse components of **s** are the spin fluctuations as they are seen from a local system of reference whose z-axis points into the instantaneous direction of the fluctuating $\mathbf{n}_B(x)$ at all points in space. We are interested in the case in which their energy $\omega \gg kT$. Therefore, we expect that their amplitude will be small. As a first approximation we may treat them in the spin wave approximation neglecting for the moment the interactions:

$$s^o(x) = \sqrt{1 - s^+(x)s^-(x)} \cong 1 - \frac{1}{2} s^+(x)s^-(x) \qquad (17)$$

In this approximation the last term of the right hand side of (12) vanishes, and we are left with a simpler action:

$$\mathcal{A}_{SW} = \frac{\rho_o}{2}\int d^2\mathbf{x} \int_0^\beta d\tau \quad s^+\left[-(\partial_\mu - i A_\mu)^2 - A_\mu^+ A_\mu^- \right]s^-$$

$$- \frac{\rho_o}{2}\int d^2\mathbf{x} \int_0^\beta dt \; \frac{1}{2}\left[\left(A_\mu^+ s^-\right)^2 + \left(A_\mu^- s^+\right)^2 \right] \qquad (18)$$

Because of the second term on the right hand side of (18) the correlation function is no longer diagonal:

$$G_{i,j}(x) = \begin{Vmatrix} \langle s^-(x)s^+(0)\rangle & \langle s^+(x)s^+(0)\rangle \\ \langle s^-(x)s^-(0)\rangle & \langle s^+(x)s^-(0)\rangle \end{Vmatrix} \qquad (19)$$

is the solution of

$$G_{i,j}(x) = G^{\,o}_{i,j}(x) + \int d^3y \;\; G^{\,o}_{i,k}(x-y)\, V_{k,l}(y)\, G_{l,j}(y) \qquad (20)$$

where

$$G^o_{i,j}(x) = \begin{Vmatrix} g(x) & 0 \\ 0 & g^*(x) \end{Vmatrix} , \qquad (21)$$

$$V_{i,j}(x) = \frac{\rho_o}{2} \left\| \begin{matrix} 0 & A^+_\mu(x)A^+_\mu(x) \\ A^-_\mu(x)A^-_\mu(x) & 0 \end{matrix} \right\| , \tag{22}$$

and

$$\left[\left(\partial_\mu - i\, A^o_\mu(x)\right)^2 + A^+_\mu(x)A^-_\mu(x) \right] g(x-y) = -\frac{2}{\rho_o}\,\delta^3(x-y) \tag{23}$$

These equations define a problem that is still very difficult. However, it may be simplified if we notice that at low temperature ξ is very large compared to the wavelength of the excitations of interest, and we have $q\xi \gg 1$. It is then sufficient to work to order $(\kappa_o/q)^2$ in working out the consequences of (19) - (23).

Equation (23) is reminiscent of the equation for the Green function of a spin-zero relativistic particle in the presence of a "vector potential", A^o_μ, and of a "scalar potential" $A^+_\mu A^-_\mu$. Using this analogy we see that the "magnetic field" asociated to the vector potential is a small quantity of the second order in the gradients and we may, to leading order in κ_o, ignore the field-induced curvature of the trajectory of the "particle" and approximate:

$$g(x-y) \cong \exp\left(i\int_{y\to x} ds.\mathbf{A}^o(s)\right)\, g^o(x-y) \tag{24}$$

where the line integral is taken along the straight path that runs from y to x, and g^o is the solution of

$$\left[(\partial_\mu)^2 + A^+_\mu(x)A^-_\mu(x) \right] g^o(x-y) = -\frac{2}{\rho_o}\,\delta^3(x-y) \tag{25}$$

The "scalar potential" can be decomposed into a thermal average and a fluctuation

$$A^+_\mu(x)A^-_\mu(x) = \left\langle\, |A^-_\mu|^2 \,\right\rangle + \left[\, |A^-_\mu|^2 - \left\langle\, |A^-_\mu|^2 \,\right\rangle \right] \tag{26}$$

By using (9) and (14) we easily see that

$$\left\langle |A^-_\mu(x)|^2 \right\rangle = \frac{1}{2}\left\langle (\partial_\mu n_B(x))^2 \right\rangle \equiv M^2 \tag{27}$$

which shows that the first term on the right hand side of (26) is of order κ_o^2. The effects of the fluctuating part of the "potential" are more difficult to treat. An estimate may be obtained by calculating the self-energy of the average Green's function. A lenghty but straightforward calculation shows that the fluctuations enter only to order κ_o^4 and may safely be neglected to our level of accuracy. Similar arguments may be applied to the off-diagonal part of the potential, (22).

It follows that, to lowest order in the gradients,

$$G_{i,j}(x,y) \cong \left\| \begin{matrix} \exp\left(i\int_{y\to x} ds.\mathbf{A}^{o}(s)\right) & 0 \\ 0 & \exp\left(-i\int_{y\to x} ds.\mathbf{A}^{o}(s)\right) \end{matrix} \right\| g^{o}(x-y) \qquad (28)$$

where $g^{o}(x-y)$ is the Fourier transform of

$$g^{o}(q,\omega_n) = \frac{2c^2}{\rho_o} \frac{1}{(qc)^2 + \omega_n^2 - M^2} \qquad (29)$$

We may summarize the effects of the slowly varying background on the spin waves in the local frame by saying that their dispersion relation is modified from $\omega(q) = cq$ to:

$$\omega(q) = c\sqrt{q^2 - M^2} \cong cq - c\frac{M^2}{2q} = cq\ \left[1 - \alpha\ (q\xi)^{-2}\right] \qquad (30)$$

and that scattering off the disorder introduces a phase shift. We'll see shortly that the latter determines the line-shape. α is a numerical factor of order one that expresses the fact that M is of order κ_0, as is evident from (27). We do not try to evaluate the coefficient α for it depends on the precise form of the correlation function. It will certainly depend on the temperature. Notice that for $q\xi \gtrsim 1$ we recover unrenormalised spin waves as we should.

IV- CORRELATION FUNCTION

Using the results of the previous section in equation (15) we obtain:

$$\langle \mathbf{n}(x).\mathbf{n}(y) \rangle = g^{o}(x-y) \left\langle \Re\left[\exp\left(i\int_{y\to x} ds_\mu \cos\theta(s)\ \partial_\mu\varphi(s)\right) \Phi(x,y)\right]\right\rangle_{n_B} \qquad (31)$$

where $\Re$ denotes the real part, and

$$\Phi(x,y) = (\cos\theta(x)\cos\varphi(x) - i\sin\varphi(x))\ (\cos\theta(y)\cos\varphi(y) + i\sin\varphi(y))$$
$$+(\cos\theta(x)\sin\varphi(x) + i\cos\varphi(x))\ (\cos\theta(y)\sin\varphi(y) + -\cos\varphi(y)) \qquad (32)$$

Notice that all reference to ψ has disappeared as it must.

It is not easy to evaluate (31) in general, neither it is worth the effort since, anyway, it is only expected to be valid at short distance. A simple calculation shows that, to second order in the gradients, (31) is equivalent to

$$\langle \mathbf{n}(x).\mathbf{n}(y) \rangle = g^{o}(x-y)\, \tilde{\Phi}(x-y) \tag{33}$$

where

$$\tilde{\Phi}(x-y) = 2 - \frac{1}{2} \langle [(\theta(x)-\theta(y))^2 + (\varphi(x)-\varphi(y))^2 \sin^2 \theta(x)] \rangle \tag{34}$$

$$\cong 2 \left[1 - \frac{1}{2} \left\langle (\mathbf{n}_B(x)-\mathbf{n}_B(y))^2 \right\rangle \right]^{1/2} \cong 2 \sqrt{\langle \mathbf{n}_B(x)\mathbf{n}_B(y) \rangle} = 2 \sqrt{\mathcal{S}_B(x-y)} \tag{35}$$

By elaborating on (3) taking into account the fact that the spectral weight of $\tilde{\Phi}$ is concentrated at low frequencies we get the imaginary part of the susceptibility

$$\chi''(q,\omega) = \int \frac{d^2q'}{(2\pi)^2} \int_{-\infty}^{\infty} \frac{d\,\omega'}{\pi} \; P(q',\omega')\, \mathfrak{Im}\, g^{o}(q-q',\omega-\omega') \tag{36}$$

where $P(q',\omega')$ is the Fourier transform in space and time of $\sqrt{\mathcal{S}_B(x-y)}$.

On general grounds one expects that the low frequency low momentum part of the structure factor will be of the form:

$$\mathcal{S}_B(q,\omega) = \text{const}\, \frac{1}{q^2 + \xi^{-2}} \; \frac{\Gamma(q)}{\omega^2 + \Gamma^2(q)} \tag{37}$$

$$\Gamma(q) \cong \Gamma_o + \Gamma_1\, (q\xi)^2 \tag{38}$$

If there is long range order at T=0 the constants Γ_o and Γ_1 vanish as $T \to 0$[10]. It is expected that, for sufficiently strong disorder, ξ stays finite at T=0 . In that case Γ will also stay finite, $\Gamma \cong c\xi^{-1}$.

At long distances and times

$$\sqrt{\mathcal{S}_B(x-y)} \cong \left(\frac{R}{\xi}\right)^{-1/4} \exp\left(-\frac{R}{2\xi}\right) \exp\left(-\Gamma_o \frac{t}{2}\right) \tag{39}$$

From (29), (36), and (39) we get

$$\chi''(q,w) = \int_{-\infty}^{\infty} \frac{d\Omega}{\pi} \; \frac{\Gamma_o/2}{\Omega^2 + [\Gamma_o/2]^2} \; \frac{|\omega-\Omega|}{\omega-\Omega} \; \mathcal{F}(q,\omega-\Omega) \tag{40}$$

where

$$\mathcal{F}(q,\omega) = \frac{\pi}{\rho_o} \int_o^\infty dR\, R \left(\frac{R}{\xi}\right)^{-1/4} J_o[qR]\, J_o[\kappa(\omega)R]\, e^{-(R/2\xi)} \tag{41}$$

and

$$\kappa(\omega) = \frac{|\omega|}{c} \left[1 + 2\alpha \left(\frac{c}{\xi\omega} \right)^2 \right]^{1/2} \tag{42}$$

The integral is standard[11] and may be expressed in terms of the hypergeometric function

$$\mathcal{F}(q,\omega) = \frac{\pi\xi^2}{\rho_o} \sum_{m=o}^{\infty} F[-m,-m,1,\zeta(q,\omega)]\; (-q\xi)^{2m} \frac{\Gamma(2m-1/4)}{m!\;\Gamma(m+1)} \tag{43}$$

with $\zeta(q,\omega) = [\kappa(\omega)/q]^2$.

This is peaked at $\zeta(q,\omega) = 1$. For $q\xi \gtrsim 1$, near the peak we have the simpler expression:

$$\mathcal{F}(q,\omega) \sim \frac{\xi/\sqrt{q\,\kappa(\omega)}}{[1 + (2\xi(q-\kappa(\omega)))^2]^{3/8}} + \frac{\xi/\sqrt{q\,\kappa(\omega)}}{[1 + (2\xi(q+\kappa(\omega)))^2]^{3/8}} \tag{44}$$

For fixed ω, along a fixed direction in q-space, we find two peaks at $q = \pm\,\kappa(\omega)$. They are remnants of the T=0 spin waves. Notice that according to (42) they are farther apart than in the ordered case, a result of the softening implicit in (30). These peaks have acquired a width of order ξ^{-1} in q reflecting that the latter is not a good quantum number when disorder is present. The corresponding peaks in ω have a width $\Gamma_\omega = c\,\xi^{-1}$. Notice that this is not the usual hydrodynamic width, proportional to q^2. This was to be expected since we are treating a limit that is quite outside the range of validity of hydrodynamics.

The predicted lineshape is manifestly non-lorenzian. This is precisely the kind of inhomogeneous broadening that one would expect form the physical picture developed in the introduction. The lineshape in the laboratory frame is simply related to this one by a convolution as follows from (40). As shown elsewhere[10], in the case of the undoped system:

$$\Gamma_o \cong c\,\xi^{-1} \sqrt{\frac{T}{2\pi\,\rho_s}} \tag{45}$$

At low temperatures this is much less than the intrinsic width given by (44) and the effect of the convolution may be neglected. In the case of the doped samples Γ_o may not vanish at T=0 and it is not clear whether or not it will remain smaller than $c\xi^{-1}$. In any event the lineshape will be qualitatively the same.

In the case of the stochiometric compounds of interest the correlation length is exceedingly large, even at room temperature . Using the experimental values of c and ξ in the expressions derived in this paper one can readily see that the corrections to the T=0 behavior should be very small throughout the whole relevant range of temperatures. This is consistent with observations.

For the doped compounds the situation is expected to be different because in this case the correlation length is limited by the presence of impurities that locally quench the direction of **n**. Detailed experimental results have been recently obtained on doped $YBa_2Cu_3O_6$[4] . It is known that the presence of holes on the CuO_2 planes induces planar disorder, i.e., in the ground state, the order parameter has no component outside the plane, and its direction within the plane is random with a correlation length that scales with the average distance between impurities. The planar situation is most easily treated by choosing a gauge where only $A_\mu^{\pm}(x)$ are non zero (cf. (7), and (8)). It is assumed that the only rôle of impurities is to limit the correlation length. The results of repeating the calculations of this section depend on the polarisation of the spin waves: the fluctuations out of the plane are frequency shifted but show no broadening; the frequency of the fluctuations within the plane is unrenormalised but they are still broadened. This behavior is in agreement with experiment [4] .

V- CONCLUSIONS

In this paper we have discussed the high energy spin waves in the disordered phase of the quantum 2-D antiferromagnet. At low temperatures these excitations interact with non-propagating relaxational modes whose dynamics is much slower. As a consequence the spin waves follow adiabatically the slow modes. We find that the spin waves are modified in two ways: there is a change in their dispersion relation and they acquire a phase shift. The space and time dependence of the phase shift determines the lineshape of the excitations. The transverse character of the coupling between fast and slow modes is reflected by a peculiar dependence of the renormalisation effects on the polarisation of the spin wave. The predicted effects are in agreement with recent experiments[4] .

We have not discussed at all in this paper the effect of spin wave interactions on our results. As it was argued in the introduction we do not expected them to be an important source of damping in the region of interest to us. It can be shown, however, that they renormalise the spin wave velocity in the manner described by CHN[8] . The results derived in this paper remain valid provided one changes everywhere the bare value of c by the renormalised one.

REFERENCES

1. Y. Endoh, K. Yamada, R.J. Birgenau, D.R. Gabbe, H.P. Jensen,M.A. Kastner, C.J. Peters, P.J. Picone, T.T. Thurston, J. Tranquada, G. Shirane, Y. Hodaka, M. Oda, Y. Enomoto, M. Suzuki, and T. Murakami, Phys.Rev.**B37**, 7443 (1988).

2. J. Rossat-Mignod, P. Burlet, M.J. Jurgens, C. Vettier, L.P. Regnault, J.Y. Henry, C. Ayache, L. Forro, H. Noel, M. Potel, P. Gougeon, and J.C. Levet, J.Phys.France C8 **12**, 2119 (1988).

3. G. Aeppli, S.M. Hayden, H.A. Mook, Z. Fisk, S.W. Cheong, D. Rytz, J.P. Remeika, G.P. Espinosa, and A.S. Cooper, Phys.Rev.Lett.**62**, 2052 (1989). See also the article by H. A. Mook in these Proceedings.

4. C. Vettier, P. Burlet, J.Y. Henry, M.J. Jurgens, G. Lapertot, L.P. Regnault, and J. Rossat-Mignod, Physica Scripta (in press). See also the article by J. Rossat-Mignod in these Proceedings.

5. S. Tyc, B.I. Halperin, and S. Chakravarty, Phys.Rev.Lett.**62**,835 (1989).

6. F. Moussat, and J. Villain, J.Phys.C: Solid State Physics **9**,4433 (1976)

7. V. Korenman, J.L. Murray, and R.E. Prange, Phys.Rev.**B16**, 4032,4048, and 4058 (1977).

8. S. Chakravarty, B.I. Halperin, and D.R. Nelson, Phys.Rev.**B39**,2344 (1989).

9. A.M. Polyakov, Phys.Lett.**B59**, 79 (1975).

10. D.R. Grempel, Phys.Rev.Lett.**61**, 1041 (1988).

11. I.S. Gradshteyn, and I.M. Ryzhik, Tables of Integrals, Series, and products, Academic Press, New York (1965).

EXACT DIAGONALIZATION STUDIES OF QUASIPARTICLES IN DOPED QUANTUM ANTIFERROMAGNETS

P. Prelovšek, J. Bonča, A. Ramšak and I. Sega

J.Stefan Institute, E.Kardelj University of Ljubljana
61111 Ljubljana, Yugoslavia

INTRODUCTION

Due to their relevance for superconducting copper oxides[1], models for strongly correlated systems have been studied extensively in last two years. In spite of a number of analytical approaches applied to these models, and to models of the CuO_2 layers in particular, some of the crucial questions lack even a qualitative answer. In such a situation exact diagonalization studies of small correlated systems have proven to be very valuable. Employing this method several groups obtained important results on various models[2-10], representing the insulating state and the low doping regime in CuO_2 layers.

The essential limitation of the exact diagonalization approach is in the smallness of systems which can be investigated. The allowed size of the system is determined by the number of quantum states N_{st}, representing the basis for the ground state (or the excited state) wavefunction $|\Psi_0\rangle$. For most efficient numerical approaches it is required that N_{st} be substantially smaller than the available computer memory. The complexity can be estimated from the total number of states $N^0_{st} = m^N$, where m is the number of quantum states per unit cell and N is the number of cells. This puts a severe restriction on the studies of models with larger m. Increasing in the complexity are thus: a) the Heisenberg model with $m = 2$ where sizes up to $N = 24$ have been reached by the Lanczos technique[2], b) the $t - J$ model with $m = 3$ and maximum size $N = 16$ (for a single hole state also $N = 18$) [3-8], c) the single band Hubbard model with $m = 4$ and typicaly $N = 12$ (using a different technique also calculations for $N = 16$ are presented at this conference)[9], d) the two (three) band model for CuO_2 layers with $m = 64$ and maximum size $N = 4$ [10]. These estimates give a clear motivation for the studies of simpler models.

In most diagonalization procedures the Lanczos method is used. Starting with a simple wave function $|\Phi_0\rangle$, a sequence of orthogonal functions $|\Phi_n\rangle$ is generated by the recursion relation :

Dynamics of Magnetic Fluctuations in High-Temperature Superconductors
Edited by G. Reiter *et al.*, Plenum Press, New York, 1991

$$H \mid \Phi_n >= b_{n-1} \mid \Phi_{n-1} > +a_n \mid \Phi_n > +b_n \mid \Phi_{n+1} >, \tag{1}$$

where $b_{-1} = 0$. Usually less than 40 Lanczos steps are needed in order to have a good convergence of the ground state energy E_0 and the wavefunction $|\Psi_0\rangle$, which is obtained by the diagonalization of the tridiagonal matrix with elements a_n, b_n. It is then straightforward to use $|\Psi_0\rangle$ for the evaluation of static expectation values or correlation functions.

The Lanczos method can be easily extended also to the calculation of dynamic response functions.[6,8] We show this on the example of the frequency dependent conductivity $\sigma(\omega)$,[8] which can be studied in a finite system as an extrapolation for $q \to 0$ of

$$\sigma(\vec{q}, \omega) = -\frac{1}{\pi\omega} ImG(\omega + i\epsilon), \tag{2a}$$

$$G(z) = \langle\Psi_0|j_{-\vec{q}}(z + E_0 - H)^{-1} j_{\vec{q}}|\Psi_0\rangle, \tag{2b}$$

In order to evaluate $G(z)$ we start the Lanczos procedure, Eq.(1), with an initial wavefunction $|\tilde{\Phi}_0\rangle = A\ j_{\vec{q}}|\Psi_0\rangle$. Then $G(z)$ can be expressed with Lanczos corresponding coefficients α_n, β_n in the form of continued fractions

$$G(z) = \cfrac{\| j_{\vec{q}}\, \Psi_0 \|^2}{z - \alpha_0 - \cfrac{\beta_1^2}{z - \alpha_1 - \cfrac{\beta_2^2}{z - \ldots}}}. \tag{3}$$

In a finite system $G(z)$ has poles on the real axis. Although the convergence of the entire spectrum is hard to reach, only a small number of Lanczos steps (< 50) is needed to reproduce essential features.[8]

In order to reduce N_{st} it is important to employ symmetry properties of the system. Whereas the total number of fermions N_f and the corresponding S^z_{tot} are easily taken into account, representations having well defined wavevector $\vec{q}$ are more difficult to implement for a larger number of holes. On the other hand, it seems that the direct inclusion of the conservation of S_{tot} is not practical.

COMPARISON OF MODELS

Let us first investigate the relation between several models proposed for the CuO_2 layers. The two band model introduced by Emery[11] assumes that only $Cu\ d_{x^2-y^2}$ orbitals and $O\ p_\sigma$ orbitals are essential for the electronic properties,

$$H_{Hubb} = -t_{pd} \sum_{(ij)s} c^\dagger_{is} c_{js} + \Delta_0 \sum_{(i\in p)} n_i + \sum_i U_i n_{i\uparrow} n_{i\downarrow} \tag{4}$$

where $c_{is}, c^{\dagger}_{is}$ represent hole operators on Cu d and O p sublattices, with a corresponding vacuum of filled shells $Cu\ d^{10}, O\ p^6$. Here, we take into account the hybridization (t_{pd}) contribution, the charge transfer (Δ_0) term and the onsite Coulomb repulsions with $U_i = U_{pp}, U_{dd}$.

For the undoped system with $\Delta_0 > 0$ and $U_{dd} \gg t_{pd}$ it is expected that the model (1) can be described well by the Heisenberg model with only spin degrees on the Cu sites. Additional holes introduced by doping (mainly) on O sites can be mobile, hopping through intermediate Cu sites (or directly by the $O-O$ hopping). For this case, the mobile holes and localized spins on Cu sites are relevant degrees of freedom and coupled hole-spin models have been derived,[12] treating t_{pd} as the smallest quantity. Models can be further symplified introducing for holes the Wannier functions[13] corresponding to the free hole hopping, but centered on Cu sites. Terms involving the antisymmetric orbitals couple only weakly to the more relevant symmetric subsystem and seem to be less important at low doping. With $\tilde{c}_{is}$ denoting the operators for these symmetric orbitals, we get the symmetrized hole-spin model,[13]

$$H_{sym} = J\sum_{<ij>} \vec{S}_i\cdot\vec{S}_j - t_0\sum_{(ij)s} \tilde{c}^{\dagger}_{is}\tilde{c}_{js} + V\sum_i \vec{\tilde{s}}_i\cdot\vec{S}_i + t_1\sum_{(ij)\,ss'} \tfrac{1}{2}\vec{\sigma}_{ss'}\cdot\vec{S}_i(\tilde{c}^{\dagger}_{is}\tilde{c}_{js'} + \tilde{c}^{\dagger}_{js}\tilde{c}_{is'}), \tag{5}$$

where some less important terms have been omitted. For large V a local singlet state, formed out of the local d hole and the symmetrized p hole, can be used as a new vacuum[14]. Neglecting higher energy triplet states, the generalized t-J model[14,13] is obtained

$$H_{tJ} = J\sum_{<ij>} \vec{S}_i\cdot\vec{S}_j - t\sum_{(ij)s} d^{\dagger}_{is}d_{js} - t'\sum_{((jk))s} d^{\dagger}_{js}d_{ks} + \varsigma\sum_{(i,j\neq k)ss'} \tfrac{1}{2}\vec{\sigma}_{ss'}\cdot\vec{S}_i d^{\dagger}_{js}d_{ks'} \tag{6}$$

with operators d_{is}, $d^{\dagger}_{is}$ acting on a subspace with no doubly occupied sites. Included are the nnn $((jk))$ hopping terms with the (t') spin independent hopping and the (ς) hopping dependent on the intermediate spin $\vec{S}_i$.

In Eq. (1) t_{pd} is not small enough compared to Δ_0 to ensure the perturbation derivation of Eq. (5), so as V is not large enough for a straighforward derivation of Eq. (6). Therefore we performed a quantitative comparison of models,[13] allowing deviations of parameters from their perturbational values. We adopted the view that effective models should reproduce as well as possible the low energy spectra of a single hole in the antiferroamgnet (AFM), i.e. a single quasiparticle (QP), of the original model, Eq. (4). Here we present results obtained by the exact diagonalization of a $d = 1$ system with $N_0 = 4$ cells.

Since the main open question is whether models are compatible in the mixed valence regime $t_{pd} \gtrsim \Delta_0$,[12] we present in Fig. 1 results in the latter regime. With the use of the appropriate renormalized parameters the agreement is even quantitative, especially for the lowest QP branch. Also we find that corrections to the simplest t-J model are small, i.e. t' and ς terms are even smaller than those derived from a single

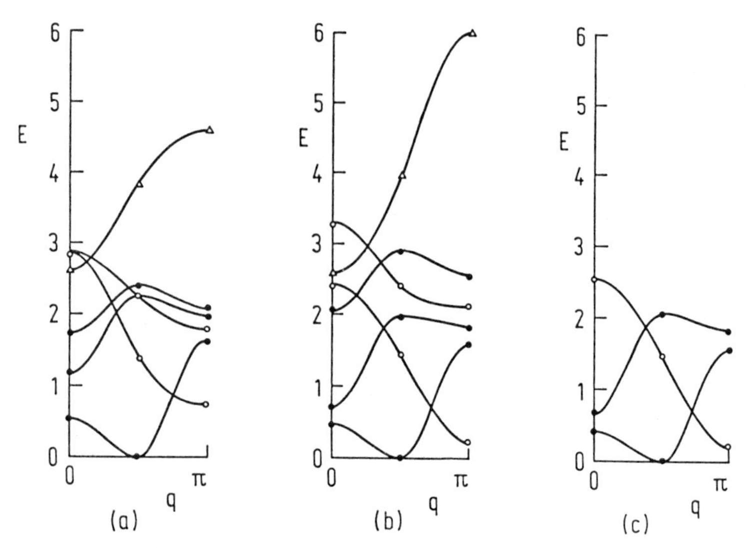

FIG. 1. The lowest lying branches for a system with a single QP on a chain of $N_0 = 4$ cells in a mixed valence regime, $\Delta_0 = 2$, $t_{pd} = 1.4$ and $U_{dd} = 7$. • represent the $S = \frac{1}{2}$, ∘ the $S = \frac{3}{2}$ and △ the $S = \frac{5}{2}$ levels, respectively. Lines are guides to the eye only. Here are: (a) two band Hubbard model, (b) symmetrized hole-spin model and (c) generalized t-J model.

band Hubbard model. It should be stressed however that the agreement between the Hubbard model and the Heisenberg model is less satisfactory in the mixed valence regime. In view of the existing experimental data[15] on magnetic properties, this could be an indication that real copper oxides are in the charge transfer regime $\Delta_0 \gg t_{pd}$.

SINGLE QUASIPARTICLE PROPERTIES

In the following we restrict our discussion to the prototype t-J model,[16] Eq. (6),

$$H = -t \sum_{<ij>s} d^\dagger_{is} d_{js} + J \sum_{<ij>} \left(\vec{S}_i \cdot \vec{S}_j - \frac{1}{4} n_i n_j \right) . \tag{7}$$

Here, we present a more detailed analysis[8] of the eigenstates of a single hole (QP) in a $d = 2$ system, which supplements the existing analytical and numerical results.[17,18,4,6] First, we calculate by the method of the diagonalization of finite $d = 2$ system the lowest branch of the QP energy dispersion $E(\vec{q})$. Whereas small 4×4 system can be diagonalized exactly, larger 8×4 systems are treated approximately, allowing only a finite number of spin flips $N_r \leq 6$ relative to the initial Néel AFM state.

In a t-J model on a $N = 4 \times 4$ system a single hole state is degenerate along the AFM Brillouin zone boundary , i.e. at $\vec{k}_0 = (\pm\pi/2, \pm\pi/2)$, $(\pi, 0)$ and $(0, \pi)$. In a larger system the lowest energy state is $\vec{k}_0 = (\pm\pi/2, \pm\pi/2)$. Still the effective mass and the mass enhancement $\underline{\mu}$ are highly anisotropic tensors,

$$\underline{\mu}^{-1} = \frac{1}{2t} \frac{\partial^2 E(\vec{k}_0 + \vec{p})}{\partial \vec{p} \partial \vec{p}} \bigg|_{\vec{p}=0} . \tag{8}$$

The enhancement is large along the AFM zone boundary (infinite for a 4×4 system), i.e. $\vec{p} \perp \vec{k}_0$ and finite $\mu_{\|} \equiv \mu$ for $\vec{p} \parallel \vec{k}_0$. Values for μ presented in Fig. 2 show an approximate $1/J$ dependence, as predicted theoretically.[17,18]

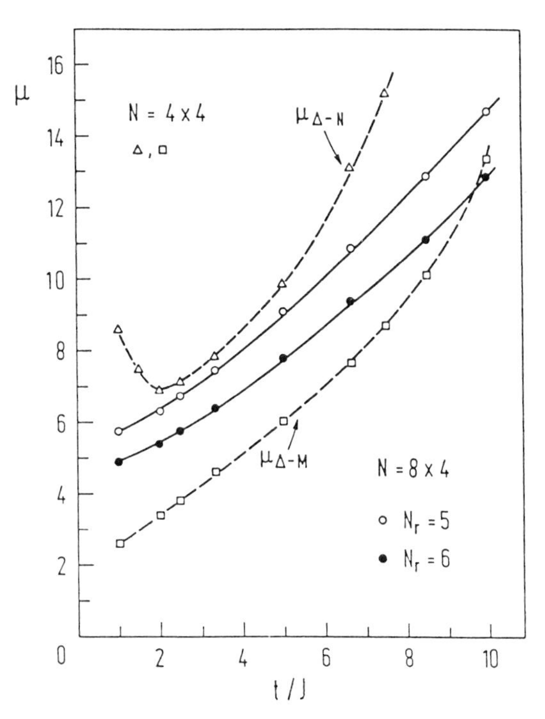

FIG. 2. Mass enhancement μ vs t/J for a) $N = 8 \times 4$ system for $N_r = 5$ and $N_r = 6$; b) for $N = 4 \times 4$ along $\vec{p} \parallel \vec{k}_0$. In the latter case the value of μ has been extracted from the variation along the $\Delta - N$ and the $\Delta - M$ line in a Brillouin zone.

The Lanczos method for the calculation of dynamical conductivity $\sigma(\omega)$ has been described in Sec.1. Since generally systems are too small for the evaluation of $q_x \to 0$, we compared two approaches:[8] a) By using the 8×4 system q_x becomes sufficiently small and an extrapolation $q_x \to 0$ can be performed for different parts of spectra. Results are however only aproximate due to restricted $N_r \lesssim 6$. b) Imposing $q = 0$ the conductivity sum rule is violated in a system with periodic boundary conditions. This can be however traced back to the disappearance of the lowest QP contribution from the spectra. Thus $q = 0$ results can be used for the investigation of the remaining part. Moreover smaller systems as 4×4 can be again used. Both methods give qualitatively similar results.

As seen in Fig. 3, where a typical plot of $\sigma(\omega)$ for finite but small q at $J/t = 0.2$ is presented, the spectra show two distinct parts: a) an undamped QP peak, which would approach $\omega \to 0$ for $q \to 0$ and is expected to broaden into a Drude peak only at $T > 0$, and b) the higher frequency part due to the incoherent hopping of the hole, where excitations have mainly the magnon character. We note that at finite J the lower magnon peaks are still well pronounced, the lowest being at $\omega \sim 2J$, in contrast to the smooth variation $\sigma \propto 1/\omega$ expected for $J \to 0$.[19] The optical sum rule is mainly exhausted by the incoherent part, since the QP peak takes only the fraction $1/\mu$ of the total intensity. It should be noted that our results are for $T = 0$, and that finite T would lead to the broadening of the coherent part and of the incoherent part. Whereas there are qualitative similarities between our and experimental results[29], measured $\sigma(\omega)$ show much broader features.

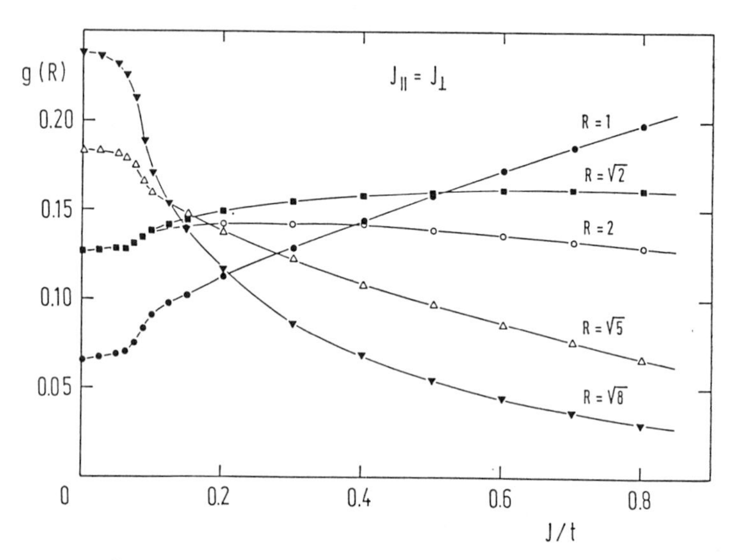

FIG. 3. Frequency dependent conductivity $\sigma(\omega)$ vs ω in units of t for $N = 8 \times 4$ and $J/t = 0.2$ ($N_r = 6$).

BINDING OF QUASIPARTICLES

We studied the binding of QP by numerically solving the t-J model with $N_h = 2$ holes on a 4×4 lattice.[4] In the whole regime of J/t that we investigated, the $N_h = 2$ ground state was found to be a spin singlet $S = 0$ and triply degenerate[4,5], corresponding to $\vec{k}_0 = (0,0)$, $(0,\pi)$ and $(\pi,0)$. In order to test the binding we calculate the binding energy of the hole pair,

$$\Delta = E_0\,(N_h = 2) - 2E_0\,(N_h = 1) + E_0\,(N_h = 0)\,, \tag{9}$$

and the hole density correlation function

$$g(\vec{R}) = \sum_i \langle \Psi_0 | n_h(\vec{R}_i) n_h(\vec{R}_i + \vec{R}) | \Psi_0 \rangle. \tag{10}$$

$g(\vec{R})$ as presented in Fig. 4 as well as Δ clearly indicate on the bound state of a hole pair at $J/t > 0.2$. Whereas $|\Delta| \simeq J$ in this regime, the hole density correlations fall off with distance so that the finite size effects seem not to be crucial. On the other hand the $S = 1$ hole pair state was found to be very weakly bound, i.e. $\Delta \sim 0$.

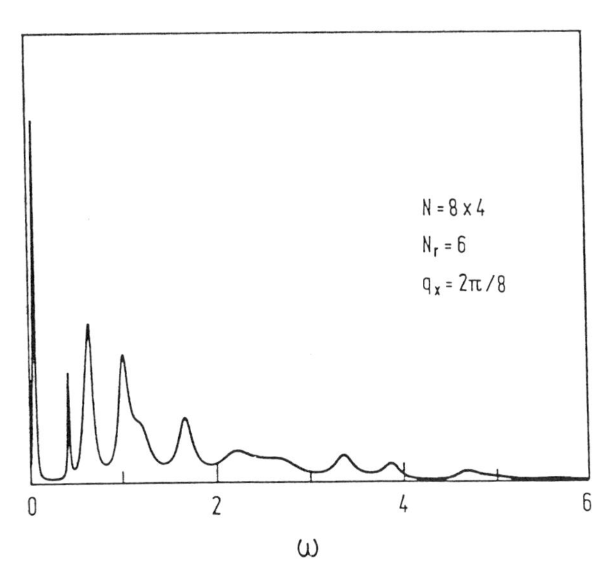

FIG. 4. Hole - density correlations $g(R)$ vs. J/t.

A gradual change to an entirely different state below $J/t < 0.1$ is evident also from nearest neighbour spin correlations $C(\vec{R})$,

$$C(\vec{R}) = \sum_i \langle \Psi_0 | \vec{S}(\vec{R}_i) \cdot \vec{S}(\vec{R}_i + \vec{R}) | \Psi_0 \rangle, \tag{11}$$

which become ferromagnetic - like for nearest neighbours below $J/t < 0.1$. Such a situation can be simply explained by the formation of two oppositely polarized ferromagnetic spin polarons which repel each other, what is consistent with the attractive-repulsive transition observed in $g(R)$.[4]

The origin of a substantial hole binding for $J/t > 0.2$ regime is still not understood. Clearly, so large effect cannot be explained by a simple broken bond argument.[21] We investigate this question further by performing the diagonalization in a restricted basis set with a finite number of reversed spins N_r (relative to the Néel state). In Fig. 5 we present the result for the density correlations $g(R)$ as a function of N_r. It is quite surprising that the correct qualitative behaviour is obtained already with small $N_r > 1$. Moreover the $N_r \geq 2$ accounts well also for the attraction - repulsion transition at $J/t \sim 0.2$. Since at $N_r = 2$ the number of involved states is not large, the binding should be quite local phenomenon clearly related to the AFM correlated background. Our results stimulate the use of analytical approaches (cumulant expansion) using t/J as an expansion parameter. Our diagonalization results as well as preliminary cumulant expansion results indicate on several contributions to the hole binding: a) the static exchange bond contribution [21], b) the quantum interference effect [17], preventing the loss of kinetic energy of two holes to the order t^2/J, c) lowering of the kinetic energy of two holes due to the reduction of the local AFM order (spin bag effect).

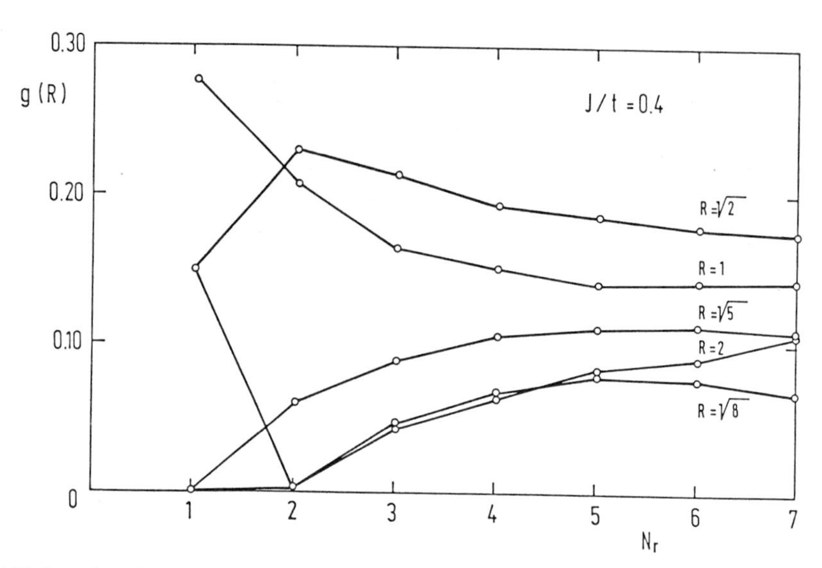

FIG. 5 Hole - density correlations $g(R)$ at fixed $J/t = 0.4$ as a function of the number of reversed spins N_r.

HIGH DOPING REGIME

In order to investigate the many-hole state and the possible SC hole pairing, we also performed the exact diagonalization of the $N = 4 \times 4$ system with $N_h = 3, 4$,[7] corresponding to concentrations $x = N_h/N < 0.25$, representing in real copper oxides the substances with highest T_c. For $N_h = 4$ we found in the whole regime the ground state to be again a spin singlet $S = 0$. A clear effect of higher doping is the reduction of spin correlations $C(R)$ as shown in Fig. 6 as a function of the concentration x. Whereas at low doping $x < \frac{1}{16}$ correlations are qualitatively consistent with $C(R)$ in a layered quantum AFM, weak AFM correlations remain essentially only among nearest neighbours $R = 1$ at high doping $x = \frac{1}{4}$. A decrease of the AFM correlation length ξ can be in our system tested also by the AFM Fourier component

$$\tilde{C}_{\mathrm{AFM}} = \sum_i e^{i\vec{q}_0 \cdot \vec{R}} C(\vec{R}), \quad \vec{q}_0 = (\pi, \pi), \tag{12}$$

which would be related to the correlation length as $\tilde{C}_{AFM} = A\xi^2$, at least for $\xi \gg 1$. From our results in Fig. 6 we get $\tilde{C}_{AFM} \propto \frac{1}{x}$ in the relevant regime $x > \frac{1}{16}$. This is consistent with experiments and with a simple argument [15] that the average distance between holes determines the AFM correlation length, i.e. $\xi \propto \frac{1}{\sqrt{x}}$.

An information on the collective state of holes in a $N_h = 4$ system can be gained

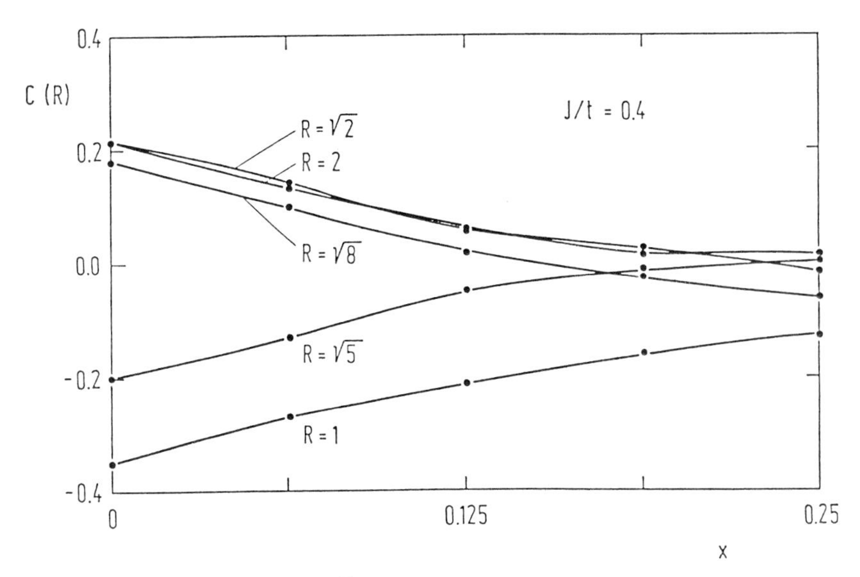

FIG. 6. Spin correlations $C(\vec{R})$ vs hole concentration x, at fixed $J/t = 0.4$.

from the four-point density correlations

$$G(\vec{R}_1, \vec{R}_2, \vec{R}_3) = \sum_i \langle \Psi_0 | n_h(\vec{R}_i) n_h(\vec{R}_i + \vec{R}_1) n_h(\vec{R}_i + \vec{R}_2) n_h(\vec{R}_i + \vec{R}_3) | \Psi_0 \rangle. \tag{13}$$

We present here G for four characteristic configurations, as shown in Fig. 7. At very large $J/t > 1.2$, G_1 gives the largest contribution what shows that the model becomes unstable against the formation of droplets in this unrealistic regime. G_1 is strongly suppressed with decreasing J/t and the G_4 correlation becomes dominant. This can be interpreted as an indication for a paired state, where pairs with the interhole distance $R = \sqrt{2}$, being the most probable in this regime, are at the largest possible interpair distance in such a small system. We note also that G_4 is the largest at the intermediate $J/t = 0.4$. Although the many hole state is in certain properties very similar to a dilute fermion system with $N_f = N_h$ [7], e.g. in hole density correlations $g(R)$, G correlations are substantially different. The main difference is in the exchanged role of G_4 and G_2, so that N_f system does not show a tendency towards pairing while $N_h = 4$ does.

Our results show that the exact diagonalization of small systems in spite of its deficiencies yields important results on the properties of a single hole in an AFM , on the nature of hole binding and on the nature of their collective state. Our analysis also shows that the effective t-J model exhibits attractive quasiparticle interactions and pairing phenomena and should be further considered as a possible model for superconductivity at high temperatures.

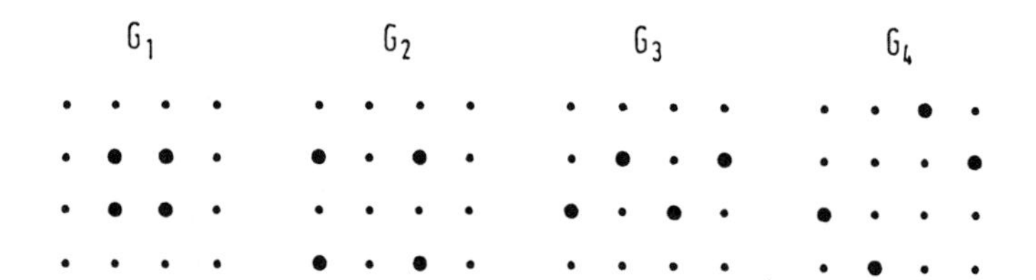

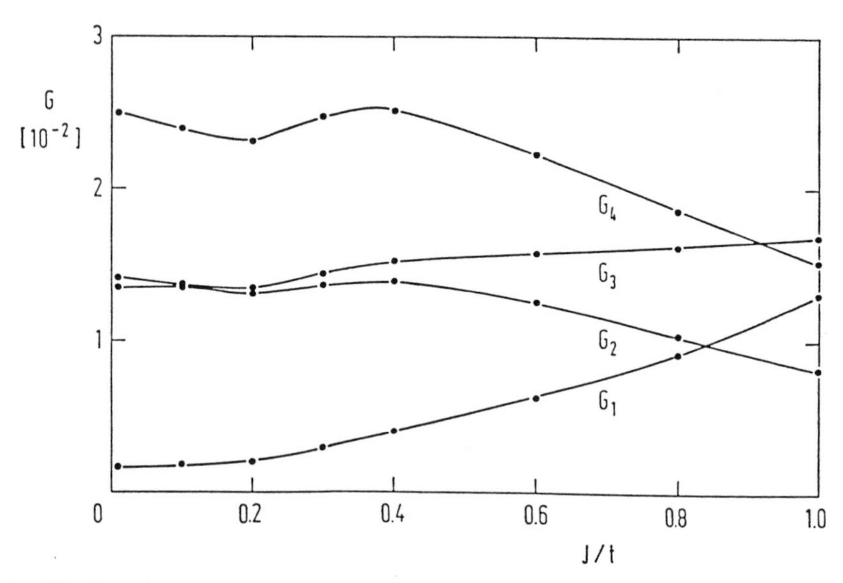

FIG. 7. The four-point density correlations for configurations G_1 to G_4 vs J/t, at fixed $x = \frac{1}{4}$.

REFERENCES

1. For an overview see e.g. Proc. of Int. Conf. on High - Temperature Superconductivity, Interlaken, eds. J. Müller and J. L. Olsen, Physica C **153-155** (1988); Proc. of Int. Conf. on High - Temperature Superconductivity, Stanford 1989.
2. J. Oitmaa and D. D. Betts, Can. J. Phys. **56** , 897 (1978); E. R. Gagliano, E. Dagotto, A. Moreo and F. C. Alcaraz, Phys. Rev. B **34**, 1677 (1986).
3. E. Kaxiras and E. Manousakis, Phys. Rev. B **37**, 656 (1988).
4. J. Bonča, P. Prelovšek and I. Sega , Phys. Rev. B **39**, 7074 (1989).
5. Y. Hasegawa and D. Poilblanc, Phys. Rev. B **40**, 9035 (1989).
6. E. Dagotto, A. Moreo, R. Joynt, S. Bacci and R. Gagliano, preprint; K. J. Szczepanski, P. Horsch, W. Stephan and M. Ziegler, preprint.
7. J. Bonča, P. Prelovšek and I. Sega, Europhys. Lett. **10**, 87 (1989).
8. I. Sega and P. Prelovšek, to be published.
9. see A. Parola, in this volume.
10. M. Ogata and H. Shiba, J. Phys. Soc. Jpn. **57**, 3074 (1988); J. E. Hirsch, S. Tang, E. Loh and D. J. Scalapino, Phys. Rev. Lett. **60**, 1688 (1988).

11. V. J. Emery , Phys. Rev. Lett. **58**, 2794 (1987).

12. P. Prelovšek, Phys. Lett. A **126**, 287 (1988); J. Zaanen and A. M. Oles, Phys. Rev. B **37**,9423 (1988); V. J. Emery and G. Reiter, Phys. Rev. B **38**, 4547 (1988).

13. A. Ramšak and P. Prelovšek, Phys. Rev. B **40**, 2239 (1989).

14. F. C. Zhang and T. M. Rice, Phys. Rev. B **37**, 3759 (1988).

15. Y. Endoh *et al.*, Phys. Rev. B **37**, 7443 (1988); Birgeneau R.J. *et al.*, Phys. Rev. B **38**, 6614 (1988).

16. for a recent review see T. M. Rice, Proc. of 9th Gen. Conf. of Condensed Matter Division of EPS, Nice, 1989.

17. S. A. Trugman, Phys. Rev. B **37**, 1597 (1988); B. I. Shraiman and E. D. Siggia , Phys. Rev. Lett. **60**, 740 (1988).

18. C. L. Kane, P. A. Lee and N. Read, Phys. Rev. B **39**, 6880 (1989).

19. T. M. Rice and F. C. Zhang, Phys. Rev. B **39**, 815 (1989).

20. G. A. Thomas *et al.*, Phys. Rev. Lett. **61** , 1313 (1988).

21. J. E. Hirsch, Phys. Rev. Lett. **59**, 228 (1987).

Copper Spin Correlations Induced By Oxygen Hole Motion

M.W. Long

School of Physics, Bath University, Claverton Down

BATH, BA2 7AY, United Kingdom

1. Introduction

The superconductivity in the perovskite systems is one of the least well understood, but most important, aspects of the Cuprates. Antiferromagnetism is another important facet of these systems, but the magnetism is very well understood in comparison to the superconductivity. The ease with which the magnetic phenomena are modelled, leads to a natural tendancy for magnetic explanations to *all* possible behaviour. We will try to develop the corresponding charge motion explanations for some of the characteristics, including even some magnetic phenomena.

The magnetism is usually modelled with static spins on the Copper atoms which interact with each other via either the Heisenberg Hamiltonian or the $x-y$ model. The systems are driven superconducting by doping of charge carriers. The way charge carriers are included in magnetic descriptions is usually with *static* inclusions to the spin Hamiltonian. The two natural inclusions are either missing spins, corresponding to electron doping, or impurity bonds, corresponding to hole doping. The physics associated with these ideas, is that actually on the impurity the spin order is distorted and the spins in the neighbourhood are then disturbed by the existing spin interactions. Unfortunately, elementary considerations indicate that this picture is difficult to justify. The exchange process between spins is a fourth order 'hopping' process whereas the delocalisation process is a second order 'hopping' process and hence would be expected to locally dominate in the vicinity of a charge carrier. The spin distortion around a charge carrier would be expected to be mediated by the charge motion and *not* the spin exchange. In this article we will develop a description for the spin correlations induced by the *dynamics* of the charge carriers.

Dynamics of Magnetic Fluctuations in High-Temperature Superconductors
Edited by G. Reiter *et al.*, Plenum Press, New York, 1991

We are describing the behaviour of charge carriers, and as such we should be developing a useful foundation with which to describe the superconductivity, which is basically coherent, correlated charge motion.

Firstly we will describe the experimental foundations and then we will discuss the present theoretical background.

1.1 Experimental Considerations

The phase diagrams of the perovskite superconductors involve several phases, all of which suggest likely physical effects at work. The first interesting phase, which seems to be found for most systems, is an antiferromagnetic insulating phase. In the simplest case, this phase is found in a stochiometric 'parent' compound and is associated with well defined charge states for each of the atoms. The superconductivity is then found when positive charge carriers are doped into this antiferromagnetic phase, and seems to occur simultaneously with the delocalisation of the charge carriers. Prior to the paramagnetic superconducting phase, there is sometimes observed a 'spin glass' phase, where 'freezing' of spins is observed with local probes and irreversible phenomena are found, but there is no discernible long range magnetic order. A simple understanding of these phases in terms of charge carrier motion is our basic physical motivation.

With the recent discovery of electron doped superconductors, it has become possible to compare the behaviour of electrons and holes as charge carriers. Although the superconductivity is directly comparable, the antiferromagnetism is quite asymmetrical. A very small hole concentration eliminates the magnetic order, but an order of magnitude larger electron concentration is required to kill the antiferromagnetism. Any decent description of the properties of these systems ought to predict this asymmetry.

There are two quite natural probes of the low energy spin excitations; Nuclear Magnetic Resonance (NMR) and inelastic neutron scattering. Both probes have produced some exciting new results.

NMR has demonstrated that, at low temperatures, the Oxygen holes and Copper spins are in strong interaction and that the low energy excitations are probably *composites* of both. The basic evidence is that below a certain temperature, both the Copper nucleus and Oxygen nucleus show the same temperature dependence of the nuclear spin relaxation rate[1], indicating that the same excitation mediates both processes. Above this temperature, the relaxation rates are seen to be rather different. Another exciting result is that there is an indication that something occurs *above* the superconducting transition temperature; precursor spin correlations. This new temperature scale seems to be just over 100K and may signal a change in the behaviour of the Copper spin system.

The most exciting development in inelastic neutron scattering has been the discovery of a low temperature *gap* in the spin excitation spectrum[2], which seems to be a precursor to the superconducting phase transition. At a similar temperature to the NMR precursor phenomenon, there is a drop in the intensity of the low energy inelastic magnetic scattering. Restricted to a region of $E < 10\text{meV}$, the result suggests the opening of a small gap in the excitation spectrum of the Copper spins and perhaps a corresponding change in the spin structure.

A theory which suggests a low temperature coupling between Oxygen holes and Copper spins seems called for, and further, a modification to the natural spin exchange interactions at low energies with new spin coherence induced would be an attractive prospect. This is precisely the physical content of our calculations, where the motion of a hole dominates many spins in its vicinity producing a strongly coupled quasi particle, combined with the prediction that the spin coherence induced by the hole is that of short range valance bonds; coherence which is expected to have a gap in the spin excitations.

1.2 Theoretical Considerations

The magnetism in the stochiometric compounds is fairly easy to understand in terms of a square lattice of magnetic strong coupling Copper atoms which interact with kinetic exchange induced antiferromagnetic Heisenberg interactions. Doping is achieved by atomic substitution or atomic vacancies in regions fairly well removed from the two dimensional layers where the conductivity is assumed to take place. The dopants are often disordered, and the resulting random electrostatic interactions with the charge carriers, are the natural explanation for the spin glass effects. Simple Anderson localisation ideas would also suggest that this disordered potential could explain the large region where the systems are doped but do not conduct. Once the doping is sufficient to wash out the randomness, then the materials become superconductors. If we ignore the random potential as being an unnecessary complication to start, then the first task of a theory is to find a microscopic model of the materials which can be used to describe the systems.

If we assume that the phenomena are basically two dimensional and confined to the CuO_2 planes, then there is only one natural tight binding model of the relevant planes, the $d-p$ model, which in its simplest form is:

$$H = T\sum_{i\sigma} d^{\dagger}_{i\sigma}d_{i\sigma} + U\sum_{i} d^{\dagger}_{i\sigma}d^{\dagger}_{i\bar{\sigma}}d_{i\bar{\sigma}}d_{i\sigma} + E\sum_{j\sigma} p^{\dagger}_{j\sigma}p_{j\sigma} + t\sum_{<ij>\sigma}(d^{\dagger}_{i\sigma}p_{j\sigma} + p^{\dagger}_{j\sigma}d_{i\sigma}) \quad (1.1)$$

where $d^{\dagger}_{i\sigma}$ creates a *hole* of spin σ (complementary spin $\bar{\sigma}$) on a Copper atom (denoted by i) and $p^{\dagger}_{j\sigma}$ creates a *hole* of spin σ on an Oxygen atom (denoted by j). There are three relevant energy scales; $\Delta = E - T$, the relative stability of a single hole on a Copper atom versus an Oxygen atom; U, the Coulombic penalty against adding a second hole to a Copper atom and, t, the hybridisation energy between orbitals on neighbouring atoms (denoted by $< ij >$). Although we have included the dominant effects in this description, direct Oxygen-Oxygen hopping, Oxygen-Oxygen onsite repulsion and nearest neighbour Copper-Oxygen repulsion have all been neglected; all of these omissions constitute arguable approximations.

The superconducting compounds are modelled with one hole per Copper atom corresponding to the parent compounds and then doping of extra Oxygen holes leading to the superconductivity. The parent compounds are insulating and antiferromagnetic, which suggests a parameterisation with $U > \Delta > | t |$.

The Hamiltonian combined with the parameterisation constitues a well defined problem, but unfortunately, this problem is much too difficult to solve and so further simplification is required. Two different approximations have so far been pursued; Firstly, the Oxygen atom topology has been argued to be irrelevant and then the Hubbard model on the square lattice might constitute a simpler alternative starting point. Secondly, the parameterisation includes some *limits* which might prove both simpler and fruitful. The most natural limit involves allowing the hopping to become infinitesimal, viz $t \mapsto 0$, and this has lead to a study of the $t - J$ model. One of the conclusions of this article is that naively taking the hopping to zero does *not* yield the $t - J$ model, but a much more interesting model which we term the $X - J$ model in this article.

The square lattice Hubbard model[3]:

$$H = -t \sum_{<ii'>\sigma} d^{\dagger}_{i\sigma} d_{i'\sigma} + U \sum_{i} d^{\dagger}_{i\sigma} d^{\dagger}_{i\bar{\sigma}} d_{i\bar{\sigma}} d_{i\sigma} \tag{1.2}$$

where $d^{\dagger}_{i\sigma}$ creates a hole of spin σ which must be carefully interpreted if a connection to the Cuprates is to be found, is a much simpler Hamiltonian with only one relevant energy scale; t/U, the relative strength of the hopping compared to onsite Coulomb penalty. The parameterisation which yields a Mott insulating antiferromagnet is $U >| t |$. This Hamiltonian and parameterisation yields a much studied problem which is *still* probably too difficult to understand. The idea of allowing the hopping to become infinitesimal may be applied to this model, and the result is in fact precisely the $t - J$ model, although this model is also suggested as the relevant limit to the $d - p$ model.

The $t - J$ Hamiltonian is:

$$H = -t \sum_{<ii'>\sigma} (1 - d^{\dagger}_{i\bar{\sigma}} d_{i\bar{\sigma}}) d^{\dagger}_{i\sigma} d_{i'\sigma} (1 - d^{\dagger}_{i'\bar{\sigma}} d_{i'\bar{\sigma}}) + \frac{J}{2} \sum_{<ii'>} \mathbf{S}_i . \mathbf{S}_{i'} \tag{1.3}$$

where $\mathbf{S}_i = (1/2) \sum_{\sigma\sigma'} d^{\dagger}_{i\sigma} \hat{\sigma}_{\sigma\sigma'} d_{i\sigma'}$ are the relevant spin operators, and the projection operators surrounding the hopping restrict attention to the subspace with no double occupancy of sites. This Hamiltonian can be fairly rigorously justified as the strong coupling limit to the Hubbard model[4], provided that $J = 4t^2/U$. Can this Hamiltonian also be justified as a limit of the $d-p$ model? Most groups seem to agree that it can be used[5], although there is still controversy[6]. It is straightforward to find the corresponding Hamiltonian for the $d-p$ model and the $X-J$ model results[7]. Our main motivation in this article is to show that the $X-J$ Hamiltonian does *not* reduce to the $t-J$ Hamiltonian in general and has an important physical effect which is not present in the $t-J$ model.

The $X-J$ model is:

$$H = -X(1+\alpha)\sum_i B_i^\dagger B_i + X(1-\alpha)\sum_i \mathbf{T}_i^\dagger.\mathbf{T}_i + \frac{J}{2}\sum_{<ii'>} \mathbf{S}_i.\mathbf{S}_{i'} \qquad (1.4)$$

in terms of the singlet and triplet pair creation operators, which have been symmetrically averaged over the four Oxygen atoms surrounding the relevant Copper atom:

$$B_i^\dagger = \frac{1}{\sqrt{2}}\sum_{<ij>\sigma} \sigma p_{j\sigma}^\dagger d_{i\bar{\sigma}}^\dagger \qquad (1.5a)$$

$$\mathbf{T}_i^\dagger = \frac{1}{\sqrt{2}}\sum_{<ij>\sigma\sigma'} p_{j\sigma}^\dagger \hat{\sigma}_{\sigma\sigma'} d_{i\bar{\sigma}'}^\dagger \qquad (1.5b)$$

where $\hat{\sigma}$ are Pauli matrices, and the $\mathbf{S}_i$ are the same Copper spin operators as before.

The reparameterisation of the system in this limit is achieved by:

$$X = \frac{t^2 U}{\Delta(U-\Delta)} \qquad (1.6a)$$

which corresponds to the t in the $t-J$ model:

$$J = \frac{4X^2}{U}(1-\alpha)^2(1+\frac{1}{\alpha}) \qquad (1.6b)$$

the analagous J, but there is a new parameter:

$$\alpha = \frac{\Delta}{U} \qquad (1.6c)$$

which has no analogue in the $t-J$ model and is the measure of whether the Oxygen level is close to the Cu^+/Cu^{2+} level ($\alpha \mapsto 0$) or to the Cu^{2+}/Cu^{3+} level ($\alpha \mapsto 1$).

It is important to realise that the superexchange *vanishes* in the absence of Cu^+ excitations and the experimental antiferromagnetism on a 250K energy scale, together with the lack of Cu^{3+} in spectroscopic experiments, strongly suggest that the experimental limit has a small value of α.

The connection between the $X-J$ Hamiltonian and the $t-J$ Hamiltonian is best attempted in the limit $\alpha \mapsto 1$, since the triplet operators, $\mathbf{T}_i^\dagger$, disappear from the Hamiltonian. The idea is that, firstly the description is already restricted to the Copper lattice, secondly that a hole in the $t-J$ description corresponds to the symmetrised singlet combination, viz $B_i^\dagger$, and thirdly that orthogonalising these operators yields an effective nearest neighbour hopping[8]. For the case of one doped hole, the mapping is exact as $\alpha \mapsto 1$[9]. The problems emerge in the opposite limit as $\alpha \mapsto 0$, and we will devote a lot of thought to this limit in section 2.

Now that we have the Hamiltonians with which to model the Cuprates, we need to evaluate whether or not they yield the basic physics, and to what extent their physical behaviour can be deduced.

It is now generally believed that the antiferromagnetic Heisenberg model on the square lattice in two dimensions has a Neel ordered ground state at zero temperature. The kinetic exchange induced Heisenberg interaction leads immediately to an explanation for the experimentally observed magnetic order. The J aspect of either the $t-J$ or $X-J$ models is the natural starting point, and therefore the Hubbard model or $d-p$ models both yield reasonable descriptions of the magnetism. The important question, when trying to compare the two models, relates to the behaviour of the holes. The holes go onto Oxygen sites and therefore the $d-p$ model and hence the $X-J$ model is the natural starting point. What does it predict and how might we compare it with the $t-J$ model or even experiment? Unfortunately, the systems yield only very qualitative behaviour for comparison. Doping destroys the Neel order and replaces it with superconductivity. Superconductivity is too subtle a correlation to deduce from the Hamiltonian and so we are left with the destruction of the Neel order. All descriptions predict the loss of antiferromagnetism, so how can we compare the $t-J$ and $X-J$ models? The answer seems to lie in the comparison between hole doped and electron doped systems.

The electron doped systems are, in principle, much easier to describe than the hole doped systems. The electrons in the planes must sit on the Copper sites and then the electrons can move around with a t^2/Δ hopping to neighbouring Copper atoms. This system is very well described by the strong coupling square lattice Hubbard model, or equivalently the $t-J$ model.

Any difference between the electron doped system and the hole doped systems would be naturally ascribed to differences between the $X-J$ and $t-J$ Hamiltonians. This is the line we will pursue, and the natural experimental difference between the two systems is the asymmetry in the antiferromagnetism. We will argue that the $X-J$ model predicts this asymmetry.

There is a magnetic explanation to the antiferromagnetic asymmetry, which arises from the fact that the doped holes reside on Oxygen sites and prefer the neighbouring Copper spins to be antiparallel. If both spins are antiparallel to the Oxygen hole, then they are ferromagnetically aligned and the implied 'ferromagnetic bond' could 'frustrate' the Neel order in a much stronger way than the simple 'dilution' caused by introducing spinless Cu^+ sites. This ferromagnetic bond has previously been argued for on exchange[10] and hybridisation[11] grounds. We will argue that the Oxygen hole is a much more vicious destroyer of Neel order than this ferromagnetic bond, which we consider to be a misleading aspect.

2. Copper Spin Correlations in a Small Cluster

We will consider the spin correlations induced by one added Oxygen hole in interaction with the Copper spins via the $X - J$ Hamiltonian in this section. There are *three* quite natural limits, each with quite particular Copper spin correlations; $\alpha = 1$ and $J = 0$ corresponds to hole motion by virtual Cu^{3+} excitations and finds a mapping of the model onto the strong coupling Hubbard model[3-4,9], for which hole motion yields Nagaoka ferromagnetism[12]; $\alpha = 0$ and $J = 0$ corresponds to hole motion by virtual Cu^+ excitations and yields a form of paramagnetism[13]; $X = 0$ corresponds to the absence of holes and yields the square lattice Heisenberg model and probably long range antiferromagnetism[14]. The correlations which are the least well understood are the paramagnetic correlations of the Cu^+ limit. This is the situation where motion on a *frustrated* topology leads to *low* spin Nagaoka coherence and we give evidence for a low spin ground state in this section together with some interpretation of the mechanism.

Our main motivation in this article is the *competition* between these three phenomena. The cluster has been chosen for its high symmetry, which facilitates calculations, and because it has an *even* number of relevant fermions, which allows total spin zero solutions. We will study the $X - J$ model by exact diagonalisation of one single cluster of seven Copper atoms and sixteen Oxygen atoms:

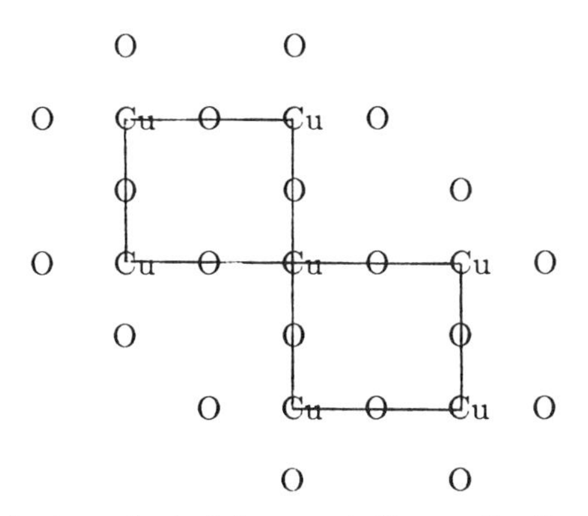

Figure 1. The cluster selected for exact diagonalisation in this article.

There are two types of state competing for the role of ground state. Firstly there is the 'High Spin' state[8-9], which is the ferromagnetic ground state to the Hubbard model limit ($\alpha \mapsto 1$). This state finds the Oxygen hole in a relative singlet

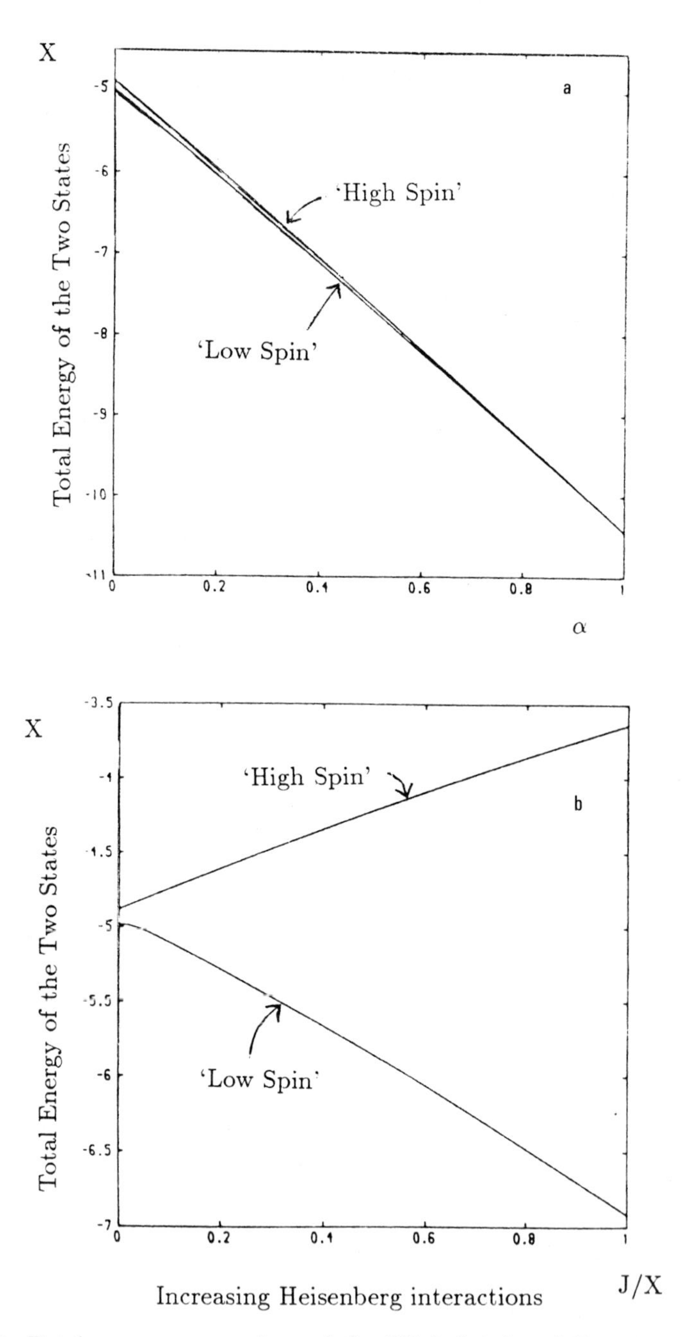

Figure 2. Total energy comparison of the 'High Spin' and 'Low Spin' States.
(a) Cu^{+} Hopping compared with Cu^{3+} Hopping.
(b) Cu^{+} Hopping compared with Heisenberg interactions.
(c) Cu^{3+} Hopping compared with Heisenberg interactions.

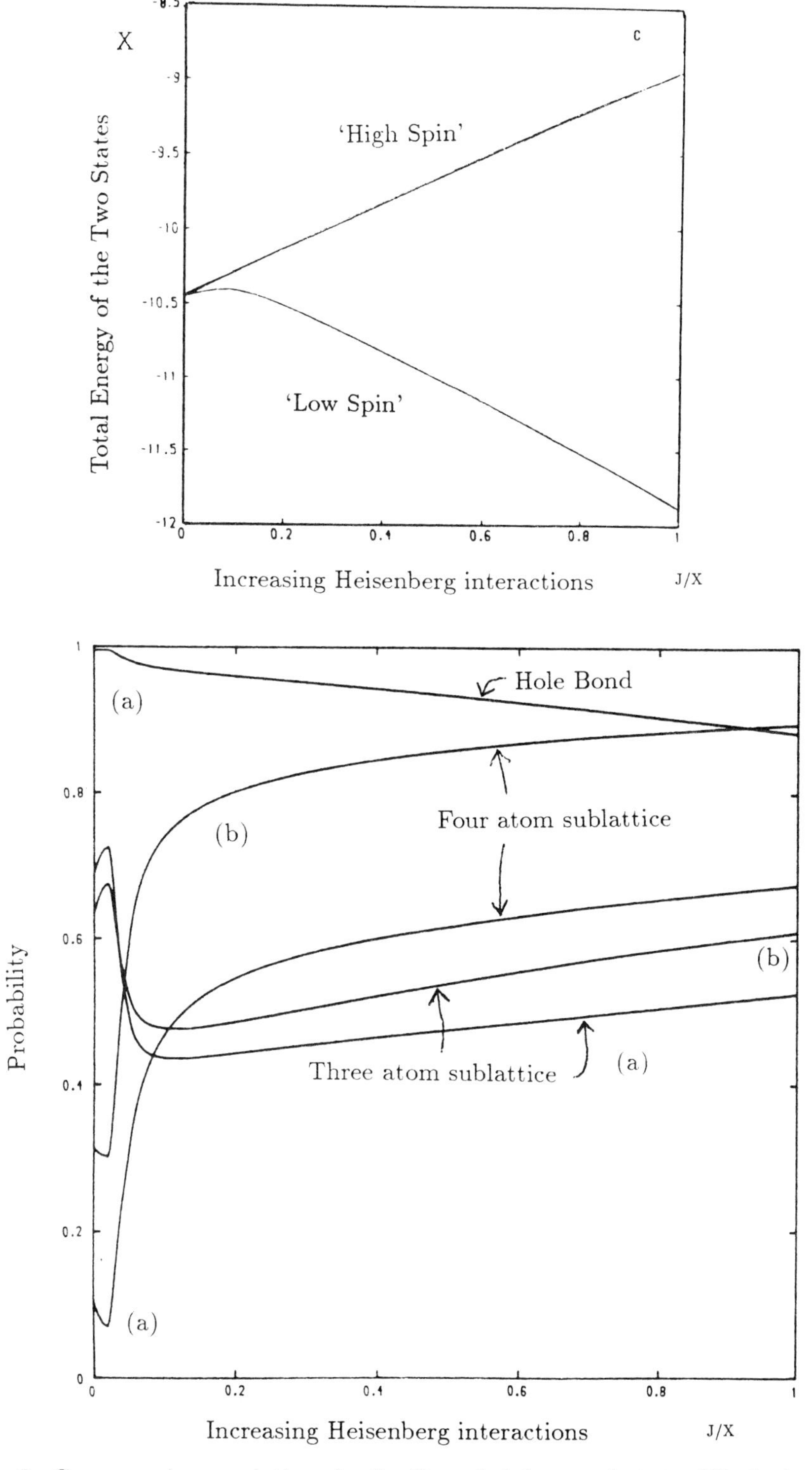

Figure 3. Copper spin correlations in the 'Low Spin' ground state. We depict the probability that the Hole Bond is triplet and the total spin on the two magnetic sublattices, normalised so that saturated ferromagnetism is unity. (a), (b) denote hole positions.

with one of its two neighbouring Copper holes and all the other Copper holes aligned ferromagnetically. The hole has uniform phase which optimises coherence around all the closed loops, satisfying the restrictions of Nagaoka[12], which apply to the present one hole case. This state also corresponds to a single magnon bound to the hole. Secondly there is a 'Low Spin' state with the minimum allowable total spin and a wavevector corresponding to the non-interacting Fermi surface.

We picture the ground state energies of the 'Low Spin' and 'High Spin' bound magnon states as the parameters are varied in figure 2. The basic physics is clear, the low spin state is vigorously favoured by both hole motion via virtual Cu^{+} excitations *and* by the Heisenberg interactions, whereas the high spin state is only favoured by hole motion by Cu^{3+} excitations. In the absence of Heisenberg interactions, the two states are nearly degenerate, but surprisingly the 'Low Spin' state is the ground state! It is stabilised by the effect which is omitted from the $t-J$ model.

Although the total energy calculations of figure 2 give a comparison which is of use in determining the phase diagram, they do not lead to insight into the types of correlations inherent in the competing low energy states. The remainder of the article will try to address the possible ways of comparing and interpreting these correlations.

It is easy to understand the difference between the high spin and low spin states, but is there a big difference between the low spin correlations induced by hopping via Cu^{+} excitations and those induced by the Heisenberg interactions?

We have elected to calculate the *square of the total spin* on each of the two sublattices as a measure of the Neel order. In the absence of hopping, the small sublattice of three sites achieves 87.5% of the maximum possible value (viz 3.75), whereas the large sublattice of four sites achieves 89.3% of the maximum value (viz 6). The values of these unit normalised sublattice moments are plotted in figure 3, for the cases where the hole neighbours the central Copper atom (denoted by (a)) and where the hole is on a square but not neighbouring the central Copper atom (denoted by (b)). Ignoring the initial behaviour, which corresponds to the irrelevant crossing of two eigenstates (the state sympathetic to Heisenberg interactions having the same quantum numbers and being $0.033X$ above the ground state), it is clear that the hopping is not inconsistent with a large local *unoriented* antiferromagnetic moment.

A second correlation which is of some interest, is the spin coherence of the pair of Copper spins which neighbour the hole. The exchange arguments of Aharony[10] and the hybridisation arguments of Emery[11] both suggest a local ferromagnetic correlation. The probability that the relevant pair of spins is in a triplet is a straightforward calculation and the result is plotted in figure 3. The local ferromagnetic correlations are clearly observed for both clusters.

There are three quite remarkable conclusions from these calculations:
(1) Low spin correlations are induced by pure hole motion in the Cu^+ limit.
(2) *Unoriented* Neel order is not inconsistent with hole motion by Cu^+.
(3) The ferromagnetic bond surrounding the hole is present, but does not seem to destroy local antiferromagnetism.
It turns out that all three of these effects are connected and can be simultaneously understood.

In figure 4, we present the spin configurations found when a hole is passed along a chain of CuO:

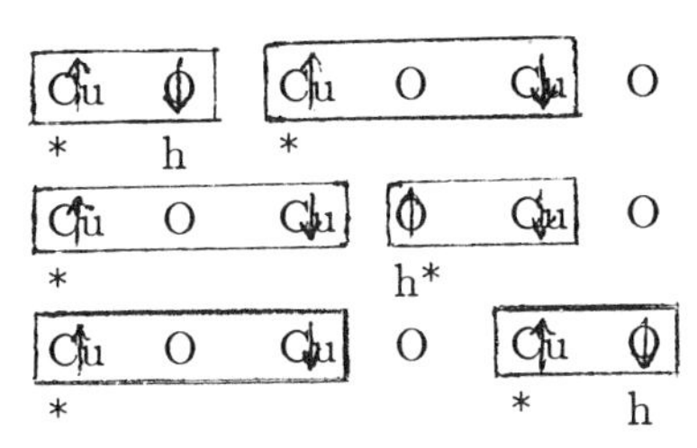

Figure 4. Hole motion via Cu^+ along a chain.

The hole is denoted by h and the spins in boxes form 'Valence Bonds'. In the first situation, the sites labelled by (*) are expected to have local triplet correlations. As the hole moves on, the triplet is seen to become a coupling between next nearest neighbour Copper atoms. *The local ferromagnetic hole bonds become Neel correlations once the hole has passed.* The second crucial feature is that the hole moves from being in a relative singlet with respect to one Copper atom, *directly* to being in a singlet with respect to a *next nearest neighbour* Copper atom. This effect suggests a next nearest neighbour hopping matrix element if a $t-J$ model is to be considered[15]. More importantly, the eventual singlet configuration is formed from a 'Valence Bond' which was originally passive in the Copper spin background. This is the source of the low spin correlations induced by hole motion. We expect a short range 'Resonating Valence Bond' state to be stabilised by this effect.

3. Conclusions

The strong coupling limit of the $d-p$ model is the $X-J$ model, *not* the $t-J$ model. There is a new physical phenomenon present in the $X-J$ model, which is not contained in any Hubbard model description. Passive spin singlets in the Copper spin background can become active singlets when a hole is present, leading to a huge gain in local hopping energy. The hole motion induces coherent 'Resonating Valence Bond' correlations in its vicinity, and so the hole quasi particles and Copper spin correlations are inter-related, as is suggested by the low energy NMR relaxation rates. It is well known that short range RVB states have a gap to spin excitations, and a gap has recently been seen in inelastic neutron scattering experiments. The low spin 'Valence Bond' correlations are complementary to *unoriented* Neel fluctuations and hole motion quite naturally leads to high spin correlations on each antiferromagnetic sublattice.

Finally, we would like to give an interpretation for the precipitous loss of Neel order in the hole doped materials, and a reason for its absence in the electron doped materials. Let us assume that holes move via the $X - J$ model and that electrons move via the $t - J$ model. Electron motion will try to generate high spin correlations locally, as suggested by the Nagaoka theorem[12]. Holes on the other hand, will try to generate the low spin correlations studied in this article. If we now consider the hole to be a 'Spin Polaron'; a charge with a surrounding cloud of sympathetic spins, then the crucial observation is that the hole polaron would be expected to be much larger than the electron polaron. This result follows from the fact that local high spin correlations are very expensive in Heisenberg energy, whereas our low spin correlations are very cheap. A simple percolation argument for localised polarons can then be invoked to explain the loss of long range Neel order. The fact that hole polarons are larger, means that the antiferromagnetism will be destroyed correspondingly quicker. Another way to understand this result comes from previous RVB analysis. It was soon realised that variational estimates of Heisenberg energies from RVB states without long range Neel order were remarkably close to the true ground state energy. The actual saving in energy from long range order is minimal, *provided the state which replaces the ordered state is a carefully chosen RVB state.* The low spin states of the present article satisfy precisely this criterion and therefore easily destroy the antiferromagnetism.

References

[1] Takigawa M; Present Volume
[2] Shirane G; Present Volume
[3] Hubbard J; 1965 Proc Roy Soc A **285** p542
[4] Hirsch JE; 1985 Phys Rev Lett **54** p1317
[5] Eg; The groups lead by PW Anderson and TM Rice.
[6] Emery VJ; Present Volume
[7] Long MW; 1988 Z Phys B: Condensed Matter **69** p409
[8] Zhang FC and Rice TM; 1988 Phys Rev B **147** p3759
[9] Zhang FC; 1989 Phys Rev B **39** p7375
[10] Aharony A, Birgeneau RJ, Coniglio A, Kastner MA and Stanley HE; 1988 Phys Rev Lett **60** p1330
[11] Emery VJ and Reiter G; 1988 Phys Rev B **38** p3759
[12] Nagaoka Y; 1966 Phys Rev **147** p392
[13] Long MW; 1988 J Phys C: Solid State Phys **21** L939
[14] Fischer MA and Young P have produced independent contributions with collaborators.
[15] Sawatzky GA; Present Volume

SPIN POLARONS IN THE $t - J$ MODEL

J M F Gunn[†] and B D Simons[‡]

[†]Rutherford Appleton Laboratory, Chilton, Didcot, Oxon OX11 0QX

[‡]Cavendish Laboratory, Madingley Road, Cambridge CB3 0HE

INTRODUCTION

Any theory of high temperature superconductors must consider charge carriers (usually holes) moving through spins with short range antiferromagnetic order. (this is demonstrated, particularly, by the results of neutron (Birgeneau *et al.* 1988a, 1988b) and Raman (Lyons *et al.* 1988) scattering experiments. The two schools of thought on magnetic mechanisms for superconductivity focus either on the short-range antiferromagnetism or the long-range paramagnetism as the underlying cause of the phenomenon. It is not clear that these points of view are really distinct.

In this article we will mainly be concerned with the first viewpoint and indeed with how a *single* hole in an antiferromagnet (treated by the $t - J$ model) affects, and is affected by, the spin system. The disruption of the magnetic order around the hole, along with the hole itself, defines a "spin polaron". We will be interested in considering its size, the nature of the magnetic order within its extent and whether it may have "internal" excited states.

HOLES IN ANTIFERROMAGNETS

The subject of added particles (electrons or holes) in antiferromagnets has an old history dating back, at least, to the work of de Gennes (1960), where he discussed the distortion of the magnetic structure to form a "spin polaron". There were three simplifying features in the situations that he examined in his treatment: firstly the electron which was causing the distortion was not part of the band that was forming the magnetic moments; secondly the spins were treated in a semiclassical manner, which is of dubious applicability in the case of interest here — spin 1/2; finally the hopping was not coupled to the spin system in the manner that it is explicitly in the Emery model and implicitly in the Hubbard model.

In the Hubbard model all these difficulties are really the same thing: the single electron on each site *is* the magnetic moment, which is thus spin 1/2 and the spins are coupled to the motion in a literal manner — when the hole moves, a spin (*i.e.* an electron)

must move in the opposite direction. In the Emery model the first of de Gennes' simplifications remains, but the others are not appropriate.

Prior to high temperature superconductivity there was much investigation of the effect of holes on magnetism in the Hubbard model, motivated in the main by the magnetic transition metal oxides. For our present purposes, there are two particularly important results from that work: firstly Nagaoka's (1966) result for one hole in an array of infinitely repulsive ($U = \infty$) electrons and secondly the results of Nagaev and co-workers (Bulaevskiĭ *et al.* 1967, for a review see Nagaev 1974) on "quasi-oscillators".

Nagaoka's theorem considers the effect of a hole on the order of the spin system in the *absence* of superexchange. The essence of the result can be appreciated in a very simple example of the Hubbard model — a triangle of three sites with one hole and two electrons, one up and one down. The electrons have an infinite U, so that there is never any double occupancy of sites (and hence no superexchange). Imagine generating the states connected by the hole motion: after moving the hole around two sites we find that the spin arrangement is the *opposite* of the initial one, despite the hole occupying its initial site (see figure 1). Thus the hole must move around the triangle *twice* (thrice for a square) to return to its initial configuration. If the hopping integral is negative, then the ground state will be the equal superposition of all of the six states generated as the hole moves around (as their energies are degenerate).

To determine the magnetic ordering induced by the hole, note that the spin configurations are superposed with the same sign: the spins are in a relative triplet, which is the Nagaoka result in this case. (Note that this triplet has $S^z = 0$ so that it points perpendicular to the "sublattice" magnetisation — the direction that we selected as the initial axis of quantisation.) We may then note two points: firstly that the excited states have singlet magnetic "order"; secondly that if the hopping integral is positive, then the *ground* state is singlet, as the sign of the superposition is changed.

Nagaoka's actual result was that the ground state of one hole in a square lattice (for infinite U) is ferromagnetic. In this case the sign of the hopping term is irrelevant, unlike in the triangle (due to an even number of hops being required to superpose spin configurations with the hole in the same place). This makes the important point that lattices which cannot be divided into two sublattices (non-bipartite) are distinct and the nature of the ground state for the triangular *lattice* (with positive hopping integral for the hole) is unknown. Secondly the result of triangular example suggests that the form of the magnetic order depended on the degree of motion (the "momentum") of the hole; this seems a largely unexplored issue (although results on the density of states for excitations were obtained by Brinkman and Rice 1970).

The simplest form of superexchange is Ising, as in that case the spins have no dynamics without the intervention of the hole. Nagaev and his collaborators are performed an *approximate* treatment of one hole in the *presence* of superexchange, the approximations being particularly appropriate in the Ising limit. They envisaged a hole placed in a Néel state leaving a trail of displaced spins as it moved. The spins are all displaced one lattice parameter, at least until the hole re-encounters its own trajectory. Since the displaced spins are on the wrong sublattice, their neighbours transverse to the motion of the hole are *ferromagnetically* aligned. Thus the energy of the spin system is increased and the "string" of flipped spins acts as a potential varying (approximately) linearly with arc-length that the hole has travelled. This tends to confine the hole to its initial position. This picture is precise in the limit of high dimensionalities as then the number of self-intersections of the hole trajectory are negligible. In reality the spins in the Cu-O planes have Heisenberg exchange; we will argue that this picture is a useful starting point for the consideration of the spin 1/2 case: the *quantum* spin polaron.

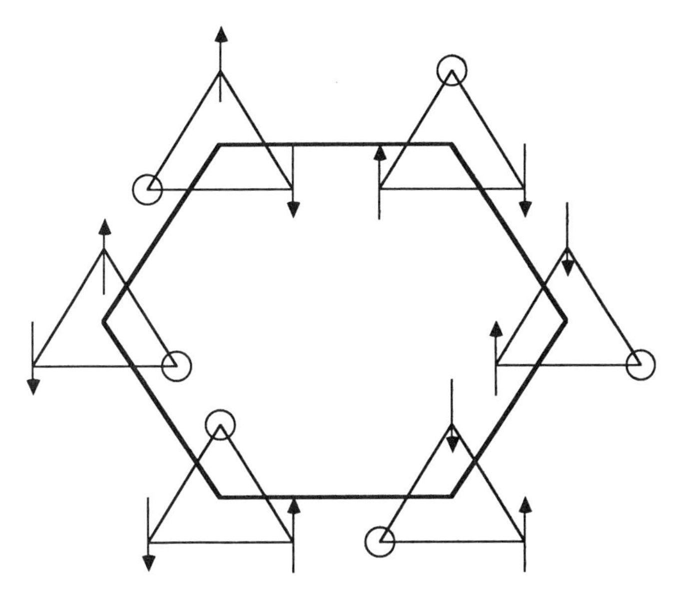

Figure 1. Generating the configuration space for the triangle with one hole.

ISING SPIN POLARONS

Although it is not immediately apparent, Nagaoka's and Nagaev's results are related; the magnetic order (in the ground state) induced by the string of flipped spins is ferromagnetic (on a bipartite lattice). In this section we will see why this is true, and deduce some results about the "internal" excitations of spin polaron (Simons and Gunn [1989]). Finally we determine the effective mass and mention the effect of long-range spin waves.

Before deducing the ground state of the Ising spin polaron, we need to be able to describe the configuration space, or the *argument* of the wavefunction, of the hole and string of flipped spins. The point to note is that one needs not merely the position of the hole, but also the trajectory that it has followed (*i.e.* the positions of the displaced spins). The construction of such states is revealed in figure 2. We construct such states from a hole placed in an initially perfect Néel state. Initially the hole may move to one of the four (we choose the square lattice) neighbours. The second "generation" of states are constructed by moving the hole to the next neighbours (moving backwards produces one of the states already considered). And so on. This procedure demonstrates that the configuration space of the hole plus spins is a Bethe Lattice. (In fact it overcounts the states, but rather weakly so.) It is important to note that this is not merely a bad representation of the two-dimensional lattice that the hole moves upon: for instance the state where the hole moves "up" and then to the right is not the same as that when the hole moves to the right *first* and then "up". If we were merely considering a single particle problem then these would be identical.

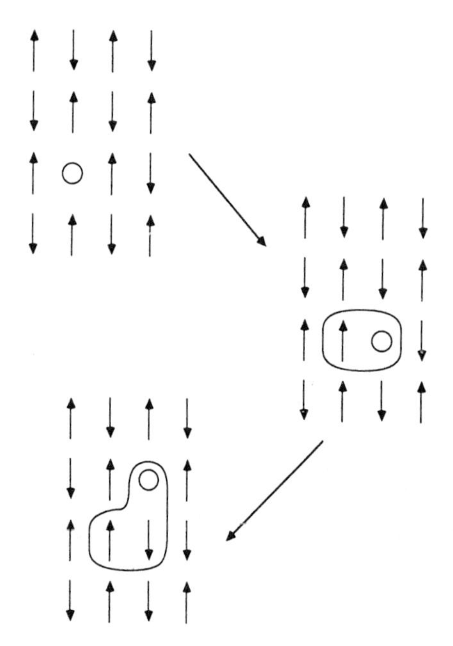

Figure 2. The construction of the hole plus spin configuration space by hole motion in an initially perfect Néel state.

Turning to the Hamiltonian, the string provides a potential which depends linearly on the "generation number" of the above construction, the coefficient being roughly $J/2$ (where J is the exchange constant). (This statement is still true *on average* when the hole encounters its trajectory; the result becomes exact in high dimensionalities.) The Hamiltonian may be written as (n is the generation number of a site on the Bethe lattice and $\mathbf{i}$ is the "trajectory" encoded in the sequence of choices made at the earlier generations of the lattice):

$$\mathcal{H} = t \sum_{n\langle\{\mathbf{i}\}\{\mathbf{j}\}\rangle} \{|n;\{\mathbf{i}\}\rangle\langle n+1;\{\mathbf{j}\}| + h.c.\} + \sum_{n\{\mathbf{i}\}} V_n |n;\{\mathbf{i}\}\rangle\langle n;\{\mathbf{i}\}| \tag{3.1}$$

Here $\langle\{i\}\{j\}\rangle$ indicate paths which up to n are equivalent. (The states $|n;\{i\}\rangle$ are naturally associated with the Bethe lattice sites, $n;\{i\}$.) We will assume that the hopping parameter, t, is positive. V_n is defined as:

$$V_n = \begin{cases} (Z-2)\frac{J}{2}n + (Z-1)\frac{J}{2} & n \geq 0, \\ 0 & n = -1 \end{cases} \tag{3.3}$$

If we assume the hopping integral, t, is negative (in fact this assumption is unecessary on the square lattice as all configurations with the hole in the same place differ by an even number of hops) then it is clear that the ground state superposes all the relevant configurations with the same phase. That is amplitudes with the hole on the same site, but the spins displaced, have the same phase. This implies that the displaced spins tend to have their *total* spin increased — *i.e.* the hole-induced order amongst the spins is ferromagnetic. This is obvious for two spins and may be deduced for the case of several spins being on the wrong sublattice by noting that:

$$\begin{aligned} \mathbf{S}_{\text{tot}}^2 &= 2\sum_{n=1}^{N} \mathbf{S}_n^2 + 2\sum_{n \neq n'} \mathbf{S}_n \cdot \mathbf{S}_{n'} \\ &= 2\sum_{n=1}^{N} \mathbf{S}_n^2 + 2\sum_{n \neq n'} \left(S_n^z S_{n'}^z + \frac{1}{2}\left[S_n^+ S_{n'}^- + S_{n'}^+ S_n^-\right] \right) \end{aligned}$$

In evaluating the expectation value of this, the only part that has off-diagonal contributions from the superposition of different spin configurations is the sum over the spin raising and lowering operators. To make this contribution maximal ('ferromagnetic') the different spin configurations must have the same phase. Since the ground state *does* have the same phase for all spin configurations, the hole — within its confined region — tends to create ferromagnetic (Nagaoka) order. This is the reconciliation between the two old results. Note that the total S^z must still be zero as the states that we are considering have $S_{\text{tot}}^z = 0$; this is a quantum analogue of canting — the quantum feature is that the symmetry in the plane normal to the sublattice direction is unbroken. (Remember that in state $S_{\text{tot}}^z = 0$, the expectation values of S^x and S^y are zero.)

In parallel with the static local distortion of the magnetic order around the hole, one expects local, or "internal", excitations of the polaron. Indeed the excitation spectrum (within the manifold of total $S^z = 0$ states) of the Nagaoka polaron may be readily calculated (Simons and Gunn [1989]). The excited states (shown in figure 3) fall into two classes: "invariant", with the same phase on configurations of the same generation (these have been discussed before by several authors *e.g.* Bulaevskiĭ *et al.* 1967), and "noninvariant", where the phases of the different members of a generation are different. The latter are a manifestation of the string not merely being the agent for the linear potential but also enlarging the configuration space from that of a hole in a nonmagnetic background. These noninvariant states are lowest lying excited states, of order J above the ground state, and determine the optical absorbtion, due to parity (Simons and Gunn [1989]). These excitations may account for the maximum in the optical absorption at *ca.* 0.3eV ($J \simeq 0.1eV$) observed by Thomas *et al.* (1988).

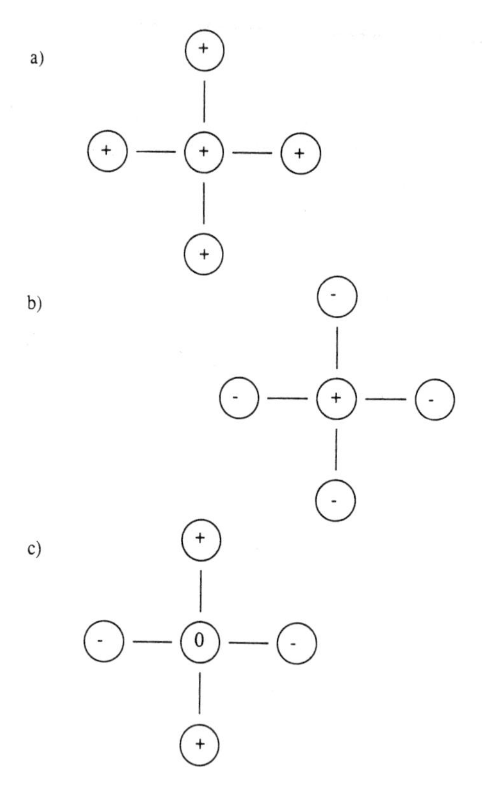

Figure 3. The phases of the different components of the wavefunction, up to the first generation, for: a) the ground state; b) an invariant excited state; c) a noninvariant excited state.

SPINLESS HEISENBERG POLARONS

The real Cu-O planes have negligible magnetic anisotropy. Unsurprisingly, the inclusion of the internal dynamics of the spin system (*i.e.* the spin waves) is not easy: one must resort to numerical work (*e.g.* Bonča *et al.* 1989 and Hasegawa and Poilblanc 1989) on small systems or to more drastic approximations (than in the Ising case) analytically (*e.g.* Kane *et al.* 1989 and Gros and Jonson 1989). For the small-scale behaviour, the most important process is that the string can dissolve *via* the transverse spin matrix element by the initial two displaced spins flipping each other. Two obvious questions are: how does this change the spin distortion around the hole from the Ising case? What is the "band structure" of the resulting quasiparticles?

Most treatments of these problems are performed in k-space (*e.g.* Kane *et al.* 1989 and Gros and Jonson 1989), but see Sachdev (1989)). The virtue of this is that it is likely to get the long distance structure (see Shraiman and Siggia [1988] and Godfrey and Gunn [1989]) correct, as the long wavelength behaviour of the spin waves are probably accurate — semiclassical approximations such as Holstein-Primakoff may be more successful there. However the short range behaviour, where one expects some coherence in the spin-flip (wave) "emission" by the hole, *i.e.* they are emitted sequentially on neighbouring sites, is rather difficult to handle in k-space (almost by definition).

A different approach (Gunn and Simons [1989]) is to treat the short-distance behaviour more accurately, whilst surrendering information about the large-scale effects. Since the purpose is to treat any coherence among the spin-flips it is natural to use the same basis as in the Ising case and then to truncate at some generation. These basis states are then associated with each site in one sublattice and an effective Hamiltonian formed by using the matrix element associated with the two displaced spins flipping by the transverse spin coupling. In the calculations of Gunn and Simons (1989) a basis set of seventeen states was used — this being the number of walks of length two and the minimum number required for a consistent treatment at the number of hops required for the basic relaxation process of the string.

Flat bands are a dominant feature of the results — see figure 4 for the case of $J/t = 0.2$. It is very easy to create spin distortions that cannot move by ensuring that there are "nodes" in the wavefunction for the configurations that are coupled to the neighbouring sites by the transverse spin coupling. Of the seventeen bands generated by the seventeen basis states, thirteen are dispersionless. (The fact that there are seventeen bands, as against one, is a warning of the *essentially* many-particle nature of the quasiparticles.) The ground state *is* dispersive, with a band width that varies as J if t is not too large. This is in qualitative agreement with the results of angle-resolved photoemission (*e.g.* Takahashi *et al.* 1989).

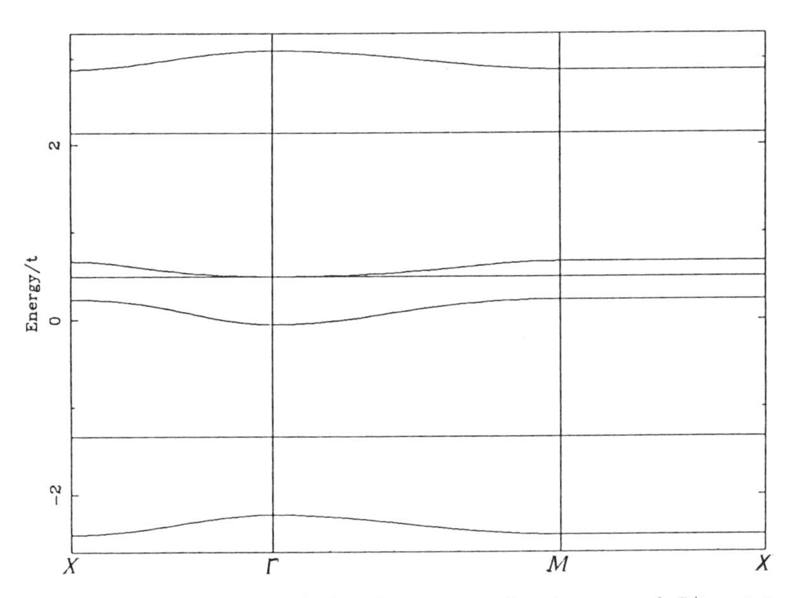

Figure 4. The quasiparticle band structure for the case of $J/t = 0.2$.

However the most striking result is that the correlations amongst pairs of spins around the hole depend on the position in the Brillouin Zone: at the band minimum (at the magnetic zone boundary) the correlations are of a singlet form, however at the centre of the zone they are triplet. It is the change in phase of the hole wavefunction as the hole moves from magnetic cell to cell that induces this change of symmetry from the Ising case. (In the Ising case the special trajectories discussed by Trugman (1988), originally discovered in the consideration of the motion of a vacancy in an ordered alloy — Elcock and McCombie (1958) and Elcock (1959), allow the hole to move from cell to cell, but in that case the band minimum is at the band centre.) Thus the polaron becomes "spinless", perhaps a microcosm of the paramagnetic (spin-liquid) state which the antiferromagnet becomes as the doping is increased.

Of course the case where $t \gg J$ is outside the domain of validity of our approximations (due to the truncated nature of our basis set) and it may be that there there is a Nagaoka polaron régime. Indeed some numerical calculations show a sudden increase in the total spin of the ground state as J/t is decreased; however it is consistent to deduce that at that point the size of the polaron is the size of the system that they consider. That is, although the polaron may be an overall singlet, there may be large regions which have high spin within it.

CONCLUSIONS

In summary we would propose that the experimental data (on energy scales somewhat larger than the superconducting gap) are consistent with holes moving through a spin system that has short-range Néel order. The resulting quasiparticles are likely to be complicated entities, due to the strong correlations between the spins and the holes. Calculations on a perfectly ordered Néel state imply that polarons with a locally enhanced singlet content are energeticaly preferred. The effective mass is proportional to the exchange constant J and is consistent with the bandwidths observed in angle-resolved photoemission. Finally the size of the spin polarons, for the estimated experimentally relevant régime is *ca.* 5 lattice sites, implying that the polarons are not "closed-packed" in superconducting samples at low doping. The nature of the superconducting state is still mysterious.

ACKNOWLEDGEMENTS

We would like to thank M W Long for many fruitful discussions.

REFERENCES

Birgeneau R.J, D.R. Gabbe, H.P. Jensen, M.A. Kastner, P.J. Picone, T.R. Thurston, G. Shirane, Y. Endoh, M. Sato, K.Yamada, Y. Hikada, M. Oda, Y. Enomoto, M. Suazuki and T. Murakami, *Phys. Rev. B* **38**, 6614 (1988).

Birgeneau R.J., Y. Endoh,, K. Kakurai, Y. Hidaka, T. Murakami, M.A. Kastner, T. R. Thurston, G. Shirane and K. Yamada 1989 *Phys. Rev. B* **39** 2868.

Bonča J, P. Prelovšek and I. Sega 1989 submitted to *Europhys. Lett.*

Brinkman W.F. and T. M. Rice, 1970 *Phys. Rev. B* **2**, 1324

Bulaevskiĭ L.N, É.L. Nagaev and D.I. Khomskiĭ, 1967 *Sov. Phys. JETP* **27**, 836

Elcock E.W., 1959 *Proc. Phys. Soc.* **73** 250

Elcock E.W. and C.W. McCombie, 1958 *Phys. Rev.* **109** 605

de Gennes P.G. 1960 *Phys. Rev.* **118**, 141

Godfrey M.J. and J.M.F. Gunn 1989 *J. Phys. Condens. Matter.* **1** 5821
Gros C. and M. D. Johnson 1989 to be published
Gunn J.M.F. and B.D. Simons 1989 submitted to *Phys. Rev. B*
Hasegawa Y and D. Poilblanc 1989 *Phys. Rev. B* **40** 9035
Kane C.L, P. A. Lee and N. Read, 1989 *Phys. Rev. B* **39**, 6880
Lyons K.B., P. A. Fleury, L. F. Schneemeyer and J. V. Waszczak, 1988 *Phys. Rev. Lett.* **60**, 732.
Nagaev É.L. 1974 *Phys. Stat. Sol. (b)* **65** 11
Nagaoka Y. 1966 *Phys. Rev.* **147**, 392
Sachdev S. 1989 *Phys. Rev. B* **39**, 12232
Shraiman B.I. and E.D. Siggia 1988 *Phys. Rev. Lett.* **61** 465
Simons B.D. and J.M.F. Gunn 1989 submitted to *Phys. Rev. B*
Takahashi T., H. Matsuyama, H. Katayama-Yoshida, Y. Okabe, S. Hosoya, K. Seki, H.Fujimoto, M. Sato and H. Inokuchi 1989 *Phys. Rev. B* **39**, 6636.
Thomas G.A., J. Orenstein, D. H. Rapkine, M. Capizzi, A. J. Millis, R. N. Bhatt, L. F. Schneemeyer and J. V. Waszczak 1988 *Phys. Rev. Lett.* **61**, 1313
Trugman S.A. 1988 *Phys. Rev. B* **37**, 1579.

ANALYTIC EVALUATION OF THE 1-HOLE SPECTRAL FUNCTION FOR THE 1-D t-J MODEL IN THE LIMIT $J \to 0$

Michael Ziegler and Peter Horsch

Max-Planck Institut für Festkörperforschung

Stuttgart, Federal Republic of Germany

Some twenty years ago Brinkman and Rice (BR) [1] studied in detail the form of the density of states (DOS) for a single hole in a half-filled one-band Hubbard model in the atomic limit , described by the effective Hamiltonian

$$H_t = -t \sum_{<i,j>,\sigma} (1 - n_{i,-\sigma})\, a^+_{i,\sigma} a_{j,\sigma} (1 - n_{j,-\sigma}) \; + \; h.c. \tag{1}$$

where the sum is over nearest neighbour pairs $< i, j >$ of electrons and σ takes on the values $+1$ (spin$\uparrow$) and -1 (spin$\downarrow$). Using a formulation introduced by Nagaoka [2] which involves the number of possible paths on a lattice and restricting these paths to the class of self- retracing ones (i.e. no closed loops), which they were able to sum to all orders, they found a reasonable approximation to the bulk of the lower Hubbard band in various background spin configurations (ferro-,antiferro-,paramagnetic). They also showed that this procedure, unlike various Green's function decoupling schemes, is equivalent to approximating the exact set of Green's functions by a simple recurrence relation which can easily be solved [3] . Moreover, in 1-D the retraceable path approximation turns out to be exact for topological reasons: it is not possible to form closed loops on a chain. In particular BR found the hole to be localized in the antiferromagnetic Néel configuration, i.e. the Green's function

$$\mathbf{G}^{AF}(k,\omega) = sgn(\omega) / \sqrt{\omega^2 - 4t^2} \tag{2}$$

is k-independent. Recently investigations of the Hubbard model in the large-U limit raised the question of how spin dynamics, described by adding to (1) a small Heisenberg AF (HAF) term H_J, affects the propagation of a single hole in the ground state of the half-filled system. Clearly one can separate the problem into two parts:
a) the effect of quantum spin fluctuations *in the ground state*,
b) inclusion of H_J *in the propagator*,
which in the limit of an infinitesimal Heisenberg coupling $J \to 0$ reduces to a).
As was pointed out by Brenig and Becker [4], who considered this case to lowest order in the spin fluctuations, these quantum deviations from the Néel state strongly influence the DOS in 1-D leading to a *dispersion of the BR continuum on the scale of the hopping energy t*. This dispersion, however, is not sharply peaked in the sense of a quasiparticle

dispersion and is not to be confused with the low-energy quasiparticle of mass $m \sim J^{-1}$ found by Kane, Lee and Read [5] in the 2-D case where propagation with H_J (case b)) becomes essential.
In the present work we focus our attention on the *exact* calculation of the propagator matrix

$$\mathbf{G}(k,\omega) = (\omega - E_0 + H_t - i\delta)^{-1} \tag{3}$$

in a *subspace* S_t of the total 1-hole Hilbert space which is a reducible representation of H_t and includes the 1-hole Néel configuration, as described below (because translational symmetry is unbroken we can define Bloch momentum-eigenstates) ; E_0 is the energy of the half-filled reference state. We obtain the matrix elements in closed form and observe that in this basis they are *momentum-independent.* In principle one can then calculate the Green's function of the hole inserted into an *arbitrary* $S_z = 0$ configuration at half-filling, subject to the constraint that the resulting 1-hole state is projected onto S_t (here an even number of sites N is assumed). In view of what was said above it is clear that in this representation all momentum dependence arises from the particular background spin configuration chosen. Finally, we demonstrate how this works for a *spin-fluctuating state*, approximating the exact spin singlet ground state of the Heisenberg antiferromagnet by Bartkowski's wave function [6] which is a coherent superposition of spin-flips in the Néel state modelled after the ground state in linear spin wave theory. In the projected one-hole Bartkowski state the only allowed configurations are those in which the spin-flips form *strings connected to the hole.* We arrive at an analytic expression of $G(k,\omega)$ (as defined in (8)) by summing a convergent series to infinite order. Even though the continuum still extends between $-2t$ and $2t$ as in BR, its overall shape is now seen to be strongly momentum-dependent due to the spin fluctuations. This result is in good agreement with exact diagonalization calculations by von Szczepanski et al. [7] obtained in the complete Hilbert space starting from the half-filled singlet ground state.
The subspace S_t is obtained by annihilating an electron with momentum k and spin σ in the zero-momentum Néel state and successively applying the hopping Hamiltonian H_t to this state. One obtains the orthonormal basis

$$\begin{aligned} |\ 0\ >_{k,\sigma} &= \sqrt{2}\, a_{k\sigma}\, |\psi_{N\acute{e}el}> \\ |\pm n >_{k,\sigma} &= 2^{-1/2}\,(\,|R_n>_{k,\sigma} \pm |L_n>_{k,\sigma}\,)\,, \qquad n = 1,...,N/2-1 \end{aligned} \tag{4}$$

The state $|R_n\,(L_n)>_{k,\sigma}$ denotes a Bloch superposition of configurations in which the hole has been moved n times to the right (left) with respect to the Néel configuration, leaving behind a string of flipped spins:

$$|R_n\,(L_n)>_{k,\sigma} = (N/2)^{-1/2} \sum_{j=1}^{N} e^{ikr_j} a_{j+(-)n,\sigma(n)} \prod_{l=0}^{n-1} S_{j+(-)l}^{-\sigma(l)} |\psi_{N\acute{e}el}> \tag{5}$$

Here $\sigma(l) \equiv (-1)^l\sigma$ and $S_l^{\pm}$ are the spin raising /–lowering operators acting at site l. For the considerations that follow it is convenient to divide S_t into two subspaces, S_t^+ and S_t^-, spanned by states $0...N/2-1$ and $-1...-(N/2-1)$ respectively.
H_t connects the subspaces only via boundary terms and thus they effectively are decoupled in the thermodynamic limit $N \to \infty$. In this limit $\tilde{H}_t \equiv P_t H_t P_t$ has block-diagonal form

$$\tilde{H}_t = \tilde{H}_t^+ \oplus \tilde{H}_t^- \quad ; \tag{6}$$

P_t projects onto S_t and the superscripts + and − refer to $S_t^{\pm}$ respectively. Each block is by itself a half-infinite dimensional tridiagonal matrix

$$\tilde{H}_t^{\pm} = \begin{pmatrix} 0 & t\eta_{\pm} & & & \\ t\eta_{\pm} & 0 & t & & \\ & t & 0 & t & \\ & & t & \ddots & \ddots \\ & & & \ddots & \end{pmatrix} \tag{7}$$

where $\eta_{\pm} = \sqrt{(3 \pm 1)/2}$. Starting from a half-filled state, $|\psi_0^{(N)}>$, and defining $|\psi^{(N-1)}>= P_t a_{k\sigma}|\psi_0^{(N)}>$ the approximate single-particle Green's function

$$G(k,\omega) \equiv \frac{1}{2} \cdot \frac{<\psi^{(N-1)}|\mathbf{G}(\omega)|\psi^{(N-1)}>}{<\psi^{(N-1)}|\psi^{(N-1)}>} \tag{8}$$

decomposes, due to (6), into the sum of two contributions $G^{\pm}(k,\omega)$ which can easily be calculated once the propagator matrices $\mathbf{G}^{\pm} \equiv P_t\left(\omega + H_t^{\pm} - i\delta\right)^{-1} P_t$ are known ($E_0^{(N)} = 0$). The calculation of the latter proceeds in two steps. First, one determines $\mathbf{G}_{00} =< 0|\mathbf{G}^{+}|0>$ as well as $\mathbf{G}_{-1-1} =< -1|\mathbf{G}^{-}|-1>$ from which, next, all other matrix elements can be obtained by explicitly solving a recurrence relation based on the tridiagonal structure of $\tilde{H}_t$. The resulting expressions are:
i)

$$\mathbf{G}_{00}(\omega) = \frac{1}{\omega - 2\,\Sigma(\omega)}, \qquad \mathbf{G}_{-1-1}(\omega) = \frac{1}{\omega - \Sigma(\omega)}, \tag{9}$$

where the self-energy

$$\Sigma(\omega) = \frac{1}{2}\left(\omega - sgn(\omega)\sqrt{\omega^2 - (2t)^2}\right) \tag{10}$$

solves the self-consistency equation $\Sigma(\omega) = t^2/(\omega - \Sigma(\omega))$. It is convenient to use (10) in order to identify

$$\mathbf{G}_{00}(\omega) \equiv \mathbf{G}^{AF}(\omega) \tag{11}$$

and relate

$$\mathbf{G}_{-1-1}(\omega) = \Sigma(\omega)/t^2 = \left(1 - (\Sigma(\omega)/t)^2\right) \cdot \mathbf{G}^{AF}(\omega) \tag{12}$$

ii) defining

$$\gamma(\omega) \equiv \Sigma(\omega)/t, \qquad \beta(\omega) \equiv \gamma^2(\omega) \tag{13}$$

a general matrix element is given by

$$\begin{aligned} \mathbf{G}^{\tau}_{\tau\cdot m,\tau\cdot m}(\omega) &= (1 + \tau\cdot\beta^m)\cdot\mathbf{G}^{AF}(\omega) \\ \mathbf{G}^{\tau}_{\tau\cdot m,\tau\cdot n}(\omega) &= (-\gamma)^{n-m}\cdot\mathbf{G}^{\tau}_{\tau\cdot m,\tau\cdot m}(\omega) \\ \mathbf{G}^{\tau}_{\tau\cdot n,\tau\cdot m}(\omega) &= \mathbf{G}^{\tau}_{\tau\cdot m,\tau\cdot n}(\omega) \\ \mathbf{G}^{+}_{0,n}(\omega) &= \sqrt{2}(-\gamma)^n\cdot\mathbf{G}^{AF}(\omega) \\ &\tau \in \{+1,-1\}, \qquad m \geq 1, n \geq m \end{aligned} \tag{14}$$

Note that the self-energy $\Sigma(\omega)$ and hence the propagator $\mathbf{G}(\omega)$ is momentum-independent. Thus all k-dependence of the single-particle Green's function arises from the particular 1-hole state $|\psi^{(N-1)}>$ chosen and is of the form

$$G(k,\omega) = g(k,\omega) \cdot \mathbf{G}^{AF}(\omega) \tag{15}$$

with $g(k,\omega)$ a nonsingular function of its arguments. The spectral weight function

$$A(k,\omega) = \begin{cases} 0 & , \quad \omega > 2t \\ \frac{1}{\pi}\Im m G(k,\omega) & , \quad \omega < 2t \end{cases} \tag{16}$$

is nonzero in the interval $-2t < \omega < 2t$ as determined by the square-root behaviour of the self-energy

$$A(k,\omega) = -\frac{1}{\pi}\frac{\Re e(g(k,\omega))}{\sqrt{4t^2 - \omega^2}} \tag{17}$$

i.e. only the real part of $g(k,\omega)$ contributes.

A good example for these dispersive effects is provided by choosing for $\psi_0^{(N)}$ the Bartkowski state

$$|\psi_B^{(N)}> = \left(exp\left(-\alpha_B \sum_{i_A,\delta} S_{i_A}^- S_{i_A+\delta}^+ \right) + h.c. \right) |\psi_{N\acute{e}el}>_{q=0} \ , \tag{18}$$

which is a variational Ansatz for the ground state of the AF Heisenberg spin Hamiltonian. The parameter α_B measures the amount of spin fluctuations, and in 1-D its optimal value is determined to be 0.355 [6]. The corresponding one-hole state, projected onto S_t, is given by

$$\begin{aligned} |\psi^{(N-1)}>_{k,\sigma} = & \frac{1}{\sqrt{2}}|0>_{k,\sigma} \\ & + \sum_{l=1}^{\infty} \alpha_B^l \cdot (\cos(k\cdot(2l-1))\,|(2l-1)>_{k,\sigma} + i\sin(k\cdot(2l-1))\,|(-(2l-1))>_{k,\sigma} \\ & + \cos(k\cdot 2l)\,|(2l)>_{k,\sigma} + i\sin(k\cdot 2l)\,|(-(2l))>_{k,\sigma} \ ) \end{aligned} \tag{19}$$

Inserting (14) and (19) in (8) yields

$$G(k,\omega) = \left(g^{(1)}(k,\omega) + g^{(2)}(k,\omega)\right) \cdot \mathbf{G}^{AF}(\omega) \tag{20}$$

with

$$g^{(1)}(k,\omega) = \left(\frac{1}{2} + \alpha_B \sum_{\delta=-1}^{+1} \phi(\delta) \frac{1-\phi(\delta)}{1+\alpha_B\phi^2(\delta)} \right) \cdot \frac{1-\alpha_B^2}{1+3\alpha_B^2}$$

$$\begin{aligned} g^{(2)}(k,\omega) = \alpha_B^2 & \left(2\frac{1-\gamma\cos(k)}{1-\alpha_B^2} \right. \\ & \left. + \frac{1}{2}\sum_{\delta=-1}^{+1} \phi(\delta) \cdot \frac{(1-\phi(\delta))^2}{1+\alpha_B\phi^2(\delta)} \cdot \left(\frac{\phi(\delta)}{1+\alpha_B\phi^2(\delta)} + \frac{2\alpha_B}{1-\alpha_B^2} \right) \right) \cdot \frac{1-\alpha_B^2}{1+3\alpha_B^2} \end{aligned}$$

with $\phi(\delta) \equiv \gamma(\omega)\, e^{ik\delta}$.

Fig. 1 depicts the spectral distribution $A(k,\tilde{\omega})$ in the interval $-2t \ \leq \tilde{\omega} \ \leq 2t$, where

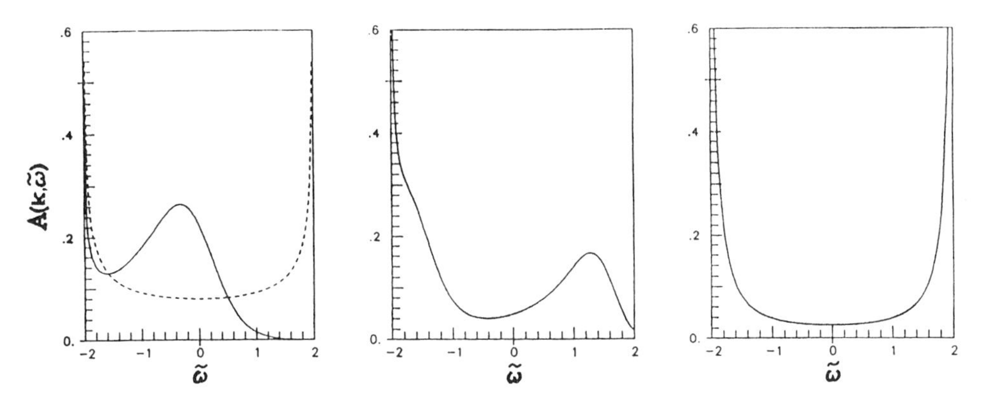

Fig. 1. Spectral function $A(k,\omega)$ in 1-D (N=∞) for $J = 0$ and various wave numbers $k = 0, \pi/4$ and $\pi/2$ (from left to right). The spectra are obtained in the restricted Hilbert space S_t using the variational state (19).

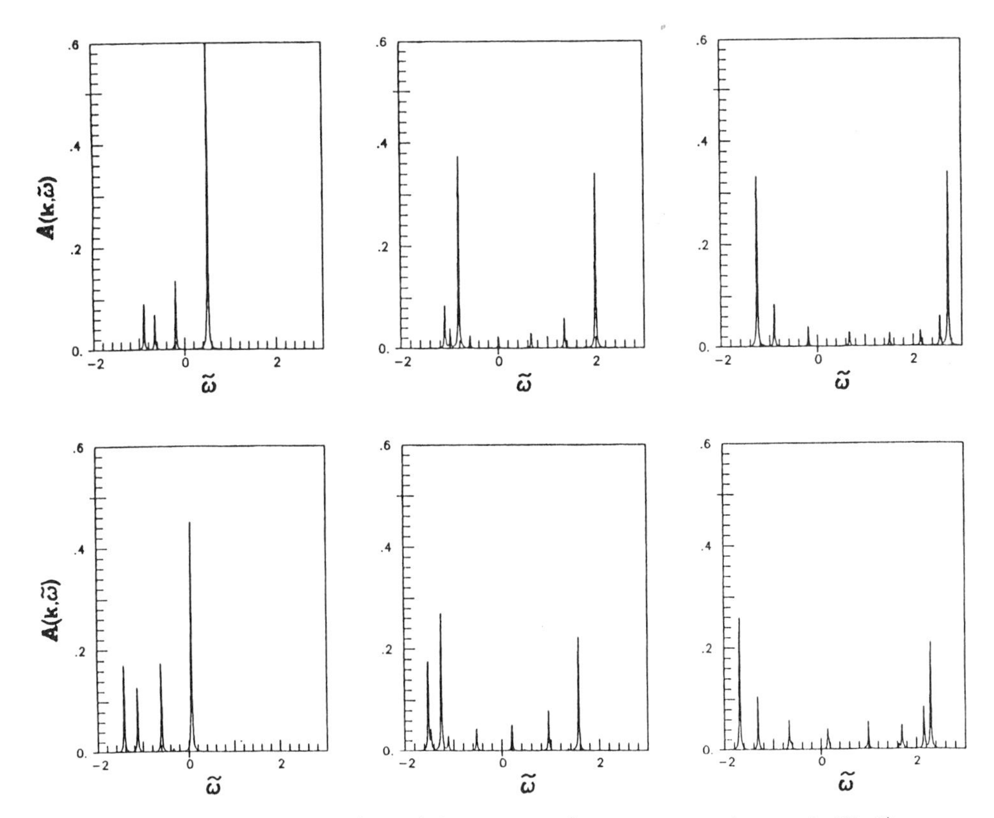

Fig. 2. $A(k,\omega)$ in 1-D (N=16) for $J = 0.2$ (same wave numbers as in Fig.1). The spectra in the upper panel (a) are obtained in the subspace S_t and in the lower panel are the corresponding results for the full Hilbert space. In both cases the hole is moving in the exact spin singlet HAF ground state.

$\tilde{\omega} \equiv -\omega$. Choosing $\tilde{\omega}$ as the relevant energy variable accounts for the fact that we are looking at the spectrum for a *hole* (lower hole energies to the left, higher hole energies to the right). For comparison we have plotted in Fig. 2 the exact diagonalization results for the t-J model ($J > 0$) both for the full Hilbert space and the subspace S_t. We observe good agreement in the finite size system between the two cases and thus we expect calculations restricted to S_t to provide a valid description of the spectral function also in the infinite system, *even for finite values of* J. Comparing Fig.1 with Fig.2 we conclude that the analytic result (20), based on the variational state (18), qualitatively reproduces the k-dependence of the single-hole spectral function as observed in the exact cluster calculation.

Acknowledgements. One of us (M.Z.) acknowledges support by the ESPRIT Program on High Temperature Superconductors. I would like to thank the Institute for Scientific Interchange (ISI) in Torino/Italy, where part of this work was completed, for its hospitality and stimulating environment.

REFERENCES

[1] W.F. Brinkman and T.M. Rice, Phys. Rev. **B 2**, 1324 (1970)

[2] Y. Nagaoka, Solid State Comm. **3**, 409 (1965); Phys. Rev. 147, 392 (1966)

[3] T.M. Rice and W.F. Brinkman, in *Critical Phenomena in Alloys, Magnets and Superconductors*, edited by R.E. Mills, E. Ascher and R.I. Jaffee, McGraw-Hill Series in Materials Science and Engineering (McGraw-Hill, 1971)

[4] W. Brenig and K.W. Becker, Z. Phys. **B 76**, 473 (1989)

[5] C.L. Kane, P.A. Lee, N. Read, Phys. Rev. **B 39**, 6880 (1989)

[6] R.R. Bartkowski, Phys. Rev. **B 5**, 4536 (1972)

[7] K.J. von Szczepanski et al., Phys. Rev. **B 41**, (Feb 1990)

DOPING EFFECTS ON THE SPIN-DENSITY-WAVE BACKGROUND

Z. Y. Weng and C. S. Ting

Department of Physics, University of Houston
Houston, Texas 77204-5504

I. INTRODUCTION

The single-band Hubbard model has been proposed[1] as a candidate to describe the high-temperature copper-oxide superconductors. There are two ways to approach such a strongly correlated electron system with the intrinsic 2D antiferromagnetic correlations. One is from the strong coupling, localized limit, where one could get a Mott insulator in the half-filled case with one electron per Cu site, described by the antiferromagnetic (AF) Heisenberg Hamiltonian. The alternative way is from the itinerant approach, where the insulating spin-density-wave (SDW) state is present at the half-filling. From the itinerant approach, the doping effects of holes on the SDW background will be discussed in the present paper.

It is noted that at half-filling, the mean-field SDW ground state will approach to the exact Néel state in the strong coupling limit. In the same limit, the collective modes obtained in the itinerant picture[5] agree with the spin wave excitations[6] in the antiferromagnetic Heisenberg model. This implies that the itinerant-perturbative approach could be also applied in the strong coupling regime with regard to the long range AF state although it could become only a metastable state upon doping.

In Section II, the ground state of one doped hole in the SDW background is studied[7]. Then, the effective interaction between two spin bags is calculated[7] in Section III. How the short-range AF state develops with doping will be studied[9] in Section IV. Finally, conclusions and discussions are presented in Section V.

II. A MOBILE SPIN BAG

The two-dimensional Hubbard Hamiltonian with nearest-neighbor hopping on a square lattice is given by

$$H = -t \sum_{<i,j>\sigma} (c_{i\sigma}^{+} c_{j\sigma} + h.c.) + U \sum_{j} n_{j\uparrow} n_{j\downarrow}, \tag{1.1}$$

which could be rewritten in two parts: $H = H_0 + H_f$ where H_0 is the mean-field part

$$H_0 = -t \sum_{<i,j>\sigma} (c_{i\sigma}^{+} c_{j\sigma} + h.c.) + U \sum_{j} (n_{j\uparrow} < n_{j\downarrow} > + < n_{j\uparrow} > n_{j\downarrow})$$
$$-U \sum_{j} < n_{j\uparrow} >< n_{j\downarrow} >, \tag{1.2}$$

and

$$H_f = U \sum_j \delta n_{j\uparrow} \delta n_{j\downarrow}. \tag{1.3}$$

where $\delta n_{j\sigma} = n_{j\sigma} - < n_{j\sigma} >$. H_f includes the fluctuations above the ground state.

In the mean-field approximation, H_f is neglected. Through the following canonical transformation[4,7]

$$c_{j\sigma} = \sum_{\mathbf{k}}{}' \frac{e^{i\mathbf{k}\cdot\mathbf{R}_j}}{\sqrt{N}} [(u_{\mathbf{k}} + \sigma p_j v_{\mathbf{k}})\alpha_{\mathbf{k}\sigma} + p_j(u_{\mathbf{k}} - \sigma p_j v_{\mathbf{k}})\beta_{\mathbf{k}\sigma}], \tag{1.4}$$

H_0 could be diagonalized as follows

$$H_0 = const. + \sum_{\mathbf{k},\sigma}{}' [(-E_{\mathbf{k}} - \mu)\alpha^+_{\mathbf{k}\sigma}\alpha_{\mathbf{k}\sigma} + (E_{\mathbf{k}} - \mu)\beta^+_{\mathbf{k}\sigma}\beta_{\mathbf{k}\sigma}], \tag{1.5}$$

where $\mu = -U/2$ at the half-filled case.

In Eq.(1.5), the $\mathbf{k}$-space operators $\alpha_{\mathbf{k}\sigma}$ and $\beta_{\mathbf{k}\sigma}$ describe respectively the lower and upper quasi-particle bands split by the SDW gap Δ and the wave vector $\mathbf{k}$ extends within the new Brillouin zone (i.e., the magnetic zone) which is reduced to the half of the original one with the nesting vector $\mathbf{Q} = (\pm\pi/a, \pm\pi/a)$ being a new reciprocal vector (see Fig. 1 of Ref.4). The coefficients $u_{\mathbf{k}}$ and $v_{\mathbf{k}}$ in Eq.(1.4) are determined by

$$u_{\mathbf{k}} = \left[\frac{1 - \varepsilon_{\mathbf{k}}/E_{\mathbf{k}}}{2}\right]^{1/2}; \tag{1.6a}$$

$$v_{\mathbf{k}} = \left[\frac{1 + \varepsilon_{\mathbf{k}}/E_{\mathbf{k}}}{2}\right]^{1/2}, \tag{1.6b}$$

in which

$$\varepsilon_{\mathbf{k}} = -2t[\cos(k_x a) + \cos(k_y a)]; \tag{1.7}$$

$$E_{\mathbf{k}} = \sqrt{\varepsilon^2_{\mathbf{k}} + \Delta^2}. \tag{1.8}$$

And the quantity p_j appearing in Eq.(1.4) is defined by

$$p_j = e^{i\mathbf{Q}\cdot\mathbf{R}_j} \tag{1.9}$$

which has only two values: ± 1 according to different lattice site $\mathbf{R}_j$. In fact, p_j determines the two sublattices of the SDW state. This can be seen from the staggered magnetization moment m_j at the lattice site j whose expression in the mean-field theory is given by

$$m_j = (2\Delta/U)p_j \tag{1.10}$$

Thus one gets two sublattices with the total spin at each site either up or down, corresponding to $p_j = +1$ and $p_j = -1$ respectively.

Therefore, the mean-field ground state $|0>$ in the half-filled case is an itinerant antiferromagnetic insulating state, with the lower quasi-particle band filled with electrons while the upper band being totally empty. When a hole is doped into such a SDW background, it is described by the quasi-hole state $|\mathbf{k}\sigma> = \alpha_{\mathbf{k}\sigma}|\mathbf{0}>$ with the energy spectrum $E_{\mathbf{k}} - \frac{U}{2}$ in the mean-field theory, which is an eigenstate of H_0. But beyond the mean-field approximation, one finds a coupling of the quasi-hole with the SDW background through H_f.

The motion of the doped hole is described by the Heisenberg equation

$$-i\frac{d}{dt}\alpha_{\mathbf{k}\sigma}(t)=(E_{\mathbf{k}}-\frac{U}{2})\alpha_{\mathbf{k}\sigma}(t)+[H_f(t),\alpha_{\mathbf{k}\sigma}(t)]. \tag{1.11}$$

where $\alpha_{\mathbf{k}\sigma}(t)$ and $H_f(t)$ are both in the Heisenberg representation. For one doped hole case, Eq.(1.11) will act on the half-filled SDW ground state. But we shall pursue a basic point of view[2] that the spin bag effect comes from the coupling of the doped hole with the *mean − field* SDW background. Thus in the following approach the half-filled true ground state will be replaced by the mean-field state $|0>$. The correction to $|0>$ due to the zero-point fluctuations will be neglected. Under this approximation, we study the renormalization effect on the quasi-hole due to its coupling with the excitations above the mean-field SDW ground state.

Hence, the doped hole's time-dependent state is determined by $\alpha_{\mathbf{k}\sigma}(t)|0>$. In the absence of the commutator $[H_f(t),\alpha_{\mathbf{k}\sigma}(t)]$, Eq.(1.11) simply gives a phase factor to the time dependence of the state $\alpha_{\mathbf{k}\sigma}(t)|0>$. The commutator $[H_f(t),\alpha_{\mathbf{k}\sigma}(t)]$ in Eq.(1.11) could be regarded as instant excitations at time t, induced by the doped hole in the SDW background. Thus a basic excitation is

$$|\phi_{\mathbf{k}\sigma}>\propto[H_f,\alpha_{\mathbf{k}\sigma}]|0>, \tag{1.12}$$

$|\phi_{\mathbf{k}\sigma}>$ involves a particle-hole pair excited around the doped hole which actually is the most important excitation state in the weak and intermediate coupling regime ($U<W=8t$).

Then a variational state for a doped hole could be constructed by $|\mathbf{k}\sigma>$ and $|\phi_{\mathbf{k}\sigma}>$ as follows

$$|\psi_{\mathbf{k}\sigma}>=\sin\theta_{\mathbf{k}}|\mathbf{k}\sigma>-\cos\theta_{\mathbf{k}}|\phi_{\mathbf{k}\sigma}>. \tag{1.13}$$

By minimizing $<\psi_{\mathbf{k}\sigma}|H|\psi_{\mathbf{k}\sigma}>$, $\theta_{\mathbf{k}}$ is determined by

$$\tan 2\theta_{\mathbf{k}}=-\frac{<\mathbf{k}\sigma|H_f|\phi_{\mathbf{k}\sigma}>+c.c.}{<\phi_{\mathbf{k}\sigma}|H|\phi_{\mathbf{k}\sigma}>-E_0-E_{\mathbf{k}}+\frac{U}{2}} \tag{1.14}$$

where $E_0=<0|H|0>$. The quasi-hole's spectrum is shifted from $E_{\mathbf{k}}-\frac{U}{2}$ to $\tilde{E}_{\mathbf{k}}$ as follows

$$\tilde{E}_{\mathbf{k}}=E_{\mathbf{k}}-\frac{U}{2}-\frac{<\mathbf{k}\sigma|H_f|\phi_{\mathbf{k}\sigma}>+c.c.}{2}\cot\theta_{\mathbf{k}}. \tag{1.15}$$

The lowest energy of the spin-bag band as a function of U/t is shown in Fig.1 by a solid curve, in contrast with $\Delta-\frac{U}{2}$, the quasi-hole's lowest energy which is represented by a dotted curve. The crossed curve in Fig.1 represents a localized spin bag solution which has been obtained in several numerical work[3] within the inhomogeneous mean-field approximation. But the spin bag in the present paper is a mobile one which gains more kinetic energy and thus has lower energy as compared to the localized spin bag solution, as Fig.1 shows. On the other hand, the "spin bag" effect not only greatly reduces the quasi-hole's energy, but also narrows the band width. For example, the reduction of the band width reaches to 50% at $U=5t$.

$|\phi_{\mathbf{k}\sigma}>$ is deduced from $\dot{\alpha}_{\mathbf{k}\sigma}|0>$ or $[H_f,\alpha_{\mathbf{k}\sigma}]|0>$. More excitation states could be obtained from $\ddot{\alpha}_{\mathbf{k}\sigma}|0>$, $\dddot{\alpha}_{\mathbf{k}\sigma}|0>$ etc., in which terms like $[H_f,[H_f,\alpha_{\mathbf{k}\sigma}]]|0>$, and so on, are involved. To improve the spin bag solution Eq.(1.13), one should include more excitation states like $|\tilde{\phi}>\propto[H_f,[H_f,\alpha_{\mathbf{k}\sigma}]]|0>$ which involves two particle-hole pairs excited from the background by the doped hole. Thus we could obtain a correction to energy as shown by the dashed curve in Fig.1. One finds that it is negligible in the regime $U<W=8t$.

Therefore, we have obtained a fairly good spin bag solution in the intermediate coupling regime $U<W=8t$, in agreement with the conjecture of Schrieffer et al.[2].

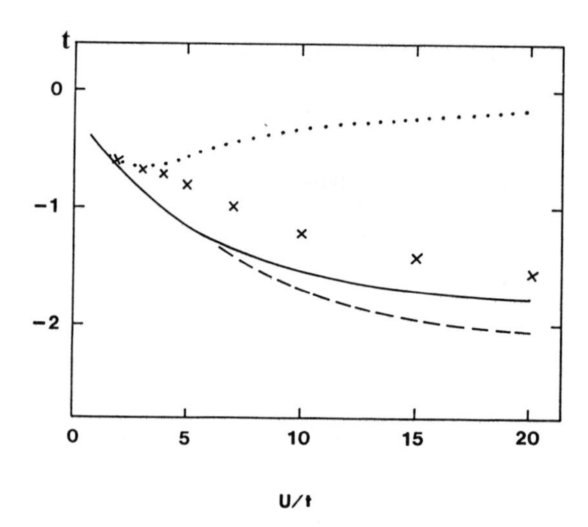

Fig.1. The lowest energy of the spin bag band versus U is presented by the solid curve, in contrast with the dotted curve, $\Delta - \frac{U}{2}$, the mean-field quasi-hole's energy along the magnetic zone boundary. The crossed curve is the localized spin bag energy obtained by the numerical approach[3]. The dashed curve gives a higher order correction to the spin bag solution.

III. THE EFFECTIVE INTERACTION

After the single spin-bag solution is obtained in the above section, one could construct a basic state for two free spin-bags in the subspace of total zero-momentum and antiparallel spin $|\mathbf{k}' \uparrow, -\mathbf{k}' \downarrow>$. The matrix element $< \mathbf{k}' \uparrow, -\mathbf{k}' \downarrow |H| \mathbf{k} \uparrow, -\mathbf{k} \downarrow>$ can be expressed as

$$< \mathbf{k}' \uparrow, -\mathbf{k}' \downarrow |H| \mathbf{k} \uparrow, -\mathbf{k} \downarrow> = (2\tilde{E}_{\mathbf{k}} + E_0)$$

$$\cdot < \mathbf{k}' \uparrow, -\mathbf{k}' \downarrow | \mathbf{k} \uparrow, -\mathbf{k} \downarrow> + H'_{\mathbf{k}',\mathbf{k}}. \tag{2.1}$$

By a lengthy but straightforward calculation, the interaction matrix element $H'_{\mathbf{k}',\mathbf{k}}$ could be reduced to the following form:

$$H'_{\mathbf{k}',\mathbf{k}} = -\frac{\cos(k'_x - k_x)a + \cos(k'_y - k_y)a}{2} \left(\frac{1}{N}\right) V, \tag{2.2}$$

under the condition of $|\varepsilon_{\mathbf{k}}|, |\varepsilon_{\mathbf{k}'}| < \Delta$. By using the following s-wave-like, p-wave-like and d-wave-like symmetry functions

$$g_s(\mathbf{k}) = \frac{1}{2}[\cos(k_x a) + \cos(k_y a)]; \tag{2.3a}$$

$$g_p(\mathbf{k}) = \frac{1}{2}[\sin(k_x a) + \sin(k_y a)]; \tag{2.3b}$$

$$g_p'(\mathbf{k}') = \frac{1}{2}[\sin(k_x a) - \sin(k_y a)]; \tag{2.3c}$$

$$g_d(\mathbf{k}) = \frac{1}{2}[\cos(k_x a) - \cos(k_y a)], \tag{2.3d}$$

one has

$$\frac{\cos(k_x' - k_x)a + \cos(k_y' - k_y)a}{2} = g_s(\mathbf{k})g_s(\mathbf{k}')+$$

$$+g_p(\mathbf{k})g_p(\mathbf{k}') + g_p'(\mathbf{k})g_p'(\mathbf{k}') + g_d(\mathbf{k})g_d(\mathbf{k}'). \tag{2.4}$$

Therefore, $H'_{\mathbf{k}',\mathbf{k}}$ in Eq.(2.2) is composed by the p-wave and s+id-wave- like components, which in fact is the direct consequence of the existence of two sublattices.

The numerical values of V as a function of U/t is shown in Fig.2 by the solid curve in the intermediate coupling regime $U \leq 8t$. One can see that the attractive potential obtained in the present paper is generally reduced in $U < W$ as compared with the RPA result[4] shown by the dashed curve in Fig.2, which corresponds to the d-wave component. But when $U > W$ the attraction potential V becomes stronger than that obtained in the RPA approach. Actually, the former behaves like $\frac{1}{U}$ in the large U limit whereas in the latter case, the potential decreases as quickly as $\frac{1}{U^3}$.

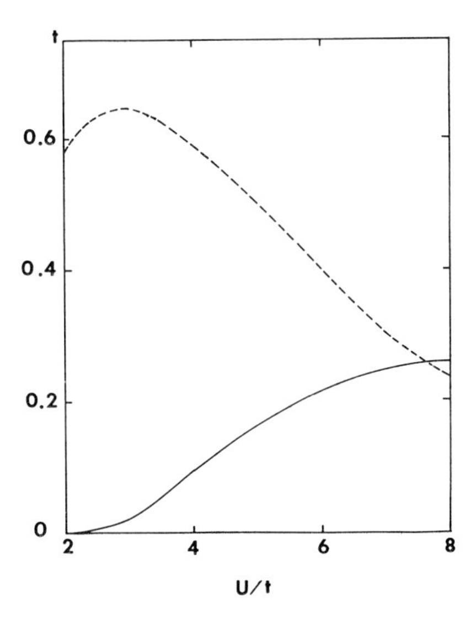

Fig.2. The attractive potential V versus U/t is shown by the solid curve. The dashed curve is the d-wave component of the attractive potential obtained in the RPA approach.

One could show in the equation-of-motion formalism that not only the RPA process of Fig.3(a), which corresponds to the exchange of the simple amplitude fluctuations, is present in the effective interaction $H'_{\mathbf{k},\mathbf{k}'}$ (Eq.(2.2)) but those processes shown in Fig.3(b) also contribute to the attractive potential which in fact is dominant over the former. It is noted that the most important contribution to the spin bag energy as discussed in Section II also comes from similar diagrams of the vertices shown in Fig.3(b). These diagrams involves the low energy spin-flip excitations. Therefore, both the strong spin-bag effect and the attractive interaction between spin bags are closely related to the low-lying spin-flip excitations. On the other hand, a repulsive contribution is also present in $H'_{\mathbf{k}',\mathbf{k}}$ or V which involves the vertex correction to the direct Coulomb interaction of the doped hole. The lowest order diagram is shown in Fig.3(c), which is not present in simple RPA approach. An intuitive way to understand this interaction is to note that, while the on-site Coulomb interaction is reduced in the SDW background due to the existence of two sublattices, it is restored within the "bag" which suppresses the local antiferromagnetic ordering or sublattices in it. Therefore, there is an extra Coulomb repulsion when two spin bags temporarily share a common bag. Due to this reason, the total attractive interaction is much reduced in the weak coupling regime as compared to the RPA result, as Fig.2 shows. On the other hand, with the increase of U, the attractive interaction is enhanced due to the contribution of the low-lying spin-flip processes appearing in the vertices.

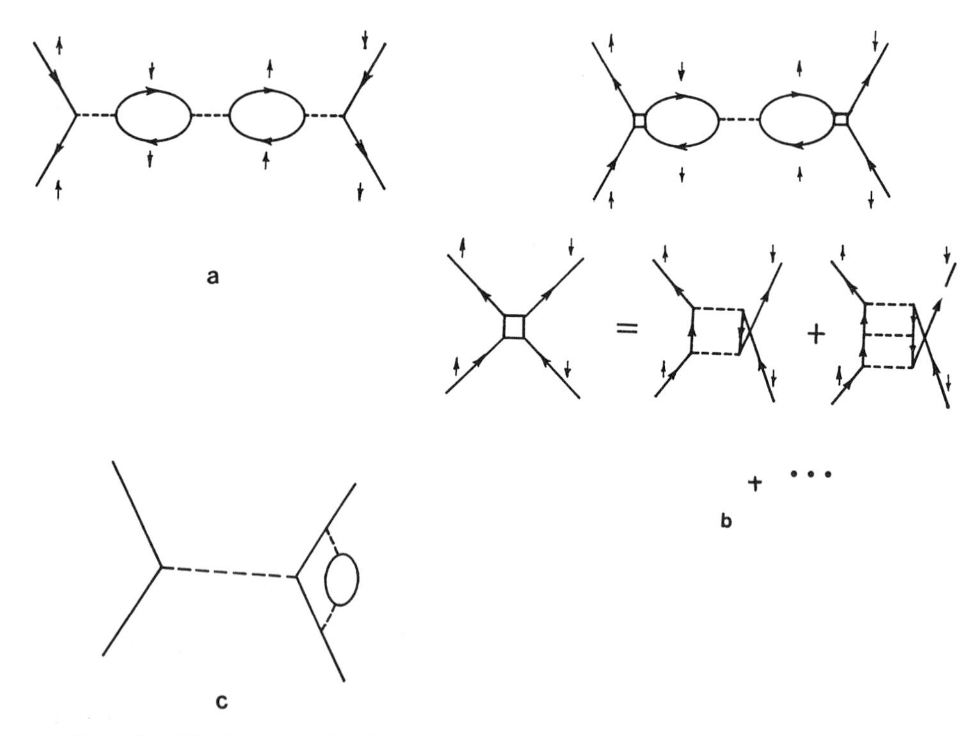

Fig.3. (a) The lowest order (actually the most important) process of the exchange of the amplitude fluctuation in the RPA. (b) A set of processes involved in the attractive potential (2.2) which in fact is dominant over the process (a). (c) The Coulomb interaction with the lowest order vertex correction.

IV. DISAPPEARANCE OF THE LONG-RANGE AF ORDER IN FINITE DOPING

Although our above calculations are carried out in the half-filling limit where the long range antiferromagnetic ordering is well-defined, the general features of the spin bags and their interaction, such as their small size and the nearest neighbor attractive coupling, enable one to expect that even in the large doping case with disappearing of the long-range order, the spin bag picture is still applicable as long as the local antiferromagnetic order exists. On the other hand, how the short-range AF background itself develops with doping will be studied in the present section[9].

Assume an unit vector $\hat{\mathbf{n}}_j$ to each lattice site in the direction, or the opposite direction, of the on-site magnetization, depending on the different sublattice. Then $\hat{\mathbf{n}}_j$ varies slowly in the lattice space when there exists a distortion of the AF order in a length scale $l \gg \xi$, where $\xi \sim v_f/\Delta$ is the SDW coherence length which determines the fundamental quantum scale of the system. $\hat{\mathbf{n}}_j$ is restricted to have a rotation in a plane, say, the z-x plane, and is determined by the angle θ from the polar axis z. The annihilation operator $\tilde{c}_{j\sigma}$ will be defined in the spin representation of $\hat{\mathbf{n}}_j \cdot \hat{\sigma}$ which is related to the original operator $c_{j\sigma}$ (in the spin representation of $\hat{\sigma}_z$) as follows

$$\tilde{c}_{j\sigma} = \sum_{\sigma'} (e^{i\hat{\sigma}_y \theta/2})_{\sigma\sigma'} c_{j\sigma'}. \tag{3.1}$$

Then, the Hubbard Hamiltonian (1.1) is transformed into the following form according to Eq.(3.1):

$$H = -t \sum_{<i,j>\sigma_1\sigma_2} [(e^{i\hat{\sigma}_y \frac{\theta_i - \theta_j}{2}})_{\sigma_1\sigma_2} \tilde{c}^+_{i\sigma_1} \tilde{c}_{j\sigma_2} + h.c.] + U \sum_j \tilde{n}_{j\uparrow}\tilde{n}_{j\downarrow} \tag{3.2}$$

The Hubbard U term is invariant under the transformation (3.1). As the deviation of $\theta_j - \theta_i$ for two nearest neighbor sites is assumed small, so that the Hamiltonian (3.2) could be rewritten in a perturbation form

$$H = \tilde{H} + \tilde{H}', \tag{3.3}$$

where $\tilde{H}$ is the unperturbed part

$$\tilde{H} = -t \sum_{<i,j>\sigma} (\tilde{c}^+_{i\sigma}\tilde{c}_{j\sigma} + h.c.) + U \sum_j \tilde{n}_{j\uparrow}\tilde{n}_{j\downarrow} \tag{3.4}$$

and the perturbation

$$\tilde{H}' = -t \sum_{<i,j>\sigma} \sigma \sin\left(\frac{\theta_i - \theta_j}{2}\right) (\tilde{c}^+_{i\sigma}\tilde{c}_{j-\sigma} + h.c.) + t \sum_{<i,j>\sigma} \left(1 - \cos\frac{\theta_i - \theta_j}{2}\right) (\tilde{c}^+_{i\sigma}\tilde{c}_{j\sigma} + h.c.). \tag{3.5}$$

Obviously, $\tilde{H}$ has the same form as the original Hubbard model H. But it describes a pre-assumed commensurate AF ordering, due to the presence of local staggered magnetization polarized along $\pm\hat{\mathbf{n}}_j$, which may become only a metastable state upon doping. By the standard mean-field approximation, $\tilde{H}$ is found to give a SDW long-range ordering in the new reference frame of $\{\hat{\mathbf{n}}_j\}$. The doped holes will induce the spin-bag effects in such a SDW background. But the spin-bag entities and the pairing correlation between them all occur in a short length scale $\sim \xi$, which is expected not to change dramatically when $\tilde{H}'$ is included which involves a large-length scale ($\gg \xi$) effect. Thus, in the following discussion of the lower-energy state of H which could have a long-range distortion in the AF order, the short-length scale spin-bag effects will be neglected for simplicity. Therefore, under the mean-field approximation discussed in section II, only the following diagonalized form is retained in $\tilde{H}$[11]

$$\tilde{H}_0 = \sum_{\mathbf{k},\sigma}{}' [-E_{\mathbf{k}} \alpha^+_{\mathbf{k}\sigma} \alpha_{\mathbf{k}\sigma} + E_{\mathbf{k}} \beta^+_{\mathbf{k}\sigma} \beta_{\mathbf{k}\sigma}] + \frac{N}{U} \Delta^2, \tag{3.6}$$

where the chemical potential μ is set as zero for simplicity comparing to Eq.(1.5). $\alpha_{\mathbf{k}\sigma}$ and $\beta_{\mathbf{k}\sigma}$ are related to $\tilde{c}_{j\sigma}$ through the canonical transformation (1.4):

$$\tilde{c}_{j\sigma} = \sum_{\mathbf{k}}{}' \frac{e^{i\mathbf{k}\cdot\mathbf{R}_j}}{\sqrt{N}} [(u_{\mathbf{k}} + \sigma p_j v_{\mathbf{k}}) \alpha_{\mathbf{k}\sigma} + p_j (u_{\mathbf{k}} - \sigma p_j v_{\mathbf{k}}) \beta_{\mathbf{k}\sigma}], \tag{3.7}$$

In the laboratory reference frame (i.e., the spin representation of $\hat{\sigma}_z$), the SDW ground state of $\tilde{H}_0$ becomes a distorted ordering state described by $\{\hat{\mathbf{n}}_j\}$ in the presence of the perturbation $\tilde{H}'$. As we shall consider the twist SDW long-range order in the scale $l \gg \xi > a$, $\theta_j \equiv \theta(\mathbf{R}_j)$ is a slowly varying function of $\mathbf{R}_j$ and could be treated as a continuous quantity, that is,

$$\theta(\mathbf{R}_i) - \theta(\mathbf{R}_j) \simeq \nabla\theta(\mathbf{R}_j) \cdot (\mathbf{R}_i - \mathbf{R}_j), \tag{3.8}$$

where $\mathbf{R}_i$ and $\mathbf{R}_j$ are the lattice vectors for the nearest neighbor sites. Introducing the following Fourier transformation:

$$\nabla\theta(\mathbf{q}) = \frac{1}{N} \sum_j e^{-i\mathbf{q}\cdot\mathbf{R}_j} \nabla\theta(\mathbf{R}_j), \tag{3.9}$$

then $\tilde{H}'$ could be transformed into the following form

$$\tilde{H}' \simeq \frac{i}{2} \sum_{\mathbf{k}'\mathbf{k}\sigma}{}' \nabla\varepsilon_{\mathbf{k}} \cdot [\sigma \nabla\theta(\mathbf{k}' - \mathbf{k}) + \nabla\theta(\mathbf{k}' - \mathbf{k} + \mathbf{Q})] \alpha^+_{\mathbf{k}'\sigma} \alpha_{\mathbf{k}-\sigma}$$

$$+ \frac{a^2}{8} (\nabla\theta)^2 \sum_{\mathbf{k}}{}' \frac{\varepsilon^2_{\mathbf{k}}}{E_{\mathbf{k}}}, \tag{3.10}$$

for a small doping concentration where the doped holes will distribute along the magnetic zone boundary within a width $< \xi^{-1}$ such that $u_{\mathbf{k}} \sim v_{\mathbf{k}} \sim \frac{1}{\sqrt{2}}$. The summation in Eq.(3.10) actually goes over only along the magnetic zone boundary within a width $< \xi^{-1}$.

In the half-filled case, only the second term on the right-hand side of Eq.(3.10) remains, which is always positive and proportional to $(\nabla\theta)^2$. The distortion $\nabla\theta$ could be only stabilized in the presence of the doped holes. One sees that the first term in $\tilde{H}'$ (Eq.(3.10)) gives an additional hopping channel of the hole which involves a 'spin-flip' in the $\{\hat{\mathbf{n}}_j\}$ reference frame. In the $\hat{\sigma}_z$ representation, this process means that the doped hole has a finite amplitude for hopping from one sublattice to the opposite one due to the existence of twist in the AF ordering .

Fig.4 shows the magnetic zone. In the mean-field theory, it is completely filled with electrons in the half-filled case and the holes will be doped close to the zone boundary. Along the four pieces of the zone boundary, one finds that $\nabla\varepsilon_{\mathbf{k}}$ points either to $\pm\mathbf{e}_a$ or to $\pm\mathbf{e}_b$ respectively, where $\mathbf{e}_a = \frac{\hat{\mathbf{x}}+\hat{\mathbf{y}}}{\sqrt{2}}$ and $\mathbf{e}_b = \frac{\hat{\mathbf{y}}-\hat{\mathbf{x}}}{\sqrt{2}}$. As $\nabla\theta(\mathbf{q})\cdot\nabla\varepsilon_{\mathbf{k}}$ determines the matrix elements in $\tilde{H}'$, the doped holes could gain additional kinetic energy in the four areas around the magnetic zone boundary. These areas (a_1, a_2 and b_1, b_2 in Fig. 4) do not cover the four corners of the zone boundary because

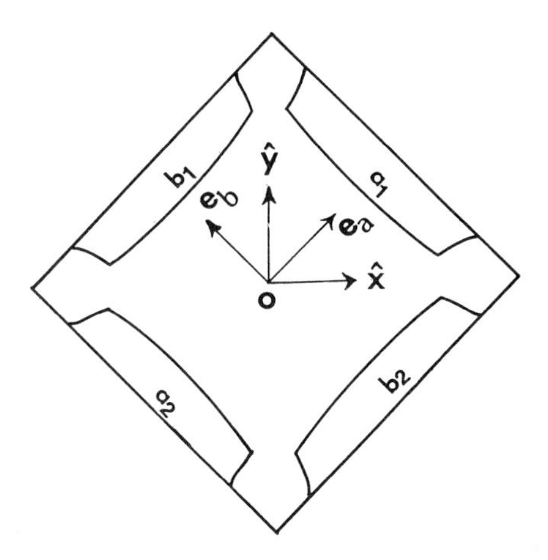

Fig.4. The square is the reduced Brillouin zone, or magnetic zone in the SDW state. When a twist state is present, holes will be distributed in the areas a_1, a_2 and b_1, b_2 with a width $< \xi^{-1}$ in the small doping case .

$|\nabla\varepsilon_{\mathbf{k}}|$ becomes vanishing around those regions and $\tilde{H}'$ has essentially no contribution there. The width of areas a's and b's is assumed less than ξ^{-1} close to the zone boundary . The doped holes will be distributed in the areas of a's and b's of Fig.4 when doping is small with the average hole-hole distance being sufficiently larger than ξ. In fact, the hole-hole distance $\sim l$ and we shall assume that $l \gg \xi$. As $\nabla\varepsilon_{\mathbf{k}}$ is almost perpendicular to the boundary line in each area, One could define

$$\nabla\theta(\mathbf{q}) = \nabla^a\theta(\mathbf{q})\mathbf{e}_a + \nabla^b\theta(\mathbf{q})\mathbf{e}_b \tag{3.11}$$

As $|\mathbf{q}| \sim l^{-1} \ll \xi^{-1}$, there is negligible hopping process between the areas a's and b's. In $\tilde{H}'$ (Eq.(3.10)), the holes in areas $a's$ and $b's$ couple with the fields $\nabla^a\theta(\mathbf{R})$ and $\nabla^b\theta(\mathbf{R})$ respectively. It is noted that the fields $\nabla^a\theta(\mathbf{R})$ and $\nabla^b\theta(\mathbf{R})$ in Eq.(3.10) are not simply independent from each other. There is an obvious constraint on them, i.e., $\nabla^b\nabla^a\theta = \nabla^a\nabla^b\theta$. However, to reduce the additional boundary energy cost, we shall assume $\nabla^b\nabla^a\theta = \nabla^a\nabla^b\theta = 0$ such that the fields $\nabla^a\theta(\mathbf{R})$ and $\nabla^b\theta(\mathbf{R})$ become one-dimensional independent functions in the directions of $\mathbf{e}_a$ and $\mathbf{e}_b$ respectively. Therefore, the holes in areas $a's$ and $b's$ are decoupled and could be treated independently and thus two areas are actually independent.

Hence, although the quasi-hole energy spectrum $E_{\mathbf{k}}$ in the unperturbed Hamiltonian $\tilde{H}_0$ is highly degenerate as $E_{\mathbf{k}} \sim \Delta$ within the width of ξ^{-1} close to the magnetic zone boundary, $\tilde{H}'$ could change the energy spectrum and thus the distribution of holes. As $\tilde{H}'$ of Eq.(3.10) describes the coupling of the doped holes with the long wave-length mean-field, the individual motion of holes will be not essential in determining the fields $\{\nabla^a\theta, \nabla^b\theta\}$ and could be averaged out. Minimizing $< H >$ with respect to $\{\nabla^a\theta, \nabla^b\theta\}$, then one could find

$$\nabla^{a,b}\theta(\mathbf{R}_j^{a,b}) = ca^{-1}n_j^{a,b}. \tag{3.12}$$

in which

$$c = \frac{v_f}{\sqrt{2}a}\left(\frac{1}{N}\sum_{\mathbf{k}}{}' \frac{\varepsilon_{\mathbf{k}}^2}{E_{\mathbf{k}}}\right)^{-1}, \tag{3.13}$$

$n_j^{a,b}$ is the hole distribution at the square lattice site j contributed by the holes from the areas $a's$ and $b's$ in Fig.4 respectively. It is noted that n_j^a (n_j^b) could have one-dimensional long wave-length statistic fluctuations along the direction of $\mathbf{e}_a$ ($\mathbf{e}_b$) which will then result in the fluctuations of the field $\nabla^a\theta$ ($\nabla^b\theta$) in this direction. If $n_j^{a,b}$ in Eq.(3.13) is replaced by its statistical average n_0, then the corresponding derivative $\nabla^{a,b}\theta_0(\mathbf{R})$ leads to a spiral long-range AF ordering with the local staggered magnetization moment rotating uniformly within the z-x plane along the direction $\mathbf{e}_a$ and $\mathbf{e}_a$. Such an incommensurate AF ordering has been also obtained in the strong-coupling, localized approach[8].

Eq.(3.12) shows that the statistic fluctuation $n_j - n_0$ leads to a fluctuation of the field $\nabla\theta(\mathbf{R})$ around $\nabla\theta_0(\mathbf{R})$. Such a statistical fluctuation will have an important effect on the long-range behavior of the spin-spin correlation function $< \mathbf{S_1} \cdot \mathbf{S_2} >$. When $|\mathbf{R}_1 - \mathbf{R}_2| \gg \xi$, one has

$$< \mathbf{S}_1 \cdot \mathbf{S}_2 > \simeq < \mathbf{S}_1 > \cdot < \mathbf{S}_2 > = \pm m^2 \cos(\theta_2 - \theta_1), \tag{3.14}$$

where $m = 2\Delta/U$ is on-site magnetization moment and the sign $\pm$ depends on whether $\mathbf{R}_1$ and $\mathbf{R}_2$ are at the same sublattice or not. A special case is $\mathbf{R}_2 - \mathbf{R}_1 \parallel \mathbf{e}_a$. Then according to Eq.(3.12), one has

$$\theta_2 - \theta_1 = \int_1^2 dR^a \cdot \nabla^a\theta(\mathbf{R}) \simeq c\sum_1^2 n_j^a. \tag{3.15}$$

That is, the deviation of $\theta_2 - \theta_1$ is utterly determined by the number of holes in the area $a's$ which are distributed along a one-dimensional chain in the range of $\{\mathbf{R}_2, \mathbf{R}_1\}$. Suppose there are totally $N\delta$ holes, then half of them will be doped into the areas $a's$ defined in Fig.4. As is noted before, n_j^a has an one-dimensional fluctuation along $\mathbf{e}_a$ and is averaged out in $\mathbf{e}_b$ direction. Then for each one-dimensional lattice chain parallel to $\mathbf{e}_a$, there are averagely $n = \sqrt{\frac{N}{2}}\delta$ holes. Under the assumption of the equal probability distribution of holes for different configurations in the chain (only the fluctuations with the scale larger than ξ will be considered), then for $|\mathbf{R}_2 - \mathbf{R}_1| \gg \xi$ one finds at last

$$< \mathbf{S}_1 \cdot \mathbf{S}_2 > \sim \pm m^2 \cos(\frac{|\mathbf{R}_2 - \mathbf{R}_1|}{\sqrt{2}a/\delta \sin c}) e^{-\frac{|\mathbf{R}_2 - \mathbf{R}_1|}{\sqrt{2}a/\delta(1-\cos c)}}. \tag{3.16}$$

Hence the AF correlation length is given by

$$l_0 = \frac{\sqrt{2}a}{1 - \cos c} \delta^{-1}. \tag{3.17}$$

which becomes finite for any non-zero doping concentration as long as $\cos c \neq 1$. The spin-spin correlation function $< \mathbf{S}_1 \cdot \mathbf{S}_2 >$ for general two sites $\mathbf{R}_1$ and $\mathbf{R}_2$ could be similarly discussed.

The correlation length l_0 defined in Eq.(3.17) has the order of the hole-hole mean distance which is assumed larger than the coherence length ξ. The correlation function in Eq.(3.16) also shows a spiral modulation to the AF order in the scale $\sim 2\sqrt{2}a\pi/(\delta \sin c)$. The incommensurate structure of the short-range AF order has been reported recently in the superconducting phase of $La_{2-x}Sr_xCuO_{4-y}$[10]. We have noted that the spiral structure of the AF order has also been obtained in the Ref.8 by the strong-coupling, localized approach which gives an incommensurate long-range AF order upon doping. But we found that the statistical fluctuations of $\nabla\theta(\mathbf{R})$ could easily break the AF correlation in the scale larger than l_0 and thus restore the global symmetry of the system.

In the present approach, the long-range AF state is found unstable against doping. But we have shown the possibility to find a short-ranged AF state with a lower energy. The essential point is that the knowledge of the dynamics and correlation of holes in the above-mentioned metastable long-range AF state might be qualitatively applied to the short-range AF state, as long as the two length scales are distinguishable, i.e., $l_0 > \xi$. The spiral structure of the short-range order determines the distribution of holes as shown in Fig.4, which would lead to a nodeless p-wave pairing of spin-bags with antiparallel spins according to the results obtained in the section II and the discussion in Ref.5. Such a spin-bag mechanism for high-temperature copper-oxide superconductivity needs further more detailed studies.

V. Conclusions

The doping effects on the AF background are studied in two length scales: $\sim \xi$ (the SDW coherence length) and $\gg \xi$. Within the scale of ξ, the doped hole will induce a distortion in the AF background which accompanies the hole moving around as an spin-bag entity. Such a spin-bag solution has been studied in the intermediate coupling regime. The effective attraction between two spin-bags with antiparallel spins has been shown to be p- and s+id-wave like. Such an attractive potential, involving the spin-flip low-lying excitations, is found stronger than that through exchanging the simple amplitude fluctuations in the RPA approach.

In the finite doping, there is a spiral structure appearing in the AF ordering and the long-range order disappears exponentially with a correlation length $l_0 \sim \delta^{-1}$. Such a short-range AF state with a spiral modulation is stabilized by the doped holes which tend to get more kinetic energy from the new channel opened in the twist state. The distribution of the doped holes in $\mathbf{k}$ space is determined by the spiral structure of the local ordering which could lead to a nodeless p-wave superconducting condensation as long as the spin-bag picture built in the long-range AF state is applicable to the short-range order background. The long-range AF ordering observed in the high-T_c copper-oxide materials at small doping region would be only a three-dimensional effect which is so sensible to the doped holes and quickly disappear with the increase of doping.

ACKNOWLEDGMENTS

The authors would like to thank Prof. T.K. Lee for stimulating discussions. The present work is supported by a grant from the Robert A.Welch Foundation and also by the Texas Center for Superconductivity at the University of Houston under the Prime Grant No.MDA-972-88-G-0002 from Defence Advanced Research Project Agency.

REFERENCES

[1]P.W. Anderson, Science **235**, 1196(1987).
[2]J.R. Schrieffer, X.-G. Wen and S.-C. Zhang, Phys. Rev. Lett. **60**, 944(1988).
[3]W.P. Su, Phys. Rev. B**37**, 9904(1988); H.Y. Choi and E.J. Mele, Phys. Rev. B**38**, 4540(1988); W.P. Su and X.Y. Chen, Phys. Rev. B**38**, 8879(1988).
[4]Z.Y. Weng, T.K. Lee and C.S. Ting, Phys. Rev. B**38**, 6561(1988); G. Vignale and K.S. Singwi, Phys. Rev. B**39**, 2956(1989).
[5]J.R. Schrieffer, X.G. Wen and S.C. Zhang, Phys. Rev. B**39**, 11663(1989).
[6]P.W.Anderson, Phys. Rev. **86**, 694(1952).
[7]Z.Y. Weng, C.S. Ting and T.K. Lee, to be published in Phys. Rev. B.
[8]B.I.Shraiman and E.D.Siggia, Phys. Rev. Lett. **62**, 1564(1989); C.L. Kane, P.A. Lee, T.K. Ng, B. Chakraborty and N. Read, preprint
[9]Z.Y. Weng, C.S. Ting, preprint.
[10]R.J. Birgeneau et al., Phys. Rev. B**39**, 2868(1989); T.R. Thurston et al., Phys. Rev. B**40**, 4585(1989); G. Shirane et al., Phys. Rev. Lett. **63** 330(1989).
[11] A generalized Fork term like $U\sum_j < \tilde{c}^+_{j\uparrow}\tilde{c}_{j\downarrow} > \tilde{c}_{j\uparrow}\tilde{c}^+_{j\downarrow} + U\sum_j \tilde{c}^+_{j\uparrow}\tilde{c}_{j\downarrow} < \tilde{c}_{j\uparrow}\tilde{c}^+_{j\downarrow} >$ should be also retained in $\tilde{H}$ for a self-consistent mean-field approach as $\tilde{H}'$ involves spin-flip process. But we found such a term actually has a negligible contribution when $l \ll \xi$.

PHASE SEPARATION IN A $t-J$ MODEL*

M. Marder[a], N. Papanicolaou[b], and G. C. Psaltakis

Department of Physics, University of Crete
and Research Center of Crete
71409 Iraklion, Greece

Abstract

We study a simple extension of the Heisenberg model which is abstracted from the large-U limit of the Hubbard Hamiltonian and includes charge fluctuations. An attempt is made to elucidate some basic features of the $T=0$ phase diagram within a suitable $1/N$ expansion. We find that the ferromagnetic boundary dictated by a naive application of the Nagaoka theorem is actually incorrect because of an instability induced by phase separation. We derive what we believe to be the correct ferromagnetic boundary for sufficiently high dimension, and provide a detailed description of the ensuing phase separation. We also find that a uniform canted antiferromangetic state is stable over a nontrivial region of the phase diagram. Potential implications of these results for the physics of the original Hubbard model are discussed briefly.

The Hubbard model has recently received renewed interest in view of potential applications to the theory of high-T_c superconductivity. Nonetheless, many fundamental aspects of its solution remain elusive, with the notable exception of an exact solution available in one dimension through the Bethe Ansatz[1]. The only rigorous result thought to be applicable to higher dimensions is a theorem due to Nagaoka[2], whose implications have often been exaggerated in the literature. As a consequence, even the gross features of the $T=0$ phase diagram have not been firmly established, except in one dimension where the ground state is a spin singlet for arbitrary electron filling.

In order to achieve a manageable theoretical framework, the original Hubbard model is simplified in two respects. First, we restrict attention to the currently popular $t-J$ model. Second, we partially alter the commutation relations of the Hubbard operators. The $t-J$ model may be obtained as the large-U limit of the Hubbard model[3], and is most conveniently formulated in terms of the Hubbard operators $\chi^{ab} = |a\rangle\langle b|$. Because double occupancy is projected out in the large-U limit, the Latin indices $a, b, \ldots$ assume only three distinct values, say $0, 1$, and 2, corresponding to a hole, a spin-up, and a spin-down electron respectively. We also use Greek indices $\mu, \nu, \ldots$ taking the two distinct values 1 and 2, and invoke the usual summation convention for repeated indices without exception. With these conventions the $t-J$ Hamiltonian is written as

$$\mathcal{H} = \mathcal{H}_1 + \mathcal{H}_2 + \mathcal{H}_3,$$
$$\mathcal{H}_1 = -\sum_{i,j} t_{ij}\chi_i^{0\mu}\chi_j^{\mu 0},$$

*Lecture delivered by N. Papanicolaou

Dynamics of Magnetic Fluctuations in High-Temperature Superconductors
Edited by G. Reiter *et al.*, Plenum Press, New York, 1991

$$\mathcal{H}_2 = \frac{1}{U} \sum_{i,j} t_{ij}^2 \left(\chi_i^{\mu\nu} \chi_j^{\nu\mu} - \chi_i^{\mu\mu} \chi_j^{\nu\nu} \right), \tag{1}$$

$$\mathcal{H}_3 = \frac{1}{U} \sum_{\substack{i,l,j \\ (i \neq j)}} t_{il} t_{lj} \left(\chi_i^{0\mu} \chi_l^{\mu\nu} \chi_j^{\nu 0} - \chi_i^{0\mu} \chi_l^{\nu\nu} \chi_j^{\mu 0} \right),$$

where we allow for arbitrary hopping constants $t_{ij} = t_{ji}$. In our explicit calculations hopping is assumed to occur only between neighboring sites with amplitude t. Hence the term $\mathcal{H}_1$ in (1) describes exchanges of holes with electrons occupying neighboring sites, whereas the three-site term $\mathcal{H}_3$ induces exchanges of holes with nearby pairs of electrons of opposite spin. The physical significance of $\mathcal{H}_2$ becomes apparent by expressing it in terms of the spin operators

$$\vec{S}_l = \frac{1}{2} \vec{\sigma}_{\mu\nu} \chi_l^{\mu\nu}, \tag{2}$$

where the $\vec{\sigma}_{\mu\nu}$ are matrix elements of the Pauli operators $\vec{\sigma} = (\sigma^x, \sigma^y, \sigma^z)$. A short calculation shows that $\mathcal{H}_2$ is essentially the usual antiferromagnetic Heisenberg Hamiltonian with exchange constant $J = 4t^2/U$. The hopping constant t and exchange constant J will be used as the independent coupling constants in all subsequent calculations. In particular, we will frequently use the dimensionless ration $t/J = U/(4t)$.

The Hamiltonian (1) must be supplemented by the local constraint

$$\chi_l^{aa} = \chi_l^{00} + \chi_l^{\mu\mu} = 1, \tag{3}$$

which expresses the fact that a site l is either empty or occupied by no more than one electron of arbitrary spin, and by the global constraint

$$\sum_l \chi_l^{\mu\mu} = N_e = n_e \Lambda, \tag{4}$$

where N_e is the total number of electrons and Λ is the total number of sites. The average density $n_e = N_e/\Lambda$ is referred to as the filling factor and will take values in $[0, 1]$. To complete the description of the Hamiltonian we must examine the commutation relations satisfied by the Hubbard operators. From their definition, $\chi^{ab} = |a\rangle \langle b|$, the Hubbard operators at any given site satisfy the $U(3)$ algebra

$$\left[\chi_l^{ab}, \chi_l^{cd} \right] = \delta^{bc} \chi_l^{ad} - \delta^{ad} \chi_l^{cb}, \tag{5}$$

where $[\ ,\ ]$ denotes the usual commutator. Although anticommutators do not occur in (5) the Fermi character of the original degrees of freedom is completely taken into account by the compactness of the unitary algebra $U(3)$. However operators at different sites may commute or anticommute depending on the specific choice of indices. For instance, $\chi_i^{0\mu}$ anticommutes with $\chi_j^{0\nu}$ if $i \neq j$ while it commutes with $\chi_j^{\mu\nu}$ or χ_j^{00}. This brings us to the second simplification of the Hubbard model; namely, operators at different sites will always be assumed to commute. Combining this simplification with the commutation relations (5), which remain intact, the general commutation relations will be written as

$$\left[\chi_i^{ab}, \chi_j^{cd} \right] = \delta_{ij} \left(\delta^{bc} \chi_i^{ad} - \delta^{ad} \chi_i^{cb} \right), \tag{6}$$

It is difficult at this point to gauge the effect of altering the commutation relations. This model defines a simple extension of the Heisenberg model which includes charge fluctuations, containing most if not all of the essential physics of the standard $t - J$ model, and perhaps of the original Hubbard model as well. We study this model in its own right. The main advantage of the partial bosonization incorporated in (6) is that we are now able to develop a complete semiclassical theory in close analogy with earlier work on quantum spin-1 systems[4].

Thus we attempt to elucidate some basic features of the phase diagram of the Hubbard model. We find that the ferromagnetic boundary dictated by a naive application of the Nagaoka theorem is actually incorrect because of an instability induced by phase separation.

We derive what we believe to be the correct ferromagnetic boundary for sufficiently high lattice dimension, and provide a detailed description of the ensuing phase separation. In the course of our investigation, we learned that general arguments in favor of phase separation had already been given in an early paper by Visscher[5], and in recent works by Ioffe and Larkin[6], and Foerster[7]. Related ideas may also be found in articles on spin polarons[8] and on spin bags[9]. But we believe that our recent work[10] contains the first attempt to develop the scenario of phase separation in correlated electron systems with reasonable completeness. Here we briefly describe the main results.

The local constraint (3) indicates that the relevant representation of $U(3)$ is the fundamental (quark) representation. To derive a sensible semiclassical theory we generalize the constraint according to

$$\chi_l^{aa} = \chi_l^{00} + \chi_l^{\mu\mu} = N, \tag{7}$$

where N, not to be confused with the electron number N_e, is arbitrary. From the algebraic point of view, (7) is equivalent to asserting that we consider an arbitrary symmetric representation of $U(3)$ by analogy with the generalization of the spin-1/2 Heisenberg model to arbitrary spin S. One may then derive a semiclassical theory based on a $1/N$ expansion, along the lines of the $1/S$ expansion extensively used in the study of the Heisenberg model, setting $N = 1$ at the end of the calculation. In order to appreciate fully the semiclassical nature of the large-N limit, one should proceed with the derivation of a phase-space path integral in terms of generalized coherent states for the symmetric representations of the unitary algebra[4]. However, for most practical purposes, the essential result is the generalized Holstein-Primakoff realization

$$\begin{aligned} &\chi_l^{00} = N - \xi_l^{\mu\star}\xi_l^{\mu}, \quad \chi_l^{\mu\nu} = \xi_l^{\mu\star}\xi_l^{\nu}, \\ &\chi_l^{0\mu} = (N - \xi_l^{\nu\star}\xi_l^{\nu})^{1/2}\xi_l^{\mu}, \quad \chi_l^{\mu 0} = \xi_l^{\mu\star}(N - \xi_l^{\nu\star}\xi_l^{\nu})^{1/2}, \end{aligned} \tag{8}$$

where the ξ_l^{μ} are Bose operators;

$$\left[\xi_i^{\mu}, \xi_j^{\nu\star}\right] = \delta_{ij}\delta^{\mu\nu}. \tag{9}$$

Inserting (8) into (1) yields a completely bosonized effective Hamiltonian which is suitable for the study of low-temperature dynamics using standard spin-wave techniques. An essential step in this approach is to determine the minimum of the effective Hamiltonian in the large-N limit, where the operators ξ_l^{μ} may be treated as classical (commuting) fields. To make the semiclassical nature of the large-N limit explicit one may introduce rescaled operators ζ_l^{μ} from

$$\xi_l^{\mu} = N^{1/2}\zeta_l^{\mu}, \quad \left[\zeta_i^{\mu}, \zeta_j^{\nu\star}\right] = \frac{1}{N}\delta_{ij}\delta^{\mu\nu}, \tag{10}$$

whose commutator vanishes for large N. Note also that N factors out nicely when the ζ_l^{μ} are introduced into the Holstein-Primakoff realization (8). Treating ζ_l as a commuting field in the large N limit, we may parameterize it so as to reveal the underlying rotational invariance by writing

$$\begin{aligned} \zeta_l^1 &= n_l^{1/2} e^{i\psi_l} \cos\left(\frac{\theta_l}{2}\right) e^{-i\phi_l/2} \\ \zeta_l^2 &= n_l^{1/2} e^{i\psi_l} \sin\left(\frac{\theta_l}{2}\right) e^{i\phi_l/2} \end{aligned}, \tag{11}$$

where n_l is the local charge density, so that the spin density at site l takes on the usual spherical parameterization

$$\begin{aligned} s_l^x &= \frac{1}{2} n_l \sin\theta_l \cos\phi_l \\ s_l^y &= \frac{1}{2} n_l \sin\theta_l \sin\phi_l \\ s_l^z &= \frac{1}{2} n_l \cos\theta_l. \end{aligned} \tag{12}$$

These expressions also reveal that to leading order the magnitude of the local spin is one half of the local charge density.

The large-N classical energy is calculated by inserting (8) and (10) into the Hamiltonian and by neglecting the ordering of the operators. In this note, we will neglect the three-site term $\mathcal{H}_3$ in the Hamiltonian (1). This restriction simplifies the analysis while preserving the essential physical picture. A discussion of the effects of $\mathcal{H}_3$ may be found in Ref. 10. Hence we obtain an explicit expression for the classical energy by inserting (8) and (10) into $\mathcal{H} = \mathcal{H}_1 + \mathcal{H}_2$ and by neglecting the ordering of the operators:

$$
\begin{aligned}
\mathcal{E} &= \mathcal{E}_1 + \mathcal{E}_2 \\
\mathcal{E}_1 &= -N^2 t \sum_{<ij>} \left[\left(1 - \zeta_i^{\mu\star}\zeta_i^{\mu}\right)\left(1 - \zeta_j^{\nu\star}\zeta_j^{\nu}\right)\right]^{1/2} \left(\zeta_i^{\lambda\star}\zeta_j^{\lambda} + \zeta_j^{\lambda\star}\zeta_i^{\lambda}\right), \\
\mathcal{E}_2 &= \frac{1}{2} N^2 J \sum_{<ij>} \left[\left|\zeta_i^{\mu\star}\zeta_j^{\mu}\right|^2 - \left(\zeta_i^{\mu\star}\zeta_i^{\mu}\right)\left(\zeta_j^{\nu\star}\zeta_j^{\nu}\right)\right].
\end{aligned} \tag{13}
$$

Here $< ij >$ denotes summation over neighboring sites, and $J = 4t^2/U$ is the exchange constant. The overall factor N^2 is important only to distinguish the semiclassical approximation from higher-order $1/N$ corrections. With this understanding, N will be set equal to the value of actual interest ($N = 1$) in the following calculations. It also proves convenient to work with the spherical variables given in (12), but the resulting expressions will not be written out explicitly here[10]. Finally, the global constraint (4) will be treated by means of a chemical potential.

The explicit calculation of the classical minimum of (13) turned out to be more difficult than we had expected and occupied most of our effort in Ref. 10. This situation should be contrasted with the corresponding one in the Heisenberg model, where the determination of the classical minimum is straightforward; the minimum occurs when all spins are alligned, in the case of a ferromagnet, or for a Néel lattice, in the case of an antiferromagnet.

We will proceed by invoking certain simplifying assumptions whose validity will be examined at later stages of our discussion. We thus assume

i. a uniform electron density; $n_l = n_e$ on all sites,

ii. uniform phases ψ_l and azimuthal angles ϕ_l,

iii. a bipartite lattice with a constant angular difference $\theta_i - \theta_j = \pm\theta_c$ between neighboring sites i and j.
The *canting angle* θ_c should reduce to $\theta_c = \pi$ at $n_e = 1$, and $\theta_c = 0$ at $n_e = 0$.

Incorporating these assumptions into (13) one finds that

$$
\mathcal{E}/\Lambda = -ztn_e\,(1 - n_e)\cos\frac{\theta_c}{2} - \frac{1}{4} zJn_e^2 \sin^2\frac{\theta_c}{2}, \tag{14}
$$

where Λ is the total number of sites, and z is the lattice coordination number.

As it turns out, the first assumption is not actually justified, but it is instructive to follow this line of reasoning through to its conclusion. Minimizing (14) with respect to θ_c we find that

$$
\cos\left(\frac{\theta_c}{2}\right) = \begin{cases} 2\frac{t}{J}\frac{1-n_e}{n_e} & \text{for } t/J \le \frac{n_e}{2(1-n_e)}; \\ 1 & \text{for } t/J \ge \frac{n_e}{2(1-n_e)}. \end{cases} \tag{15}
$$

Therefore a critical line develops in the $n_e - t/J$ plane, given by

$$
\frac{t}{J} = \frac{n_e}{2\,(1 - n_e)}, \tag{16}
$$

and drawn as a dashed line in Fig. 1. Above the critical line, the system orders ferromagnetically. The destruction of perfect ferromagnetism at the line is consistent with the usual interpretation of Nagaoka's original calculation. Below the line, the system forms a canted phase which interpolates smoothly betwen ferromagnetism ($\theta_c = 0$) and antiferromagnetism ($\theta_c = \pi$) at half filling ($n_e = 1$).

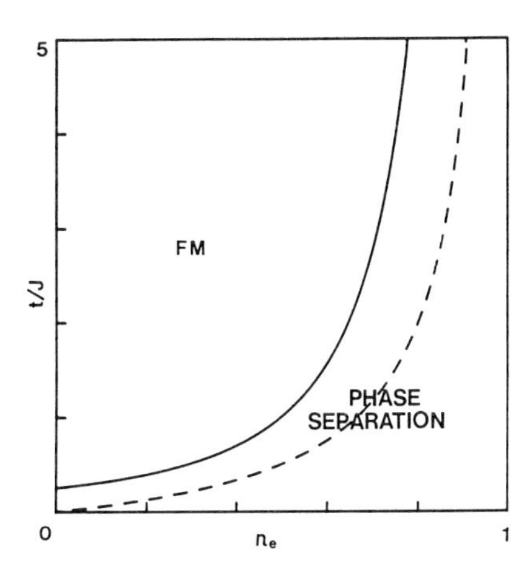

FIGURE 1. $T = 0$ phase diagram in the absence of $\mathcal{H}_3$. The dashed line indicates the limit of metastability defined in Eq. (16), whereas the solid line is the true critical line of Eq. (18). Above the critical line the system orders ferromagnetically (FM), while it undergoes phase separation in the remaining region.

The ground state energy is given by

$$\mathcal{E}/zt\Lambda = \begin{cases} -\frac{t}{J}(1-n_e)^2 - \frac{J}{4t}n_e^2, & \text{for } t/J \leq \frac{n_e}{2(1-n_e)}; \\ -n_e(1-n_e) & \text{for } t/J \geq \frac{n_e}{2(1-n_e)}, \end{cases} \tag{17}$$

and is a continuous function of n_e across the critical line. The chemical potential, $\mu = \mathcal{E}'(n_e)/\Lambda$, is also continuous. However the second derivative, $\mathcal{E}''$ is discontinuous, a fact of importance to which we will return shortly.

In order to decide whether or not this classical configuration is a reasonable candidate for the ground state, we examine its stability. As it turns out, the issue of stability can be settled by simple arguments based on the convexity of the ground state energy $\mathcal{E} = \mathcal{E}(n_e)$, which appeared in Eq. (17). Our subsequent analysis will show that the canted phase is actually everywhere unstable against long wavelength fluctuations. And, while the ferromagnetic phase is locally stable, it becomes globally unstable over a nontrivial region of the phase diagram. The true critical line will be shown to be given by

$$\frac{t}{J} = \frac{1}{4(1-n_e)^2}, \tag{18}$$

instead of Eq. (16), and is depicted by a solid line in Fig. 1. Above the true critical line, the system is indeed realized in a ferromagnetic phase which is both locally and globally stable. Below the true critical line, the system undergoes phase separation. In the space between solid and dashed lines in Fig. 1, a ferromagnetic phase would be metastable.

The simplest route to Eq. (18) proceeds with an examination of the second derivative of the ground state energy $\mathcal{E} = \mathcal{E}(n_e)$ of Eq. (17);

$$\mathcal{E}''(n_e)/zt\Lambda = \begin{cases} -\frac{2t}{J} - \frac{J}{2t}, & \text{for } t/J \leq \frac{n_e}{2(1-n_e)}; \\ 2 & \text{for } t/J \geq \frac{n_e}{2(1-n_e)}. \end{cases} \tag{19}$$

The second derivative is negative for $t/J < n_e/2(1-n_e)$. This guarantees that the energy of Eq. (17) is concave in the canted phase. An argument going back to Maxwell and Gibbs[11]

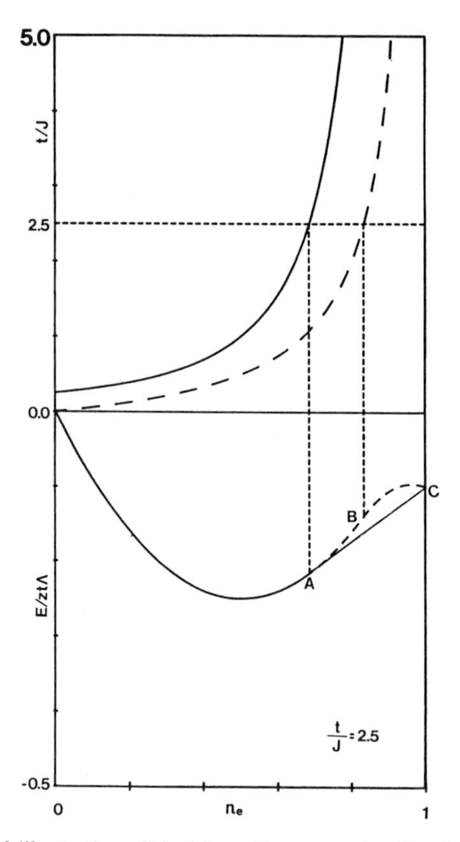

FIGURE 2. Geometrical illustration of the Maxwell construction for phase separation; see the text for further explanation.

asserts that the free energy of a system must always be a convex function of density; otherwise, one can always construct a phase-separated state of lower free energy, and whose free energy is (marginally) convex.

An explicit demonstration of Maxwell's argument, as applied to the present case, is summarized by the geometrical construction of Fig. 2. The lower half of the figure shows the energy, Eq. (17), as a function of n_e at fixed t/J. In our particular example, we take $t/J = 2.5$. The energy (17) is depicted in part by a dashed line, curve ABC, where it is concave. Point B shows where the ferromagnetic phase should end according to the early estimate, Eq. (16). Let us compare the energy of a uniform charge density state that sits upon this dashed curve with one that involves macroscopic phase separation. If a fraction m of the system has charge density n_+, and a fraction $1-m$ of the system has density n_-, with $n_+ > n_-$ and

$$mn_+ + (1-m)n_- = n_e, \tag{20}$$

then the energy of a phase separated state is

$$\mathcal{E}_{ps} = m\mathcal{E}(n_+) + (1-m)\mathcal{E}(n_-), \tag{21}$$

where $\mathcal{E}(n_\pm)$ is the value of the energy (17) at $n_e = n_\pm$. Note that the interface energy is ignored in (21), as is appropriate in the thermodynamic limit.

Viewed as a function of n_e, the energy of the phase-separated configuration given in Eq. (21) is a straight line connecting the points $(n_-, \mathcal{E}(n_-))$ and $(n_+, \mathcal{E}(n_+))$. One chooses n_+ and n_- so that the straight line connecting these points lies as low as possible. This construction is illustrated by the solid line connecting points A and C in Fig. 2. One sees that $n_+ = 1$, so a fraction $m = (n_e - n_-)/(1 - n_-)$ is purely antiferromagnetic. Again inspecting Fig. 2, one sees that at point A, the solid and dashed lines must be tangent to one another. This means that

$$\mathcal{E}'(n_-) = \frac{\mathcal{E}(n_+) - \mathcal{E}(n_-)}{n_+ - n_-}, \tag{22}$$

with $n_+ = 1$, while n_- lies somewhere in the ferromagnetic region. Expressing the energies appearing in (22) in units of $zt\Lambda$, Eq. (17) yields $\mathcal{E}(n_+) = \mathcal{E}(1) = -J/4t$, $\mathcal{E}(n_-) = -n_-(1 - n_-)$ and $\mathcal{E}'(n_-) = 2n_- - 1$. Inserting these expressions into Eq. (22) and solving for n_- we find that

$$n_+ = 1$$

$$\begin{cases} n_- = 1 - \frac{1}{2}[J/t]^{1/2} & \text{for } t/J > 1/4; \\ n_- = 0 & \text{for } t/J < 1/4, \end{cases} \tag{23}$$

which are the densities that correspond to optimal phase separation. Hence the ground state configuration consists of a purely antiferromagnetic component with density $n_+ = 1$, and a purely ferromagnetic component of density n_-. Although the canted phase is the best configuration with uniform charge density for certain values of n_e, it is everywhere unstable against phase separation. However, the canted phase is stabilized to some extent by the inclusion of the three-site term in the Hamiltonian (1), a point discussed in ref. 10. The second equation in (23) is shown in the upper half of Fig. 2 as a solid line, and gives the true critical line announced earlier in Eq. (18) and in Fig. 1. To complete the picture, we must update the expressions for the ground state energy given earlier in Eq. (17);

$$\mathcal{E}/zt\Lambda = \begin{cases} \left(1 - [J/t]^{1/2}\right) n_e - \left(1 - \frac{1}{2}[J/t]^{1/2}\right)^2, & \text{for } t/J \leq \dfrac{1}{4(1-n_e)^2}; \\ -n_e(1 - n_e) & \text{for } t/J \geq \dfrac{1}{4(1-n_e)^2}, \end{cases} \tag{24}$$

Although reasoning based upon the Maxwell construction establishes that the phase-separated configuration is lower in energy than the canted state, it falls short of proving that a configuration of still lower energy does not exist. An analytical proof that the phase separated configuration is indeed the absolute minimum of the classical energy (13) has been possible in one dimension[10]. We have not been able to obtain a similar proof in higher dimensions, but numerical experiments point to the same conclusion. By a numerical experiment we mean a direct numerical minimization of the classical energy (13). Such task is not entirely straightforward because of the large number of variables involved, especially for large lattice dimension. Hence most of our numerical calculations were based on a simplified form of the Hamiltonian, which assumes that the overall phase ψ_l and the azimuthal angle ϕ_l, defined in (11), are uniform at the minimum; this assumption implies, in particular, that all spins are contained in a plane. We have thus confirmed the heuristic picture derived above and obtained some additional insight concerning the nature of the interface separating the ferromagnetic and antiferromagnetic domains.

We have found that the classical energy possesses a multitude of local minima which correspond to formation of ferromagnetic or antiferromagnetic bubbles of varying size. Generically, these minima are metastable because the absolute minimum is achieved when the (positive) interface energy is minimized; this corresponds to the formation of exactly two domains, one ferromagnetic, the other antiferromagnetic. Bubbles attract each other with a strong but very short ranged force and lower their energies when they unite. Examples of such bubbles are displayed in Fig. 3, and were obtained by a numerical minimization of (13) on a two-dimensional square lattice. Although it is difficult to discern from Fig. 3 the detailed nature of the interface, a more careful analysis shows that the interface is sharp on the antiferromagnetic side while it develops an exponential tail on the ferromagnetic side. One should also keep in mind that Fig. 3 was produced on the assumption that all spins lie in a plane.

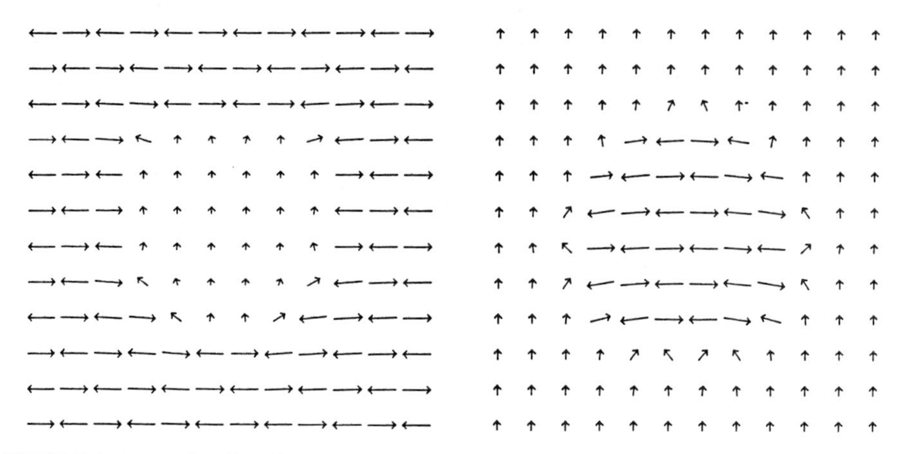

FIGURE 3. Examples of antiferromagnetic and ferromagnetic bubbles obtained with a numerical minimization of the classical energy (13) on a square lattice.

When this assumption is relaxed, spins on opposite sides of the interface might prefer to be perpendicular to each other. A more detailed understanding of the interface is available in one dimension [10].

The formation of bubbles is reminiscent of earlier discussions of spin polarons[8] and spin bags[9]. In particular, holes are expelled from the antiferromagnetic region and prefer to cluster together within a ferromagnetic region of reduced charge density. Therefore, although an attraction between holes is clearly at work, it leads to the formation of macroscopic domains rather than to a conventional pairing. In this respect, the picture emerging here might be unfavorable for applications of the $t - J$ model to the theory of high-T_c superconductivity.

Note that the phase diagram constructed above is independent of lattice dimension. One should recall that a similar situation arises in the ordinary Heisenberg model, where the Néel state is the lowest energy semiclassical configuration in all dimensions. On the other hand, the Néel state is certainly destroyed by quantum fluctuations in one dimension, while almost certainly being preserved in three dimensions. The situation in two dimensions is currently under debate, but the available evidence[12] suggests that long-range order is present at $T = 0$. Of course, it would be difficult to draw a precise analogy with the current model, but a few general comments can be made. It is reasonable to assume that the phase separation obtained here through semiclassical arguments is an accurate description of the true (quantum) ground state at sufficiently large dimension D, perhaps as low as $D = 3$. At the other extreme, $D = 1$, quantum fluctuations do destroy magnetic order and lead to a ground state which is a spin singlet; it is unlikely that phase separation would occur within a singlet ground state[10]. Needless to say, we do not know at this point whether or not the picture derived here is applicable in two dimensions. Nevertheless, the overall picture suggests that phase separation is a generic feature of the $t - J$ model, just as the Néel state is the generic ground state of the Heisenberg model. Thus we summarize our main conclusions:

First, we will return to the partial bosonization incorporated in the commutation relations (6). Although we cannot gauge at this point the full effect of this simplification, we should reemphasize that our model does share with the usual $t - J$ model many essential physical characteristics. In particular, Nagaoka's arguments go through without modification. Thus our conclusions concerning the true ferromagnetic boundary should have impact on the corresponding question in the usual $t - J$ model as well as in the Hubbard model.

Second, one must question whether a $t - J$ model of any kind is an appropriate description of the Hubbard model. Recall that the $t - J$ model is derived as the large-U limit of the

Hubbard model, the limit being taken at any fixed electron density. Hence if a ferromagnetic boundary actually exists, the limit would have to be taken through a phase boundary. Thus it is not obvious that the details of the phase diagram below the ferromagnetic boundary coincide with those of the original Hubbard model. Of course this criticism need not apply in low dimensions, e.g. $D = 1$, where the phase diagram of the quantum model is known to be featureless.

Let us assume for the sake of argument that the potential pitfalls described above prove harmless, and that phase separation is indeed predicted by the Hubbard model. One should then question whether or not phase separation is generic; in particular, whether this picture is stable against variations of the Hubbard Hamiltonian. One worrisome objection is that we have described macroscopic aggregation of charge without considering the long-range effects of Coulomb repulsion. Coulomb repulsion may limit the growth of phase separating bubbles, or else screening may allow them to grow to arbitrary size. In the latter case, one contemplates the important effects of screening charges which until now have not even been included in the problem. Either scenario suggests that phase separation may push the Hubbard model beyond its limit of applicability. Such a conclusion could prove disastrous not only for potential applications of the simple Hubbard model to high T_c superconductivity, but also for more conventional applications to metallic magnetism. We end with almost as many questions as when we began:

Ο γυαλός είναι στραβός, ή στραβά αρμενίζουμε?

Acknowledgements

This work was supported in part by a research grant from the E.E.C. (ESPRIT-3041). The work of N.P. was also supported by the U.S. Department of Energy, while G.C.P. acknowledges support also from the Greek Secretariat for Science and Technology (contract No. 87ED215). M.P.M. is grateful to the members of the Research Center of Crete for welcoming him to join their work during his summer on Crete. We thank E.N. Economou, P. Spathis, and L. Kleinman for valuable discussions.

References

(a) Present address: Center for Nonlinear Dynamics, and Department of Physics, University of Texas, Austin TX 78712, U.S.A.

(b) Also at the Department of Physics, Washington University, St. Louis, MO 63130, U.S.A.

1. E.H. Lieb and F.Y. Wu, *Phys. Rev. Lett.* **20**, 1445 (1968).
2. Y. Nagaoka, *Phys. Rev.* **147**, 392 (1966).
3. A.B. Harris and R.V. Lange, *Phys. Rev.* **157**, 295 (1967).
4. N. Papanicolaou, *Nucl. Phys. B* **240**, 281 (1984); **305**, 386 (1988) .
5. P.B. Visscher, *Phys. Rev. B* **10**, 943 (1974).
6. L.B. Ioffe and A.I. Larkin, *Phys. Rev. B* **37**, 5730 (1988).
7. D. Foerster, *Z. Phys. B–Condensed Matter* **74**, 295 (1989).
8. W.F. Brinkman and T.M. Rice, *Phys. Rev. B* **2**, 1324 (1970).
9. J.R. Schrieffer, X.-G. Wen and S.-C. Zhang, *Phys. Rev. Lett.* **60**, 944 (1988).
10. M. Marder, N. Papanicolaou, and G.C. Psaltakis, preprint (1989).
11. J.C. Maxwell,*Nature*, **11**, (1874) 53; also in *Collected Works*, v. 2, (Dover, New York, 1952) p. 424; J.W. Gibbs, "On the Equilibrium of Heterogeneous Substances," *Trans. Conn. Acad. Arts. Sci.* **3**, 108 (1878); also in *The Scientific Papers of J.W. Gibbs*, (Dover, New York, 1961).
12. S. Tang and J.E. Hirsch, *Phys. Rev. B* **39**, 4548 (1989).

SPIRAL MAGNETIC PHASES AS A RESULT OF DOPING IN HIGH Tc COMPOUNDS

Marc Gabay

Université Paris-Sud, Centre d'Orsay

91405 Orsay, France

INTRODUCTION

The exciting discovery of high Tc superconductivity in copper oxides has generated a lot of experimental and theoretical activity but, so far, the underlying mechanism responsible for this superconductivity remains elusive. Yet several experimental features have emerged which any realistic theoretical model should take into account. (i) NMR shifts of Y^{89} and O^{17} on the YBCO material [1] indicate that the Cu(3d)-O(2p) hole orbitals are of σ character and that strong correlations between the Cu(3d) and O(2p) holes do occur. (ii) Superconductivity in electron doped compounds [2,3] suggests that there exists a symmetry with respect to adding or removing electrons from the half filled band. (iii) Magnetism and unusual magnetic fluctuations are found in most copper oxides [4,5]. (iv) photoemission data and band structure calculations allow to compute the relevant energies of the problem [6] : they yield a factor of 2-3 for the ratio between the site copper energy and the bandwidth so that the electronic states of the CuO_2 sheets are neither totally atomic-like nor fully band-like.

In the atomic limit, two models are popularly invoked : the two band model introduced by Emery [7] and the single band $t-J$ model proposed by Zhang and Rice [8]. Although there is a current theoretical and experimental [9] debate on whether the former or the latter model is better qualified to describe these cuprates, we

Dynamics of Magnetic Fluctuations in High-Temperature Superconductors
Edited by G. Reiter *et al.*, Plenum Press, New York, 1991

will use the $t-J$ model to study the magnetic fluctuations of these systems; our choice - in addition to experimental evidences - is based on results of numerical diagonalization of Cu_5O_{16} clusters. They reveal that charge and spin excitation energies are separated by a gap of order $2.5eV$ and that the lowest energy levels (corresponding to the spin fluctuations) are well reproduced by a J, J', t, t' model where $J'/J \approx 3.10^{-2}$ ($J \approx 0.13eV$) and $|t'|/t \approx 10^{-1}$ ($t \approx 0.4eV$). t' and J' represent respectively diagonal hopping terms and spin exchange terms on a square lattice. In view of their smallness the $t-J$ model seems a valid starting point.

In section A we will briefly recall the procedure we used to obtain an effective spin $S = 1/2$ hamiltonian away from half filling. Then, in section B we shall analyze the ground state of our model in the classical ($S \to \infty$) limit. Quantum fluctuations are handled in section C in a variational fashion and *a la* Chakravarty Halperin and Nelson (CHN) [11]. Results of the scaling theory are then compared to predictions made within the context of flux phases [12] and to the dynamical spin fluctuations seen by neutron scattering experiments on single crystals of $La_{2-\delta}Sr_\delta CuO_4$ [13].

A- THE EFFECTIVE SPIN HAMILTONIAN

Magnetic fluctuations of a single layer of CuO_2 are described by the following $t-J$ model on a square lattice:

$$H = \mathcal{H}_{hop} + \mathcal{H}_1 + \mathcal{H}_2 \tag{1}$$

$$\mathcal{H}_{hop} = -t \sum_{i,j,\sigma} P c^\dagger_{i,\sigma} c_{j,\sigma} P^\dagger \tag{1a}$$

$$\mathcal{H}_1 = J/2 \sum_{i,j,\sigma} (\vec{S}_i \cdot \vec{S}_j - n_i \cdot n_j/4) \tag{1b}$$

$$\mathcal{H}_2 = J/4 \sum_{i,j,k,\sigma} P[c^\dagger_{i,\sigma} c^\dagger_{j,-\sigma} c_{j,\sigma} c_{k,-\sigma} - c^\dagger_{i,\sigma} n_{j,-\sigma} c_{k,\sigma}] P^\dagger \tag{1c}$$

where $J = 4t^2/U$, P is a Gutzwiller operator prohibiting double occupancy of the sites in the large U limit, and $\vec{S}_i$ is a spin 1/2 operator on site i. Equation (1c) describes the motion of holes on the same sublattice with (2nd term) and without (1st term) spin flip.

At half filling, H reduces to $\mathcal{H}_1$. This hamiltonian is believed to yield antiferromagnetic long range order [14]. The effect of $\mathcal{H}_{hop}$ and $\mathcal{H}_2$ is to destabilize this order by inducing longer range frustrating interactions in the spin part of H [15]. To see that we consider the exactly soluble case of three sites, denoted by a, b, c, and two electrons. We call $|C>$ the state where the hole sits on the central site b, and $|E>$ the state where the hole resides on the end sites a and c. The ground state $|\Psi_{gs}>$ is a coherent superposition of $|C>$ and $|E>$. Now, since $\mathcal{H}_2$ can only operate on $|E>$, $<\Psi_{gs}|\mathcal{H}_2|\Psi_{gs}>$ is proportional to $<E|\mathcal{H}_2|E>$. If we introduce the operator T which transfers a hole from b to a and c – that is, such that $T|C>=|E>$ – we can write $<E|\mathcal{H}_2|E>=<C|\tilde{\mathcal{H}}_2|C>$. This operator can easily be constructed in the three site case, and we find that $\tilde{\mathcal{H}}_2 = T^{\dagger}\mathcal{H}_2 T$ acts only on $|C>$ and is given by

$$\tilde{\mathcal{H}}_2 = J(1-n_b)(\vec{S}_a \cdot \vec{S}_c - n_a n_c/4) \tag{2}$$

If the hole is mobile, and if the time scale of its motion is much shorter than the time scale of spin fluctuations,we can use a Born-Oppenheimer type of approximation to study these degrees of freedom separately. The spin part will then read

$$\mathcal{H}_{eff} = J/2(\vec{S}_a \cdot \vec{S}_b + \vec{S}_b \cdot \vec{S}_c) + J_{eff}\vec{S}_a \cdot \vec{S}_c \tag{3}$$

where J_{eff} is obtained by averaging n_b for a fixed spin background. With these same assumptions, we have derived an effective spin hamiltonian for the lattice, in the presence of a **finite** concentration δ of holes (in the case of a single hole the second assumption appears to break down [16]). To lowest order in δ we find [15]

$$\mathcal{H}_{eff} = J_1 \sum_{i,j} \vec{S}_i \cdot \vec{S}_j + J_2 \sum_{i,k} \vec{S}_i \cdot \vec{S}_k + J_3 \sum_{i,l} \vec{S}_i \cdot \vec{S}_l \tag{4}$$

k (resp l) is a second (resp third) neighbor on the square lattice and $J_1 = J/2(1 - 5\delta) - |t|\delta/2$, $J_2 = J\delta$, $J_3 = J\delta/2$.

At this stage a remark is in order; the Hilbert space of $\mathcal{H}_{eff}$ seems to pertain to an SU_2 symmetry whereas, in the presence of holes, it should be larger. In fact our effective hamiltonian (4) no longer posess the naive SU_2 symmetry due to non nearest-neighbor interactions, but a larger symmetry as well.

B-THE CLASSICAL LIMIT

For $S \rightarrow \infty$ we can study the classical ground state of Hamiltonian (4) :

• So long as $2J_2 + 4J_3 \leq J_1$,

the antiferromagnetic structure characterized by a wavevector $\vec{Q} = (\pi, \pi)$ is stable.

• When $2J_2 + 4J_3 > J_1$,

an incommensurate spiral structure, characterized by two pairs $\pm\vec{Q}_1, \pm\vec{Q}_2$ develops:

⋆⋆ For $2J_3 < J_2$ we have

$$\vec{Q}_1 = (\pi, q_0), \vec{Q}_2 = (q_0, \pi), \quad \cos q_0 \equiv -\frac{(J_1 - 2J_2)}{4J_3}$$

⋆⋆ For $2J_3 > J_2$ we have

$$\vec{Q}_1 = (q_0', q_0'), \vec{Q}_2 = (-q_0', q_0'), \quad \cos q_0' \equiv -\frac{J_1}{2J_2 + 4J_3}$$

⋆⋆ For $2J_3 = J_2$, which is precisely the situation we have from hamiltonian (4), we find a continuous manifold of wavevectors given by

$$\cos q_x + \cos q_y = -\frac{J_1}{2J_2} \tag{5}$$

the Neel and spiral phases meet at a Lifshitz point; with the values of t and J quoted above, this occurs for $\delta_c \approx 0.09$. Naturally, for such a value, the expansion in δ has to be pushed to higher than first order.

C-QUANTUM GROUND STATES AND FLUCTUATIONS

Including quantum fluctuations, we look for ground states with long range order as well as spin liquid states. To perform analytical calculations, we construct the long range ordered states as the semi-classical extensions of the classical states determined in section **B**. As for the disordered state we use a short range RVB model, and only retain the most degenerate configurations (those which lower the energy most) [14] .

We have also studied the antiferromagnetically correlated and the spin liquid phases by constructing variational Jastrow wavefunctions *a la* Kalmeyer and Laughlin [17,18]. We chose them of the form $\Psi(\vec{r}_1, .., \vec{r}_i, \vec{r}_{\frac{N}{2}}) \sim \exp^{-\sum_{i,j} V(\vec{r}_i - \vec{r}_j)}$.

V is chosen as a hard core short range repulsive potential to describe the long range ordered state and as a screened one component plasma of logarithmicaly interacting charges to represent the R.V.B state. We then find that, increasing δ, we switch from an antiferromagnetic order to a spin liquid phase for $\delta_c \approx 0.067$, then to a modulated phase. To make further predictions, we write the non linear σ models corresponding to hamiltonian (4).

Starting from the Neel phase, we get

$$\mathcal{Z}_N = \int \mathcal{D}\vec{n} \exp \frac{-1}{2g_N^*} \int_0^{\beta\hbar c_N^* \Lambda} \int d\tau d^d x [(\vec{\nabla}\vec{n})^2 + (\partial\vec{n}/\partial\tau)^2] \tag{6}$$

This expression is identical with that obtained by CHN except for the fact that the coupling constant g_N^* and the spin wave velocity c_N^* are complicated functions of J_2/J_1. Jolicoeur and Le Guillou [19] have computed the correlation length of this model as a function of temperature for $S = 1/2$ and $\delta = 0.06$; they find

$$\xi(\delta = 0.06) \sim 1.8 \exp(\frac{42\pi}{T})$$

to be compared with the $\delta = 0$ value $\xi_0 \sim 0.5 \exp(\frac{480\pi}{T})$. The agreement with the experiments of Birgeneau et al and of Endoh et al [20] is quite good, down to $T = 200K$. Below that temperature the growth of ξ is cut off and tends to saturate to the $T = 0K$ value proportional to $\delta^{-1/2}$ [20].

Starting from the modulated phase, we arbitrarily select one of the possible directions of the incommensurate spiral – obtained from equation (5) – and obtain

$$\mathcal{Z}_B = \int \mathcal{D}\vec{n} \exp -\frac{1}{2g_B^*} \int_0^{\beta\hbar c_B^* \Lambda} \int d\tau d^d x d^d x' [\vec{n}(\vec{x}) G^{-1} \vec{n}(\vec{x}') + (\partial\vec{n}/\partial\tau)^2] \tag{7}$$

where the Fourier transform of G^{-1} reads

$$G^{-1}(\vec{q}) = (\cos q_x + \cos q_y + \frac{J_1}{2J_2})^2 \tag{8}$$

Thus, near the Lifshitz point, equation (7) takes the simpler form

$$\mathcal{Z}_B = \int \mathcal{D}\vec{n} \exp -\frac{1}{2g_B^*} \int_0^{\beta\hbar c_B^* \Lambda} \int d\tau d^d x [(\{\vec{\nabla}^2 + \vec{Q}_0^2\}\vec{n})^2 + (\partial\vec{n}/\partial\tau)^2] \tag{9}$$

which is the action of the Brazovskii model [21]. We note that the gapless spinwave excitations correspond to those obtained by slave fermion approaches on the t-J model [22]. At $T = 0K$, actions (6) and (9) pertain to $d + 1 = 3$ dimensional hamiltonians. Thus the Lifshitz point at $Q_0 = 0$ only occurs for $g_B^* = 0$, that is for $S \to \infty$. Away from this point, equation (9) leads to a line of first order phase transitions in the (g_B, δ) plane. The incommensurate structures which can order at low g^B are the single spiral or the double spiral (with two orthogonal wavevectors)[23]. The resulting phase diagram is shown in figure 1. The straight line ($g_b = g_b^*$) corresponds to the physical case $S = 1/2$. In the vicinity of the Lifshitz point, on the modulated side, we see that the zero temperature state corresponds to a disordered phase. the spin dynamics in that region will be well described by a massive version of action (9) that is by pseudo spinwaves of the form $\omega^2 = r + (\vec{q}^2 - \vec{Q}_0^2)^2$ with $r \propto \delta - \delta_c$. Also we notice that, contrary to the situation relevant to the non modulated Neel phase, the correlation length will always remain finite even at $T = 0$. Experimentally speaking, this effect will be hard to see since anyway ξ is limited by the average hole distance.

At that stage we can try to establish a contact between our approach at $T = 0K$, based on fluctuations **from a broken symmetry spin state**, and RVB or flux phases approaches, which start from a disordered spin liquid state. In standard classical theories where flux phases come about, such as in the frustrated XY model, or for a network of Josephson junctions, frustration causes the symmetry of the relevant problem to be different from the naive U(1) group one might have assumed [24]. This is most easily revealed by looking for a long wavelength continuous theory of these systems. One then finds that the hallmark of frustration is to introduce (at least) two competing wavevectors. In the case of staggered frustration with two chiralities, one gets precisely two wavevectors. The action describing the critical fluctuations of such models will be identical to the high "temperature" (here the high g) version of our action (9), but for the peculiar expression of the bare propagator (8).
So we would like to propose that the double spiral state represents a staggered flux phase, as proposed by Kane et al [23]. Furthermore, since we found, using a variational method, that the RVB phase touched the Neel phase, we can infer that the " high g " phase of our model also corresponds to the RVB phase.

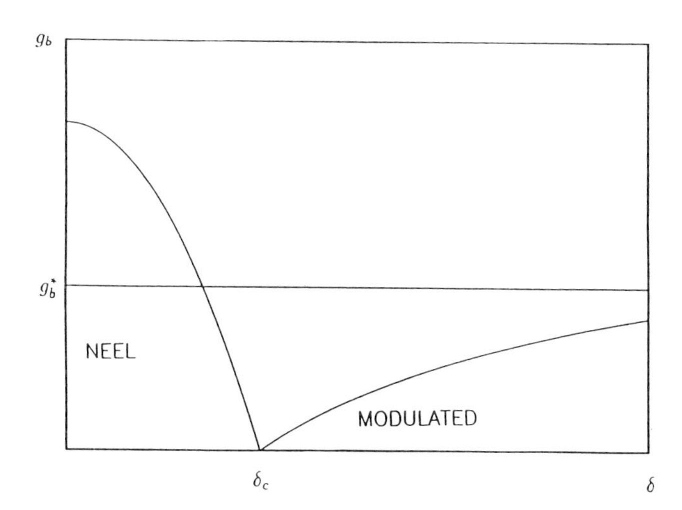

Fig. 1 The g_b versus δ phase diagram at $T = 0K$

ACKNOWLEDGMENTS

Fruitful discussions with Drs T. Garel, D. Poilblanc and J. Tranquada are gratefuly acknowleged. I warmly thank Dr G. Reiter for putting this workshop together, and NATO for financial support.

REFERENCES

1. H. Alloul, preprint and APS March Meeting, St Louis USA (1989) M. Takigawa, preprint and APS March Meeting, St Louis USA (1989)

2a. Y. Tokura, H. Takigi and S. Uchida, Nature **337**, 345 (1989)

2b. H. Takagi, Y. Tokura, S. Uchida, Phys. Rev. Lett. **62**, 1197 (1989)

3. V.J. Emery, Nature **337**, 306 (1989)

4. R.J. Birgeneau and G. Shirane, *Physical properties of HTc Superconductors*, edited by D.M Ginsberg, World Scientific Publishing, 1989)

5. G.M. Luke, Nature **338**, 49 (1989)

6a. A.K. McMahan, R.M. Martin and S. Satpathy, Phys. Rev. B**38**, 6650 (1989)

6b. M.S. Hybertsen, M. Schluter and N.E. Christensen, Physica C**153-155**, 1217 (1988) and Phys. Rev. B**39**, 9028 (1989)

7. V. J. Emery, Phys. Rev. Lett. **58**, 2794 (1987)

8. F.C. Zhang and T.M. Rice, Phys. Rev. B**37**, 3759 (1988) and T.M. Rice, J. Magn. and Magn .Mat. **76-77**, 542 (1988)

9a. See T.M Rice, this Conference and V.J. Emery, this Conference

9b. See M. Takigawa, this Conference and C. Berthier, this Conference

10. M.S. Hybertsen, E.B. Stechel, M. Schluter and D.R. Jennison, preprint (1989)

11. S. Chakravarty, B.I. Halperin and D.R. Nelson, Phys. Rev. Lett. **60**, 1057 (1988) and Phys. Rev. B**39**, 2344 (1989)

12. I. Affleck, J.B Marston, Phys. Rev. B**37**, 3774 (1988), P.W Anderson, B. Shastry, D. Hristopoulos, preprint (1989), P. Lederer, D. Poilblanc, T.M. Rice, preprint (1989), R.B. Laughlin, Z. Zou, preprint (1989) , D.

Poilblanc, Phys. Rev. B**40**, 7376 (1989), D. Poilblanc and Y. Hasegawa, to appear in Phys. Rev. B (1990), D. Poilblanc, to appear in Phys. Rev B Rapid Comm. (1990)

13. G. Shirane, R.J. Birgeneau, Y. Endoh, P. Gehring, M.A. Kastner, K. Kitazawa, H. Kojima, I. Tanaka, T.R Thurston and K. Yamada, Phys. Rev. Lett . **63**, 330 (1989)
14. S. Liang, B. Doucot and P.W. Anderson, Phys. Rev. Lett. **61**, 365 (1988)
15. M. Gabay, S. Doniach, M. Inui, Physica C**153-155**, 1277 (1988) M. Inui, S. Doniach and M. Gabay, Phys. Rev. B**38**, 6631 (1988); see also L.P. Gor'kov and A.V. Sokol, J.E.T.P. Lett. **46**, 420 (1987), L.P. Gor'kov, preprint (1989)
16. P. Horsch, W. Stephan, M. Ziegler and K. von Szczepanski, preprint (1989) and this Conference; see also E. Dagotto, A. Moreo, R. Joynt, S. Bacci and E. Gagliano, preprint (1989), S.A. Trugman, preprint (1989)
17. S. Doniach, M. Inui, V. Kalmeyer and M. Gabay, Europhys. Lett. **6**, 663 (1988)
18. V. Kalmeyer, R.B. Laughlin, Phys. Rev. B**39**, 11879 (1989)
19. T. Jolicoeur and J.C. Le Guillou, preprint (1989)
20. Y. Endoh, Phys. Rev. B**37**, 7443 (1988)
21. S.A. Brazovskii, J.E.T.P. **41**, 85 (1975)
22. D. Yoshioka, J. Phys. Soc. Japan **58**, 32 (1989), ibid. **58**, 1516 (1989)
23. C.L. Kane, P.A. Lee and T.K. Ng, B. Chakraborty and N. Read, preprint (1989), Y. Hasegawa, D. Poilblanc preprint Oct. (1989)
24. T. Garel, S. Doniach, J. Phys C**13**, L887 (1980), M.Y. Choi and S. Doniach, Phys. Rev. B**31**, 4516 (1985), D.K Campbell, P. Kumar and S.E. Trullinger, Springer Series Solid State Sci. **69**, 361 (1987)

INDEX